# Asymptotic Characteristics of Entire Functions and Their Applications in Mathematics and Biophysics

Second Edition, Revised and Enlarged

*by*

L.S. Maergoiz
*The Krasnoyarsk State Academy of Architecture and Civil Engineering, Institute of Computational Modelling, Krasnoyarsk, Russia*



SPRINGER-SCIENCE+BUSINESS MEDIA, B.V.

A C.I.P. Catalogue record for this book is available from the Library of Congress.

ISBN 978-90-481-6360-1 ISBN 978-94-017-0807-4 (eBook)
DOI 10.1007/978-94-017-0807-4

*Printed on acid-free paper*

# Asymptotic Characteristics of Entire Functions and Their Applications in Mathematics and Biophysics

Mathematics and Its Applications

Managing Editor:

M. HAZEWINKEL

*Centre for Mathematics and Computer Science, Amsterdam, The Netherlands*

Volume 559

To the memory of my teacher
**Valentin Konstantinovich Ivanov**

Abstract

**Maergoiz L.S. Asymptotic Characteristics of Entire Functions and Their Applications in Mathematics and Biophysics**

This book is the revised and enlarged second edition of the following monograph published in Russian:

**Маергойз Л. С. Асимптотические характеристики целых функций и их приложения в математике и биофизике. Новосибирск: Наука, Сибирское отделение, 1991. - 272 с.**

The book is devoted to asymptotic methods in the theory of entire and plurisubharmonic functions, and to applications of these methods. Traditional and new asymptotic characteristics of entire functions of one or several complex variables are studied, and then applied to various topics of complex analysis (Borel-Laplace transforms and their modifications; Mittag-Leffler functions and their natural gene- ralizations; integral methods of summing the power series; Riemann surfaces; series expansions of analytic functions in domains of special form, etc.). In an appendix written especially for the second edition, many of these topics are examined for the class of entire functions of one variable with a given proximate order of growth.

In a separate chapter, biophysical applications are treated. The entire functions are applied to dynamical testing of homeostasis processes. We present and investigate in detail certain algorithms of analysis of homeostasis systems under external impacts (modifications of Prony's algorithm). Along with biophysics, such algorithms may be applied in other fields of natural science and technology.

The book is addressed to experts in theoretical and applied mathematics and mathematical biophysics, and to students (of graduate as well as of undergraduate level) interested in complex analysis and its applications.

Illustrations - 14. References - 248.

# TABLE OF CONTENTS

## From the Editors of the Russian Edition

Theory of functions of complex variable finds immediate use in many branches of science. Applications to the problems of aeromechanics, and hydromechanics, elasticity theory, electrostatics, to calculations of electric circuits, etc. are well known and adequately represented in textbooks and monographs. These applications employ the apparatus of Cauchy-type integrals, conformal and quasiconformal mappings, Laplace transforms and so on. Meanwhile, the applied aspects of the theory of entire functions of one or several variables (an important section of complex analysis) have not been adequately exposed as yet. In the Russian literature, applications of entire functions were considered in two monographs by Ya. I. Khurgin and V. P. Yakovlev: "Methods of the Theory of Entire Functions in Radiophysics, Communication Theory and Optics" (M., 1962) and "Finite Functions in Physics and Technology".

The fields of application treated in these books are evident from the titles. Applications to biophysics have never been touched upon in the monographic literature. In this respect the present study is pioneering. The theoretical part of the book (devoted to the intrinsic problems of the theory of entire functions) is focused on the topics related to the Borel transformation and its generalizations, and to asymptotic growth characteristics of entire functions. The author of the book took active part in elaboration of these topics, and the book presents certain results of himself (some of which are published for the first time). Among these results, two groups should be mentioned. The first deals with indicator diagrams. In 1929 G. Polya found that for an entire function $f$ of exponential type the indicator $h(\theta; f)$ allows a simple geometric interpretation: up to reflection in the real axis, this is the support function for the convex hull of the set of singular points for the Borel transform of $f$. Several generations of mathematicians were trying to find an analog of the Polya theorem for entire functions of an arbitrary finite order and to interprete the corresponding indicators geometrically. This was achieved in the 60s for $h^+(\theta; f)$ only. The author was the first who solved this old problem completely. This required new approaches pertaining to minimal $\rho$-trigonometrically convex functions and to construction of special Riemann surfaces. The methods developed by the author made it possible to obtain an analog of the Polya theorem for the entire functions of several variables with nonnegative indicator.

The other group of results is related to new growth characteristics (introduced by the author) of entire functions of several variables. Considerable difficulties arise here compared to the case of one variable. Usually, the growth of an entire function of several variables is described by a point in $\mathbb{R}_+$ or in $\mathbb{R}_+^n$. The author employs a function $\mathbb{R}^n \to \mathbb{R}$ instead, which is fundamentally novel and, at first sight, unexpected. To understand this viewpoint, attention should be paid to the fact that, in the case of one variable, usual growth characteristics of entire functions of normal type and finite order are associated with the coefficients of the linear function representing an asymptote for a certain plane curve determined by $M(r; f)$. In the book the new notions are studied in detail, their relationship with the known concepts is traced, and important applications are indicated.

A characteristic feature of the book is the interrelation of the methods of complex and convex analysis. For the problems in question, this interrelation is exploited for the first time in such depth and scope. On the whole, the book is quite a valuable study.

Professor A. A. Goldberg

It was already Wigner who wrote about the "unfathomable efficiency of mathematics in science". A prominent example of this is the expansion of quite a common class of functions, namely, the entire functions, to biophysical research. The author of the book demonstrates an efficient application of entire functions by the example of the topical biophysical problem of identifying relaxation characteristics for homeostatic processes. Mathematical applications of entire functions to biophysics are considered in a special chapter. Its subject is concerned with an important applied field, relaxation spectroscopy, the basic idea of which is to present results of experiments as a sum of exponentials to be split by mathematical processing. The life of some mathematical ideas can be divided into two periods: modest existence until their applied value is comprehended, which is followed by expansion and prosperity. This is the case of the relaxation spectroscopy. M. Eigen won the Nobel prize for its development and application to chemistry, but the idea itself dates back to A. M. Lyapunov (Lyapunov series), whereas the methods date back to the work done by Prony as early as in the 19th century, and to the works of some other authors.

The monograph develops the method of Prony and presents its detailed application to the analysis of experimental data for homeostasis systems. Publication of the book can be expected to help in propagation of this elegant and useful method.

Professor R. G. Khlebopros

## Foreword to the Russian Edition

The book is an attempt to set forth the fundamentals of the *growth theory* for entire functions of one or several variables, and to consider its applications to biophysical research. The framework of the theory in question has encompassed certain problems the interest to which arose in the late 50s - early 60s at the seminar on the methods of mathematical physics held by V. K. Ivanov (Correspondence Member of the Russian Academy of Sciences) at the Urals University. A series of these problems were associated with the result of Polya on the relationship between the indicator diagram and the Borel transform of an entire function of exponential type. Participants of the seminar were amused by the broad range of applications of that result to various fields of pure mathematics and, surprisingly, to some applied research. In particular, the author was greatly impressed by the solution of a certain geophysical problem in which V. K. Ivanov employed the spatial analog of the Polya theorem. So, the effort of many researches in the USSR and abroad towards extension of the "Polya theorem effect" to other classes of entire functions in one or several variables was perceived as indisputably topical. Moreover, many of us also took active part in that work.

Another series of problems (of immediate interest for the author at that time) were related to the growth theory of entire functions of several variables. The order, type, and indicator of an entire function of *one* variable are growth characteristics well recognized in the mathematical literature. At the time described, various multidimensional analogs used to arise in abundance, which persuaded V. K. Ivanov to pose the question of detecting the "right" growth characteristics in the case of several variables that would be as ample and convenient as the one-variable notions mentioned above.

We pass to a more detailed overview of the content of the book.

In Chapter 1, some elementary facts are discussed: the basics of convex analysis and of the growth theory of entire functions, the theory of trigonometrically $\rho$-convex, subharmonic, and plurisubharmonic functions, etc. To make the presentation more self-contained, a part of this material is given with proofs. Other necessary information can be found in standard university textbooks on complex analysis, topology, and functional analysis.

Chapter 2 is devoted to applications of the theory of entire functions to mathematical analysis of relaxation characteristics of homeostasis processes (in the Introduction below, we dwell upon this in more detail). A prominent example of such characteristics is the so-called sugar curves. The author does not rule out the possibility that the results of that chapter can be obtained by other methods more usual for the experts in differential equations. To read the chapter, it suffices to know some facts discussed in Sections 3, 4, 10 of Chapter 1. If the reader is interested only in the formulation of the algorithm of mathematical prediction for a homeostasis system behavior under external impacts, it suffices to know the initial definitions in Section 10 of Chapter 1. For the reader's convenience, the references to Chapter 2 are given on a separate list.

In Chapters 3-5, we develop the construction of the plane indicator diagram for an entire function of one variable of order $\rho > 0$ and with indicator $h$ of general form. For such entire functions, the generalized Borel transformation is introduced. These notions make it possible to represent the spaces $[\rho, h(\theta))$ and $[\rho, h(\theta)]$ as the duals to certain spaces of functions holomorphic on so-called $(\rho, \alpha)$-convex sets in $\mathbb{C}$. This result is illustrated by the following applications:

expansion of functions holomorphic in the vicinity of a closed bounded $(\rho, \alpha)$-convex domain $D$ in a series that involves a certain specific system of entire functions;

continuability inwards of a system of entire functions that is representing for the class $H(D)$ of functions holomorphic in $D$ (specifically, continuability to a subdomain $G$ of $D$);

analytic realization of operators of generalized differentiation and of generalized convolution;

description of $(\rho, \alpha)$-convex supports of analytic functionals;

analytic continuation of power series.

In Chapter 3 the case of $h \geq 0$ is treated separately.

In Chapters 6-8 we study growth characteristics of functions plurisubharmonic in $\mathbb{C}^n$, and of entire functions of several variables. These characteristics are the order function, the system of type functions, and the system of regularized indicators. The above growth indices constitute a multidimensional analog of the well-known one-variable triad: order, type, indicator. The background for these indices is provided by geometric analysis (see Chapter 6) of the asymptotics for functions plurisubharmonic in $\mathbb{C}^n$ and convex in $\mathbb{R}^n$. The idea of the geometric approach to growth characteristics can be illustrated by the case of one variable as follows.

If an entire function $f(z)$ has finite order $\rho$ and normal type $\sigma$, then

$$\overline{\lim_{r\to\infty}}\ [V_f(u) - \rho u - \ln\sigma] = 0, \text{ where } V_f(u) = \ln^+ \ln^+ M_f(e^u),$$

and

$$M_f(r) = \sup\{|f(re^{i\theta})| : \theta \in \mathbb{R}\}.$$

Geometrically this means that the plane curve $y = V_f(u)$ has the "upper" asymptote $y = \varphi(u)$, where $\varphi(u) = \rho u + \ln\sigma$. The function $\exp\{\exp\{\varphi(\ln r)\}\}$ is an element of the well-known growth scale for the entire functions of one variable.

The results of Chapter 6 are used in Chapter 7 to study similar growth characteristics of *entire* functions of several variables. These characteristics are employed in the analysis of comparative growth of certain functions associated with a given entire function $f$, namely, the maximum of the modulus $M_f$, the maximal term $\mu_f$, and the Nevanlinna characteristic

$$m(r;f) = \frac{1}{(2\pi)^n} \int_0^{2\pi} \cdots \int_0^{2\pi} \ln^+ |f(r_1 e^{i\theta_1}, ..., r_n e^{i\theta_n})| d\theta_1 ... d\theta_n.$$

In Chapter 8 the notion of an indicator diagram is introduced and analyzed for entire functions of several variables with a given system of positively conjugate orders and with nonnegative indicator corresponding to this system. For this class of entire functions, an analog of the Polya-Martineau-Ehrenpreis theorem (see Theorem $A2$) is obtained, and then used to show the existence of an entire function with indicator positive in $\mathbb{C}^n \setminus 0$ and corresponding to the given system of positively conjugate orders. Among other applications, we mention the description of the generalized Borel polygon for a system of integral summation methods for $n$-fold power series with $n > 1$.

In Appendix A, we study the construction of the Riemann surface for the inverse to the mapping

$$\alpha(z) = z^\rho \left[1 + \frac{a_1}{z} + \cdots + \frac{a_n}{z^n}\right]$$

with nonintegral $\rho > 0$; this Riemann surface is used to introduce the plane indicator diagram of an entire function of fractional order.

Bibliographical references are mostly given in the Comments at the end of the book. Quite often, exercises represent some facts required in the subsequent exposition; the proofs of such facts are immediate from the definitions and are left to the reader. The beginning and the end of every proof are marked with the signs ◀, ▶. Theorems, lemmas, propositions, properties, corollaries, exercises, definitions, proofs, chapters are often referred to as T., L., P., Prop., Cor., Ex., Def., Pr., Ch. respectively; enumeration is chapterwise and unique for all types of statements. All statements (and also the formulas) are enumerated by three indices; we prefix the first index by the letter A or B when referring to an appendix A or B, respectively. For example, "Theorem 1.9.11" means the 11th theorem in Section 9 of Chapter 1, and "Theorem A.2.9" means the 9th theorem in Section 2 of Appendix A. Next, we enumerate sections by two indices and subsections by three indices. For example, "Section 5.1" is the first section of Chapter 5, and "Subsection 3.1.1" is the first subsection in Section 3.1. For the matters of enumeration, we treat the introduction to the book as "Chapter 0" and the introductory section to each chapter as "Section 0". Thus, "formula (3.0.1)" refers to the first formula in the introduction to Chapter 3, etc.

If a function under consideration may take infinite values, this is indicated explicitly. The lower bound of an empty set of real numbers is $\infty$. Unions, intersections, suprema, and infima of families are sometimes written as follows:

$$\overline{\lim} = \limsup; \underline{\lim} = \liminf; \bigcup_{\alpha\in A} U_\alpha = \bigcup U_\alpha | \alpha \in A; \quad \bigcap_{\alpha\in A} U_\alpha = \bigcap U_\alpha | \alpha \in A;$$

$$\sup_{\alpha\in A} V_\alpha = \sup\{V_\alpha : \alpha \in A\}; \inf_{\alpha\in A} V_\alpha = \inf\{V_\alpha : \alpha \in A\}.$$

* *

*

Contacts with V. K. Ivanov have greatly influenced the subject of the book. A. A. Goldberg kindly agreed to edit the main "theoretical" part. His benign criticism, and valuable remarks of the reviewer L. I. Ronkin considerably improved the content of the first version of the manuscript. Contacts with R. G. Khlebopros who edited the "applied" part of the book, and with V. P. Nefedov excited the author's interest in the problems of mathematical biophysics. O. V. Epifanov wrote the proof of Theorem 10.8 in Ch. 1. Individual useful remarks about the manuscript were made by A. V. Abanin, Yu. E. Anikonov, B. N. Apanasov, A. N. Gorban, M. A. Evgrafov, Yu. A. Kazmin, Yu. F. Korobeinik, I. F. Krasichkov, M. A. Mkrtchyan, V. N. Novosel- tsev, A. K. Tsikh, N. M. Chernykh, A. P. Yuzhakov. It is my pleasant duty to express my sincere gratitude to these people and all mathematicians who, at various seminars and conferences, discussed the author's reports about individual results of this study.

* *

*

For the second edition, the text was revised thoroughly. Many proofs were refined and simplified, and some were replaced by new ones (this is especially true for Chapter 4 and Section 5.1). Appendix B of the first edition, devoted to a software implementation of the algorithm of mathematical analysis for homeostasis systems, was withdrawn as obsolete from the viewpoint of software designers. It was replaced with a survey of results (presented with proofs) about the relationship between the indicator and conjugate diagrams of an entire function of a given proximate order of growth. To Chapter 2, a section was added 2.5 about the Prony harmonic expansion algorithm and its modifications, and the proof of Property 2.4.2′ was added to Section 2.4. A new version of Sections 2.4 and 2.5 was written by the author together with B. N. Varava. Section 3.1 (describing the theory of $\rho$-convex sets) was extended substantially. Finally, Section 8.2 now contains more detailed proofs (as compared to the first edition).

The author expresses his deep gratitude to V.A. Ouskin who prepared the first version of the translation, and to T. G. Belokopytova, I. M. Fedotova, and T. O. Kochetkova, who did a great deal of AMS-LATEX keyboarding work. Besides, the author is sincerely gratefull to S. V. Kislyakov for linguistic editing of the translation. This research was done with the aid of the Russian Foundation for Basic Research (grants no. 96-15-96266, no. 00-15-96140, no. NSH-1212.2003.1 for support of leading research schools and grant no. 03-01-00460).

# Introduction

The theory of entire functions of one or several variables is a popular section of complex analysis; it has numerous applications to adjacent topics, and to other fields of pure and applied mathematics.

Applications of entire functions to biophysical research are considered in the second chapter of this book. A usual method of investigating homeostasis systems of living organisms is to study the response of characteristics of these systems to an external impact. The response of certain system parameters (that are called relaxation characteristics) to this impact, and the regain of equilibrium (called relaxation) by these parameters forms the basis to judge the functional capacities of the system in question, and to disclose the presence of one or another pathology in it. The book presents a modification of a method (dating back to Prony) for identifying relaxation characteristics of homeostasis processes whose dynamics is described by differential equations of finite order with unknown constant coefficients. The numerical data are assumed to be measured in uniform time lapses. The methods of the theory of entire functions are employed for a detailed analysis of the conditions of existence, uniqueness, and stability of the identification algorithm in question; this algorithm is designed on the basis of the study of similar questions related to the inverse problem for the differential equation mentioned above.

The proof of the consistency of the above algorithm leans upon two fundamental notions associated with a broad field of research in the theory of entire functions. These are the *Borel transform* $F(z) = \sum_{k=0}^{\infty} a_k/z^{k+1}$ *of an entire function* $f(z) = \sum_{k=0}^{\infty} a_k z^k/k!$ *of exponential type* and *its indicator diagram* $I_h$. The latter is the compact convex subset of $\mathbb{C}$ the support function of which is equal to the indicator $h(\theta) = \overline{\lim_{r\to\infty}}\, r^{-1}\ln|f(re^{i\theta})|$, $\theta \in \mathbb{R}$, of $f$.

In Chapters 3-8 these concepts are extended to classes of entire functions of finite order, which leads to new notions in the theory of growth of entire functions of one and several variables. Various applications of these notions in complex analysis are discussed.

The notion of an indicator diagram was introduced by G. Polya who employed it in 1929 to prove the following theorem.

THEOREM $A_1$. *In the above notation, let $K$ be the conjugate diagram of a function $f$, i.e., the smallest compact convex set in $\mathbb{C}$ outside of which $F$ admits analytic continuation. Then $K = \{z \in \mathbb{C} :\ \overline{z} \in I_h\}$.*

G. Polya found a number of spectacular applications of this theorem to analytic continuation problems; see the excellent exposition in the well-known monograph by L. Bieberbach. A multidimensional analog of Theorem $A_1$ related to a realization of the dual to the space of functions holomorphic in a convex domain $D \subset \mathbb{C}^n$ dates back to A. Martineau and L. Ehrenpreis.

THEOREM $A_2$. *Let $D$ be a bounded convex domain in $\mathbb{C}^n$. We denote by $H(D)$ the space of all functions holomorphic in $D$ and by $H^*(D)$ the dual space endowed with the strong topology. Putting $\langle\zeta, z\rangle = \sum_{k=1}^{n} \zeta_k z_k$, we introduce the "support function" of $D$ by the formula $h(\zeta) = \sup\{\langle\zeta, z\rangle : z \in D\}$. Let $[\mathbb{I}, h) = \{f\}$ be the space of entire functions of exponential type and such that*

$$\overline{\lim_{w\to\zeta}}\ \overline{\lim_{r\to\infty}}\ r^{-1} ln|f(rw_1, ..., rw_n)| < h(\zeta), \quad \zeta \in \mathbb{C}^n.$$

*We endow this space with the standard inductive limit topology. Then the Laplace transformation, defined on $H^*(D)$ by the formula*

$$f(z) := \mu(e^{\langle\zeta,z\rangle}),\ z \in \mathbb{C}^n, \quad \mu \in H^*(D),$$

*establishes topological isomorphism between the spaces $H^*(D)$ and $[\mathbb{I}, h)$.*

The Polya theorem was a kind of a bridge for the methods of the theory of entire functions of exponential type to penetrate into other sections of complex analysis, such as, e.g., summation of power series, complex analysis in convex domains (completeness of systems of exponentials, expansion of holomorphic functions in exponential series, and so on), interpolation of entire functions, differential equations of infinite order, convolution equations, etc. For $n = 1$, these problems were considered by A. O. Gelfond, A. F. Leont'ev, B. Ya. Levin, A. I. Markushevich, M. A. Evgrafov, Yu. F. Korobeinik, I. F. Krasichkov-Ternovsky, Yu. N. Frolov, O. V. Epifanov, and for $n > 1$ by B. Malgrange, L. Ehrenpreis, A. Martineau, V. V. Napalkov, L. Gruman, V. V. Morzhakov and other authors.

The scope of such applications can be enlarged substantially by employing the full-scale theory of growth of entire functions of finite order, which is well developed and plentiful of results. Theorem $A_1$ made topical the problem of constructing an analog of the Polya theorem for entire functions of finite order $\rho \neq 1$. This has led to the problem of constructing a geometric image in $\mathbb{C}$ of the indicator of an entire function of order $\rho \neq 1$ (the indicator diagram), and also of constructing a transform of this function (an analog of the Borel transform) such that the corresponding "conjugate diagram" coincide with the indicator diagram.

Contributions of M. F. Subbotin, V. Bernstein, A. Macintyre, A. I. Markushevich, M. M. Dzhrbashyan, M. A. Evgrafov, Yu. A. Kazmin made it possible to introduce and study the generalized Borel transformation for vast classes of entire functions. For instance, A. I. Markushevich considered "the inverse Borel transformation corresponding to an arbitrarily fixed entire function $K$ of two variables". With each function $F$ holomorphic in an open neighborhood $U = U(F)$ of the point $\infty$ in $\mathbb{C}$ and such that $F(\infty) = 0$, this operator associates the entire function

$$f(\lambda) = \frac{1}{2\pi i} \int\limits_{|z|=\sigma} K(\lambda, z)F(z)dz, \quad \lambda \in \mathbb{C}. \tag{1}$$

Here $\sigma > 0$ is a number such that $\{z \in \mathbb{C} : \ |z| \geq \sigma\} \subset U$. The best-studied case is that of $K(\lambda, z) = A(\lambda z)$, where $A$ is an entire function of a special form with positive Taylor coefficients. The generalized Borel transformation has found the widest application in the particular case where there exists an inverse integral representation of $F$ in terms of $f$ (see (1)); for instance, this is so if

$$A(u) = E_\rho(u; \mu) = \sum_{k=0}^{\infty} \frac{u^k}{\Gamma(\mu + k/\rho)}, \qquad u \in \mathbb{C}, \ \mathrm{Re}\, \mu > 0,$$

is the Mittag-Leffler function. For example, in the years 30s-60s of the XXth century, M. F. Subbotin, I. F. Lokshin, A. E. Avetisyan, and M. M. Dzhrbashyan obtained an analog of Theorem $A_1$ for the class of entire functions of order $\rho > 0$, of normal type, and with nonnegative indicator. Moreover, the indicator diagram in this case is a so-called $\rho$-convex compact set. These results have formed the basis for considerable progress made by Soviet mathematicians in complex analysis in $\rho$-convex domains.

Multidimensional analogs of the Polya theorem were studied by V. K. Ivanov, A. Martineau, L. Ehrenpreis, S. G. Gindikin, L. I. Ronkin, C. O. Kiselman and other authors. Only the results of A. Martineau and L. Ehrenpreis (see Theorem $A_2$) are related to the geometric image of the indicator in $\mathbb{C}^n$. In essence, they considered the case of the convex regularized radial indicator of an entire function of exponential type. It is well known that in the general case this indicator may be nonconvex and even discontinuous. Furthermore, the concept of the radial indicator (which receives widespread attention) is based on the definition (given by Borel at the turn of the XXth century) of the order of an entire function $f(z_1, ..., z_n)$, $n > 1$, with respect to the totality of the variables $z_1, ..., z_n$ :

$$\rho(f) = \varlimsup_{R \to \infty} (\ln R)^{-1} \ln^+ \ln^+ M_f(R, ..., R),$$

where $M_f(r)$ is the maximum of the modulus of $f$ in the polydisk $\{z \in \mathbb{C}^n : |z_j| \leq r_j, \ j = 1, ..., n\}$.

The order $\rho(f)$ coincides with the order of the trace of the function

$$\Phi_f(r) = \ln^+ M_f(r), \quad r \in \mathbb{R}_0^n := \{r \in \mathbb{R}^n : r_1 > 0, ..., r_n > 0\}$$

on the "radial ray" $\Gamma = \{r \in \mathbb{R}_0^n : r_1 = \cdots = r_n\}$ *). It is natural to consider other directions of growth of $\Phi_f$ at infinity. They can be described with the

*) The trace of the function $\Phi_f$ on every other ray $L \subset \mathbb{R}_0^n$ with vertex at $0 \in \mathbb{R}^n$ has the same order of growth (see Lemma 6.2.8).

help of a system of parabolic rays in $\mathbb{R}_0^n$. Then the growth of $\Phi_f$ is characterized by the order function

$$\rho_f(u) = \overline{\lim_{t\to\infty}} (\ln t)^{-1} \ln^+ \Phi_f(t^{u_1}, ..., t^{u_n}), \ u \in \mathbb{R}^n .$$

If $0 < \rho(f) < \infty$, then the totality of directions $\{u \in \mathbb{R}^n : \rho_f(u) > 0\}$ of nonzero order of growth for $f$ is a concave cone in $\mathbb{R}^n$. From this viewpoint, if we aim at the general problems of the theory of entire functions rather than at specific applied questions, special attention to the radial indicator looks merely like a tribute to tradition. In this connection, the problem arises of investigating "nontraditional" growth directions for an entire function of several variables, and of introducing new indicators of its asymptotic behavior with respect to the whole collection of growth directions. These studies were initiated by L. Baumgartner, J. Valiron, M. M. Dzhrbashyan, L. I. Ronkin, and A. A. Goldberg.

The book is devoted to traditional and new growth characteristics for entire and plurisubharmonic functions. All these characteristics are designed with the aid of the geometric theory of functions of complex variable and convex analysis. In one or another way, the applications to mathematics and biophysics treated in the monograph are related to Theorems $A_1$ and $A_2$.

# Chapter 1

# Preliminaries

## 1.1. On the growth of nondecreasing functions of one variable

Let $\Lambda = \{\varphi(t)\}$ be the class of nonnegative monotone nondecreasing functions defined on the semiaxis $t \geqslant 0$. We consider basic concepts associated with the simplest growth scale for $\Lambda$ (a *growth scale* is a set of functions, which are used to measure the growth).

**Definition 1.1.1.** Let $\varphi \in \Lambda$. The number

$$\rho = \rho[\varphi] = \overline{\lim_{t\to\infty}}\ (\ln t)^{-1} \cdot \ln^+ \varphi(t)$$

is called the *order* of $\varphi$. If $\rho < \infty$, then the number

$$\sigma = \sigma[\varphi] = \overline{\lim_{t\to\infty}}\ t^{-\rho} \cdot \varphi(t)$$

is called the *type* of $\varphi$. In the cases where $\sigma = 0$, $0 < \sigma < \infty$, and $\sigma = \infty$, the function $\varphi$ is said to have *minimal, normal, and maximal* type, respectively.

**Example.** Let $\varphi_1(t) = t$ for $0 \leqslant t \leqslant e$ and $\varphi_1(t) = (\ln t)^{-1} \cdot t$ for $t \geqslant e$. Define $\varphi_2(t) = At$, $t \geqslant 0$, and $\varphi_3(t) = t \ln^+ t$, $t \geqslant 0$. We have:

$$\varphi_i \in \Lambda,\ \rho[\varphi_i] = 1, i = 1, 2, 3;\ \sigma[\varphi_1] = 0,\ \sigma[\varphi_2] = A,\ \sigma[\varphi_3] = \infty.$$

The geometric meaning of these concepts was mentioned in the Foreword. Let $N \subset \Lambda$ be the subclass of functions of finite nonzero order and of normal type. As the growth scale for $N$, we choose the set $Q = \{\psi(t) = \tau t^\gamma : \tau, \gamma > 0\}$.

**Proposition 1.1.2.** *Let $\varphi \in N$. In the growth scale $Q$ there exists a unique function $\psi_0$ asymptotically equivalent to $\varphi$ in the following sense:*

$$\overline{\lim_{t\to\infty}}\ \varphi(t)/\psi_0(t) = 1,$$

*and $\psi_0(t) = \sigma t^\rho$, where $\rho$ and $\sigma$ are the order and the type of $\varphi$.*

The growth of a function $\varphi \in \Lambda$ can also be characterized by the behavior (convergence or divergence) of the integral

$$J(\alpha) = \int_1^\infty \varphi(t)/t^{\alpha+1} dt, \quad \alpha > 0.$$

**Proposition 1.1.3.** *Suppose that a function $\varphi \in \Lambda$ has order $\rho < \infty$; put $\mu = \inf\{\alpha > 0 : J(\alpha) < \infty\}$. Then $\mu = \rho$.*

◀ Let $\gamma > \rho$. For some $t_0 > 0$ we have $\varphi(t) < t^\gamma \ \forall\, t > t_0$. Therefore, the integral $J(\alpha)$ converges for $\alpha > \gamma$ and $\mu \leqslant \rho$. On the other hand, if the integral $J(\alpha)$ converges for some $\alpha > 0$, then for every $\varepsilon > 0$ there exists $r_0 = r_0(\varepsilon) \geqslant 1$ such that

$$\frac{\varphi(r)}{\alpha r^\alpha} = \varphi(r) \cdot \int_r^\infty \frac{dt}{t^{\alpha+1}} \leqslant \int_r^\infty \frac{\varphi(t)dt}{t^{\alpha+1}} < \varepsilon$$

for all $r > r_0$, or

$$\varphi(r) < \varepsilon \alpha r^\alpha \quad \forall\, r > r_0. \tag{1.1.1}$$

Now we conclude that $\rho \leqslant \mu$. Thus, $\rho = \mu$ ▶

**Definition 1.1.4.** A function $\varphi \in \Lambda$ of finite order $\rho$ is said to *belong to the convergence class or to the divergence class* depending on whether or not the integral $\int_1^\infty \varphi(t)/t^{\rho+1}dt$ converges or diverges.

Inequality (1.1.1) implies the following assertion.

**Proposition 1.1.5.** *If a function $\varphi$ in $\Lambda$ belongs to the convergence class, it is of minimal type. The converse is wrong* (consider the function $\varphi_1$ in the above example).

**Corollary 1.1.6.** *The functions of normal or maximal type belong to the divergence class.*

## 1.2. Semicontinuous functions

**Definition 1.2.1.** A function $V$ defined on a set $S \subset \mathbb{R}^n$ is said to be *lower semicontinuous* if $V(S) \subset (-\infty, \infty]$ and

$$\varliminf_{u \to x; u \in S} V(u) = V(x) \quad \forall\, x \in S, \tag{1.2.1}$$

where

$$\varliminf_{u \to x; u \in S} V(u) = \sup_{\varepsilon > 0}\{\inf\{V(u) : u \in S; |u - x| < \varepsilon\}\}.$$

Next, $V$ is said to be *upper semicontinuous* on $S$ if $-V$ is lower semicontinuous.

We present a geometric characterization of lower semicontinuous functions.

**Theorem 1.2.2.** *Let $V$ be a function that maps $\mathbb{R}^n$ to $(-\infty, \infty]$. The following properties are equivalent:*

1) *the function $V$ is lower semicontinuous on $\mathbb{R}^n$;*
2) *the epigraph* epi $V$ *of $V$ (see Notation) is closed in $\mathbb{R}^{n+1}$;*
3) *the level set $T_\alpha := \{u \in \mathbb{R}^n : V(u) \leqslant \alpha\}$ of $V$ is closed for every $\alpha \in \mathbb{R}$.*

◀ 1)→2). Suppose $(u_m, \mu_m) \in \operatorname{epi} V$, $(u_m, \mu_m) \to (x, \mu)$ as $m \to \infty$. For every $\varepsilon > 0$ we have

$$\mu + \varepsilon > \mu_m \geqslant V(u_m) \quad \forall\, m > m_0 = m_0(\varepsilon). \tag{1.2.2}$$

Identity (1.2.1) for $S = \mathbb{R}^n$ implies that

$$V(u_m) > V(x) - \varepsilon \quad \forall\, m > m_1 = m_1(\varepsilon).$$

Now from inequality (1.2.2) we conclude that $V(x) < \mu + 2\varepsilon \ \forall\, \varepsilon > 0$ or $V(x) \leqslant \mu$, i.e., $(x, \mu) \in \operatorname{epi} V$.

2)→3). Let $\Gamma_\alpha = \{(u, u_{n+1}) \in \mathbb{R}^n \times \mathbb{R} : u_{n+1} = \alpha\}$, $a \in \mathbb{R}$. Then $\Gamma_\alpha \bigcap \operatorname{epi} V$ and, consequently, the $T_\alpha$ are closed sets.

3)→1). Suppose $x \in \mathbb{R}^n$, and denote $L_\alpha = \mathbb{R}^n \setminus T_\alpha$. Then $x \in L_\alpha \ \forall\, \alpha < V(x)$. But $L_\alpha$ is open in $\mathbb{R}^n$ for every $\alpha \in \mathbb{R}$. This implies the lower semicontinuity of $V$ at $x$ (see (1.2.1)) ▶

**Exercise 1.2.3.** The sum of a finite number of lower semicontinuous functions is lower semicontinuous.

**Exercise 1.2.4.** A lower semicontinuous function $V$ defined on a compact set $K$ is lower bounded on $K$ and there exists a point $x \in K$ such that $\inf\{V(u) : u \in K\} = V(x)$.

**Exercise 1.2.5.** The supremum of every family of lower semicontinuous functions defined on a set $S \subset \mathbb{R}^n$ is a lower semicontinuous function on $S$.

**Exercise 1.2.6.** The lowest upper bound $V(u)$ of any monotone nondecreasing sequence $\{V_m(u)\}_1^\infty$ of lower semicontinuous functions on a set $S \subset \mathbb{R}^n$ is a function lower semicontinuous on $S$, and for all $x \in S$ we have

$$V(x) = \sup_{\varepsilon > 0; k > 0} \inf\{V_m(u) : |u - x| < \varepsilon;\ u \in S;\ m > k\}.$$

**Exercise 1.2.7.** If a function $F(r, y)$ is continuous in the variable $r$ on a set $\Omega \subset \mathbb{R}^n$ for every $y \in K$, where $K$ is a compact set in $\mathbb{R}^n$, and is lower semicontinuous with respect to the totality of variables on the set $\Omega \times K$, then $\Phi(r) = \sup_{y \in K} F(r, y)$ is a continuous function on $\Omega$.

We mention the following general fact; surely, it is also true, in particular, for semicontinuous functions.

**Exercise 1.2.8.** Let $X, Y$ be sets in $\mathbb{R}^n$, and let $F(x, y)$ be a function defined on $X \times Y$. Then

$$\sup_{x \in X} \sup_{y \in Y} F(x, y) = \sup_{y \in Y} \sup_{x \in X} F(x, y).$$

**Definition 1.2.9.** Consider a function $V : S \to [-\infty, \infty]$. The *lower semicontinuous envelope* of $V$ is the function $\operatorname{cl} V(x) = \inf\{\lambda \in R : (x, \lambda) \in \overline{\operatorname{epi} V}\}$ defined on $\mathbb{R}^n$.

**Exercise 1.2.10.** The function $\operatorname{cl} V$ is the largest lower semicontinuous minorant of $V$, and (see (1.2.1))

$$\operatorname{cl} V(x) = \varliminf_{u \to x : u \in S} V(u) \quad \forall\, x \in \overline{S}; \quad \operatorname{cl} V(x) = \infty \quad \forall\, x \in \mathbb{R}^n \setminus \overline{S}.$$

## 1.3. Convex sets and associated functions

### 1.3.1. Some properties of convex sets.

**Definition 1.3.1.** A set $M \subset \mathbb{R}^n$ is said to be *affine (convex)* if the straight line $\Pi = \{\lambda x + (1-\lambda)y : \lambda \in R\}$ (the segment $\Gamma = \{\lambda x + (1-\lambda)y : \lambda \in [0,1]\}$) is included in $M$ for arbitrary points $x, y \in M$ such that $x \neq y$.

**Exercise 1.3.2.** If $M$ is a convex set, then it consists of all *convex combinations* of its points, i.e., $M = \bigcup M_m | m \in \mathbb{N}$, where

$$M_m = \left\{ \sum_{i=1}^{m} \lambda_i x_i : \sum_{i=1}^{m} \lambda_i = 1;\ \lambda_i \geqslant 0;\ x_i \in M;\ i = 1, \dots, m \right\}.$$

**Exercise 1.3.3.** Every affine set $M$ in $\mathbb{R}^n$ is a shift of a unique linear subspace $L \subset \mathbb{R}^n$, namely, $L = M - M$, where $M - M = \{x - y :\ x, y \in M\}$.

**Definition 1.3.4.** The *affine (convex) hull* $\operatorname{aff} M$ (respectively, $\operatorname{conv} M$) of a set $M \subset \mathbb{R}^n$ is the smallest affine (convex) set including $M$.

**Exercise 1.3.5.** For every $M \subset \mathbb{R}^n$ the set $\operatorname{conv} M$ consists of all convex combinations of points of $M$.

**Definition 1.3.6.** The *dimension of an affine set* $M \subset \mathbb{R}^n$ is the dimension of the linear subspace $L = M - M$ (see Exercise 1.3.3): $\dim M := \dim L$. The *dimension of a convex set* $M \subset \mathbb{R}^n$ is the dimension of its affine hull $\operatorname{aff} M : \dim M := \dim \operatorname{aff} M$. If $\dim M = n - 1$, then the affine set $M$ is called a *hyperplane*.

**Exercise 1.3.7.** Let $\Delta_k = \operatorname{conv}\{x_0, \dots, x_n\}$ with $x_0, \dots, x_n \in \mathbb{R}^n$. If $\dim \Delta_k = l$, then $k \geqslant l$ (if $l = k$, the set $\Delta_k$ is called a *k-dimensional simplex*).

The following assertion shows that in Exercise 1.3.2 it suffices to take convex combinations of at most $n+1$ points of $M$.

**Theorem 1.3.8.** *If $M$ is a set in $\mathbb{R}^n$, then*

$$\operatorname{conv} M = \left\{ \sum_{i=0}^{n} \lambda_i x_i : (\lambda_0, \dots, \lambda_n) \in B; x_0, \dots, x_n \in M \right\},$$

*where*

$$B = \left\{ (\lambda_0, \dots, \lambda_n) \in \mathbb{R}^{n+1} : \sum_{i=0}^{n} \lambda_i = 1;\ \lambda_i \geqslant 0 \right\}. \tag{1.3.1}$$

◀ Let $x \in \operatorname{conv} M$. Suppose $k$ is the smallest integer such that

$$x = \sum_{i=0}^{k} \lambda_i x_i, \text{ where } \sum_{i=0}^{k} \lambda_i = 1, \ \lambda_i > 0, \text{ and } x_i \in M, \ i = 0, 1, \ldots, k, \quad (1.3.2)$$

$\Delta_k = \operatorname{conv}\{x_0, \ldots, x_k\}$, and $l = \dim \Delta_k$ (see Exercise 1.3.5). Since $l \leqslant n$, the theorem will be proved if we show that $l = k$. By Exercise 1.3.7, $k \geqslant l$. Suppose $k > l$. Then the vectors $x_i - x_0$, $i = 1, \ldots, k$, are linear dependent. Therefore, there exist numbers $\alpha_0, \ldots, \alpha_k$ not all equal to zero and such that

$$\sum_{i=0}^{k} \alpha_i x_i = 0; \quad \sum_{i=0}^{k} \alpha_i = 0. \quad (1.3.3)$$

The set $J = \{i \in \{0, 1, \ldots, k\} : \alpha_i < 0\}$ is nonempty. There exists an index $j \in J$ with $\max\limits_{i \in J} \dfrac{\lambda_i}{\alpha_i} = \dfrac{\lambda_j}{\alpha_j} < 0$ (see (1.3.2)). By formulas (1.3.3), (1.3.2), we have

$$x = \sum_{i=0}^{k} \mu_i x_i, \text{ where } \mu_i = \lambda_i - \frac{\lambda_j}{\alpha_j}\alpha_j, \text{ and } \sum_{i=0}^{k} \mu_i = 1. \quad (1.3.4)$$

If $\alpha_i \geqslant 0$, then $\mu_i > 0$, because $\alpha_j < 0$. If $\alpha_i < 0$, i.e., $i \in J$, then $\mu_j \geqslant 0$, because $\dfrac{\lambda_i}{\alpha_i} \leqslant \dfrac{\lambda_j}{\alpha_j}$. But $\mu_j = 0$. This contradicts the choice of the number $k$ (see (1.3.4), (1.3.2)). So, $k = l$ ▶

**Theorem 1.3.9.** *The convex hull* conv $K$ *of every compact set* $K$ *in* $\mathbb{R}^n$ *is a compact set* .

◀ We consider the continuous mapping of the topological product $A = B \times \underbrace{K \times \cdots \times K}_{n+1}$ (see (1.3.1)) to $\mathbb{R}^n$ defined by the formula $((\lambda_0, \ldots \lambda_n), x_0, x_1, \ldots, x_n) \to \sum\limits_{i=0}^{n} \lambda_i x_i$. The set conv $K$ is the image of $A$ under this mapping (see Theorem 1.3.8) and, since $A$ is a compact set, conv $K$ is also compact. ▶

**Exercise 1.3.10.** Let $K, L$ be convex sets in $\mathbb{R}^n$. Then

$$\operatorname{conv}(K \cup L) = \bigcup_{0 \leqslant \lambda \leqslant 1} (\lambda K + (1 - \lambda) L).$$

**Exercise 1.3.11.** Let $\mathcal{D}$ be an $n$-dimensional closed convex set in $\mathbb{R}^n$ with nonempty boundary $\Gamma$. If $\mathcal{D}$ is not a half-space, then conv $\Gamma = \mathcal{D}$.

**Theorem 1.3.12.** *Let* ri $M$ *be the interior of a convex set* $M \subset \mathbb{R}^n$ *with respect to* aff $M$ $(\dim M \geqslant 1)$. *Then* $\overline{\operatorname{ri} M} = \overline{M}$, *and* ri $M$ *is convex.*

◀ There is no loss of generality in assuming that $\dim M = n$, $\operatorname{ri} M = \overset{\circ}{M}$. Clearly, $\overset{\circ}{M} \neq \varnothing$: $M$ always contains some $n$-dimensional simplex. Suppose

$y \in \overline{M}$, $x \in \overset{\circ}{M}$, $S = \{u \in \mathbb{R}^n : |u| \leqslant 1\}$, and $\lambda \in [0,1)$. Then $y \in M + \varepsilon S \ \forall \ \varepsilon > 0$, because $\overline{M} = \bigcap\limits_{\varepsilon>0} (M + \varepsilon S)$. On other hand, we see that $x \in \overset{\circ}{M}$ and, consequently, $x + \delta S \subset M$ for some $\delta > 0$. Therefore, for sufficiently small $\varepsilon > 0$ we obtain:

$$(1-\lambda)x + \lambda y + \varepsilon S \subset (1-\lambda)x + \lambda(M + \varepsilon S) + \varepsilon S =$$

$$= (1-\lambda)[x + \varepsilon(1+\lambda)(1-\lambda)^{-1}S] + \lambda M \subset (1-\lambda)M + \lambda M = M,$$

i.e., $(1-\lambda)x + \lambda y \in \overset{\circ}{M} \quad \forall \ \lambda \in [0,1),\ y \in \overline{\overset{\circ}{M}}$. Thus $\overline{M} \subset \overline{\overset{\circ}{M}}$. The inverse inclusion $\overline{\overset{\circ}{M}} \subset \overline{M}$ is obvious ▶

**1.3.2. The Minkowski functional and the support function.**

**Definition 1.3.13.** Let $M$ be a star-like set in $\mathbb{R}^n$ (i.e., if $x \in M$, then $\lambda x \in M \ \forall \ \lambda \in [0,1]$). The function $p_M(x) = \inf\{\alpha > 0; \alpha^{-1} \cdot x \in M\}$ is called the *Minkowski functional* of $M$.

So, if $x \neq 0$ and the ray with vertex at 0 and containing the point $x$ intersects the boundary of $M$ at the only point $y \neq 0$, then $p_M(x) = |x| \cdot |y|^{-1}$.

The next fact is an immediate consequence of Definition 1.3.11.

**Proposition 1.3.14.** *The Minkowski functional has the following properties:*

1) $p_A(\lambda x) = \lambda p_A(x)$ *and* $p_{\lambda A}(x) = \lambda^{-1} p_A(x)$ *for all* $\lambda > 0$ *and* $x \in \mathbb{R}^n$;

2) *if* $A \subset B$, *then* $p_A(x) \geqslant p_B(x)$ *for all* $x \in \mathbb{R}^n$;

3) *if* $A = \bigcap A_i | i \in I$, *then* $p_A(x) = \sup\{p_{A_i} x : i \in I\}$, $x \in \mathbb{R}^n$;

4) *if* $A = \bigcup A_i | i \in \mathbb{N}$, *where* $\{A_i,\ i \in \mathbb{N}\}$ *is an increasing sequence of sets, i.e.,* $A_i \subset A_{i+1}$, $i \in \mathbb{N}$, *then* $p_A(x) = \inf\{p_{A_i}(x) : i \in \mathbb{N}\}$, $x \in \mathbb{R}^n$. *Here* $A$, $B$, *and* $A_i, i \in I$, *are nonempty star-like sets in* $\mathbb{R}^n$.

The subsequent exposition will require some notions of the theory of partially ordered sets. Let $M, M'$ be partially ordered sets. A mapping $\varphi : M \to M'$ is called an *(anti)homomorophism* if $a \leqslant b$ in $M$ implies $\varphi(a) \leqslant \varphi(b)$ (respectively, $\varphi(b) \leqslant \varphi(a)$) in $M'$. A one-to-one (anti)homomorophism is called an *(anti)isomorphism.*

On the set of functions defined on $\mathbb{R}^n$ (that may also take infinite values), we introduce a natural order relation: $f \leqslant g$ if $f(x) \leqslant g(x) \ \forall \ x \in \mathbb{R}^n$. We order the sets in $\mathbb{R}^n$ by inclusion.

The Minkowski functional helps to relate certain classes of functions and sets (Minkowski duality).

**Theorem 1.3.15.** *Let* $P_1 = \{u : \mathbb{R}^n \to [0,\infty]\}$ *be the class of lower semicontinuous positively homogeneous functions of degree* 1, *i.e.,* $p(\lambda x) = \lambda p(x)$ *for all* $\lambda > 0$ *and* $x \in \mathbb{R}^n$. *We put*

$$P_2 = \{p \in P_1 : p(x) > 0 \quad \forall \ x \in \mathbb{R}^n \setminus \{0\}\}.$$

*Let $X_1$ be the class of closed star-like sets in $\mathbb{R}^n$, and let $X_2$ be the subclass of compact sets in $X_1$. If the sets $P_1$ and $X_1$ are partially ordered as indicated above, then the mapping*

$$\Phi : P_1 \to X_1, \ \Phi(p) = \Omega_p = \Omega(p) := \{x \in \mathbb{R}^n : p(x) \leqslant 1\},$$

*is an antiisomorphism with the following properties:* 1) $\Phi^{-1}(K) = p_K \ \forall \ K \in X_1$, *where $p_K$ is the Minkowski functional of $K$;* 2) $\Phi(P_2) = X_2$.

◄ By the properties of the functions in $P_1$ and by Theorem 1.2.2, we have $\Phi(P_1) \subset X_1 \ \forall \ p \in P_1$. Definition 1.3.13 and the positive homogeneity of any $p \in P_1$ show that $p = p_\Omega$, where $p_\Omega$ is the Minkowski functional of $\Omega_p$. This implies that $\Phi$ is a monomorphism.

Let $K \in X_1$. Obviously, $x_0 \in \Omega(p_K)$. Definition 1.3.13 implies that $\dfrac{x_0}{p_K(x_0)+\varepsilon} \in K \ \forall \ \varepsilon > 0$. Consequently, $x_0 \in K$ if $p_K(x_0) < 1$. If $p_K(x_0) = 1$, then $x_0/(1+\varepsilon) \in K \ \forall \ \varepsilon > 0$. Taking into account that $K$ is closed, again we find: $x_0 \in K$. Thus, $K = \Omega(p_K)$. Recalling that $p_K$ is positively homogeneous of degree 1 (see P. 1.3.14, 1)) and applying Theorem 2.2, we conclude that $p_K \in P_1$, i.e., $\Phi^{-1}(K) = p_K$. Consequently, $\Phi$ is an epimorphism. By property 2) in Proposition 1.3.14, $\Phi$ is an antiisomorphism.

Let $m = \inf\{p(x) : x \in \mathbb{R}^n; |x| = 1\}$. If $p \in P_2$, then $m > 0$ (see Exercise 1.2.4). Therefore, $p(x) \geqslant m \cdot |x| \ \forall \ x \in \mathbb{R}^n$ and $\Omega_p \subset \{x \in \mathbb{R}^n : |x| \leqslant 1/m\}$, i.e., $\Omega_p \in X_2$. On the other hand, if $K \in X_2$, then $p_K \in P_2$. Indeed, if a point $x_0 \in \mathbb{R}^n \setminus \{0\}$ with the property $p_K(x_0) = 0$ exists, then by the positive homogeneity of the function $p$, the ray $\{tx_0 : t > 0\}$ is included in $\Omega(p_K) = K$. This contradicts the fact that $K$ is a compact set. Thus, $\Phi(P_2) = X_2$ ►

An entirely similar theorem can be stated about the Minkowski duality between the class $\{p : \mathbb{R}^n \to [0,\infty]\}$ of sublinear functions (i.e., of the functions with the properties

$$p(\lambda x) = \lambda p(x) \ \forall \ \lambda > 0, x \in \mathbb{R}^n, \ p(x+y) \leqslant p(x) + p(y) \ \forall \ x, y \in \mathbb{R}^n,$$

and $p(0) = 0$) and the class of convex sets in $\mathbb{R}^n$ containing 0. This theorem is based on the following well-known result.

**Theorem 1.3.16.** *Let $K$ be a convex set in $\mathbb{R}^n$ containing 0. Then the Minkowski functional $p_K(x)$ is a sublinear function with values in $[0,\infty]$.*

**Theorem 1.3.17.** *Let $K$ be a closed convex set in $\mathbb{R}^n$ such that $0 \in \overset{\circ}{K}$, and $K \neq \mathbb{R}^n$. Put $T = \{x \in \mathbb{R}^n : p_K(x) < 1\}$. Then*

$$\overset{\circ}{K} = T; \quad \partial K = \{x \in \mathbb{R}^n : P_K(x) = 1\}. \tag{1.3.5}$$

◄ If $x \in \overset{\circ}{K}$, then for sufficiently small $\varepsilon > 0$ we have: $(1-\varepsilon)^{-1} \cdot x \in K$ and $p_K(x) < 1$. So, $\overset{\circ}{K} \subset T$. Suppose $y \in T$ and $\varepsilon \in (0, 1 - p_K(y))$. Then $z := [p_K(y) + \varepsilon]^{-1} \cdot y \in K$. But $0 \in \overset{\circ}{K}$. Therefore (see the proof of Theorem

1.3.12), $\lambda z \in \overset{\circ}{K}$ $\forall\ \lambda \in [0, 1)$. Consequently, $y \in \overset{\circ}{K}$ and $T \subset \overset{\circ}{K}$, i.e., $T = \overset{\circ}{K}$. Now (1.3.5) follows from the formula $K = \{x \in \mathbb{R}^n : p_K(x) \leqslant 1\}$ (see Theorem 1.3.15) ►

**Definition 1.3.18.** The *support function of a set* $M \subset \mathbb{R}^n$ is the function $h_M(u) = \sup\{\langle u, x\rangle : x \in M\}$, $u \in \mathbb{R}^n$.

If $|u| = 1$, then $d = |h_M(u)|$ is the distance from the origin in $\mathbb{R}^n$ to the hyperplane $\{x \in \mathbb{R}^n : \langle u, x\rangle = h_M(u)\}$. This is a support hyperplane to the set $M$ (see the remark to Theorem 1.3.21 below) with direction vector $u$ if $d < \infty$.

**Theorem 1.3.19.** *Let $K$ be a compact convex set in $\mathbb{R}^n$, and let $h$ be a finite sublinear function on $\mathbb{R}^n$. We put $T(h) = \{x \in \mathbb{R}^n : \langle u, x\rangle \leqslant h(u)\ \forall\ u \in \mathbb{R}^n\}$. Then $h$ coincides with the support function $h_K$ of $K$ if and only if $K = T_h$.*

◄ *The "only if" part.* Suppose $K \neq T(h)$ and $x_0 \in T(h) \setminus K$ (the inclusion $K \subset T(h)$ is obvious under the assumptions of T. 1.3.19). By Theorem 1.3.22 (see below), there exists a hyperplane $\Gamma = \{x \in \mathbb{R}^n : \langle u_0, x\rangle = c\}$ such that $h_K(u_0) < c < \langle u_0, x_0\rangle$. This contradicts the fact that $x_0 \in T(h)$. So, $K = T(h)$.

*The "if" part.* By the corollary to the Hahn-Banach theorem, for every point $u \in \mathbb{R}^n$ there exists a linear functional $l$ on $\mathbb{R}^n$ such that $l(u) = h(u)$ and $l(v) \leqslant h(v)$, $v \in \mathbb{R}^n$. Clearly, q$l(v) = \langle v, x_0\rangle\ \forall\ v \in \mathbb{R}^n$ for some $x_0 \in \mathbb{R}^n$. Then $x_0 \in T(h)$ and, consequently, $h \leqslant h_K$. The reverse inequality is obvious under the assumptions of the theorem. Thus, $h = h_K$ ►

**Remark.** Theorem 1.3.19 remains valid if in its notation $K$ is a closed convex set in $\mathbb{R}^n$ and $h : \mathbb{R}^n \to (-\infty, \infty]$ is a lower semicontinuous sublinear function (see [157, p. 130]).

**Exercise 1.3.20.** Let $A, B$ be compact convex sets, and let $C \doteq A + B$. Then $h_C(u) = h_A(u) + h_B(u)\ \forall\ u \in \mathbb{R}^n$.

In conclusion we formulate separation theorems for convex sets. Let $K, T$ be two nonempty sets in $\mathbb{R}^n$. A hyperplane $H$ *separates* (*strongly separates*) $K$ and $T$ if $K$ is contained in one of the closed (open) half-spaces bounded by $H$, and $T$ is contained in the other closed (open) half-space.

**Theorem 1.3.21 (separation theorem).** *Let $K$ and $T$ be nonempty convex sets in $\mathbb{R}^n$, and let* $\operatorname{ri} K \cap \operatorname{ri} T = \varnothing$. *Then there exists a hyperplane $H$ separating $K$ and $T$.*

**Remark.** The assumption of the theorem is true, in particular, when $\dim K = n$, and $T = \{t\}$ is a boundary point for $K$. Then the hyperplane $H$ is called *supporting for the set* $K$.

**Theorem 1.3.22 (strong separation theorem).** *If $K$ and $T$ are nonempty convex sets in $\mathbb{R}^n$ satisfying the condition $\overline{K} \cap \overline{T} = \varnothing$, then there exists a hyperplane $H$ separating $K$ and $T$ strongly.*

## 1.4. Convex functions

**1.4.1. Some properties of convex functions.**

**Definition 1.4.1.** A function $V : S \to [-\infty, \infty]$, where $S \subset \mathbb{R}^n$, is said to be *convex* if its epigraph epi $V$ is a convex set in $\mathbb{R}^{n+1}$. A function $V$ is said to be *concave* on $S$ if $-V$ is convex.

We note that under the assumptions of Definition 1.4.1 the set $\operatorname{dom} V = \{u \in \mathbb{R}^n : V(u) < \infty\}$ is convex. Convex functions that do not take infinite values of opposite signs can be defined by the following inequality.

**Proposition 1.4.2.** *Suppose $S$ is a convex subset of $\mathbb{R}^n$ and $R = (-\infty, \infty]$ or $[-\infty, \infty)$. A function $V : S \to R$ is convex if and only if*

$$V(\lambda x + (1-\lambda)y) \leqslant \lambda V(x) + (1-\lambda)V(y) \quad \forall\, \lambda \in (0,1) \tag{1.4.1}$$

*for all $x, y \in S$.*

**Exercise 1.4.3.** If $V$ is a finite convex function defined on a convex set $S \subset \mathbb{R}^n$, then

$$V\left(\sum_{i=0}^{m} \lambda_i x_i\right) \leqslant \sum_{i=0}^{m} \lambda_i V(x_i) \tag{1.4.2}$$

for all $\lambda_0, \dots \lambda_m \geqslant 0$, $\sum\limits_{i=0}^{m} \lambda_i = 1$, and $x_0, \dots, x_m \in S$, $m \in \mathbb{N}$.

**Theorem 1.4.4.** *A finite convex function $V$ defined on a set $S \subset \mathbb{R}^n$ is continuous on the set* ri $S$ *if* $\dim S \geqslant 1$.

◀ The set $S$ is convex. There is no loss of generality in assuming that $\dim S = n$, $\operatorname{ri} S = \overset{\circ}{S}$. Let $x \in \overset{\circ}{S}$. We consider the convex function $g(y) = V(x+y) - V(x)$ defined on the convex set $T = \overset{\circ}{S} - x$ (see T. 1.3.12). Let $A \subset \overset{\circ}{\Delta} \subset \Delta \subset T$, where $A$ is a ball with center at 0, and $\Delta$ is an $n$-dimensional simplex. Together with Exercise 1.3.5 and formula (1.3.1), inequality (1.4.2) implies that

$$a := \max\{g(y_0), g(y_1), \dots, g(y_n)\} \geqslant g\left(\sum_{i=0}^{n} \lambda_i y_i\right) \quad \forall\, \lambda \in B,$$

where $\{y_0, \dots, y_n\}$ are the vertices of the simplex $\Delta$. Since $0 \in \Delta$, we have $g(0) = 0 \leqslant a$. For every $\varepsilon > 0$ we choose $\delta \in (0,1)$, $\delta a < \varepsilon$, and take the ball $U = \delta A$. Using the convexity of the function $g(y)$ and inequality (1.4.2), we obtain:

$$g(y) = g\left((1-\delta)\cdot 0 + \delta \frac{y}{\delta}\right) \leqslant \delta g\left(\frac{y}{\delta}\right) \leqslant \delta a < \varepsilon \quad \forall\, y \in U;$$

$$0 = 2g(0) \leqslant g(y) + g(-y) \quad \forall\, y \in U.$$

These two inequalities imply that $g(y) \geqslant -\delta a > -\varepsilon$, $|g(y)| < \varepsilon \ \forall\, y \in U$, i.e., the function $V$ is continuous at the point $x \in \overset{\circ}{S}$ ▶

**Proposition 1.4.5.** *Let $V : S \to [-\infty, \infty)$ be a convex function. If $V(x) = -\infty$ for some $x \in S$, then $V(u) \equiv -\infty$ for $u \in \operatorname{ri} S$.*

◀ As above we assume that $\dim S = n$. For every fixed $u \in \overset{\circ}{S}$, if $\lambda \in (0,1)$ is sufficiently small, then $z = (1-\lambda)^{-1} \times [u - \lambda x] \in S$, and in this case $u = \lambda x + (1-\lambda)z$. Taking (1.4.1) into account, we find

$$V(u) \leqslant \lambda V(x) + (1-\lambda)V(z) = -\infty, \quad V(u) = -\infty \ ▶$$

We consider a convexity criterion for functions defined on a convex domain $\mathcal{D} \subset \mathbb{R}^\backslash$. Let $\Gamma = \{xt + b \in \mathbb{R}^n : t \in \mathbb{R}\}$ (where $x \in \mathbb{R}^n \setminus \{0\}$, $b \in \mathbb{R}^n$) be a straight line in $\mathbb{R}^n$ having a nonempty intersection with $\mathcal{D}$.

**Theorem 1.4.6.** *Let $V : \mathcal{D} \to \mathbb{R}$. Denote by $A = \{l\}$ the class of affine functions on $\mathbb{R}$ (a function is affine on $\mathbb{R}$ if it is of the form $l(t) = \gamma t + \delta$, $t \in \mathbb{R}$, where $\gamma, \delta \in \mathbb{R}$, $v(t) = V(xt + b)$, $t \in \Delta$). The following conditions are equivalent:*

*1) $V$ is a convex function;*

*2) if $B_l = \{t \in \Delta : v(t) < l(t)\} \neq \emptyset$, then $B_l$ is a connected set for every admissible $x$, $b$, $l$.*

◀ 1) → 2). The function $v$ is convex on the interval $\Delta$. The function $\Phi(t) = v(t) - l(t)$, $t \in \Delta$, possesses the same property. Therefore, the set $B_l = \{t \in \Delta : \ \Phi(t) < 0\}$ is convex and, consequently, connected.

2) → 1). Let $u, b \in \mathcal{D}$, $u \neq b$. We denote $x = u - b$, $v(t) = V(xt+b)$, $t \in \Delta$, and take $\varepsilon > 0$. The set $A$ contains a unique function $l$ determined by the conditions $l(0) = V(b) + \varepsilon$, $l(1) = V(u) + \varepsilon$. Then $\{0,1\} \subset B_l$ and, by assumption, $[0,1] \subset B_l$, i.e.,

$$V[(1-t)b + tu] = v(t) < l(t) = (1-t)l(0) + tl(1) =$$

$$= (1-t)V(b) + tV(u) + \varepsilon \quad \forall\, t \in [0,1].$$

Calculating the limit as $\varepsilon \to 0$, we verify the convexity of $V$ ▶

**Exercise 1.4.7.** If $V(t)$ is a monotone nondecreasing convex function defined on $\mathbb{R}$ and bounded from above, then $V(t) \equiv const$, $t \in \mathbb{R}$.

**Exercise 1.4.8.** Let $V(t)$ be a convex function for $t > 0$. Then the function $\varphi(t) = t \cdot V(1/t)$ is convex for $t > 0$.

**Exercise 1.4.9.** Let $V$ be a convex function on a set $S \subset \mathbb{R}^n$, and let $\varphi$ be a monotone nondecreasing convex function on a set $T \subset \mathbb{R}$ such that $T \supset V(S)$. Then $\varphi \circ V$ is a convex function on $S$.

**Definition 1.4.10.** The *convex envelope* $\operatorname{conv} V$ *of a function* $V : S \to (-\infty, \infty]$ is defined to be the following function on $\mathbb{R}^n$:

$$\operatorname{conv} V(u) = \inf\{\mu \in \mathbb{R} : (u, \mu) \in \operatorname{conv} \operatorname{epi} V\}.$$

This is the largest convex function dominated by $V$. A nontrivial example to this concept is indicated below.

**Example.** Suppose $K = \{u \in \mathbb{R}^2 : |u| \leqslant 1\}$ and $V(u) = 0$, $u \in \partial K \setminus \{1,0\}$, $V(1,0) = -1$. By applying formula (5.4) in §5, it can be proved that

$$\operatorname{conv} V(u) = \frac{1}{2} \cdot \frac{u_1^2 + u_2^2 - 1}{1 - u_1}, \quad u \in \overset{\circ}{K},$$

$$(\operatorname{conv} V)|_{\partial K} = V, \quad \operatorname{conv} V(u) = \infty, \quad u \in \mathbb{R}^2 \setminus K.$$

Exercise 1.3.2 implies the following result.

**Proposition 1.4.11.** *We have*

$$\operatorname{conv} V(x) = \inf\left\{\sum_{i=0}^{m} \lambda_i V(x_i) : \ m \in \mathbb{N}, \ x_i \in S, \ \lambda \in U_m\right\},$$

*where* $\lambda = (\lambda_0, \lambda_1, \ldots, \lambda_m)$ *and*

$$U_m = \left\{\lambda : \ \sum_{i=0}^{m} \lambda_i x_i = x, \ \sum_{i=0}^{m} \lambda_i = 1, \ \lambda_i \geqslant 0\right\}.$$

**Exercise 1.4.12.** If a function $V : S \to (-\infty, \infty)$ is lower bounded on a set $S$, then the function $\operatorname{conv} V$ possesses the same property.

**Exercise 1.4.13.** Let $S$ be a closed convex set in $\mathbb{R}^n$ with $\dim S \geqslant 1$. Suppose $T$ is a convex set on the boundary $\partial S$ (with respect to $\operatorname{aff} S$), and $V : S \to (-\infty, \infty]$ is a function convex on $T$. Then

$$\operatorname{conv} V(x) = V(x) \quad \forall\, x \in T. \tag{1.4.3}$$

**1.4.2. Quasiconvex and logarithmically convex functions.**

**Definition 1.4.14.** A function $V : S \to [-\infty, \infty]$, where $S$ is a convex set, is said to be *quasiconvex* if

$$V(\lambda x + (1-\lambda)y) \leqslant \max\{V(x), V(y)\} \ \forall\, \lambda \in [0,1]; \ x, y \in S.$$

**Example.** Let $V$ be a convex function on $\mathbb{R}^n$. Then $U = \ln^+ V$ is a quasiconvex function.

Obviously, every convex function is quasiconvex. The converse is not true (consider the function $V(t) = \ln t$, $t > 0$). Another example of a quasiconvex function is any monotonic function on $\mathbb{R}$.

**Exercise 1.4.15.** A function $V$ defined on $\mathbb{R}^n$ is quasiconvex if and only if the level set $\{u \in \mathbb{R}^n : \ V(u) \leqslant \alpha\}$ is convex for every $\alpha \in \mathbb{R}^n$.

The relationship between the quasiconvex and convex functions is described by the following theorem.

**Theorem 1.4.16.** *A function $V: S \to \mathbb{R}$ (where $S$ is a convex set in $\mathbb{R}^n$) is convex if and only if the function $V(u)+\langle u, \alpha\rangle$ is quasiconvex on $S$ for every $\alpha \in \mathbb{R}^n$.*

◀ The "only if" part is explained by the fact that every convex function is quasiconvex.

*The "if" part.* We fix $x, y \in S$, $x \neq y$. There exists a point $b \in \mathbb{R}^n$ such that

$$V(x) + \langle b, x\rangle = V(y) + \langle b, y\rangle. \tag{1.4.4}$$

Since $T(u) = V(u) + \langle b, y\rangle$ is a quasiconvex function, by (1.4.4) for all $\lambda \in [0,1]$ we have:

$$T(\lambda x + (1-\lambda)y) \leqslant T(x) = \lambda T(x) + (1-\lambda)T(y),$$

or $V(\lambda x + (1-\lambda)y) \leqslant \lambda V(x) + (1-\lambda)V(y)$, i.e., $V(u)$ is a convex function ▶

**Remark.** Theorem 1.4.16 implies that a sum of quasiconvex functions (even the sum of a quasiconvex function and a linear function) may fail to be quasiconvex.

**Exercise 1.4.17.** Suppose $V$ is a quasiconvex function defined on a convex set $S$ and $\varphi$ is a monotone nondecreasing function on an interval $(a,b) \supset V(S)$. Then $\varphi \circ V$ is a quasiconvex function on $S$.

The class of quasiconvex functions contains the logarithms of convex functions (see the example), and the following class of functions coincides with the set of exponentials of all convex functions.

**Definition 1.4.18.** A nonnegative function $V(u)$ defined on a convex set $S$ is said to be *logarithmically convex* if $\ln V(u)$ is a convex function on $S$ ($\ln 0 = -\infty$).

Proposition 1.4.5 implies the following result.

**Proposition 1.4.19.** *If a logarithmically convex function $V: S \to [0,\infty)$ is equal to $0$ at least at one point $x$ of a convex set $S$, then $V(u) \equiv 0$ for $u \in \operatorname{ri} S$.*

**Theorem 1.4.20.** *A positive function $V$ defined on a convex set $S$ is logarithmically convex if and only if the function $V_a(u) = V(u)e^{\langle u,a\rangle}$ is convex on $S$ for all $a \in \mathbb{R}^n$.*

◀ The "only if" part follows from Exercise 1.4.9, because $\ln V_a(u) = \ln V(u) + \langle u, a\rangle$ is a convex function on $S$.

*The "if" part.* The function $\ln V_a(u)$ is quasiconvex for every $a \in \mathbb{R}^n$ (see the example). By Theorem 1.4.16, $\ln V(u)$ is a convex function on $S$ ▶

### 1.4.3. Subdifferential of a convex function.

**Definition 1.4.21.** Let $V: \mathbb{R}^n \to (-\infty, \infty]$ be a convex function, and let $x \in \operatorname{int}(\operatorname{dom} V)$. The *subdifferential of $V$ at $x$* is defined to be the set

$$\partial V(x) = \{y \in \mathbb{R}^n : V(u) \geqslant V(x) + \langle u - x, y\rangle \ \forall\, u \in \mathbb{R}^n\}. \tag{1.4.5}$$

The geometric meaning of the subdifferential is as follows: if $|V(x)| < \infty$, then $\partial V(x)$ is the set of direction vectors of *nonvertical support hyperplanes* *) to the set epi $V$ at the point $(x, V(x))$. For instance, if $V(u) = |u|$, then

$$\partial V(0) = \{y \in \mathbb{R}^n : |y| \leqslant 1\},$$

$$\partial V(x) = \{x/|x|\} = \text{grad } V(x), \quad x \neq 0$$

(the unit ball is a strictly convex set and $|u|$ is its support function, see Theorem 1.3.19). We show that the subdifferential is an analog of the gradient of a differentiable convex function. First, let $n = 1$.

**Proposition 1.4.22.** *Let $\varphi$ be a finite convex function on an interval $(\alpha, \beta)$. Then at every point $\tau \in (\alpha, b)$ the finite left and right derivatives $\varphi'_-(\tau)$ and $\varphi'_+(\tau)$ of $\varphi$ exist. These derivatives are monotone nondecreasing functions on $(a, b)$, and*

$$\varphi'_-(\tau) \leqslant \varphi'_+(\tau), \quad \partial\varphi(\tau) = [\varphi'_-(\tau), \varphi'_+(\tau)]. \tag{1.4.6}$$

◀ Since $\varphi$ is convex, the quantity $(t-\tau)^{-1} \times [\varphi(t) - \varphi(\tau)]$ is monotone nondecreasing on the set $(\alpha, \beta)\backslash\{\tau\}$ as a function of $t$. Therefore, the derivatives $\varphi'_-(\tau), \varphi'_+(\tau)$ exist and $\varphi'_-(\tau) \leqslant \varphi'_+(\tau)$ $\forall\, \tau \in (\alpha, \beta)$. Let $y \in \partial\varphi(\tau)$. We have $(t-\tau)^{-1} \times [\varphi(t) - \varphi(\tau)] \geqslant y$ for $t > \tau$. Therefore, $\varphi'_+(\tau) \geqslant y$. Arguing as in the case of $t < \tau$, we obtain $\varphi'_-(\tau) \leqslant y$. Now let $y \in [\varphi'_-(\tau), \varphi'_+(\tau)]$. Then

$$y \leqslant \varphi'_+(\tau) \leqslant (t-\tau)^{-1}[\varphi(t) - \varphi(\tau)], \quad \varphi(t) \geqslant \varphi(\tau) + y \cdot (t - \tau)$$

for $t > \tau$. Precisely in the same way we verify that the latter inequality is true also for $t \leqslant \tau$, i.e., $y \in \partial\varphi(\tau)$ ▶

**Proposition 1.4.23.** *Let $V$ be a convex function on $\mathbb{R}^n$. Then its subdifferential $\partial V(x)$ is a nonempty compact convex set for every point $x \in$* int (dom $V$).

◀ The support hyperplane $\Pi$ to the set epi $V$ in $\mathbb{R}^{n+1}$ passes through the point $(x, V(x))$. This hyperplane cannot be vertical ($V$ is a finite continuous function near $x$). Therefore, the hyperplane $\Pi$ admits the representation $\Pi = \{(u, u_{n+1}) \in \mathbb{R}^n \times \mathbb{R} : u_{n+1} = \langle u - x, y\rangle + V(x)\}$ for some $y \in \mathbb{R}^n$. Obviously, $y \in \partial V(x)$, i.e., $\partial V(x) \neq \varnothing$. Definition 1.4.21 implies that $\partial V(x)$ is a closed convex set. Finally, precisely as in (1.4.6), we find

$$\partial V(x) \subset [V'_{-1}(x), V'_{+1}(x)] \times \cdots \times [V'_{-n}(x), V'_{+n}(x)], \tag{1.4.7}$$

where $V'_{-i}(x)$ and $V'_{+i}(x)$ are the left and the right derivative of the function $V$ at the point $x$ in the variable $u_i$, $i = 1, \ldots, n$ (see P. 1.4.2). So, $\partial V(x)$ is a compact set ▶

---

*) I.e., hyperplanes in $\mathbb{R}^{n+1}$, the equation of each of them has the form $u_{n+1} = \langle u, a\rangle + c$, $(u, u_{n+1}) \in \mathbb{R}^n \times \mathbb{R}$, where $a \in \mathbb{R}^n$, $c \in \mathbb{R}$, and the vector $a$ is called the *direction*.

**Definition 1.4.24.** Let $V$ be a convex function on $\mathbb{R}^n$, and let $x \in \operatorname{int}(\operatorname{dom} V)$. The *Gateaux differential of* $V$ *at* $x$ is the function

$$\delta V(x;a) = \lim_{\lambda \to +0} \lambda^{-1}[V(x+\lambda a) - V(x)], \quad a \in \mathbb{R}^n. \tag{1.4.8}$$

**Remark.** The function under the limit sign in (1.4.8) decreases as $\lambda \to +0$ (see the proof of T. 1.4.22). Consequently, this limit exists, and $\delta V(x;a) < \infty$. So, $\delta V(x;a)$ is the one-sided derivative of $V$ at $x$ in the direction $a$ if $a \neq 0$.

The Gateaux differential is closely related to the subdifferential.

**Theorem 1.4.25.** *Under the assumptions of Definition* 1.4.24, *the Gateaux differential* $\delta V(x;a)$ *of* $V(u)$ *at* $x$ *is the support function of the subdifferntial* $\partial V(x)$ *at* $x$.

◀ Taking $u = x + \lambda a$, $\lambda > 0$, and $a \in \mathbb{R}^n$ in formula (1.4.5), passing to the limit as $\lambda \to +0$, and reversing the above arguments, we find

$$\partial V(x) = \{y \in \mathbb{R}^n : \langle y, a\rangle \leqslant \delta V(x;a) \ \forall\ a \in \mathbb{R}^n\} \tag{1.4.9}$$

We have $\partial V(x) \neq \varnothing$ (see P. 1.4.23). The function $\delta V(x;a)$ is the limit of a converging sequence of functions convex in the variable $a$; therefore, this is a finite sublinear function in the variable $a$ on $\mathbb{R}^n$ (see the remark to Definition 1.4.24, Theorem 1.3.15, (1.4.8)). Therefore, by (1.4.9) and Theorem 1.3.19, we have

$$\sup\{\langle y, a\rangle : y \in \partial V(x)\} = \delta V(x;a) \ ▶ \tag{1.4.10}$$

**Theorem 1.4.26.** *For a convex function* $V$ *on* $\mathbb{R}^n$, *the subdifferential* $\partial V(x)$ *at a point* $x \in \operatorname{int}(\operatorname{dom} V)$ *is a singleton if and only if* $\operatorname{grad} V(x)$ *exists at* $x$; *moreover, in this case* $\partial V(x) = \operatorname{grad} V(x)$.

◀ Let $\partial V(x) = \{y\}$. By formula (1.4.10), we have $\delta V(x;a) = \langle y, a\rangle$. Hence, taking $a = \pm e_i$, $i = 1, \ldots, n$, we find $y_i = V_i^+(x) = V_i^-(x)$, i.e., $y = \operatorname{grad} V(x)$. The converse follows from formula (1.4.7): $\partial V(x) \neq \varnothing$ (see P. 1.4.23) ▶

To conclude the section, we mention the following fact which will be useful later on.

**Theorem 1.4.27.** *Let* $\mathcal{D}$ *be a convex domain in* $\mathbb{R}^n$, *and let* $\{V_n : \mathcal{D} \to \mathbb{R}\}$ *be a sequence of convex differentiable functions converging to a differentiable function* $V : \mathcal{D} \to \mathbb{R}$. *Then*

$$\lim_{n\to\infty} \operatorname{grad} V_n(x) = \operatorname{grad} V(x), \quad x \in \mathcal{D},$$

*and even without the differentiability assumption the sequence* $\{V_n\}$ *converges to* $V$ *uniformly on all compact subsets of* $\mathcal{D}$.

## 1.5. Duality of convex functions

We consider functions defined on $\mathbb{R}^n$ with values in $[-\infty, \infty]$. Note that if $V : S \to [-\infty, \infty]$ is a convex function on $S \subset \mathbb{R}^n$, then the function $\tilde{V}$ defined by $\tilde{V}(u) = V(u)$ for $u \in S$ and $\tilde{V}(u) = \infty$ for $u \in \mathbb{R}^n \setminus S$ is a convex extension of $V$ to $\mathbb{R}^n$.

### 1.5.1. Young transformation and its geometric meaning.

**Definition 1.5.1.** The *Young transform* of a function $V: \mathbb{R}^n \to [-\infty, \infty]$ (or the *function conjugate* to $V$) is the function

$$V^*(y) = \sup_{u \in \mathbb{R}^n} [\langle u, y \rangle - V(u)] = \sup_{u \in \mathrm{dom}\, V} [\langle u, y \rangle - V(u)]$$

defined on $\mathbb{R}^n$. The function $V^{**} := (V^*)^*$ is called the *double Young transform* of $V$.

**Example 1.** Consider the *indicator function* of a compact convex set $K$: $\delta_K(u) = 0$ for $u \in K$, $\delta_K(u) = \infty$ for $u \in \mathbb{R}^n \setminus K$. Then $\delta_K^*(y)$ is the support function of $K$ (see Def. 1.3.18); next, $\delta_K^{**}(u) = \delta_{\tilde{K}}(u)$, where $\tilde{K} = \{u \in \mathbb{R}^n : \langle u, y \rangle \leqslant \delta_K^*(y) \ \forall\, y \in \mathbb{R}^n\} = K$ (see Theorem 1.3.19).

**Example 2.** Let $\varphi(t) = Ae^{\rho t}$, $t \in R$, where $A, \rho > 0$. Then

$$\varphi^*(\lambda) = \frac{\lambda}{\rho} \ln \frac{\lambda}{Ae\rho} \quad \forall\, \lambda > 0; \ \varphi^*(0) = 0; \ \varphi^*(\lambda) = \infty \ \forall\, \lambda < 0; \ \varphi^{**} = \varphi.$$

We explain the geometric meaning of the Young transformation.

**Definition 1.5.2.** Let $\Pi$ be a nonvertical hyperplane in $\mathbb{R}^m$ for $m \geqslant 2$ (see the remark to Def. 1.4.21), and let $M$ be a set contained in one of the closed half-spaces determined by $\Pi$. The hyperplane $\Pi$ is called a *barrier* for $M$. If there are arbitrarily small translates of $\Pi$ that fail to be barriers for $M$, then $\Pi$ is called a *support hyperplane for* $M$. A function $V$ defined on $\mathbb{R}^m$ is said to be *proper* if its epigraph has at least one barrier.

**Proposition 1.5.3.** *A function $V \not\equiv \infty$ is proper if and only if* $\mathrm{dom}\, V^* \neq \varnothing$. *Moreover, the hyperplane*

$$B = \{(u, u_{n+1}) \in \mathbb{R}^n \times \mathbb{R} : u_{n+1} = \langle u, y \rangle - c\}. \tag{1.5.1}$$

*is a barrier (support hyperplane) for the epigraph* $\mathrm{epi}\, V$ *of $V$ if and only if* $(y, c) \in \mathrm{epi}\, V^*$ *(respectively, $c = V^*(y)$).*

The geometric meaning of Proposition 1.5.3 is shown in Fig. 1.

◀ If $B$ (see (1.5.1)) is a barrier for the set $\mathrm{epi}\, V$, then this set lies in the half-space

$$\Pi = \{(u, u_{n+1}) \in \mathbb{R}^n \times \mathbb{R} : \ \langle u, y \rangle - u_{n+1} \leqslant c\}. \tag{1.5.2}$$

Consequently, $\langle u, y \rangle - V(u) \leqslant c \ \forall\, u \in \mathbb{R}^n$, $c \geqslant V^*(y)$, and $\mathrm{dom}\, V^* \neq \varnothing$. Other statements of Proposition 1.5.3 are obvious ▶

**Corollary 1.5.4.** *Let $V$ be a finite convex function on $\mathbb{R}^n$, and let $D = \{u \in \mathbb{R}^n : V(u) \leqslant 0\}$, $\dim D = n$. If $\Gamma$ is a support hyperplane for $D$, then $\Gamma = \{u \in \mathbb{R}^n : \langle u, y \rangle = c\}$, where $y \in \mathrm{dom}\, V^*$, $c = V^*(y)$.*

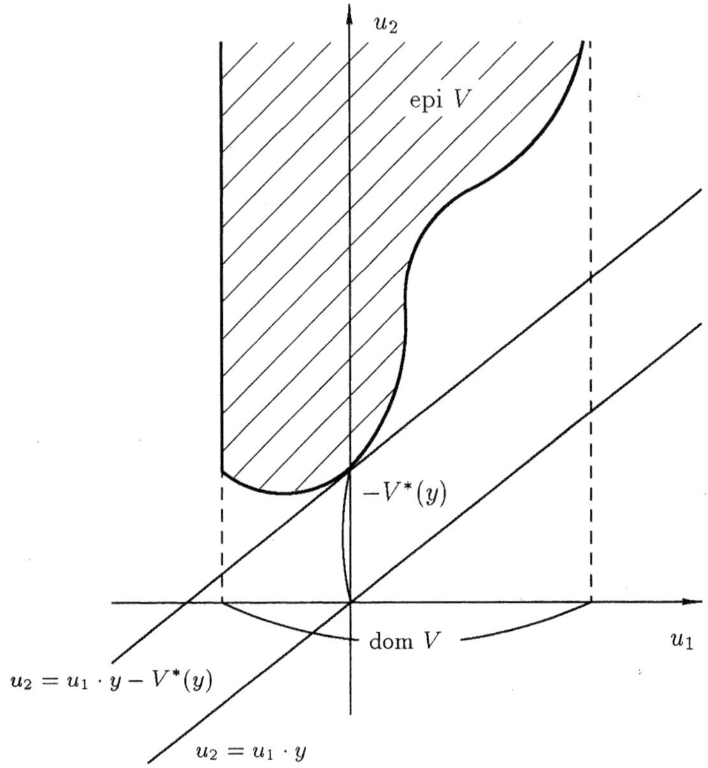


Fig. 1

◀ We consider an embeding $E : \Gamma \to \mathbb{R}^{n+1}$ such that

$$E(\Gamma) = \{(u, u_{n+1}) \in \mathbb{R}^{n+1} : \; u \in \Gamma, \; u_{n+1} = 0\}.$$

Theorem 1.3.21 (separation theorem) and the relation $\operatorname{dom} V = \mathbb{R}^n$ imply that the support hyperplane $B$ for $\operatorname{epi} V$ passes through the affine set $E(\Gamma)$. It only remains to use Proposition 1.5.3 ▶

**Corollary 1.5.5.** *If $V$ is a proper function, then*

$$\operatorname{epi} V^* = \bigcap \{(y, c) \in \mathbb{R}^n \times \mathbb{R} : \langle x, y \rangle - x_{n+1} \leqslant c\} | (x, x_{n+1}) \in \operatorname{epi} V,$$

*and $V^*$ is lower semicontinuous and convex on $\mathbb{R}^n$* (see (1.5.2)).

**1.5.2. Basic theorems.** In Examples 1, 2 the Young transformation is self-dual, i.e., $V^{**} = V$. For applications it is important to describe the class of functions possessing this property.

**Theorem 1.5.6.** *For every function $V$ defined on $\mathbb{R}^n$ we have $V^{**} \leqslant V$; moreover, equality occurs if and only if $V$ is lower semicontinuous and convex on $\mathbb{R}^n$.*

First, we prove several lemmas.

**Lemma 1.5.7.** 1) *Every convex lower semicontinuous function $V : \mathbb{R}^n \to (-\infty, \infty]$, $V \not\equiv \infty$, is proper;* 2) *if $V$ is a proper function, then so is $V^*$.*

◀ The set $M = \operatorname{epi} V$ is closed and convex (see T. 1.2.2). Suppose $(v, \alpha) \in \operatorname{epi} V$, and $m = (v, V(v) - \varepsilon)$. Then $m \notin M \ \forall\, \varepsilon > 0$. By Theorem 1.3.22 (the strong separation theorem), there exists a hyperplane $T = \{(u, u_{n+1}) \in \mathbb{R}^n \times \mathbb{R} : \ f(u, u_{n+1}) = c\}$, where $f(u, u_{n+1}) = \langle u, y \rangle - \gamma u_{n+1}$, separating strongly the set $M$ and the point $m$. In particular, $f(v, \alpha) < c$, $f(m) > c$. Hence, $\gamma \neq 0$, i.e., $T$ is a barrier for the set $\operatorname{epi} V$, and $V$ is a proper function. Statement 2) follows from Corollary 1.5.5 and statement 1) ▶

**Lemma 1.5.8.** *Let $M$ be a closed convex subset of $\mathbb{R}^{n+1}$. If $M$ has at least one barrier, then $M$ coincides with the intersection $\Pi$ of all closed half-spaces containing $M$ and bounded by barriers of $M$.*

◀ Clearly, $M \subset \Pi$. Let $m \in \mathbb{R}^{n+1} \setminus M$. There exists a linear function $f$ on $\mathbb{R}^{n+1}$ and a number $c$ such that

$$T = \{(u, u_{n+1}) \in \mathbb{R}^n \times \mathbb{R} : f(u, u_{n+1}) = c\} \tag{1.5.3}$$

is a hyperplane separating strongly the set $M$ and the point $m$ (see T. 1.3.22), moreover, $f(m) > c$, $f(M) < c$ *). If $T$ is a barrier for $M$, then $m \notin \Pi$. Suppose $T$ is a vertical hyperplane. Under the assumption of the lemma, there exists a linear function $g$ on $\mathbb{R}^{n+1}$ and a number $b$ such that $\{(u, u_{n+1}) \in \mathbb{R}^n \times \mathbb{R} : g(u, u_{n+1}) = b\}$ is a barrier for $M$, and $g[(0, \ldots, 0); 1] \neq 0$, $g(M) \leqslant b$.

For $t > 0$, let $h = g + tf$. By the above, $h[(0, \ldots, 0); 1] \neq 0$, and $h(M) \leqslant ct + b$ for every $t > 0$ (see (1.5.3)). Consequently, the hyperplane $X_t = \{(u, u_{n+1}) \in \mathbb{R}^n \times \mathbb{R} : h(u, u_{n+1}) = ct + b\}$ is a barrier for $M$ for every $t > 0$. Since $f(m) > c$, then there exists $t_0 > 0$ such that $h(m) = g(m) + tf(m) > ct + b \ \forall\, t > t_0$. Therefore, the hyperplane $X_t$ separates strongly the point $m$ and the set $M$, i.e., again $m \notin \Pi$. Thus, $\Pi \subset M$, $\Pi = M$ ▶

**Lemma 1.5.9.** *If $V$ is a proper function, then*

$$\operatorname{epi} V^{**} = \overline{\operatorname{conv} \operatorname{epi} V}, \quad V^{**} = \operatorname{cl} \operatorname{conv} V \tag{1.5.4}$$

(see Definitions 1.3.4, 1.2.9, 4.10).

◀ By Corollary 1.5.5 and Lemma 1.5.7, we have

$$\operatorname{epi} V^{**} = \bigcup \{(u, u_{n+1}) : \langle u, y \rangle - c \leqslant u_{n+1}\} | (y, c) \in \operatorname{epi} V^*. \tag{1.5.5}$$

Therefore (see P. 1.5.3), $\operatorname{epi} V^{**}$ is the intersection of all closed half-spaces containing the set $\operatorname{epi} V$ and bounded by its barriers. But every such barrier is

*) Here and below this means that $f(x) < c \ \forall\, x \in M$.

also a barrier for the set conv epi $V$. Now, by formula (1.5.5) and Lemma 1.5.8, we obtain formula (1.5.4) ▶

**Proof of Theorem 1.5.6.** Formula (1.5.4) and Theorem 1.2.2 imply that $V^{**} = V$ if $V$ is a proper lower semicontinuous convex function. Applying Lemma 1.5.7 and Corollary 1.5.5 to the function $V^*$, we deduce that the converse is true for a proper function $V$. Finally, the inequality $V^{**} \leqslant V$ follows from the inequality

$$V(u) \geqslant \langle u, y\rangle - V^*(y) \quad \forall\ (u, y) \in \mathbb{R}^n \times \mathbb{R}^n.$$

By (1.5.4) and Theorem 1.5.6 (see P. 1.4.11), we prove the following fact.

**Proposition 1.5.10.** *If $V$ is a proper function on $\mathbb{R}^n$, then $V \subset \operatorname{dom} V^{**} \subset \overline{\operatorname{conv} \operatorname{dom} V}$.*

Finally, Theorems 1.4.4, 1.5.6, Proposition 1.5.3, and (1.5.5) imply the next statement.

Corollary 1.5.11. *Let $V$ be a finite convex function on $\mathbb{R}^n$, and let $\mathfrak{M} = \{\psi\}$ be the class of convex lower semicontinuous functions defined on $\mathbb{R}^n$. Then there exists a unique function $\psi \in \mathfrak{M}$ such that $\psi^* = V$; moreover, $\psi = V^*$. In other words, there is only one function $\psi \in \mathfrak{M}$ such that the graph of $V$ is the upper envelope of the family of hyperplanes*

$$\{(u, u_{n+1}) \in \mathbb{R}^n \times \mathbb{R} :\ u_{n+1} = \langle u, y\rangle - \psi(y),\ \ y \in \operatorname{dom} \psi\}$$

*associated with $\psi$ (this is the family of all support hyperplanes for the epigraph of $V$).*

We mention another important fact of the duality theory.

**Theorem 1.5.12.** *Let $\{V_\alpha,\ \alpha \in A\}$ be a family of functions defined on $\mathbb{R}^n$. Suppose $V := \inf\{V_\alpha : \alpha \in A\}$ is a proper function. Then $V^* = \sup\{V_\alpha^* : \alpha \in A\}$.*

◀ The hyperplane $B$ (see (1.5.1)) is a barrier for epi $V$ if and only if it is a barrier for epi $V_\alpha$ for every $\alpha \in A$. It only remains to apply Proposition 1.5.3 (see also Cor. 1.5.5, Ex. 1.2.8) ▶

The Young transformation helps to investigate differential properties of convex functions.

**Proposition 1.5.13.** *Let $V : \mathbb{R}^n \to (-\infty, \infty]$, $V \not\equiv \infty$, be a convex function, and let $T_x = \{y \in \mathbb{R}^n : \langle x, y\rangle = V(x) + V^*(y)\}$. Then the subdifferential $\partial V(x)$ is included in $\operatorname{dom} V^*$, and $\partial V(x) = T_x$ for every $x \in \operatorname{dom} V$.*

◀ Let $\partial V(x) \neq \varnothing$, and let $y \in \partial V(x)$. Then $\langle y, u\rangle - V(u) \leqslant \langle y, x\rangle - V(x)\ \forall\ u \in \mathbb{R}^n$. Consequently, $V^*(y) \leqslant \langle y, x\rangle - V(x)$ and $y \in \operatorname{dom} V^*$. On the other hand, by the definition of the function $V^*$, we have $V^*(y) \geqslant \langle y, x\rangle - V(x)$, i.e., $\partial V(x) \subset T_x$. The inclusion $T_x \subset \partial V(x)$ is proved by reversing the arguments ▶

**Proposition 1.5.14.** *Let $V : \mathbb{R}^n \to (-\infty, \infty)$ be a strictly convex function, i.e., a function satisfying the inequality*

$$V(\lambda x + \mu y) < \lambda V(x) + \mu V(y), \ \lambda, \mu > 0, \ \lambda + \mu = 1, \ x \neq y, \tag{1.5.6}$$

*where $x \neq y$, and let $D = \operatorname{int}(\operatorname{dom} V^*) \neq \varnothing$. Then $V^*$ is a differentiable function on $D$.*

◀ Since $V$ is a continuous function (see T. 1.4.4), Theorem 1.5.6 yields $V = V^{**}$. Hence, applying Theorem 1.5.13 to the function $V^*$, we find

$$\partial V^*(y) = \{x \in \mathbb{R}^n : V(x) + V^*(y) = \langle x, y \rangle\} \quad \forall\, y \in D. \tag{1.5.7}$$

We fix $y \in D$ and take $x_i \in \partial V^*(y)$, $i = 1, 2$, $x_1 \neq x_2$. Then (see Def. 1.4.21) the segment $[x_1, x_2]$ is also included in $\partial V^*(y)$. Now, by (1.5.7), using the notation in (1.5.6), we obtain:

$$V(\lambda x_1 + \mu x_2) = \langle \lambda x_1 + \mu x_2, y \rangle - V^*(y) = \lambda V(x_1) + \mu V(x_2).$$

This contradicts inequality (1.5.6). Consequently, $V^*(y)$ is a singleton (see P. 1.4.23). It only remains to use Theorem 1.4.26 ▶

The geometric meaning of Proposition 1.5.14 is expressed by the following statement.

**Corollary 1.5.15.** *Under the assumptions of Proposition* 1.5.14, *let $x$ be any point in $\mathbb{R}^n$, and let $P$ be the support hyperplane for the epigraph* epi $V$ *at $(x, V(x))$ with the direction vector $y$, i.e.,*

$$\Pi = \{(u, u_{n+1}) \in \mathbb{R}^n \times \mathbb{R} : \ \langle u, y \rangle - u_{n+1} = c\} \tag{1.5.8}$$

*Then $P$ does not pass through other points of the graph of $V$, and $x =$* grad $V^*(y)$ (see T. 1.4.26).

◀ Note that $P$ is a nonvertical support hyperplane (see the remark to Def. 1.4.21). Therefore, formula (1.5.8) is true, and $c = V^*(y)$ (see P. 1.5.3). Consider the set $T := \{u \in \mathbb{R}^n : \langle u, y \rangle = V(u) + V^*(y)\}$, containing the point $x$. Since $V = V^{**}$ (see T. 1.5.6), by Proposition 1.5.13 we have $T = \partial V^*(y)$. Now, Proposition 1.5.14 yields $T = \{x\}$ (see T. 1.4.26) ▶

## 1.6. Asymptotic properties of convex functions

**Proposition 1.6.1.** *Let $S$ be an unbounded closed convex set in $\mathbb{R}^m$. Then*

*i) for every point $x \in S$ there exists a ray $L$ starting at $x$ and lying entirely in $S$;*

*ii) if a ray $\Gamma$ has vertex at 0 and $\Gamma + x \subset S$ for some $x \in S$, then $\Gamma + u \subset S$ $\forall\, u \in S$;*

*iii) the collection of all cones $X$ in $\mathbb{R}^m$ with vertex at 0 and satisfying $X + u \subset S \quad \forall\, u \in S$ possesses the largest element $A(S)$; moreover,*

$$A(S) = \{u \in \mathbb{R}^n : S + u \subset S\}. \tag{1.6.1}$$

**Definition 1.6.2.** The cone $A(S)$ (see P. 1.6.1) is called the *asymptotic (or recessive) cone* of the unbounded closed convex set $S \subset \mathbb{R}^m$.

**Proposition 1.6.3.** *The asymptotic cone $A(S)$ of an unbounded closed convex set $S$ is a closed convex set coinciding with the set $T(S)$ of the limits of all possible sequences of the form $\lambda_1 x_1, \lambda_2 x_2, \ldots$, where $x_i \in S$, $\lambda_i > 0$, and $\lambda_i \to 0$ as $i \to \infty$.*

◀ 1. Formula (1.6.1) implies that $A(S)$ is closed. Take $x, y \in A(S)$ and $0 < \lambda < 1$. By (1.6.1) and the formula $S = \lambda S + (1-\lambda)S$, we have $S + \lambda x + (1-\lambda)y = \lambda(S+x) + (1-\lambda)(S+y) \subset S$, i.e., $A(S)$ is a convex set.

2. We fix $a \in S$, $x \in A(S)$. Since $A(S)$ is a cone, it follows that (see (1.6.1)) $a + \mu x \in S \quad \forall\, \mu \geqslant 0$. For an arbitrary sequence $\{\lambda_i > 0,\ i = 1, 2, \ldots\}$ with $\lambda_i \to 0$ as $i \to \infty$, we denote $\mu_i = \lambda_i^{-1}$, $x_i = a + \mu_i x$. Then $x = \lim\limits_{i\to\infty} \lambda_i x_i$, where $x_i \in S$. So, $A(S) \subset T(S)$.

Suppose $x \in T(S)$, $x \neq 0$. Let $\lambda_i > 0$ and $x_i \in S$, $i = 1, 2, \ldots$ be two sequences such that $\lim\limits_{i\to\infty} \lambda_i x_i = x$ and $\lim\limits_{i\to\infty} \lambda_i = 0$. Since $S$ is a convex set, we have

$$a(1-\lambda_i) + \lambda_i x_i = a + \lambda_i(x_i - a) \in S \quad \forall\, a \in S,\ \lambda_i < 1.$$

Passing to the limit as $i \to \infty$ and recalling that $S$ is closed, we find: $a + x \in S$ $\forall\, a \in S$, i.e. (see (1.6.1)), $x \in A(S)$. So, $T(S) \subset A(S)$, $T(S) = A(S)$ ▶

Let $V$ be a lower semicontinuous convex function defined on $\mathbb{R}^n$ and not inedtically equal to zero. Its epigraph epi $V$ is an unbounded closed convex set in $\mathbb{R}^{n+1}$ (see T. 1.2.2), having the asymptotic cone $A(\text{epi } V)$.

**Definition 1.6.4.** The *asymptotic (or recessive) function* for $V$ is the following function defined on $\mathbb{R}^n$:

$$VO^+(x) = \inf\{\lambda \in \mathbb{R} : (x, \lambda) \in A(\text{epi } V)\}, \tag{1.6.2}$$

i.e., epi $VO^+ = A(\text{epi } V)$.

In the following statement we collect some formulas for the asymptotic function.

**Theorem 1.6.5.** *Let $V$ be a lower semicontinuous function defined on $\mathbb{R}^n$ and such that $V(u) \not\equiv \infty$, and let $x \in \mathbb{R}^n$. We introduce the functions*

$$\rho(x) = \lim_{t\to\infty} t^{-1} V(xt + u),\ u \in \text{dom } V, \tag{1.6.3}$$

$$\gamma(x) = \sup\{V(u+x) - V(u) :\ u \in \text{dom } V\}, \tag{1.6.4}$$

$$\delta(x) = \sup\{\langle x, y\rangle :\ y \in \text{dom } V^*\}, \tag{1.6.5}$$

*Then the function $\rho$ does not depend on $u$ and the asymptotic function $VO^+(x)$ of $V$ is a lower semicontinuous sublinear function, which can be calculated by any of the following formulas:*

$$VO^+(x) = \rho(x) = \gamma(x) = \delta(x) = \varliminf_{u\to x,\, t\to\infty} t^{-1} V(ut),\ x \in \mathbb{R}^n. \tag{1.6.6}$$

◀ 1. The lower semicontinuity and convexity of $VO^+$ follow from Proposition 1.6.3 and Theorem 1.2.2. Since $A(\text{epi } V)$ is a cone with vertex at 0, we have $VO^+(\lambda x) = \lambda VO^+(x)$ $\forall\, \lambda > 0$, $x \in \mathbb{R}^n$.

2. We fix an arbitrary $u \in \text{dom } V$. Since $V(xt+u)$ is a convex function in the variable $t$, the function $t^{-1}[V(xt+u) - V(u)]$ is monotone nondecreasing for $t > 0$. Therefore, the limit in (1.6.3) exists and

$$\rho(x) \geqslant V(x+u) - V(u) \quad \forall\, u \in \text{dom } V. \tag{1.6.7}$$

If $x \in \text{dom } \rho$, this means that $(u, V(u)) + (x, \rho(x)) \in \text{epi } V$. Consequently (see (1.6.1) and (1.6.2)), $(x, \rho(x)) \in A(\text{epi } V) = \text{epi } VO^+$, i.e., $\rho(x) \geqslant VO^+(x)$.

Now, let $x \in \text{dom } VO^+$, and let $\Gamma = \{(xt, tVO^+(x)) : t > 0\}$. Then (see item 1, (1.6.1), and (1.6.2)) $\Gamma \subset \text{epi } VO^+(x) = A(\text{epi } V)$ and $(u, V(u)) + \Gamma \subset \text{epi } V$. It follows that $V(u + xt) \leqslant V(u) + tVO^+(x)$ $\forall\, t > 0$. Consequently, $\rho(x) \leqslant VO^+(x)$. So, $\rho = VO^+$.

3. Let $x \in \text{dom } \rho$. By Definition 1.5.1 and formula (1.6.7), we find

$$V^*(y) \leqslant \sup\,\{\langle u, y\rangle - V(x+u) + \rho(x) : u \in \text{dom } V\} =$$
$$\sup\,\{\langle u+x, y\rangle - V(x+u) : u \in \text{dom } V\} + \rho(x) - \langle x, y\rangle \leqslant$$
$$\leqslant V^*(y) + \rho(x) - \langle x, y\rangle, \quad \text{i.e., } \langle x, y\rangle \leqslant \rho(x) \quad \forall\, y \in \text{dom } V^*.$$

Thus, $\delta \leqslant \rho$ (see (1.6.5)).

4. Since $V = V^{**}$ (see T. 1.5.6), by Definition 1.5.1 for all $x \in \text{dom } V$ we have

$$V(u+x) = \sup\,\{\langle u+x, y\rangle - V^*(y) : y \in \text{dom } V^*\} \leqslant \delta(x) + V(u).$$

It follows that $\gamma \leqslant \delta$ (see (1.6.4)).

5. Let $(u, u_{n+1}) \in \text{epi } V$, and let $x \in \text{dom } \gamma$. Since $u \in \text{dom } V$, by the inequality $V(u) + \gamma(x) \geqslant V(x+u)$ we obtain: $(u, u_{n+1}) + (x, \gamma(x)) \in \text{epi } V$. Consequently (see (1.6.1) and (1.6.2)), $(x, \gamma(x)) \in A(\text{epi } V) = \text{epi } VO^+$, i.e., $VO^+(x) \leqslant \gamma(x)$; $VO^+ \leqslant \gamma$.

The same arguments as above show that $VO^+ = \gamma - \delta - \rho$. Now, recalling that $V$ is a proper function (see Lemmaee 1.5.7), by formula (1.6.3) and Proposition 1.6.3 we verify formula (1.6.6) ▶

## 1.7. Minkowski theorem on convex bodies

The following assertion allows us to estimate the number of integral points (i.e., of points with integral coordinates) in a convex set of given volume.

**Minkowski Theorem.** *Let $D$ be a convex domain in $\mathbb{R}^n$ symmetric with respect to* 0 (i.e., if $x \in D$, then also $-x \in D$) *the volume $V$ of which is strictly greater than $m2^n$, where $m$ is a natural number (the case of $V = \infty$ is not excluded). Then the domain $D$ contains at least $m$ different pairs of integral points $\pm\mathbf{u_j}$, $1 \leqslant j \leqslant m$, not coinciding with 0.*

The proof is based on the following lemma.

**Lemma 1.7.1.** *Let $m$ be a natural number, and let $G$ be a domain in $\mathbb{R}^n$ the volume $V$ of which is strictly greater than $m$ (the case of $V = \infty$ is not excluded). Then there exist $m+1$ different points $\boldsymbol{x}_1, \dots, \boldsymbol{x}_{m+1}$ in $G$ such that all differences $\boldsymbol{x}_i - \boldsymbol{x}_j$ are integral points.*

◀ Let $\mathbb{Z}^n$ be the collection of all integral points in $\mathbb{R}^n$. Denote

$$G_u = \{y \in G : u_i \leqslant y_i < u_i + 1,\ 1 \leqslant i \leqslant n\}, \quad E_u = G_u - u,\ u \in \mathbb{Z}^n,$$

and let $V_u$ be the volume of $E_u$. Obviously,

$$\sum V_u \,|\, u \in \mathbb{Z}^n = V > m. \tag{1.7.1}$$

Since all sets $E_u$, $u \in \mathbb{Z}^n$, lie in the unit cube $K = \{y \in \mathbb{R}^n : 0 \leqslant y_i < 1\}$, by formula (1.7.1) we conclude that there exists at least one point $\nu \in K$ and integral points $\boldsymbol{u}_1, \dots, \boldsymbol{u}_{m+1}$ such that $\nu \in E_{uj}, j = 1, \dots, m+1$. Then $\boldsymbol{x}_j = \nu + \boldsymbol{u}_j \in G$, $j = 1, \dots, m+1$ (and, moreover, the points $\boldsymbol{x}_i - \boldsymbol{x}_j = \boldsymbol{u}_i - \boldsymbol{u}_j$, $i, j = 1, \dots, m+1$ are integral) ▶

**Proof of the Minkowski Theorem**. We apply Lemma 1.7.1 to the set $G = \frac{1}{2}D$ which has volume $2^{-n}V > m$. Let $\boldsymbol{x}_1, \dots, \boldsymbol{x}_{m+1} \in G$ be as in Lemma 1.7.1. The integral points $\boldsymbol{u}_j = \boldsymbol{x}_j - \boldsymbol{x}_{m+1}$, $j = 1, \dots, m$, are different, and $-\boldsymbol{x}_{m+1} \in G$ because $G$ is symmetric. Since $G$ is convex, we have $\frac{1}{2}\,\boldsymbol{u}_j = \frac{1}{2}\,\boldsymbol{x}_j + \frac{1}{2}(-\,\boldsymbol{x}_{m+1}) \in G$, $\boldsymbol{u}_j \in D$, $j = 1, \dots, m$. It is obvious that the integral points $-\,\boldsymbol{u}_j$ also belong to $D$, $j = 1, \dots, m$.

## 1.8. Plurisubharmonic functions

**Definition 1.8.1.** Let $D$ be a domain in $\mathbb{C}$. A function $u : D \to [-\infty, \infty)$ is said to be *subharmonic* in $D$ if i) it is upper semicontinuous in $D$; ii) we have

$$u(z) \leqslant \frac{1}{2\pi}\int_0^{2\pi} u(z + re^{i\theta})\, d\theta \tag{1.8.1}$$

for every $z \in D$ and for all sufficiently small $r > 0$.

We use the symbol $H(D)$ to denote the class of functions subharmonic in $D \subset \mathbb{C}$.

**Definition 1.8.2.** Let $D$ be a domain in $\mathbb{C}^n$, $n > 1$. A function $u : D \to [-\infty, \infty)$ is said to be *plurisubharmonic* in $D$ if i) it is upper semicontinuous in $D$; ii) for every $z \in D$ and every complex line $\{z + a\lambda = (z_1 + a_1\lambda, \dots, z_n + a_n\lambda), \lambda \in \mathbb{C}\}$, where $a \in \mathbb{C}^n \setminus \{0\}$, the function $\varphi(\lambda) = u(z + a\lambda)$ is subharmonic in every connected component of the set $\{\lambda \in \mathbb{C} : z + a\lambda \in D\}$.

We denote by $PSH_n(D)$ the class of functions plurisubharmonic in the domain $D \subset \mathbb{C}^n$. For every function $f$ holomorphic in $D$, this class contains $\ln|f|$ and $\ln^+|f| = \max\{0, \ln|f|\}$. Obviously, $PSH_1(D) = SH(D)$.

### 1.8.1. Basic theorems

**Theorem 1.8.3** (Maximum Principle). *If $u \in PSH_n(D)$, $n \geqslant 1$, $u \not\equiv$ const , then $u$ cannot attain its maximum inside $D$.*

**Exercise 1.8.4.** Let $\{u_\alpha,\ \alpha \in A\}$ be a family of functions in the class $PSH_n(D)$, $n \geqslant 1$, and let $u = \sup\ \{u_\alpha,\ \alpha \in A\}$ be an upper semicontinuous function in $D$; moreover, let dom $u = D$. Then $u \in PSH_n(D)$.

**Theorem 1.8.5.** *Let $\Phi \in PSH_n(D)$, $n \geqslant 1$; let $\{f_j, j = 1, 2, \ldots, n\}$ be functions holomorphic in a domain $G \subset \mathbb{C}^m$, $m \geqslant 1$. Then $u(z) = \Phi[f_1(z), \ldots, f_n(z)]$, $z \in G$, is a function belonging to $PSH_m(G)$.*

**Theorem 1.8.6** (parameter integration). *Let $u(z,\tau)$ be an upper semicontinuous function in the domain $D \times G \subset \mathbb{C}^n \times \mathbb{R}^m$. Suppose that for every fixed $\tau \in G$, the function $u(.,\tau)$ is plurisubharmonic in $D$. We denote by $K$ a compact subset of $G$. Then the function $V(z) = \int\limits_K u(z,\tau)d\mu_\tau$, where $\mu_\tau$ is a nonnegative measure on $G$, is plurisubharmonic in $D$.*

**Theorem 1.8.7** (local summability). *If $u \in PSH_n(D)$ and $u \not\equiv -\infty$, then $u$ is summable on all compact subsets of $D$.*

### 1.8.2. Sequences of plurisubharmonic functions

**Theorem 1.8.8** (Hartogs lemma). *Let $\{u_t,\ t > 0\}$ be a family of functions in $PSH_n(D)$ uniformly bounded from above on all compact subsets of $D$. Assume that there exists a constant $C$ such that $\overline{\lim\limits_{t\to\infty}}\, u_t(z) \leqslant C$, $z \in D$. Then for every $\varepsilon > 0$ and every compact set $K \subset D$, there is a constant $t_0(\epsilon)$ such that $u_t(z) \leqslant C + \varepsilon \quad \forall\ z \in K,\ t > t_0(\epsilon)$.*

**Proposition 1.8.9.** *Under the assumptions of Theorem 1.8.8, the regularization*

$$u(z) = \overline{\lim_{\omega\to z}}\ \overline{\lim_{t\to\infty}}\, u_t(\omega), \quad z \in D,$$

*of the sequence $\{u_t,\ t > 0\}$ belongs to $PSH_n(D)$.*

The following statement is related to Exercise 1.2.6.

**Exercise 1.8.10.** The limit of a monotone decreasing sequence of functions of class $PSH_n(D)$ belongs to $PSH_n(D)$

**Theorem 1.8.11.** *Let $\mathfrak{P} := PSH_n(\mathbb{C}^n)$. For every function $u \in \mathfrak{P}$ there exists a monotone decreasing sequence $\{u_j\}_1^\infty$ of inifinitely differentiable functions of class $\mathfrak{P}$ such that $u_j \downarrow u$. If, moreover, a) $u(z) \equiv u(|z_1|, \ldots, |z_n|)$, $z \subset \mathbb{C}^n$ or b) $u(tz_1, \ldots, tz_n) - t^\rho u(z)$ $\forall\ t > 0, z \in \mathbb{C}^n$ for some $\rho > 0$, then the functions $u_j$ can be chosen so that they possess the same property, and in case b) we may ensure that $u_j \in C^\infty(\mathbb{C}^n \setminus \{0\})$, $j = 1, 2, \ldots$.*

### 1.8.3. Plurisubharmonic functions and convex functions

**Proposition 1.8.12.** *If $u$ is a finite convex function in a domain $D \subset \mathbb{C}^n$, then $u \in PSH_n(D)$.*

◀ Let $z \in D$, and let $a \in \mathbb{C}^n \setminus \{0\}$. For sufficiently small $r > 0$ and every $\theta \in \mathbb{R}$, the point $z + re^{i\theta}a$ belongs to $D$. By the convexity of $u$, we obtain the formula

$$\frac{1}{2\pi}\int\limits_0^{2\pi} u(z+re^{i\theta}a)\,d\theta = \frac{1}{\pi}\int\limits_0^{\pi} \frac{u(z+re^{i\theta}a)+u(z-re^{i\theta}a)}{2}\,d\theta \geqslant u(z).$$

It only remains to use Definitions 1.8.1, 1.8.2 and Theorem 1.4.4 ▶

**Theorem 1.8.13.** *A finite function $V$ defined on a convex domain $B\subset\mathbb{R}^n$ is convex if and only if the function*

$$u(\omega) = V(\mathrm{Re}\,\omega_1,\ldots,\mathrm{Re}\,\omega_n) \tag{1.8.2}$$

*is plurisubharmonic in the tubular domain $D = B + i\mathbb{R}^n$.*

◀ *The "only if" part* follows from Proposition 1.8.12.

*The "if" part.* Let $b\in B$, and let $x\in\mathbb{R}^n\setminus\{0\}$. We put $\Delta=\{t\in\mathbb{R}: xt+b\in B\}$ and $v(t)=V(xt+b)$, $t\in\Delta$. Let $l$ be an affine function of the variable $t$. By Theorem 1.4.6, the claim will be proved if we show that the set $G=\{t\in\Delta : v(t)<l(t)\}$ is connected (if it is nonempty) for all admissible $x,b,l$. Suppose the contrary. Since $v$ is upper semicontinuous, the set $G$ is open. There exist $x,b,l$ such that the nonempty set $G$ has at least two components $\Delta_1,\Delta_2$. Let $t_i\in\Delta_i$, $i=1,2$, $t_1<t_2$, $\varepsilon\geqslant 0$, and let $l_\varepsilon(t)$ be an affine function defined by the conditions $l_\varepsilon(t_j)=l(t_j)+\varepsilon$, $j=1,2, l_0(t)=l(t)$. The function $\Phi(\varepsilon)=\sup\{v(t)-l_\varepsilon(t) : t\in[t_1,t_2]\}$ is continuous and monotone decreasing to $-\infty$ as $\varepsilon\to\infty$ (see Ex. 2.7). By assumption, $\Phi(0)\geqslant 0$. Therefore, there exists $\delta\geqslant 0$ such that $\Phi(\delta)=0$. We take the smallest number $t_0$ in the nonempty set $\{t\in[t_1,t_2]\colon v(t)=l_\delta(t)\}$ (see Ex. 2.4). Combining this with the previous arguments, we obtain:

$$v(t)<l_\delta(t),\ t\in(t_1,t_0);\ v(t_0)=l_\delta(t_0);\ v(t)\leqslant l_\delta(t),\ t\in(t_0,t_2). \tag{1.8.3}$$

Consider the functions $\varphi(\tau)=v(\mathrm{Re}\,\tau)$, $\psi(\tau)=l_\delta(\mathrm{Re}\,\tau)$, $\tau\in\mathbb{C}$.

Since $\psi$ is harmonic in $\mathbb{C}$, from (1.8.3) we deduce that

$$\frac{1}{2\pi}\int\limits_0^{2\pi}\varphi(t_0+re^{i\theta})\,d\theta < \frac{1}{2\pi}\int\limits_0^{2\pi}\psi(t_0+re^{i\theta})\,d\theta = \psi(t_0)=\varphi(t_0)$$

for a sufficiently small $r>0$. But this contradicts the subharmonicity of $\varphi$ near $t_0$ (see Def. 1.8.1 and 1.8.2) ▶

We mention the following property of the class $\mathfrak{P}$ of functions plurisubharmonic in $\mathbb{C}^n$.

**Theorem 1.8.14.** *Let $\Phi$ be a finite real function defined on $\mathbb{R}^n_+$. If the function $u(z)=\Phi(|z_1|,\ldots,|z_n|)$ belongs to $\mathfrak{P}$, then $\Phi$ is monotone non-decreasing in every variable, is upper semicontinuous on $\mathbb{R}^n_+$, and is also convex with respect to the variables $\ln r_1,\ldots,\ln r_n$ on $\mathbb{R}^n_0$.*

◀ Since $\Phi(r) = \max\{u(z) : |z_j| \leqslant r_j,\ j = 1,\dots,n\}$, the maximum principle (see T. 1.8.3) implies that the function in question is monotone nondecreasing in each variable. The upper semicontinuity of $\Phi$ is a consequence of a similar property for $u$ (see Def. 1.8.2). We consider the function $V(u) = \Phi(e^{u_1},\dots,e^{u_n})$, $u \in \mathbb{R}^n$. By Theorem 1.8.5, the function $F(\omega) = u(e^{\omega_1},\dots, e^{\omega_n})$ belongs to the class $\mathfrak{P}$. Since $F(\omega) = V(\mathrm{Re}\,\omega_1,\dots,\mathrm{Re}\,\omega_n), \omega \in \mathbb{C}^n$, Theorem 1.8.13 shows that $V$ is a convex function on $\mathbb{R}^n$. Hence, the final assertion of Theorem 1.8.14 follows ▶

**Corollary 1.8.15.** *If $u \not\equiv -\infty$ is a function of class $\mathfrak{P}$, then at every point $z \in \mathbb{C}^n$ its mean value*

$$\mathfrak{M}_u = \frac{1}{(2\pi)^n}\int_0^{2\pi}\dots\int_0^{2\pi} u(z_1 + r_1e^{i\theta_1},\dots,z_n + r_ne^{i\theta_n})\,d\theta_1\dots d\theta_n$$

*possesses the properties described in Theorem* 1.8.14.

◀ It suffices to observe the following identity:

$$\mathfrak{M}(z;|\omega|) \equiv \frac{1}{(2\pi)^n}\times\int_0^{2\pi}\dots\int_0^{2\pi} u(z_1 + \omega_1e^{i\theta_1},\dots,z_n + \omega_ne^{i\theta_n})\,d\theta_1\dots d\theta_n,$$

and to use Theorems 1.8.6, 1.8.7 and 1.8.14 ▶

The Poisson inequality is true for plurisubharmonic functions in polydisks.

**Theorem 1.8.16.** *Let $\Pi = \{z \in \mathbb{C}^n : |z_j| < R_j,\ j = 1,\dots,n\}$. If $u$ is a continuous function in $\overline{\Pi}$ and $u \in PSH_n(\Pi)$, then*

$$u(z) \leqslant \frac{1}{(2\pi)^n}\cdot\int_0^{2\pi}\dots\int_0^{2\pi} u(\zeta)\cdot\prod_{j=1}^{n}\frac{R_j^2 - |z_j|^2}{|\zeta_j - z_j|^2}\,d\theta_1\dots d\theta_n \quad \forall\ z \in \Pi, \tag{1.8.4}$$

*where $\zeta_j = R_je^{i\theta_j}$, $j = 1,\dots,n$.*

The proof is based on the Poisson formula.

**Theorem 1.8.17.** *If $u$ is a function continuous on the boundary $\Gamma$ of the disk $D = \{\zeta \in \mathbb{C} : |\zeta| < R\}$, then the Poisson integral*

$$h(z) = \frac{1}{2\pi}\int_0^{2\pi} u(\zeta)\cdot\frac{R^2 - |z|^2}{|\zeta - z|^2}\,d\theta,\ \zeta = Re^{i\theta} \tag{1.8.5}$$

*is a harmonic function in $D$ that extends to $\overline{D}$ by continuity, and $h|_\Gamma = u$.*

**Proof of Theorem 1.8.16.** 1. Let $n = 1$, and $h$ be the harmonic function defined by formula (1.8.5). Then $v := u - h$ is a continuous function in $\overline{D}$ such that $v \in SH(D)$, and $v|_\Gamma = 0$ (see T. 1.8.17). By Theorem 1.8.3, we obtain $v \leqslant 0$, $u \leqslant h$.

2. Let $n > 1$. In each variable $z_j$, $u$ is subharmonic in the disk $|z_j| < R_j$ for fixed values of the other variables. Therefore, (1.8.4) is obtained by $n$-fold application of the one-dimensional result.

Later we shall need the following well-known fact.

**Theorem 1.8.18.** *A two times continuously differentiable function $u$ defined on a domain $D \subset \mathbb{C}^n$ belongs to $PSH_n(D)$ if and only if*

$$\sum_{j,k=1}^{n} \frac{\partial^2 u}{\partial z_j \partial \bar{z}_k} s_j \bar{s}_k \geqslant 0 \quad \forall\, z \in \Omega,\ s \in \mathbb{C}^n \setminus \{0\}.$$

*For $n = 1$ this condition is equivalent to the inequality $\Delta u \geqslant 0$, where $\Delta$ is the Laplace operator, and $u \in SH(\Omega) = PSH_1(\Omega)$.*

## 1.9. Trigonometrically $\rho$-convex functions

### 1.9.1. Definition and basic properties

**Definition 1.9.1.** Let $h$ be a finite function defined on a finite or infinite interval $(\alpha, \beta) \subset \mathbb{R}$. For $\rho > 0$, we denote $M = \{l(\theta) = a\cos\rho\theta + b\sin\rho\theta,\ \theta \in \mathbb{R} :\ a, b \in \mathbb{R}\}$. Let $\theta_1, \theta_2$ be any fixed pair of points in $(\alpha, \beta)$ such that $0 < \rho(\theta_2 - \theta_1) < \pi$. The function $h$ is called *trigonometrically $\rho$-convex* if the condition $h(\theta_j) \leqslant l(\theta_j)$, $j = 1, 2$, implies the inequality $h(\theta) \leqslant l(\theta)\ \forall\, \theta \in [\theta_1, \theta_2]$ for every $l \in M$ and every pair $\theta_1, \theta_2$ as above.

We note that every function $l \in M$ is determined by its values at the points $\theta_1, \theta_2$ and

$$l(\theta)\sin\rho(\theta_2 - \theta_1) = l(\theta_1)\sin\rho(\theta_2 - \theta) + l(\theta_2)\sin\rho(\theta - \theta_1), \tag{1.9.1}$$

where $\theta \in \mathbb{R}$. This formula allows us to give an equivalent definition of a trigonometrically $\rho$-convex function in the following form.

**Definition 1.9.2.** In the notation of Definition 1.9.1, a function $h$ is said to be *trigonometrically $\rho$-convex* if for all $\theta \in [\theta_1, \theta_2]$

$$h(\theta)\sin\rho(\theta_2 - \theta_1) \leqslant h(\theta_1)\sin\rho(\theta_2 - \theta) + h(\theta_2)\sin\rho(\theta - \theta_1). \tag{1.9.2}$$

Let $L = (0, \infty) \times \mathbb{R}$ be the Riemann surface of the logarithm, the complex structure of which is defined by the local homeomorphism $\pi :\ L \to \mathbb{C},\ \pi(r, \theta) = re^{i\theta}\ \forall\, (r, \theta) \in L$. We denote the coordinates (the modulus and the argument) of an element $Z \in L$ as follows: $Z = (|Z|, \operatorname{Arg} Z)$. For $\rho > 0$, let $Z^\rho = |Z|^\rho \exp\{i\rho \operatorname{Arg} Z\}$. For every pair of points $Z_1, Z_2 \in L$ satisfying the condition

$$0 \leqslant \rho(\theta_2 - \theta_1) < \pi,\ \theta_j = \operatorname{Arg} Z_j,\ j = 1, 2, \tag{1.9.3}$$

there exists a unique element $Z := Z_1 \otimes Z_2 \in L$ such that

$$\theta_1 < \operatorname{Arg} Z < \theta_2; \quad Z^\rho = Z_1^\rho + Z_2^\rho. \tag{1.9.4}$$

**Definition 1.9.3.** Let $L(\alpha,\beta)=\{Z\in L:\ \alpha<\text{Arg } Z<\beta\}$. A function $H:\ L(\alpha,\beta)\to\mathbb{R}$ is said to be *$\rho$-sublinear* (cf. T. 1.3.15) if it satisfies the conditions

$$H(tZ)=t^\rho H(Z),\ \ tZ=(t|Z|,\text{Arg } Z),\ t>0,\ Z\in L,$$

$$H(Z_1\otimes Z_2)\leqslant H(Z_1)+H(Z_2), \tag{1.9.5}$$

where $(Z_1,Z_2)$ is any pair of points in $L(\alpha,\beta)$ satisfying (1.9.3).

The relationship between $\rho$-sublinear, trigonometrically $\rho$-convex, and convex functions is clarified by the following statement.

**Proposition 1.9.4.** *Consider a function $h:(\alpha,\beta)\to\mathbb{R}$. Suppose $\rho>0$, $\gamma_\rho=\min\{\pi,\pi/2\rho\}$, and $\varepsilon\in(0,\gamma_\rho)$. Let $B=\{\Delta\}$ b e the system of all intervals in $(\alpha,\beta)$ of the form $\Delta=(\theta-\varepsilon,\theta+\varepsilon)$, where $\theta$ is an arbitrary point in $(\alpha,\beta)$. We put $\Pi_\varepsilon(\Delta)=\{\omega=te^{i\psi}\in\mathbb{C}:\ t>0,\ \psi/\rho\in\Delta\}$. The following conditions are equivalent.*

i) *$h$ is trigonometrically $\rho$-convex;*

ii)

$$H(Z):=r^\rho h(\theta),\ Z=(r,\theta)\in L(\alpha,\beta) \tag{1.9.6}$$

*is a $\rho$-sublinear function in $L(\alpha,\beta)$* (see Def. 1.9.3)*;*

iii) *for every interval $\Delta\in B$ the function $G(\omega)=th(\psi/\rho)$, $\omega=te^{i\psi}\in\Pi_\varepsilon(\Delta)$ is convex.*

◀ ii) → i) We put $\theta=\text{Arg } Z$, $r=|Z|$, $r_j=|Z_j|$, and $\theta_j=\text{Arg } Z_j$, $j=1,2$. Condition (1.9.4) implies the identities

$$r^\rho\cos\rho\theta=r_1^\rho\cos\rho\theta_1+r^\rho\cos\rho\theta_2 \text{ and } r^\rho\sin\rho\theta=r_1^\rho\sin\rho\theta_1+r_2^\rho\sin\rho\theta_2.$$

Consequently,

$$r^\rho\sin\rho(\theta-\theta_1)=r^\rho\sin\rho(\theta_2-\theta_1) \text{ and } r_1^\rho\sin(\theta_2-\theta_1)=r^\rho\sin\rho(\theta_2-\theta).$$

Now, applying (1.9.3) and (1.9.4), we verify inequality (1.9.2).

ii) → iii). The mapping

$$\Phi:\ L(\theta-\varepsilon,\theta+\varepsilon)\to\Pi_\varepsilon;\ \Phi(Z)=Z^\rho$$

is biholomorphic (see Def. 1.9.3). The function $G$ can be represented as follows: $G(\omega)=H\circ\Phi^{-1}(\omega)$, $\omega\in\Pi_\varepsilon$, where $H$ is defined by formula (1.9.6). Since $H$ is a $\rho$-sublinear function, formula (1.9.5) implies that $G$ is sublinear, and, therefore, convex in $\Pi_\varepsilon$ (see T. 1.3.15).

The implications i) → ii), iii) → ii) are proved by reversing the arguments ▶

We examine several properties of a trigonometrically $\rho$-convex function $h:(\alpha,\beta)\to\mathbb{R}$.

**Property 1.9.5.** *The function $h$ is continuous.*

◀ We use the notation of Proposition 1.9.4 and of its proof. Let $\theta \in (\alpha, \beta)$. By Proposition 1.9.4, $G$ is sublinear and, therefore, convex in $\Pi_\varepsilon(\Delta)$. Since $G$ is continuous (see T. 1.4.4), so is the function $H = G \circ \Phi$ defined on $L(\theta-\varepsilon, \theta+\varepsilon)$. Therefore, $h(\varphi) = H(1, \varphi)$ is also a continuous function on $(\theta - \varepsilon, \theta + \varepsilon)$ and, consequently, on $(\alpha, \beta)$ ▶

The following fact is a consequence of (1.9.2) and Proposition 1.9.5.

**Property 1.9.6.** *The function $h$ satisfies the inequality*

$$h(\theta) + h(\theta + \pi/\rho) \geqslant 0 \quad \forall\, \theta,\, \theta + \pi/\rho \in (\alpha, \beta). \tag{1.9.7}$$

**Property 1.9.7.** *If $[\psi_1, \psi_2] \subset (\alpha, \beta)$, $\psi_2 - \psi_1 > \pi/\rho$, and*

$$h(\psi) \leqslant 0, \quad \psi \in (\psi_1, \psi_2), \tag{1.9.8}$$

*then $h(\psi) \equiv 0$, $\psi \in [\psi_1, \psi_2]$.*

◀ Let $\Delta \in [\psi_1, \psi_2 - \pi/\rho] \cup [\psi_1 + \pi/\rho, \psi_2]$. By (1.9.7) and (1.9.8), we find $h(\psi) \equiv 0$, $\psi \in \Delta$. If $\psi_2 - \psi_1 \geqslant 2\pi/\rho$, then $\Delta = [\psi_1, \psi_2]$. Let $\psi_2 - \psi_1 < 2\pi/\rho$, $\psi \in (\psi_2 - \pi/\rho, \psi_1 + \pi/\rho)$. Taking $\theta_1 = \psi_1$, $\theta = \psi_2 - \pi/\rho$, and $\theta_2 = \psi$ in (1.9.2), we obtain $h(\psi) \geqslant 0$. Together with (1.9.8), this implies Property 1.9.7 ▶

**Property 1.9.8.** *Suppose $[\psi_1, \psi_2] \subset (\alpha, \beta)$, $\psi_2 - \psi_1 > \pi/\rho$, and put $l(\theta) = a \cos \rho\theta + b \sin \rho\theta$, $\theta \in \mathbb{R}$. If $h(\psi) \leqslant l(\psi)$, $\psi \in [\psi_1, \psi_2]$, then $h(\psi) \equiv l(\psi)$, $\psi \in [\psi_1, \psi_2]$.*

◀ It suffices to observe that the difference $h - l$ is a trigonometrically $\rho$-convex function on $(\alpha, \beta)$ (see (1.9.1) and (1.9.2)) and to use Property 1.9.7 ▶

**Property 1.9.9.** *Let $0 < \rho \leqslant 1/2$, and let $h : \mathbb{R} \to \mathbb{R}$ be a trigonometrically $\rho$-convex $2\pi$-periodic function. Then $h(\theta) \geqslant 0\ \forall\, \theta \in \mathbb{R}$, and for $\rho < 1/2$ either $h(\theta) \equiv 0$, $\theta \in \mathbb{R}$, or $h(\theta) > 0\ \forall\, \theta \in \mathbb{R}$.*

◀ For $\rho = 1/2$ the assertion follows from (1.9.7). Let $\rho \in (0, 1/2)$. Taking $\theta_1 = \psi$, $\theta = \psi + \pi$, and $\theta_2 = \psi + 2\pi$ in (1.9.2), we have: $h(\psi) \geqslant h(\psi + \pi) \cos \pi\rho\ \forall\ \psi \in \mathbb{R}$. Applying this inequality once again and recalling that $h(\psi) = h(\psi + 2\pi)$, we obtain $h(\psi) \geqslant 0$, $\psi \in \mathbb{R}$. If $h(\psi_0) > 0$ for some $\psi_0 \in \mathbb{R}$, then, again using (1.9.2) with $\theta_1 = \psi$, $\theta = \psi_0$, and $\theta_2 = \psi + 2\pi$, where $\psi < \psi_0 < \psi + 2\pi$, we find $h(\psi) > 0\ \forall\ \psi \in \mathbb{R}$ ▶

We present a criterion of trigonometric $\rho$-convexity.

**Property 1.9.10.** *Consider a function $h : (\alpha, \beta) \to \mathbb{R}$. For $\rho > 0$, put $\mathfrak{B} = \{\Delta = (\theta_1, \theta_2) : \Delta \subset (\alpha, \beta),\ \rho(\theta_2 - \theta_1) < \pi\}$, and $M = \{l = \operatorname{Re} pe^{i\rho\theta}, \theta \in \mathbb{R} : p \in \mathbb{C}\}$* (cf. Def. 1.9.1). *The following conditions are equivalent:*

i) *$h$ is a trigonometrically $\rho$-convex function;*

ii) *for every admissible $\Delta \in \mathfrak{B}$ and $l \in M$, if the set $B_l(\Delta) = \{\theta \in \Delta : h(\theta) < l(\theta)\}$ is not empty, then it is connected.*

The proof is similar to the proof of Theorem 1.4.6 and is based on formulas (1.9.1) and (1.9.2).

The following result explains the relationship between trigonometrically $\rho$-convex functions and subharmonic functions.

**Theorem 1.9.11.** *For* $0 < \beta - \alpha < 2\pi$, *denote* $M(\alpha, \beta) = \{z = re^{i\theta} \in \mathbb{C} : r > 0,\ \theta \in (\alpha, \beta)\}$, *and let* $h : (\alpha, \beta) \to \mathbb{R}$. *The function* $h$ *is trigonometrically* $\rho$*-convex on* $(\alpha, \beta)$ *if and only if* $K(z) = r^\rho h(\theta)$, $z \in M(\alpha, \beta)$, *is a subharmonic function in the angle* $M(\alpha, \beta)$.

◀ *The "only if" part.* We consider the mapping $\pi : L(\alpha, \beta) \to M(\alpha, \beta)$, $\pi[(r, \theta)] = re^{i\theta}$. Since $\beta - \alpha < 2\pi$, it follows that $\pi$ is a homeomorphism, and $K = H \circ \pi^{-1}$ (see (1.9.6)). Let $z_0 = r_0 e^{i\theta_0} \in M(\alpha, \beta)$. We use the notation of Proposition 1.9.4: $K = G \circ W$, where $W = \Phi \circ \pi^{-1}(z) = z^\rho$, $z \in M(\theta_0 - \varepsilon, \theta + \varepsilon)$. Since $G$ is convex in the angle $\Pi_\varepsilon$, containing the point $\omega_0 = r_0^\rho e^{i\rho\theta_0}$, we see that $G$ is also subharmonic there (see P. 1.8.12). But subharmonicity is preserved under holomorphic changes of variables (see T. 1.8.5), therefore, the function $K$ is also subharmonic in the angle $M(\theta_0 - \varepsilon, \theta_0 + \varepsilon)$. Consequently, $K \in SH[M(\alpha, \beta)]$ (the definition of a subharmonic function is of local nature, see Def. 1.8.1).

*The "if" part.* Let $K \in SH[M(\alpha, \beta)]$. Suppose $h$ is not trigonometrically $\rho$-convex on $(\alpha, \beta)$. Then, by Property 1.9.10, there exists an interval $\Delta$ in $\mathfrak{B}$ and a function $l \in M$ such that the set $B_l(\Delta)$ is not connected. Using the pattern of the proof of the "if" part in Theorem 1.8.13, we arrive at a contradiction (the function $\mathrm{Re}\, pz^\rho$ is harmonic in $M(\alpha, \beta)$) ▶

The following result is true for functions defined on $\mathbb{C}$.

**Theorem 1.9.12.** *Let* $H : \mathbb{C} \to \mathbb{R}$ *be a positively homogeneous function of order* $\rho > 0$, *i.e.,* $H(tz) = t^\rho \cdot H(z)$ $\forall\, t \geqslant 0$, $z \in \mathbb{C}$. *Then the following assertions are equivalent:*

i) $h(\theta) := H(e^{i\theta})$, $\theta \in \mathbb{R}$, *is a* $2\pi$*-periodic trigonometrically* $\rho$*-convex function;*

ii) $H$ *is subharmonic in* $\mathbb{C}$.

*In the case of* $\rho = 1$ *the following assertion is added to ii):*

iii) $H$ *is sublinear in* $\mathbb{C}$ (see T. 1.3.15).

To prove Theorem 1.9.12, we need the folowing fact.

**Property 1.9.13.** *Under the assumptions of assertion i) of Theorem 1.9.12, we have* $\int_0^{2\pi} h(\theta)\, d\theta \geqslant 0$.

◀ Suppose that $\sum_- := \{\theta \in \mathbb{R} : h(\theta) < 0\} \neq \emptyset$, which is possible for $\rho > 1/2$ (see Prop. 1.9.9). The components $\{(\alpha_i, \beta_i) : i \in \mathbb{Z}\}$ and $\{[\beta_i, \alpha_{i+1}], i \in \mathbb{Z}\}$ of the sets $\sum_-$ and $\sum_+ := \{\theta \in \mathbb{R} : h(\theta) \geqslant 0\}$, respectively, interlace ($\alpha_i < \beta_i < \alpha_{i+1}\ \forall\, i \in \mathbb{Z}$), and (see Property 1.9.7 and (1.9.2)) $\beta_i - \alpha_i \leqslant \pi/\rho$, $\alpha_{i+1} - \beta_i \geqslant \pi/\rho\ \forall\, i \in \mathbb{Z}$. Therefore, $(\alpha_i + \pi/\rho, \beta_i + \pi/\rho) \subset [\beta_i, \alpha_{i+1}]\ \forall\, i \in \mathbb{Z}$. Since $h$ is a $2\pi$-periodic function, we have $\alpha_0 + 2\pi = \alpha_k$ for some integer $k \geqslant 1$.

Combining this with (1.9.7), we find

$$\int_0^{2\pi} h(\theta)\,d\theta = \int_{\alpha_0}^{\alpha_k} h(\theta)\,d\theta = \sum_{i=0}^{k-1}\left(\int_{\alpha_i}^{\beta_i} h(\theta)\,d\theta + \int_{\beta_i}^{\alpha_{i+1}} h(\theta)\,d\theta\right) \geqslant$$

$$\geqslant \sum_{i=0}^{k-1}\left(\int_{\alpha_i}^{\beta_i} h(\theta)\,d\theta + \int_{\alpha_i+\pi/\rho}^{\beta_i+\pi/\rho} h(\theta)\,d\theta\right) =$$

$$\sum_{i=0}^{k-1}\int_{\alpha_i}^{\beta_i} [h(\theta) + h(\theta+\pi/\rho)]\,d\theta \geqslant 0 \blacktriangleright$$

**Proof of Theorem 1.9.12.** i)$\to$ ii). By Theorem 1.9.11, $H$ is a subharmonic function on $\mathbb{C}\setminus\{0\}$. Property 1.9.13 means that (1.8.1) is true for $H$ if $z=0$, i.e., $H$ is subharmonic in $\mathbb{C}$. Theorem 1.9.11 implies that ii) $\to$ i).

i) $\to$ iii). Let $z_j = r_j e^{i\theta_j} \in \mathbb{C}\setminus\{0\}$, $j=0,1$. We may assume that $|\theta_2-\theta_1| \leqslant \pi$. If $|\theta_2-\theta_1|<\pi$, then Proposition 1.9.4 implies the inequality

$$H(z_1+z_2) \leqslant H(z_1) + H(z_2). \tag{1.9.9}$$

If $|\theta_2-\theta_1| = \pi$ and, for example, $r_1 \geqslant r_2$, $\theta_2 = \theta_1+\pi$, then $z_1+z_2 = (r_1-r_2)e^{i\theta_1}$. In this case inequality (1.9.9) is a consequence of (1.9.7).

The implication iii) $\to$ i) is a consequence of Proposition 1.9.4 ▶

We mention a useful criterion of trigonometric $\rho$-convexity.

**Theorem 1.9.14.** *Let $\gamma$ be any fixed number in the interval $(0,\pi/\rho)$. If a function $h:(\alpha,\beta)\to\mathbb{R}$ satisfies inequality (1.9.2) for $0<\theta_2-\theta_1<\gamma$, then $h$ is trigonometrically $\rho$-convex.*

◀ 1. We verify the following condition:

Let $\theta_0 \in (\alpha,\beta)$, $\delta>0$, and let $\Delta(\theta_0) = [\theta_0-\delta, \theta_0+\delta] \subset (\alpha,\beta)$. If $h(\theta) \leqslant l(\theta)$, $\theta\in\Delta(\theta_0)$, and $h(\theta_0) = l(\theta_0)$, then for sufficiently small $\delta>0$ we have $h(\theta) \equiv l(\theta)$, $\theta\in\Delta(\theta_0)$.

Suppose the contrary. Then there exists $\delta\in(0,\gamma/2)$ such that, e.g., $h(\theta_0-\delta) < l(\theta_0-\delta)$, $h(\theta_0+\delta) \leqslant l(\theta_0+\delta)$. By inequality (1.9.2), for $\theta_1=\theta_0-\delta$, $\theta_2=\theta_0+\delta$, $\theta=\theta_0$ we find: $h(\theta_0)<l(\theta_0)$, a contradiction.

2. We note that under the assumptions of the theorem, $h$ is a continuous function. The proof of this is similar to the case of $\gamma=\pi/\rho$ (see the proof of Prop. 1.9.5). Suppose $h$ is not trigonometrically $\rho$-convex. Then there exist points $\theta_1<\theta_0<\theta_2$ in $(\alpha,\beta)$ and a number $\varepsilon>0$ such that $\gamma<\theta_2-\theta_1<\pi/\rho$, and

$$l_\varepsilon(\theta_0) = h(\theta_0),\ h(\theta) \leqslant l_\varepsilon(\theta),\ \theta\in[\theta_0,\theta_2],\ h(\theta) < l_\varepsilon(\theta),\ \theta\in[\theta_1,\theta_0),$$

where $l_\varepsilon$ is the function in $M$ (cf. Def. 1.9.1) determined by the conditions $l_\varepsilon(\theta_j) = h(\theta_j)+\varepsilon$, $j=1,2$. We see this using the method of the proof of

Theorem 1.8.13 (see (1.8.3)) and taking into account that $h$ is continuous. This contradicts the property of $h$ proved in item 1 ▶

The following assertion provides some information about the behavior of a trigonometrically $\rho$-convex function on the boundary of its closed domain.

**Property 1.9.15.** (cf. Prop. 1.9.5). *A trigonometrically $\rho$-convex function $h : [\alpha, \beta] \to \mathbb{R}$ is upper semicontinuous on $[\alpha, \beta]$.*

◀ Taking $\theta_1 = \alpha$, $\theta_2 \in (\alpha, \beta)$ in (1.9.2) and passing to the limit in (1.9.2) as $\theta \to \alpha + 0$, we find $\overline{\lim\limits_{\theta \to \alpha+0}} h(\theta) \leqslant h(\alpha)$. Similarly we obtain $\overline{\lim\limits_{\theta \to \beta-0}} h(\theta) \leqslant h(\beta)$ ▶

Inequality (1.9.2) allows us to extend the definition of a trigonometrically $\rho$-convex function to the functions defined on a closed interval $[\alpha, \beta]$ with possible infinite values of one sign.

**Property 1.9.16.** *If $h : [\alpha, \beta] \to [-\infty, \infty)$ is a trigonometrically $\rho$-convex function such that $\{\theta \in (\alpha, \beta) : h(\theta) = -\infty\} \neq \emptyset$, then $h(\theta) \equiv -\infty$, $\theta \in (\alpha, \beta)$.*

◀ Let $h(\theta_0) = -\infty$, where $\theta_0 \in (\alpha, \beta)$, and let $\Delta = \{\theta \in \mathbb{R} : |\theta - \theta_0| < \pi/\rho\}$. In a standard way, by using (1.9.2), we find $h(\theta) = -\infty \ \forall\, \theta \in \Delta \cap (\alpha, \beta)$. Now, since $(\alpha, \beta)$ can be covered by intervals of length $< \pi/\rho$, we obtain Property 1.9.16 ▶

In what follows, we consider the trigonometrically $\rho$-convex functions with values in $(-\infty, \infty]$. Inequality (1.9.7) does not follow from trigonometric $\rho$-convexity of a function $h : [\alpha, \beta] \to (-\infty, \infty]$ if there is a point $\theta \in [\alpha, \beta]$ such that

$$\theta + \pi/\rho \in [\alpha, \beta], \ \ \theta, \ \theta + \pi/\rho \in \operatorname{dom} h, \ (\theta, \theta + \pi/\rho) \subset [\alpha, \beta] \setminus \operatorname{dom} h.$$

Therefore, for the functions with infinite values the following modification of Definition 1.9.2 is appropriate.

**Definition 1.9.2′.** *A function $h : [\alpha, \beta] \to (-\infty, \infty]$ is said to be trigonometrically $\rho$-convex if it satisfies inequality (1.9.2) for all pairs $\theta_1, \theta_2 \in [\alpha, \beta]$ such that $0 < \rho(\theta_2 - \theta_1) < \pi$ and inequality (1.9.7) for every $\theta \in [\alpha, \beta - \pi/\rho]$ under the condition $\beta - \alpha \geqslant \pi/\rho$.*

A similar remark can be made concerning the $\rho$-sublinear functions with infinite values. Consider the closed angle

$$\Delta(\alpha, \beta) = \{Z \in L : \alpha \leqslant \operatorname{Arg} Z \leqslant \beta\}$$

on the Riemann surface of the logarithm. We take its one-point topological extension $\Delta_0(\alpha, \beta) = \Delta(\alpha, \beta) \cup \{\Lambda\}$, i.e., we add the "vertex" $\Lambda$ to the angle $\Delta(\alpha, \beta)$. Moreover, we agree that $\pi(\Lambda) = 0$ and (see (1.9.4)) $\Lambda = Z_1 \oplus Z_2$ if $|Z_1| = |Z_2|$ and $\rho\theta_2 = \rho\theta_1 + \pi$.

**Definition 1.9.3′.** A function $H : \Delta_0(\alpha, \beta) \to (-\infty, \infty]$ is said to be *$\rho$-sublinear* if

i) $H(\Lambda) = 0$;

ii) the function $H$ satisfies (1.9.5), where $Z_1, Z_2$ are arbitrary points in $\Delta(\alpha, \beta)$ such that (cf. (1.9.3))

$$0 \leqslant \rho(\text{Arg } Z_2 - \text{Arg } Z_1) \leqslant \pi. \tag{1.9.3'}$$

**Remark.** Proposition 1.9.4 admits a complete analog reflecting the relationship between trigonometrically $\rho$-convex functions and $\rho$-sublinear functions in the sense of Definitions 1.9.2′ and 1.9.3′.

**1.9.2. Differential and geometric properties.** We present a simple criterion of trigonometric $\rho$-convexity.

**Theorem 1.9.17.** *Let $h : (\alpha, \beta) \to \mathbb{R}$ be a two times continuously differentiable function. Then $h$ is trigonometrically $\rho$-convex on $(\alpha, \beta)$ if and only if $h''(\theta) + \rho^2 h(\theta) \geqslant 0 \ \ \forall\, \theta \in (\alpha, \beta)$.*

◀ Let $\theta_0 \in (\alpha, \beta)$. We choose $\varepsilon \in (0, \pi)$ so that $(\theta_0 - \varepsilon, \theta_0 + \varepsilon) \subset (\alpha, \beta)$. Consider the function $K(z) = K(re^{i\theta}) = r^\rho h(\theta)$ defined in the angle $M(\theta_0 - \varepsilon, \theta_0 + \varepsilon) = \{z \in \mathbb{C} : |\text{Arg } Z - \theta_0| < \varepsilon\}$. It is subharmonic in $M(\theta_0 - \varepsilon, \theta_0 + \varepsilon)$ if and only if $h$ is trigonometrically $\rho$-convex on $(\theta_0 - \varepsilon, \theta_0 + \varepsilon)$ (see T. 1.9.11). It only remains to apply Theorem 1.8.18 to $K$, writing the Laplace operator in the polar coordinates

$$\Delta u = \frac{1}{r}\frac{\partial}{\partial r}\left(r\frac{\partial u}{\partial r}\right) + \frac{1}{r^2}\frac{\partial^2 u}{\partial \theta^2}$$

and recalling that $\theta_0$ is an arbitrary point in $(\alpha, \beta)$ ▶

**Theorem 1.9.18.** *A trigonometrically $\rho$-convex function $h : (\alpha, \beta) \to \mathbb{R}$ has finite right and left derivatives $h'_+(\theta)$, $h'_-(\theta)$ at every point $\theta \in (\alpha, \beta)$, and*

$$h'_-(\theta) \leqslant h'_+(\theta) \quad \forall\, \theta \in (\alpha, \beta)$$

◀ For $\varphi \in (\alpha, \beta) \setminus \{\theta\}$ with $\sin \rho(\varphi - \theta) \neq 0$, we obtain

$$\frac{h(\varphi) - h(\theta)}{\varphi - \theta} = \left\{A(\varphi) + \frac{h(\theta)[\cos \rho(\varphi - \theta) - 1]}{\sin \rho(\varphi - \theta)}\right\} \cdot \frac{\sin \rho(\varphi - \theta)}{\varphi - \theta},$$

where

$$A(\varphi) = [h(\varphi) - h(\theta) \cos \rho(\varphi - \theta)] \cdot [\sin \rho(\varphi - \theta)]^{-1}. \tag{1.9.10}$$

Therefore, the derivatives $h'_-(\theta)$, $h'_+(\theta)$ exist only if the following limits exist:

$$A(\theta - 0) = \lim_{\varphi \to \theta - 0} A(\varphi), \quad A(\theta + 0) = \lim_{\varphi \to \theta + 0} A(\varphi)$$

. Moreover,

$$h'_-(\theta) = \rho A(\theta - 0), \qquad h'_+(\theta) = \rho A(\theta + 0). \tag{1.9.11}$$

From inequality (1.9.2), we deduce that $A$ is monotone nondecreasing (see (1.9.10)) on each interval of length at most $\pi/\rho$ that contains $\theta$ in its closure. Consequently, the required unequality is fulfilled, moreover

$$\begin{aligned} \rho A(\varphi) &\leqslant h'_-(\theta), \qquad \max\{\alpha, \theta - \pi/\rho\} < \varphi < \theta; \\ \rho A(\varphi) &\geqslant h'_+(\theta), \qquad \theta < \varphi < \min\{\beta, \theta + \pi/\rho\}, \end{aligned} \tag{1.9.12}$$

i.e., the numbers $h'_-(\theta)$ and $h'_+(\theta)$ are finite ▶

**Property 1.9.19.** *Under the assumptions of Theorem* 1.9.18, *if* $\theta_0 \in (\alpha, \beta)$ *is a point of maximum or minimum for* $h$, *then*

$$h(\theta) \geqslant h(\theta_0) \cos \rho(\theta - \theta_0), \quad \text{for } |\theta - \theta_0| < \pi/\rho, \quad \theta \in (\alpha, \beta).$$

At the point $\theta_0$, either the derivative $h'(\theta_0)$ exists and is equal to 0, or $h'_-(\theta_0) \leqslant 0 \leqslant h'_+(\theta_0)$. Now the statement follows from inequality (1.9.12).

The following assertion characterizes the set where a trigonometrically $\rho$-convex function is differentiable.

**Property 1.9.20.** *Under the assumptions of Theorem* 1.9.18, *the function* $h$ *is continuously differentiable on* $(\alpha, \beta)$ *with the exception of an at most countable set.*

The following statement describes support properties of trigonometrically $\rho$-convex functions.

**Theorem 1.9.21.** *Let* $\Delta(\theta) = \{\varphi \in (\alpha, \beta) : \ |\varphi - \theta| < \pi/\rho\}$. *Under the assumptions of Theorem* 1.9.18, *the set*

$$\begin{aligned} Dh(\theta) = \{p \in \mathbb{C} : Re\ pe^{i\rho\theta} = h(\theta); \\ Re\ pe^{i\rho\varphi} \leqslant h(\varphi) \quad \forall\ \varphi \in \Delta(\theta)\} \end{aligned} \tag{1.9.13}$$

*is nonempty for every* $\theta \in (\alpha, \beta)$, *and*

$$\begin{aligned} Dh(\theta) = \{p \in \mathbb{C} : \rho p e^{i\rho\theta} = \rho h(\theta) - \\ i[(1-\lambda)h'_-(\theta) + \lambda h'_+(\theta)], 0 \leqslant \lambda \leqslant 1\}. \end{aligned} \tag{1.9.14}$$

◀ Let $B_\theta$ be the right hand side of inequality (1.9.14), and let $p \in B_\theta$. Then $p = [h - it/\rho]e^{-i\rho\theta}$, where $t \in [h'_-(\theta), h'_+(\theta)]$ (see (1.9.9). Using inequality (1.9.12), we immediately see that $p \in Dh(\theta)$ (see (1.9.13)). So, $B_\theta \subset Dh(\theta)$. The inclusion $Dh(\theta) \subset B_\theta$ is proved by reversing the arguments ▶

We consider the class $P_\rho = \{h : \mathbb{R} \to (-\infty, \infty]\}$ of trigonometrically $\rho$-convex $2\pi$-periodic lower semicontinuous functions. By Property 1.9.15, every function $h \in P_\rho$ is continuous in dom $h$. We examine geometric properties of functions belonging to $P_\rho$.

The structure of the set dom $h$ for a function $h \in P_\rho$ is described by the following statement.

**Property 1.9.22.** *For* $h \in P_\rho$ *with* $h \not\equiv \infty$, *denote by* $B_h$ *the collection of all components of the set* $\mathbb{R} \setminus \text{dom}\ h$. *Then:*

*i) if* $\rho \leqslant 1/2$, *then* dom $h = \mathbb{R}$*;*

*ii) if* $\rho > 1/2$, dom $h \neq \mathbb{R}$, *then every element* $\Delta \in B_h$ *is of length* $\alpha \geqslant \pi/\rho$*;*

*iii) if* $\rho < 1$, dom $h \neq \mathbb{R}$, *then there are no other elements in the set* $B_h$ *between its elements* $\Delta, \Delta + 2\pi$*; for* $\rho = 1$ *this is valid if* $\overset{\circ}{\Delta} \neq \emptyset$.

◀ Assume that dom $h \neq \mathbb{R}$. The components of dom $h$ and the elements of $B_h$ interlace. Therefore, inequality (1.9.2) shows that $\alpha \geqslant \pi/\rho$. Since $h$ is a $2\pi$-periodic function, we have $\Delta + 2\pi \in B_h$. Since $h \not\equiv \infty$, it follows that $\pi/\rho \leqslant \alpha < 2\pi$. This is possible only if $\rho > 1/2$. Assertion iii) is proved similarly ▶

Let $F_h$ be the collection of all components of dom $h$, let $\Delta \in F_h$ and, say, $\Delta = [\alpha, \beta]$. In these notation the following supplements to Theorems 1.9.18, 1.9.21 are valid.

**Theorem 1.9.18′.** *At the points* $\alpha, \beta$ *there exist right and left derivatives* $h'_+(\alpha)$ *and* $h'_-(\beta)$ *of* $h$, *and* $-\infty \leqslant h'_+(\alpha) < \infty$, $-\infty < h'_-(\beta) \leqslant \infty$.

**Theorem 1.9.21′.** *Let* $Dh(\alpha) = \{p \in \mathbb{C} :\ \mathrm{Re}\, pe^{i\rho\alpha} = h(\alpha),\ \mathrm{Re}\, pe^{i\rho\varphi} \leqslant h(\varphi),$ *where* $\alpha \leqslant \varphi \leqslant \min\{\alpha + \pi/\rho, \beta\}\}$. *If* $h'_+(\alpha) = -\infty$, *then* $Dh(\alpha) = \emptyset$. *If* $h'_+(\alpha) > -\infty$, *then* $Dh(\alpha) = \{p \in \mathbb{C} :\ \rho p e^{i\rho\alpha} = \rho h(\alpha) - it,\ t \leqslant h'_+(\alpha)\}$. *The same is true for* $\beta$.

**Remark.** For $\theta \in$ dom $h$, we put $\Pi = \{w \in \mathbb{C} :\ w = te^{i\psi},\ t \geqslant 0$ and $|\psi - \rho\theta| < \pi/2\}$ and $G(w) = th(\psi/\rho)$, $w \in \Pi$. Then $G$ is a sublinear function in the half-plane $\Pi$ (see the proof of Prop. 1.9.5). The subdifferential $\partial G(e^{i\rho\theta})$ of $G$ at $e^{i\rho\theta}$ (see Def. 1.4.21) coincides with the set $Dh(\theta)$. For this reason, the set $Dh(\theta)$ is called the *subdifferential* of $h$ at the point $\theta \in \mathbb{R}$, and the multivalued mapping $Dh : \mathbb{R} \to \mathbb{C}$ is said to be *subdifferential* .

For the sake of simplicity, we illustrate the geometric properties of the functions in $P_\rho$ by the example of the subclass $P'_\rho = \{h \in P_\rho :\ \text{dom}\ h = \mathbb{R}\}$.

Suppose $\rho = 1$, $h \in P'_1$. Then $H(z) = rh(\theta)$, $z = re^{i\theta} \in \mathbb{C}$, is a continuous sublinear function (see T. 1.9.12 and T. 1.4.5). Therefore (see Def. 1.3.18, T. 1.3.19, T. 1.5.6, and Example 1 in Section 1.5), $H$ is the support function of the compact convex set

$$I_h = \{q \in \mathbb{C} :\ \mathrm{Re}\, qe^{-i\theta} \leqslant h(\theta) \quad \forall\, \theta \in \mathbb{R}\}. \tag{1.9.15}$$

Let $h \in P'_\rho$. With $h$, we associate the family of straight lines

$$\Omega_\rho = \{l_\varphi :\ \varphi \in \mathbb{R}\};\ l_\varphi = \{p \in \mathbb{C} : \mathrm{Re}\, pe^{i\rho\varphi} = h(\varphi)\}. \tag{1.9.16}$$

For $\rho = 1$ the set $\Omega_1$ is the family of support straight lines to the compact convex set

$$K_h = \{p \in \mathbb{C} :\quad \mathrm{Re}\, pe^{i\theta} \leqslant h(\theta) \quad \forall\, \theta \in \mathbb{R}\}$$

($K_h$ is called the "conjugate diagram"). If there is $\theta \in \mathbb{R}$ such that $h(\theta) + h(\theta + \pi) = 0$ (cf. (1.9.7)), then $l_\theta = l_{\theta+\pi}$ and dim $K_h < 2$. If $\inf\{h(\varphi) + h(\varphi + \pi) :\ \varphi \in \mathbb{R}\} > 0$, then $K_h$ is the closure of the convex domain in $\mathbb{C}$ bounded the closed Jordan curve $\Gamma_h$, which is the "envelope" of the family $\Omega_1$.

In general, for $\rho \neq 1$ the "envelope" of the family $\Omega_\rho(h)$ (see (1.9.16)) is a piecewise locally convex curve $\Gamma_h$ in $\mathbb{C}$ possessing a certain cyclicity property. In the movement along $\Gamma_h$, the support straight line $l_\varphi$ rotates in one direction.

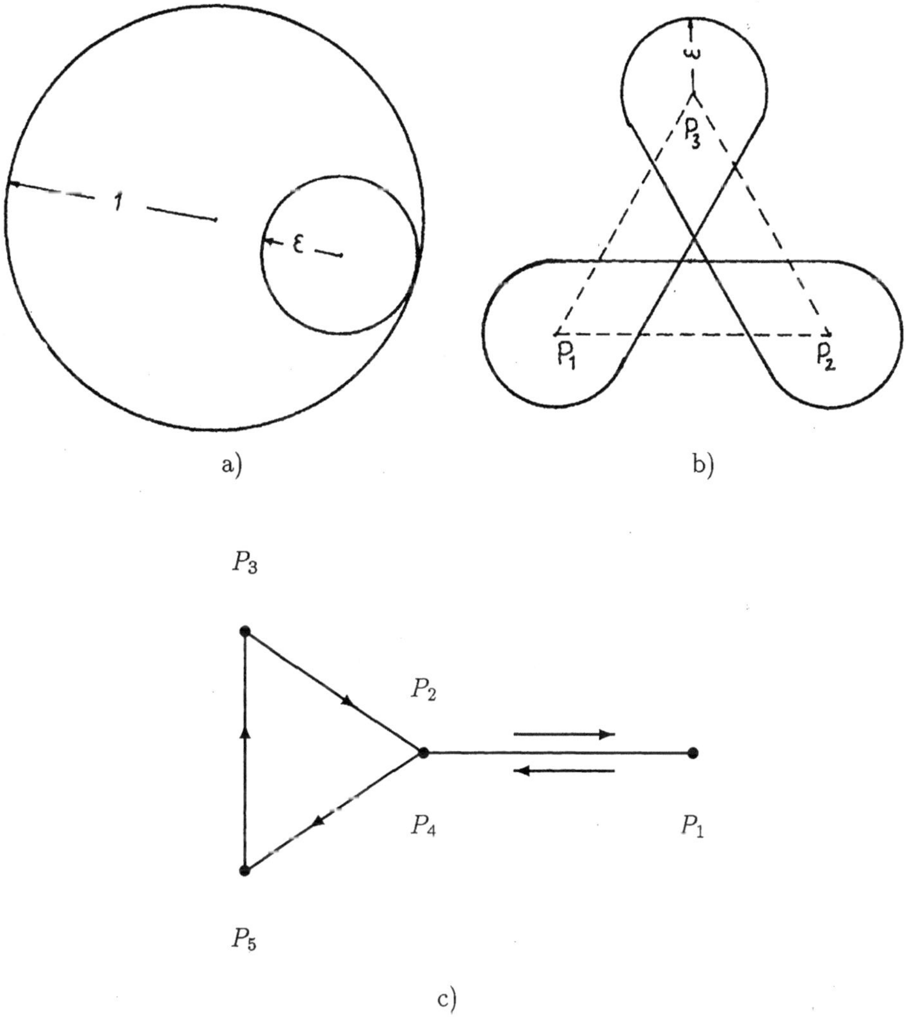


Fig. 2

**Example 1.** Fig. 2 shows examples of curves associated (in the way described above) with certain functions of class $P_\rho$ for $\rho = 2$ and 3, namely, of the $2\pi$-periodic extensions to $\mathbb{R}$ of the following functions:

$$\text{a)}\quad h_\varepsilon(\theta) = \varepsilon + (1-\varepsilon)\cos 2\theta, \quad 0 \leqslant \theta \leqslant \pi;$$
$$h_\varepsilon(\theta) = 1, \quad -\pi \leqslant \theta \leqslant 0;$$
$$\text{b)}\quad h_\varepsilon(\theta) = \operatorname{Re} p_j e^{i2\theta} + \varepsilon, \quad \theta \in [\theta_{j-1}, \theta_j], \quad j = 1, 2, 3;$$
$$\theta_j = -\pi + \frac{2}{3}\pi j, \qquad p_j = \exp\{i\frac{2}{3}\pi(j-2)\},$$

where $\varepsilon > 0$ is a sufficiently small number;

$$\text{c)}\quad h(\theta) = \operatorname{Re} p_j e^{i3\theta}, \quad \theta \in [\theta_{j-1}, \theta_j], \quad j = 1, 2, 3, 4, 5;$$
$$p_1 = -\frac{1}{2} + i\frac{\sqrt{3}}{2}, \quad p_2 = p_4 = 1; \quad p_3 = 2;$$
$$p_5 = \frac{1}{2} - i\frac{\sqrt{3}}{2}; \quad \theta_0 = -\pi; \quad \theta_1 = -\frac{7}{9}\pi;$$
$$\theta_2 = -\frac{\pi}{6}; \quad \theta_3 = \frac{\pi}{6}; \quad \theta_4 = \frac{7\pi}{9}; \quad \theta_5 = \pi.$$

Let $h$ be a two times continuously differentiable function in $P'_\rho$ satisfying the condition $h''(\theta) + \rho^2 h(\theta) > 0 \;\; \forall\, \theta \in \mathbb{R}$ (cf. T. 1.9.17). Elementary calculations show that there exists a unique smooth curve

$$\Gamma_h = \{p = \Phi(\theta) \in \mathbb{C} : \quad \theta \in \mathbb{R}\}$$

in $\mathbb{C}$ determined by the conditions

$$\operatorname{Re} \Phi(\theta) e^{i\rho\theta} = h(\theta), \quad \operatorname{Re} \Phi'(\theta) e^{i\rho\theta} = 0 \quad \forall\, \theta \in \mathbb{R},$$

and possessing the property $\operatorname{Im} \Phi'(\theta) e^{i\rho\theta} < 0 \;\; \forall\, \theta \in \mathbb{R}$. Specifically,

$$\Phi(\theta) = e^{-i\rho\theta}[h(\theta) - ih'(\theta)/\rho], \quad \theta \in \mathbb{R}. \tag{1.9.17}$$

Geometrically, these conditions mean that $e^{-i\rho\theta}$ is the normal to $\Gamma_h$ at the point $\Phi(\theta)$, the tangent vector is on the right of $e^{-i\rho\theta}$, i.e., the tangent to $\Gamma_h$ rotates continuously in one direction when moving along $\Gamma_h$. Any curve locally coinciding with the graph of a strictly convex function (in a suitable coordinate system in $\mathbb{R}^2$) possesses this property: the derivative of this function strictly increases (see P. 1.4.22).

In the case under consideration formula (1.9.17) leads to an explicit construction of the curve $\Gamma_h$ associated with a function $h \in P'_\rho$.

**Example 2.** Let $h(\theta) = 2\cos^3\theta$ for $|\theta| \leqslant \pi/2$ and $h(\theta) = 0$ for $\pi/2 < |\theta| \leqslant \pi$. Then the $2\pi$-periodic extension of this function to $\mathbb{R}$ belongs to $P'_3$ (see Prop. 1.9.17), and the function $\Phi$ (see (1.9.17)) is the $2\pi$-periodic extension to $\mathbb{R}$ of the function

$$p = \begin{cases} e^{-i2\theta}(1 + \cos 2\theta), & |\theta| \leqslant \pi/2; \\ 0, & \pi/2 < |\theta| \leqslant \pi. \end{cases}$$

So, the range of the path $\Phi: \mathbb{R} \to \Gamma_h$ is a cardioid, and in the movement along it the support straight line $l_\theta$ turns about the singular point $p=0$ of the cardioid through an angle of $3\pi$.

We describe the curve $\Gamma_h$ analytically in the case of an arbitrary function $h \in P'_\rho$. For this, we need the following fact.

**Property 1.9.23.** *Suppose $g \in P'_\rho$. For $\theta \in \mathbb{R}$ and $\delta \in (0,\pi/\rho)$, put $\Delta_\theta(\delta) = (\theta-\delta, \theta+\delta)$. Then:*

*1) if $g(\theta)=0$ and $g(\varphi) \geqslant 0 \ \ \forall \varphi \in \Delta_\theta(\delta)$, then $g(\varphi) \geqslant 0 \ \ \forall \varphi \in \Delta_\theta(\pi/\rho)$;*

*2) if $\psi \in \Delta_\theta(\pi/2\rho)$ and $g(\psi) = g(\theta-\pi/2\rho) = g(\theta+\pi/2\rho) = 0$, and $g(\varphi) \geqslant 0 \ \ \forall \varphi \in \Delta_\theta(\pi/2\rho)$, then $g(\varphi) \equiv 0$, $\varphi \in \Delta_\theta(\pi/2\rho)$.*

Both assertions follow directly from (1.9.2).

Let $\theta \in \mathbb{R}$. We put

$$K_\theta = \{p \in \mathbb{C}: \ \operatorname{Re} pe^{i\rho\varphi} \leqslant h(\varphi), \quad \varphi \in \bar{\Delta}_\theta(\pi/2\rho)\}. \tag{1.9.18}$$

1. If $h(\theta-\pi/2\rho) + h(\theta+\pi/2\rho) > 0$, then $K_\theta$ is a closed convex domain in $\mathbb{C}$. We denote

$$\Pi = \{w = te^{i\psi} \in \mathbb{C}: \ t \geqslant 0, \ |\psi - \rho\theta| \leqslant \pi/2\},$$
$$G(w) = \begin{cases} th(\psi/\rho), & w = te^{i\psi} \in \Pi, \\ \infty, & w \in \mathbb{C} \setminus \Pi, \end{cases} \tag{1.9.19}$$

and $H(u) = G(\bar{u})$, $u \in \mathbb{C}$. Proposition 1.9.4 and Property 1.9.5 imply that $H$ is a lower semicontinuous sublinear function on $\mathbb{C}$. Combining this with the remark to Theorem 1.3.19 and using the notation in that theorem, we conclude (see (1.9.18)) that $K_\theta = T(H)$ and $H$ is the support function for $K_\theta$. Consequently,

$$h(\varphi) = \sup\{\operatorname{Re} pe^{i\rho\varphi}: \ p \in K_\theta\}, \quad \varphi \in \Delta_\theta(\pi/2\rho). \tag{1.9.20}$$

Therefore, the family of the curves (see (1.9.15))

$$\{l_\varphi: \ \varphi \in \Delta_\theta(\pi/2\rho)\}, \ l_\varphi = \{p \in \mathbb{C}: \ \operatorname{Re} pe^{i\rho\varphi} = h(\varphi)\} \tag{1.9.21}$$

is the family of support straight lines to the set $K_\theta$; moreover, the set

$$\Gamma_h^\theta = \bigcup (K_\theta \cap l_\varphi) \mid \varphi \in \Delta_\theta(\pi/2\rho) \tag{1.9.22}$$

is a convex arc on the boundary of $K_\theta$, which is the "envelope" of this family.

Suppose

$$\psi \in \Delta_\theta(\pi/2\rho), \ q \in K_\theta \cap l_\psi, \ g(\varphi) = h(\varphi) - \operatorname{Re} qe^{i\rho\varphi}, \tag{1.9.23}$$

where $\varphi \in \mathbb{R}$. Then, by (1.9.18) and (1.9.20), we have: $g(\psi) = 0$ and $g(\varphi) \geqslant 0$ $\forall \varphi \in \Delta_\theta(\pi/2\rho)$. Therefore, near the point $\psi$ the function $g$ satisfies the conditions of assertion 1) of Property 1.9.23. Consequently, $g(\varphi) \geqslant 0$, $|\varphi - \psi| < \pi/\rho$, and in the notation of formula (1.9.13) we obtain: $q \in Dh(\psi)$, $K_\theta \cap l_\psi = Dh(\psi)$. So (see (1.9.22)),

$$\Gamma_h^\theta = \bigcup Dh(\varphi) \mid \varphi \in \Delta_\theta(\pi/2\rho) \tag{1.9.24}$$

is a convex arc containing the point $q$.

2. Let $h(\theta + \pi/2\rho) + h(\theta - \pi/2\rho) = 0$, and let condition (1.9.23) be true. Combining this with (1.9.20), we conclude that assertion 2) of Property 1.9.23 is valid for $g$. Consequently,

$$h(\varphi) = \operatorname{Re} q e^{i\rho\varphi}, \qquad \varphi \in \bar{\Delta}_\theta(\pi/2\rho); \tag{1.9.25}$$

$K_\theta$ is the ray with the origin at $q$ and parallel to the vector $e^{-i\rho\theta}$, and the support straight line $l_\varphi$ (see (1.9.15) and (1.9.21)) rotates about $q$ at least for $\varphi \in [\theta - \pi/2\rho, \theta + \pi/2\rho]$. Furthermore, the straight lines $l_{\theta-\pi/2\rho}$ and $l_{\theta+\pi/2\rho}$ coincide. If $\Delta_\theta(\pi/2\rho)$ is the maximal interval on which equality (1.9.25) is true, then the curve $\Gamma_h$ has a "cuspidal edge", i.e., a segment passed twice in opposite directions. This fact is illustrated by Example 1, c, where the cuspidal edge is the segment $[1, 2]$ (see Fig. 2, $c$). In the general case for the straight line $l_\varphi$ (see Fig. 2, $b$) the angle of rotation about $q$ is larger than one complete turn. This happens in the same example, in particular, for $\varphi$ ranging through the intervals $(-7\pi/9, -\pi/6)$ and $(\pi/6, 7\pi/9)$, when the straight line $l_\varphi$ rotates about 1. In such cases the curve $\Gamma_h$ may fail to be convex near $q$ (see Examples 1, $c$ and 2).

Formula (1.9.25) implies that in this case formula (1.9.17) is valid: $p = \Phi(\varphi) \equiv q$, $\varphi \in \Delta_\theta(\pi/2\rho)$. Consequently, in the notation of Theorem 1.9.18, we have $\{q\} = Dh(\varphi)\ \forall\ \varphi \in \Delta_\theta(\pi/2\rho)$ and formula (1.9.24) is true again (see (1.9.22)); moreover, $\Gamma_h^\theta = \{q\}$.

Thus, the curve $\Gamma_h$ (the "envelope" of the family of straight lines $\Omega_\rho(h)$, see (1.9.15)) admits the following description:

$$\Gamma_h = \bigcup Dh(\varphi) \mid \varphi \in \mathbb{R}, \tag{1.9.26}$$

where the set $Dh(\varphi)$ is defined constructively by formula (1.9.14).

**Definition 1.9.24.** The subdifferential mapping $Dh : \mathbb{R} \to \mathbb{C}$, where $Dh(\varphi)$ is the subdifferential of the function $h$ at the point $\varphi \in \mathbb{R}$, is called the *path associated with the function* $h \in P'_\rho$, and the range (or support) of $Dh$, i.e,. $\Gamma_h = Dh(\mathbb{R})$, is called the *curve in* $\mathbb{C}$ *associated with* $h$ (see the remark to T. 1.9.21′, and (1.9.20)).

In contrast to the curve $\Gamma_h$, the mapping $Dh$ completely determines a function $h$ in $P'_\rho$, and, in particular, provides information about the intervals where $h$ is trigonometric. Examples 1, 2 graphically illustrate this fact. In addition, the paths associated with two different functions of class $P'_\rho$ may have equal supports. For instance, this is so for the family of functions

$$\{h_k(\theta) = h(\theta + 2\pi k/\rho),\ \theta \in \mathbb{R};\ k \in \mathbb{Z}\}, \quad \text{where } h \in P_\rho,$$

because for every $k \in \mathbb{Z} \setminus \{0\}$ one and the same system of the straight lines of the form (1.9.16) corresponds to the pair of functions $h_k$ and $h_0$. A similar phenomenon may occur for functions belonging to different classes. For example, the function $h(\theta) \equiv \sigma$, $\theta \in \mathbb{R}$, belongs to $P_\rho$ for all $\rho > 0$, and $\Gamma_h = \{p \in \mathbb{C} : \ |p| = \sigma\}$.

By the arguments preceding Definition 1.9.24, we proved that the path $Dh$ is *locally convex* in the following sense: for every $\theta \in \mathbb{R}$ there exists $\delta > 0$ such that the set $\bigcup Dh(\varphi)\ |\varphi \in (\theta - \delta, \theta + \delta)$ is either a convex set in $\mathbb{C}$ of dimension $< 2$ (a point or a segment), or a convex arc in $\mathbb{C}$, i.e., an arcwise connected part of the boundary of a convex domain in $\mathbb{C}$. This and several other properties of the path $Dh$ are presented in the following theorem.

**Theorem 1.9.25.** *Let $h \in P_\rho$, and let $Dh : \mathbb{R} \to \mathbb{C}$ be the path associated with $h$* (see Def. 1.9.24). *Then the graph*

$$\operatorname{gr} Dh = \{(\theta, p) \in \mathbb{R} \times \mathbb{C} : \ \theta \in \mathbb{R}; \quad p \in Dh(\theta)\}$$

*is an arcwise connected set in $\mathbb{R} \times \mathbb{C}$. The cyclic group $G$ with generator*

$$B_\rho(\theta, p) = (\theta - 2\pi, pe^{i2\pi\rho}), \quad \theta \in \mathbb{R}; \quad p \in Dh(\theta) \tag{1.9.27}$$

*operates on this set. Moreover, $Dh$ is a locally convex path possessing the following properties:* 1) *if $h$ is a differentiable function, then*

$$Dh(\theta) = e^{-i\rho\theta}\,[h(\theta) - ih'(\theta)/\rho], \quad \theta \in \mathbb{R}; \tag{1.9.28}$$

2) *the curve $\Gamma_h$ associated with $h$* (see (1.9.26)) *is a bounded set in $\mathbb{C}$, and*

$$\sup\{|p| : \ p \in \Gamma_h\} = \sigma, \quad \text{where} \quad \sigma = \sup\{h(\theta) : \ \theta \in \mathbb{R}\}. \tag{1.9.29}$$

◀ Formulas (1.9.27) and (1.9.28) immediately follow from (1.9.13) and (1.9.14). For instance, recalling that $h$ is a $2\pi$-periodic function, by (1.9.13) we have: $pe^{i2\pi\rho} \in Dh(\theta - 2\pi)$ if $p \in Dh(\theta)$, $\theta \in \mathbb{R}$.

If $\theta \in \mathbb{R}$, $q \in Dh(\theta)$, then (see (1.9.13), (1.9.29)) $\operatorname{Re} qe^{i\rho\varphi} \leqslant \sigma$, $|\varphi - \theta| \leqslant \pi\rho$. Consequently, $|q| \leqslant \sigma \ \ \forall\, q \in \Gamma_h$. Since $h$ is a $2\pi$ periodic continuous function (see Prop. 1.9.5), there exists $\theta_0 \in \mathbb{R}$ such that $h(\theta_0) = \sigma$. Then, by (1.9.13), we find: $\sigma \leqslant |q|$, $q \in Dh(\theta_0)$. This implies formula (1.9.29). The other assertions of the theorem were proved earlier. ▶

**Remark.** Formula (1.9.27) implies that the cyclic group $G$ of rotations about 0 in $\mathbb{C}$ through an angle of $2\pi\rho$ operates on $\Gamma_h$. By using these rotations, we may form the curve $\Gamma_h$ from its part

$$E_h = \bigcup Dh(\varphi) \mid \varphi \in [\theta, \theta + 2\pi), \tag{1.9.30}$$

where $\theta \in \mathbb{R}$ is an arbitrary fixed number. If $\rho = m/n$ is a rational number, then (see (1.9.27)) $G$ is a finite group of order $n$ and $\Gamma_h$ is a closed curve, in the movement along which the support straight line $l_\varphi$ (see (1.9.15)) makes $m$ complete turns if the parameter $\varphi$ ranges through an interval of length $2\pi n$. If $\rho$ is an irrational number, then $G$ is an infinite group and $\Gamma_h$ is a nonclosed curve.

We list a series of properties of the plane curve $\Gamma_h$ associated with a function $h \in P'_\rho$.

**Property 1.9.26.** The *curve* $\Gamma_h$ *is locally rectifiable, and the length of its part* $E_h$ (see (1.9.30)) *does not depend on* $\theta$ *and is equal to* $\int_0^{2\pi} \rho h(\varphi)\, d\varphi$.

◀ For simplicity, we restrict ourselves to the case when $h$ is a two times continuously differentiable function. Let $S(E_h)$ be the length of $E_h$. We denote $\Phi(\varphi) = Dh(\varphi)$, $\varphi \in \mathbb{R}$ (see (1.9.14)). Then, by (1.9.28), $\Phi$ is continuously differentiable and

$$S(E_h) = \int_\theta^{\theta+2\pi} |\Phi'(\varphi)|\, d\varphi = \int_0^{2\pi} [\rho h(\varphi) + h''(\varphi)/\rho]\, d\varphi = \int_0^{2\pi} \rho h(\varphi)\, d\varphi,$$

because $h'$ is $2\pi$-periodic ▶

**Property 1.9.27.** *Let* $h \in P_\rho$. *We put* $H(z) = r^\rho h(\theta)$ *and* $\sigma = \sup\{h(\theta) : \theta \in \mathbb{R}\}$. *Let* $z_j = r_j e^{i\theta_j}$, $j = 1, 2$, *be an arbitrary pair of points in* $\mathbb{C}$ *satisfying the condition* $\rho|\theta_2 - \theta_1| < \pi$. *Then*

$$|H(z_1) - H(z_2)| \leqslant \sigma |z_1^\rho - z_2^\rho|, \qquad z^\rho = r^\rho e^{i\rho\theta}. \tag{1.9.31}$$

◀ We can find $\theta \in \mathbb{R}$ such that $\theta_1, \theta_2 \in \Delta_\theta(\pi/2\rho)$, where $\Delta_\theta(\delta) = (\theta - \delta, \theta + \delta)$. If $h(\theta - \pi/2\rho) + h(\theta + \pi/2\rho) = 0$, then (1.9.25) is valid. Therefore, $h(\theta_j) = \operatorname{Re} q e^{i\rho\theta_j}$, $j = 1, 2$, where $q \in Dh(\theta) \subset \Gamma_h$. From this we immediately deduce inequality (1.9.31), taking into account that $|q| \leqslant \sigma$ (see (1.9.29)). Next, let $h(\theta - \pi/2\rho) + h(\theta + \pi/2\rho) > 0$. In the notation of formulas (1.9.18) and (1.9.22) we have:

$$h(\varphi) = \sup\{\operatorname{Re} p e^{i\rho\varphi} : p \in K_\theta\} \geqslant \operatorname{Re} q e^{i\rho\varphi} \qquad \forall\, q \in \Gamma_h^\theta$$

for all $\varphi \in \Delta_\theta(\pi/2\rho)$. In particular, $h(\theta_2) \geqslant \operatorname{Re} q e^{i\rho\theta_2}$ if $q \in Dh(\theta_1)$. On the other hand, $h(\theta_1) = \operatorname{Re} q e^{i\rho\theta_1}$ (see (1.9.13)). So (see (1.9.29)),

$$H(z_1) - H(z_2) \leqslant \operatorname{Re} q(z_1^\rho - z_2^\rho) \leqslant \sigma |z_1^\rho - z_2^\rho|.$$

Interchanging the roles of $z_1$ and $z_2$ in the above arguments, we arrive at (1.9.31) ▶

**Property 1.9.28.** *Let $K$ be a compact set in $\mathbb{C}$ with $0 \in K$. Then under the assumptions and in the notation of Property 1.9.27, the following inequality is true for sufficiently large $R$:*

$$|H(z+\tau) - H(z)| \leqslant \sigma\rho[1+\varepsilon(R)]|\tau||z|^{\rho-1}, \ \tau \in K, \tag{1.9.32}$$

*where $|z| > R$ and $\varepsilon(R) \to 0$ (as $R \to \infty$) uniformly in the variable $\tau \in K$.*

◀ Suppose $z = re^{i\theta}$. Put $\Pi_\varepsilon = \{w = te^{i\varphi} \in \mathbb{C} : \ t > 0, \ |\varphi - \theta| < \varepsilon\}$, where $\varepsilon < \min\{\pi/2, \pi\rho\}$. Then $w^\rho = t^\rho e^{i\rho\varphi}$ is a function defined on $\Pi_\varepsilon$. The Newton-Leibniz formula yields

$$|(z+\tau)^\rho - z^\rho| = \left| \int\limits_z^{z+\tau} \rho t^{\rho-1}\, dt \right| \leqslant \rho|\tau| \max\{|t|^{\rho-1} : \ t \in [z, z+\tau]\}.$$

Together with (1.9.31), this implies (1.9.32) ▶

## 1.10. Selected facts about entire functions of one variable

We collect some facts of the growth theory of entire functions of one variable. These facts will be required in the subsequent chapters.

**1.10.1. Growth characteristics of entire functions of one variable.** A function $f$ of a complex variable is said to be *entire* if it is representable by a power series convergent everywhere on $\mathbb{C}$:

$$f(z) = \sum_{k=0}^{\infty} c_k z^k. \tag{1.10.1}$$

Usually, the growth of an entire function $f$ is compared to the growth of the maximum of its modulus

$$M_f(r) = \max_{|z|=r} |f(z)|. \tag{1.10.2}$$

By definition, the *order and type* of an entire function $f$ are the order and type (see Def. 1.1.1) of the function $\varphi(r) = \ln^+ M_f(r)$. The radial growth of an entire function $f$ of order $\rho < \infty$ is characterized by its *indicator*

$$h_f(\theta) = \overline{\lim_{r\to\infty}}\ r^{-\rho} \cdot \ln|f(re^{i\theta})|, \ \theta \in \mathbb{R}. \tag{1.10.3}$$

**Proposition 1.10.1.** *The maximal value of the indicator of an arbitrary entire function of finite order $\rho > 0$ coincides with the type of this function.*

**Theorem 1.10.2.** *The indicator of an arbitrary entire function of order $\rho > 0$ is a $2\pi$-periodic trigonometrically $\rho$-convex function. The class $P'_\rho$ of* *all finite functions possessing the last-mentioned properties coincides with the class of indicators of entire functions of order* $\rho > 0$ *and of finite type.*

The relationship between growth characteristics of an entire function and its Taylor coefficients is explained by the following theorem.

**Theorem 1.10.3.** *Let* $\{c_k\}_0^\infty$ *be the Taylor coefficients of an entire function* $f$ (see (1.10.1)). *Then*

*i) the order* $\rho \geqslant 0$ *of* $f$ *can be calculated by the formula*

$$\rho = \varlimsup_{k\to\infty} \frac{k \ln k}{-\ln |c_k|}; \tag{1.10.4}$$

*ii) if* $0 < \rho < \infty$, *then the type* $\sigma \geqslant 0$ *of* $f$ *can be calculated by the formula*

$$(\sigma e \rho)^{1/\rho} = \varlimsup_{k\to\infty} k^{1/\rho} \cdot \sqrt[k]{|c_k|} \tag{1.10.5}$$

**1.10.2. Indicator and conjugate diagrams of an entire function of exponential type.** Let $h = h_f$ be the indicator of an entire function $f$ of the first order and of finite type. The indicator diagram $I_h$ of $f$ defined by formula (1.9.15) (see T. 1.10.2) serves as a geometric image of the indicator $h$. It plays an important role in a large body of research. We present the proof of Theorem $A_1$ related to notion of an indicator diagram (see the Introduction). First, we mention a result by E. Borel (Theorem $A_1$ is an extension of it).

An *entire function of exponential type* is an entire function $f$ satisfying the inequality

$$|f(z)| < A\exp\{a|z|\}, \quad z \in \mathbb{C}, \tag{1.10.6}$$

for some constants $A = A_f > 0$ and $a = a_f > 0$.

**Proposition 1.10.4.** *Suppose*

$$f(z) = \sum_{k=0}^{\infty} \frac{a_k}{k!} z^k \tag{1.10.7}$$

*is an entire function of exponential type, and* $\sigma = \varlimsup_{r\to\infty} r^{-1} \cdot \ln M_f(r)$. *Let* $\gamma$ *be the radius of the smallest disk with center at 0 outside of which the Borel transform*

$$F(z) = \sum_{k=0}^{\infty} \frac{a_k}{z^{k+1}} \tag{1.10.8}$$

*of* $f$ *is holomorphic. Then* $\sigma = \gamma$.

◀ If $f$ is of order $\rho = 1$, then $\sigma$ is the type of $f$. Invoking the Stirling formula $\sqrt[k]{k!} \simeq k/e$, $k \to \infty$, by (1.10.5) and (1.10.7) we find

$$\sigma = \varlimsup_{k\to\infty} \sqrt[k]{|a_k|}. \tag{1.10.9}$$

But if $\rho < 1$, then $\sigma = 0$, and at the same time the limit in (1.10.9) is equal to 0, because by (1.10.4), (1.10.7) and the Stirling formula for every $\varepsilon \in (\rho, 1)$ we have

$$\sqrt[k]{|a_k|} < k^{(\varepsilon-1)/\varepsilon}, \quad k > k_0(\varepsilon).$$

Now, from (1.10.9), (1.10.8) and the Cauchy-Hadamard formula for the radius of convergence of a power series, we deduce Proposition 1.10.4 ▶

So, the conjugate diagram of a function $f$ (see the Introduction, Theorem $A_1$) lies in the disk of radius $\sigma$ in $\mathbb{C}$ with center at 0. To prove Theorem $A_1$ we need the following integral representations relating the functions $f$ and $F$ (see (1.10.7) and (1.10.8)).

**Proposition 1.10.5.** *Let $f$ be an entire function of exponential type, and let $F$ be the Borel transform of $f$. We put $\sigma = \overline{\lim\limits_{r\to\infty}}\, r^{-1} \cdot \ln M_f(r)$, and for $A > \sigma$ denote $\Pi(\theta; A) = \{z \in \mathbb{C} : \operatorname{Re} z e^{i\theta} > A\}$, $\theta \in \mathbb{R}$. Then*

$$F(z) = \int_0^{\infty(\operatorname{Arg}\zeta=\theta)} e^{-z\zeta} f(\zeta) d\zeta \quad \forall\, z \in \Pi(\theta; A), \quad \theta \in \mathbb{R}, \tag{1.10.10}$$

$$f(z) = \frac{1}{2\pi i} \int_\Gamma e^{z\zeta} F(\zeta)\, d\zeta \quad \forall\, z \in \mathbb{C}, \tag{1.10.11}$$

*where $\Gamma$ is a closed positively oriented rectifiable Jordan contour enclosing the conjugate diagram $K$ of $f$.*

◀ 1. Inequality (1.10.6) shows that the integral in (1.10.10) converges uniformly in $z$ on every compact subset of $\Pi(\theta; A)$ and, consequently, is a holomorphic function in $\Pi(\theta; A)$. Because of the uniqueness theorem, it suffices to prove formula (1.10.10) for $z \in \Pi(\theta; 3A)$. The following inequality (see (1.10.7)) is obvious:

$$|R_n(z, \zeta)| < e^{-3Ar} \cdot \sum_{k=n}^{\infty} \frac{|a_k|}{k!} r^k \quad \forall\, z \in \Pi(\theta; 3A),\ \zeta \in \Gamma_\theta, \tag{1.10.12}$$

forall $n \geqslant 0$, where

$$r = |\zeta|, \ \Gamma_\theta = \{\zeta = re^{i\theta} : r \geqslant 0\}, \ R_n(z, \zeta) = e^{-z\zeta} \cdot \sum_{k=n}^{\infty} \frac{a_k}{k!} \zeta^k.$$

By (1.10.9), we have

$$|a_k| < B_1 A^k \quad \forall\, k \geqslant 0, \text{ where } B_1 = B_1(f) > 0. \tag{1.10.13}$$

Since (see Example 2 in 1.5)

$$\max_{r \geqslant 0} r^k e^{-2Ar} = \left(\frac{k}{2Ae}\right)^k \quad \forall\, k \geqslant 0$$

and $k! > B_2(k/e)^k \ \forall\, k \geqslant 0$, where $B_2 > 0$, by (1.10.12) and (1.10.13) we obtain

$$|R_n(z;\zeta)| < 2^{1-n} B_1 B_2 e^{-A|\zeta|} \quad \forall\, \zeta \in \Gamma_\theta,\ z \in \Pi(\theta; 3A).$$

Therefore, the series $\sum\limits_{k=0}^{\infty} e^{-z\zeta} a_k \zeta^k/k!$ converges uniformly in $\zeta \in \Gamma_\theta$ for every fixed $z \in \Pi(\theta; 3A)$, and it can be integrated termwise over the ray $\Gamma_\theta$. But

$$\int\limits_{\Gamma_\theta} e^{-z\zeta} \zeta^k \, d\zeta = \frac{k!}{z^{k+1}}$$

for $\operatorname{Re} ze^{i\theta} > 0,\ k \geqslant 0$. This yields (1.10.10).

By Proposition 1.10.4, the compact set $T = \{\zeta = Ae^{i\theta} : |\theta| \leqslant \pi\}$ is included in the domain of convergence of the series $F$ (see (1.10.8)). Therefore, the series $F$ converges uniformly in $\zeta \in T$ and the series $e^{-z\theta} F(\zeta)$ can be integrated termwise in $\zeta \in T$ (see (1.10.7)):

$$\frac{1}{2\pi i} \int\limits_{T} e^{z\zeta} F(\zeta) \, d\zeta = f(z)$$

($T$ is oriented positively). To deduce formula (1.10.11), it only remains to apply the Cauchy theorem ▶

**Proof of Theorem $A_1$.** In the notation of Proposition 1.10.4, if $\sigma = 0$, then $h(\theta) \equiv 0$, $\theta \in \mathbb{R}$ (see P. 1.10.1), and Theorem $A_1$ follows from Proposition 1.10.4. Let $\sigma > 0$, i.e., $f$ is of normal type. Since

$$|f(re^{i\theta})| < C_\varepsilon \exp\{r[h(\theta) + \varepsilon]\} \quad \forall\, r \geqslant 0,\ C_\varepsilon > 0$$

for every $\varepsilon > 0$, if follows that the integral in (1.10.10) is a holomorphic function in the half-plane $\Pi_h(\theta) = \{z \in \mathbb{C} : \operatorname{Re}\, ze^{i\theta} > h(\theta)\}$, where $\theta \in \mathbb{R}$. Therefore, by (1.10.10), the series (1.10.8) admits holomorphic continuation to $\bigcup \Pi_h(\theta) | \theta \in \mathbb{R}$. Let $K$, $I_h$ be the conjugate and the indicator diagrams of $f$. By the above arguments , we have (see (1.9.15))

$$K \subset T_h, \quad \text{where} \quad T_h = \{z : \bar{z} \in I_h\}. \tag{1.10.14}$$

For $\varepsilon > 0$, we put $S_\varepsilon = \{z \in \mathbb{C} : |z| \leqslant \varepsilon\}$. Let $O_\varepsilon(K) = K + S_\varepsilon$ be the $\varepsilon$-hull of $K$. We put $\Gamma = \partial O_\varepsilon(K)$ in (1.10.11) and find an upper bound for $|f|$. The holomorphic function $F$ is continuous on the compact set $\Gamma$; so $|F(\zeta)| < M$ $\forall\, \zeta \in \Gamma$ for some $M > 0$. Then, by Exercize 1.3.20, we obtain $\sup\{\operatorname{Re} \zeta e^{i\theta} : \zeta \in \Gamma\} \leqslant k(-\theta) + \varepsilon \ \forall\, \theta \in \mathbb{R}$, where $k$ is the support function of $K$. Combining this with (1.10.11), we conclude that

$$|f(re^{i\theta})| < M \exp\{r[k(-\theta) + \varepsilon]\} \quad \forall\, z = re^{i\theta} \in \mathbb{C}.$$

Consequently, $h(\theta) \leqslant k(-\theta)\ \forall\, \theta \in \mathbb{R}$, i.e. (see (1.10.14)), $T_h \subset K$, $T_h = K$.

**1.10.3. The spaces $H(E), H^*(E), E \subset \overline{\mathbb{C}}$, and the dual spaces of entire functions.** For an open set $E$ in $\overline{\mathbb{C}}$, let $H(E)$ denote the vector space of functions holomorphic in $E$. Taking an increasing sequence $\{K_n\}_1^\infty$ of compact subsets of $E$ such that $\bigcup\limits_{n=1}^{\infty} K_n = E$, $K_n \subset \overset{\circ}{K}_{n+1}$ $\forall\, n \in \mathbb{N}$, we endow $H(E)$ with the topology determined by the seminorms $p_n(f) = \max\{|f(\zeta)| : \zeta \in K_n\}$, $n = 1, 2, \ldots$. A sequence $f_n$ in $H(E)$ converges to $f$ in this topology if and only if it converges uniformly on every compact subset of $E$ as $n \to \infty$. It is known that $H(E)$ is an $F$-space, i.e., a complete metrizable locally convex space.

Now, let $E$ be a compact subset of $\overline{\mathbb{C}}$, and let $\mathbf{O}^E$ be the collection of all open sets in $\overline{\mathbb{C}}$ containing $E$. We denote $A_E = \bigcup H(O)|O \in \mathbf{O}^E$. In $A_E$, we introduce an equivalence relation $(R)$ as follows: $f \sim g$ if $f|_O = g|_O$ for some $O \in \mathbf{O}^E$. The elements of the set $H(E) = A_E/R$ are called *locally analytic functions on* $E$. We endow $H(E)$ with a vector space structure in a natural way. There always exists a decreasing sequence $\{O_n\}_1^\infty$ with $O_n \in \mathbf{O}^E$, $\overline{O}_{n+1} \subset O_n$ $\forall\, n \in \mathbb{N}$, such that $\bigcap O_n | n \subset \mathbb{N} = E$. We put $F_n = \overline{O}_n$. Let $H_n$ be the space of functions holomorphic on $O_n$ and continuous on $F_n$; we equip this space with the norm $||f||_n = \max\{|f(\zeta)| : \zeta \in F_n\}$, $n = 1, 2, \ldots$ . Then $H(E) = \bigcup\limits_{n=1}^{\infty} H_n$.

We treat $H(E)$ as the inductive limit of the sequence $\{H_n\}_1^\infty$, i.e., the topology in $H(E)$ is defined as the strongest among the locally convex topologies for which ali inclusions $\varphi_n : H_n \to H(E)$, $n = 1, 2, \ldots$, are continuous. A sequence $\{\widetilde{f}_n\}_1^\infty$ of locally analytic functions on $E$ converges to zero only if for some $O \in \mathbf{O}^{\mathbf{E}}$ each equivalence class $\widetilde{f}_n$ contains a function $f_n \in H(O)$, $n \in \mathbb{N}$, and $\lim\limits_{n\to\infty} f_n = 0$ in the topology of $H(O)$.

If $E$ is an open or compact subset of $\mathbb{C}^n$, $n > 1$, the space $H(E)$ is defined similarly. We shall need the following property of inductive limits.

**Proposition 1.10.6.** *Let $H$ be a vector space endowed with the inductive limit topology of a sequence $\{H_n\}_1^\infty$ of locally convex spaces, and let $\varphi_n : H_n \to H$ be the inclusion mapping. Suppose $F$ is a locally convex space, and $\Phi : H \to F$ is a linear mapping. The mapping $\Phi$ is continuous only if $\Phi \circ \varphi_n : H_n \to F$ is a continuous mapping for every $n \in \mathbb{N}$.*

Let $E \neq \emptyset$ be an open or closed set in $\overline{\mathbb{C}}$ such that $\infty \in E$. We denote by $H_0(E)$ the subspace of $H(E)$ consisting of functions equal to 0 at the point $\infty$.

Let $D$ be a domain in $\mathbb{C}$. We put $E = \overline{\mathbb{C}} \setminus D$, and fix $\widetilde{g} \in H_0(E)$. Suppose $O$ is an element of $\mathbf{O}^{\mathbf{E}}$ with oriented Jordan piecewise smooth boundary $\Gamma$ and such that, moreover, there exists an element $O_1 \in \mathbf{O}^{\mathbf{E}}$ satisfying $O \subseteq \overline{O} \subset O_1$. By the Cauchy theorem, the number

$$\mu_{\widetilde{g}}(f) := \langle f, \widetilde{g} \rangle = \frac{1}{2\pi i} \int\limits_{\Gamma} f(z) g(z)\, dz, \quad f \in H(D),$$

where $g \in \widetilde{g}$ and $g \in H(O_1)$, does not depend on the choice of elements in

$g, O, O_1$ possessing the above properties.

**Theorem 1.10.7.** i) *In the above notation, let $H^*(D)$ be the space dual to $H(D)$ and endowed with the strong topology. Then the operator*

$$\Phi : H_0(E) \to H^*(D)\ (E = \overline{\mathbb{C}} \setminus D)$$

*defined by the formula $\Phi(\widetilde{g}) = \mu_{\widetilde{g}}$, is a topological isomorphism.*

ii) *A similar operator establishes topological isomorphism between $H_0(\overline{\mathbb{C}} \setminus D)$ and $H^*(D)$ if $D$ is a compact subset of $\mathbb{C}$.*

The above properties of the spaces $H(E)$, $H^*(E)$ will be used to prove Theorem $A_2$ for $n = 1$. First, we consider several auxiliary assertions.

Let $D$ be a bounded convex domain in $\mathbb{C}$, and let $\{K_n\}_1^\infty$ be an increasing sequence *) of compact convex sets in $\mathbb{C}$ exhausting $D$ and such that $K_n \subset \overset{\circ}{K}_{n+1} \subset D \quad \forall\, n \in \mathbb{N}$. For $\theta \in \mathbb{R}$ we put

$$h(\theta) = \sup\{\text{Re } pe^{i\theta} : p \in D\}, \quad h_n(\theta) = \sup\{\text{Re } pe^{i\theta} : p \in K_n\}.$$

Let $[1, h(\theta))$ be the space of entire functions $f$ on $\mathbb{C}$ possessing the property $\overline{\lim_{r\to\infty}}\, r^{-1} \cdot \ln|f(re^{i\theta})| < h(\theta)\ \forall\, \theta \in \mathbb{R}$. We endow this space with the topology of inductive limit for the sequence $M_n$ of Banach spaces of entire functions equiped with the following norms:

$$||f||_n = \sup\{|f(z)| \cdot \exp\{-rh_n(\theta)\} : z = re^{i\theta} \in \mathbb{C}\}, \quad n \in \mathbb{N}.$$

Consider also the space $H_0(E)$ of functions locally analytic on $E := \overline{\mathbb{C}} \setminus D$ and equal to 0 at the point $\infty$. This space is endowed with the topology of inductive limit for the sequence of Banach spaces $H_n = A(\overline{G}_n)$, $n = 1, 2, \ldots$, of functions holomorphic in $G_n := \overline{\mathbb{C}} \setminus K_n$, continuous in $\overline{G}_n$, and equal to 0 at $\infty$. The norm in $H_n$ is defined by the formula $||g||_n = \max\{|g(\zeta)| : \zeta \in G_n\}$.

**Theorem 1.10.8.** *Let $B$ be the operator defined on the space of entire functions of exponential type by the formula $B(f) = F$, where $F$ is the Borel transform of $f$* (see (1.10.7) and (1.10.8)). *Then $B$ establishes topological isomorphism between the spaces $[1, h(\theta))$ and $H_0(\overline{\mathbb{C}} \setminus D)$.*

◄ Algebraic isomorphism of the above spaces follows from Theorem $A_1$. It only remains to prove that this isomorphism is topological.

1. Fixing $n \in \mathbb{N}$, we show that $B : M_n \to H_{n+1}$ is a continuous mapping. Let $f \in M_n$. Then $F = Bf$ extends analytically to $G_n = \overline{\mathbb{C}} \setminus K_n$, and $F(\infty) = 0$ (see (1.10.14)). Let $\rho_n = \inf\{h_{n+1}(\theta) - h_n(\theta) : |\theta| \leqslant \pi\}$. Since $h_n$ and $h_{n+1}$ are continuous functions (see Ex. 2.3.7), we see that $\rho_n > 0$. Observe that $\overline{G}_{n+1} \subset G_n$ and $\overline{G}_{n+1} = \bigcup \Pi(\theta; h_{n+1}) | \theta \in [-\pi, \pi]$, where $\Pi(\theta; h_{n+1}) = \{z \in \mathbb{C} : \text{Re } ze^{i\theta} \geqslant h_{n+1}(\theta)\}$ (see the above proof of T. $A_1$ in §10). Since $f \in M_n$, we have $|f(\zeta)| \leqslant ||f||_n \cdot \exp\{|\zeta| \cdot h_n(\arg \zeta)\}\ \forall\, \zeta \in \mathbb{C}$. Now, we estimate the

*) For example, $K_n = \omega + (1 - 1/n)(\overline{D} - \omega), \quad \forall\, n \in \mathbb{N}$, where $\omega \in D$.

norm of $F$ in $H_{n+1}$ by using the integral representation of $F$ (see (1.10.10)):

$$||F||_{n+1} = \max\left\{\left|\int_0^{\infty(\arg\zeta=\theta)} f(\zeta)e^{-z\zeta}\,d\zeta\right| : z \in \Pi(\theta; h_{n+1}),\ |\theta| \leqslant \pi\right\}$$

$$\leqslant ||f||_n \max\left\{\int_0^{\infty} \Phi_n(z,\theta)\,d|\zeta| : z \in \Pi(\theta; h_{h+1}),\ |\theta| \leqslant \pi\right\}$$

$$\leqslant ||f||_n \cdot \int_0^{\infty} e^{-\rho_n|\zeta|}\,d\zeta = \rho_n^{-1} \cdot ||f||_n,$$

where $\Phi_n(z,\theta) = \exp\{|\zeta|(h_n(\theta) - \operatorname{Re} ze^{i\theta})\}$.

Thus, $B : M_n \to H_{n+1}$ is a continuous operator. Then the operator $B : M_0 \to H_0(E)$, where $E := \mathbb{C} \setminus D$, is continuous: by the definition of the inductive limit topology, the inclusion maps of $\varphi_{n+1} : H_{n+1} \to H_0(E)$, $n \in \mathbb{N}$, are continuous. So, the linear operator $B : [1, h(\theta)) \to H_0(E)$ is also continuous (see P. 1.10.6).

2. We show the continuity of the inverse operator $B^{-1}$. The above arguments imply that it suffices to establish the continuity of the mapping $B^{-1} : H_n \to M_n$ for every $n \in \mathbb{N}$. For $n \in \mathbb{N}$, let $F \in H_n$, and let $f = B^{-1}(F)$. We estimate the norm of the operator $B^{-1}$, defined on $H_n$. Using (1.10.11) and taking $\Gamma = \partial K_n$ in that formula (which is possible by the Cauchy theorem and the continuity of $F$ in $\overline{G}_n$, where $G_n = \overline{\mathbb{C}} \setminus K_n$), we obtain

$$|f(re^{i\theta})| \leqslant \frac{1}{2\pi} \cdot \max\{|F(\zeta)| : \zeta \in \partial K_n\} \cdot \int_\Gamma \exp\{rh_n(\theta)\}\,d|\zeta| \leqslant$$

$$\leqslant A_n ||F||_n \cdot \exp\{rh_n(\theta)\} \quad \forall\ re^{i\theta} \in \mathbb{C}.$$

Here $A_n = \text{const} > 0$. Consequently, $||B^{-1}F||_n \leqslant A_n||F_n||_n$, $n \in \mathbb{N}$ ▶

The proof of Theorem $A_2$ for the case where $D \subset \mathbb{C}$ immediately follows from Theorems 1.10.7 and 1.10.8, because

$$f = B_{-1} \circ \Phi_{-1}(\mu) \quad \forall\ \mu \in H^*(D) \quad (f(z) = \mu_\lambda(e^{z\lambda})).$$

**Remark.** Arguing as in the proofs of Theorems $A_2$ and 1.10.8, we extend these results to the case where $D$ is an unbounded domain in $\mathbb{C}$: a sequence of compact convex sets exhausting $D$ and possessing the above properties also exists under this condition.

Let $h : \mathbb{R} \to \mathbb{R}$ be a $2\pi$-periodic trigonometrically 1-convex function. We denote by $[1, h(\theta)]$ the space of entire functions $f$ such that

$$\overline{\lim_{r\to\infty}}\, r^{-1} \cdot \ln|f(re^{i\theta})| \leqslant h(\theta) \quad \forall\ \theta \in \mathbb{R};$$

the topology in this space is induced by the countable system of seminorms

$$||f||_n = \sup\{|f(z)|\exp\{-[h(\theta)+1/n]r\} : z \in \mathbb{C}\},\ n \in \mathbb{N}. \qquad (1.10.15)$$

**Theorem 1.10.9.** *For a compact convex set $K$ in $\mathbb{C}$, let $h(\theta) = \sup\{\mathrm{Re}\, pe^{i\theta} : p \in K\}$. We denote by $H(K)$ the space of locally analytic functions endowed with the inductive limit topology, and by $H^*(K)$ its dual space endowed with the strong topology. Let $[1, h(\theta)]$ be the space of entire functions of exponential type introduced above. Then the Laplace transformation*

$$\mu : H^*(K) \to [1, h(\theta)],\ \text{where}\ \mu_\zeta(e^{z\zeta}) = f(z),\ z \in \mathbb{C}$$

*establishes topological isomorphism between the spaces $H^*(K)$ and $[1, h(\theta)]$.*

Theorem 1.10.9 is a consequence of statement ii) of Theorem 1.10.7 and of the following analog of Theorem 1.10.8 for compact sets.

**Theorem 1.10.8′.** *Let $K$ be a compact convex set in $\mathbb{C}$. Put $h(\theta) = \sup\{\mathrm{Re}\, pe^{i\theta} : p \in K\}$. Then the above operator* (see Theorem 1.10.8, (1.10.7), and (1.10.8))

$$B : [1, h(\theta)] \to H_0(\overline{\mathbb{C}} \setminus K),\quad \text{where}\ Bf = F,$$

*is a topological isomorphism.*

◀ Let $G_n = K + S(1/n)$, where $S(\varepsilon) = \{z \in \mathbb{C} : |z| < \varepsilon\}$, $\varepsilon > 0$. We denote $K_n = \overline{\mathbb{C}} \setminus G_n\ \forall\ n \in \mathbb{N}$, and $E = \overline{\mathbb{C}} \setminus K$. Then $\{K_n\}_1^\infty$ is a sequence of compact sets exhausting the domain $E \subset \overline{\mathbb{C}}$. As in the beginning of Subsection 3, we introduce a topology in $H(E)$ and in its subspace $H_0(E) = \{g \in H(E) : g(\infty) = 0\}$ with the help of the sequence of seminorms $||g||_n = \max\{|g(\zeta)| : \zeta \in K_n\}$, $n = 1, 2, \ldots$.

We show that $B^{-1}$ is a continuous mapping. Take $F \in H_0(E)$, and denote $f = B^{-1}F$. We fix $n \in \mathbb{N}$ and put $\Gamma = \partial K_n$ in (1.10.11). Estimating $f$ much as in item 2 of the proof of Theorem 1.10.8, we obtain the following inequality (see (1.10.15)): $||f||_n \leqslant A_n \cdot ||F||_n$, where $A_n = \text{const} > 0$. Now, the continuity of $B^{-1}$ is a consequence of the following well-known continuity criterion ▶

**Proposition 1.10.10.** *Let $X, Y$ be locally convex spaces, and let $P = \{p\}$, $Q = \{q\}$ be systems of seminorms determining the topology in $X$ and $Y$, respectively. A linear operator $T : X \to Y$ is continuous if and only if for every seminorm $q \in Q$ there exists a seminorm $p \in P$ and the number $A > 0$ such that $q(Tx) \leqslant Ap(x)\ \forall\ x \in X$.*

◀ Since $H_0(E)$ and $[1, h(\theta)]$ are $F$-spaces, by the Banach theorem the continious linear operator $B^{-1} : H_0(E) \to [1, h(\theta)]$ is an open mapping. Therefore, $B$ is continuous ▶

**1.10.4. Entire functions of completely regular growth.** Let $E$ be a measurable set of positive numbers. Denote by $m$ the Lebesgue measure on

$(0, \infty)$. The set $E$ is called a *set of relative measure zero* or a *$E^0$-set* if

$$\lim_{r\to\infty} r^{-1} m[E \cap (0, r)] = 0.$$

**Definition 1.10.11.** An entire function $f$ of order $\rho > 0$ and of normal type is called a *function of completely regular growth (c.r.g.) on the ray* arg $z = \theta$, where $\theta \in \mathbb{R}$, if the limit

$$h_f(\theta) = \lim_{r\to\infty} r^{-\rho} \ln |f(re^{i\theta})| \tag{1.10.16}$$

exists along $r \in \mathbb{R}_+ \backslash E$ for some $E^0$-set $E$. If in (1.10.16) the convergence to the limit as $r \notin E$ is uniform in the variable $\theta \in \mathbb{R}$ and $E$ is a $E^0$-set independent of $\theta$, then $f$ is called *an entire function of completely regular growth.*

**Theorem 1.10.12.** *If $f$ is an entire function of completely regular growth on every ray in $\mathbb{C}$ with origin at 0, then $f$ is an entire function of completely regular growth.*

The existence of an entire function of completely regular growth with given asymptotic properties is shown by the following statement.

**Theorem 1.10.13.** *Let $\rho > 0$, and let $H$ be a finite function in the class $P_\rho$. There exists an entire function $L$ of order $\rho$ such that the indicator of $L$ is equal to $H$ and the following conditions are satisfied:*

i) *$L$ is a function of completely regular growth,*

ii) *all zeros $\{\lambda_k\}_1^\infty$ of $L$ are simple, and $0 < |\lambda_1| \leqslant |\lambda_2| \leqslant \dots$,*

iii) $\overline{\lim_{k\to\infty}} \, [-|\lambda_k|^{-\rho} \ln |L'(\lambda_k)| + H(\arg \lambda_k)] \leqslant 0.$

The distribution of zeros of entire functions (see assertion 2 of Theorem 1.10.13) is related to the growth characteristics.

**Theorem 1.10.14.** *If $f$ is an entire function of order $\rho > 0$ and of normal type, then $\overline{\lim_{k\to\infty}} \, |\lambda_k|^{-\rho} \cdot k < \infty$.*

### 1.10.5. Lagrange interpolation series.

Let $L$ be an entire function satisfying condition ii) of Theorem *1.10.13*. Suppose $\{b_k\}_1^\infty$ is a sequence in $\mathbb{C}$. If the series

$$F(z) = \sum_{k=1}^{\infty} \frac{b_k L(z)}{L'(\lambda_k)(z - \lambda_k)} \tag{1.10.17}$$

converges uniformly on every compact subset of $\mathbb{C}$, then its sum $F$ is an entire function possessing the interpolation property: $F(\lambda_k) = b_k \ \forall \ k \in \mathbb{N}$. This series is called the *Lagrange interpolation series.* In the notation of Theorem 1.10.13, we formulate a sufficient condition for the representability of an entire function by a Lagrange interpolation series.

**Theorem 1.10.15.** *Let $L$ be a function the existence of which is stated in Theorem* 1.10.13. *Every entire function $f$ of order $\rho$ the indicator of which* *satisfies the inequality* $h_f(\theta) < H(\theta)\ \forall\ \theta \in \mathbb{R}$ *is representable by the series*

$$f(\lambda) = \sum_{k=1}^{\infty} \frac{f(\lambda_k)L(\lambda)}{L'(\lambda_k)(\lambda - \lambda_k)}, \qquad \lambda \in \mathbb{C}, \tag{1.10.18}$$

*convergent uniformly on every compact subset of* $\mathbb{C}$.

**1.10.6. Growth characteristics of functions holomorphic inside an angle in $\mathbb{C}$.** For a function $f$ holomorphic inside an angle and continuous up to its boundary, the notions of the order of growth, type, and indicator are introduced much in the same way as in the above case of entire functions. In the sequel we shall need two results for holomorphic functions of exponential type inside an angle, namely, the theorems of F. Carlson and V. Bernstein.

**Theorem 1.10.16.** *If a holomorphic function* $f$ *of exponential type in the half-plane* $\operatorname{Re} z \geqslant 0$ *is equal to* $0$ *at the integral points* $n = 1, 2, \ldots$ *and* $h_f(-\pi/2) + h_f(\pi/2) < 2\pi$, *where* $h_f$ *is the indicator of* $f$, *then* $f \equiv 0$.

**Theorem 1.10.17.** *Let* $f$ *be a holomorphic function of exponential type inside the angle* $\Delta_\alpha = \{z \in \mathbb{C} : |\arg z| < \pi\alpha/2\}$, *where* $0 < \alpha \leqslant 1$. *If for some* $\lambda > 0$ *the function* $f$ *satisfies the Hölder condition of the order* $\lambda$ *at the point* $0$, *i.e.*,

$$|f(z) - f(0)| = O(|z|^\lambda), \quad z \to 0, \quad z \in \Delta_\alpha,$$

*then the Laplace transform* $\int\limits_0^\infty e^{-pt} f(t)\, dt$ *admits analytic continuation up to a function* $F$ *on the set*

$$\bigcup \{p \in \mathbb{C} : \operatorname{Re} pe^{i\theta} > h_f(\theta)\} |\ |\theta| < \pi\alpha/2.$$

*Moreover, the straight line* $\{p \in \mathbb{C} : \operatorname{Re} pe^{i\theta} = h_f(\theta)\}$ *contains singular points of* $F$ *for every* $\theta \in (-\pi/2\alpha, \pi/2\alpha)$.

Theorem 1.10.17 is an analog of the Polya theorem (see the Introduction, T. $A_1$) for functions of exponential type inside an angle.

Chapter 2

# A Method of Identifying Homeostasis Relaxation Characteristics

## 2.1. Homeostasis system relaxation characteristics and the problem of their identification

A way of investigating medico-biological, biophysical, physico-chemical or other dynamically balanced systems is to study the relaxation of certain system parameters ("variables") after an external impact. Provided the changes in the systems are not pathological, the variables either regain their original levels or pass to new (adaptation) levels. A glowing example of this is homeostasis systems of living organisms. It is well known that homeostasis, i.e., the ability of an organism to sustain permanence of its internal medium under disturbances, is the basis of self-preservation of living systems.

The response of variables of a balanced system to external disturbance makes it possible to judge the functional state of the system. The curves reflecting the transition process (behavior dynamics) for these variables are called *relaxation characteristics.* An organism's response to the impact of external medium is verified against a number of homeostatic parameters by load tests. An apt illustration is furnished by the sugar curve, which reflects the dynamics of glucose content in blood in the glucose-tolerance test. The relaxation characteristics of homeostasis parameters make it possible to judge the human health. For instance, it is well known that the sugar curves of healthy people differ from those of diabetic patients in the mode of regaining the steady-state level. Other examples of relaxation characteristics of homeostasis processes are the perfusion parameters in isolated organs observed under various impacts. Among them are the parameters of respiration and hemopoiesis under hypoxia. In bio-engineering systems designed to perfuse isolated organs, homeostasis conditions are created and maintained by engineering facilities. Analysis of the behavior of perfusion parameters subject to disturbance can help to gain a more penetrating insight into the mechanisms of physiological processes running in an integral organism.

The mathematical analysis of homeostasis relaxation characteristics is reduced to the problem of identifying them, i.e., to constructing models, optimal in a certain sense, that describe the behavior of these characteristics, and to evaluating the parameters of these models. The fact that, in general, model parameters are not so numerous as laboratory data is a major advantage of modeling. For instance, this alleviates substantially the investigation of homeostasis in a homogenous group of objects. In this case for initial laboratory data it is possible to take group-averaged data of measured parameters of the processes under study. Difficulties arise when a model partially accounts for the internal structural properties of a homeostasis system, but does not reflect possible pathological changes in it, or the influence of the external medium on the processes that run inside the system (for example, when the mechanisms of this influence are not fully understood). If, even in the case of a pathology or under different ecological conditions, a homeostatic process behaves quantitatively in the same way, it is appropriate to apply the known algorithms of automatic classification to the space of the distinguished parameters of the model. The analysis of the resulting model processes may help to recognize the main types of behavior of the initial process.

Among the most common classes of models used to study relaxation characteristics (and to solve many other problems arising in science and in engineering practice), we name the class of models describing the dynamics of these characteristics by means of an ordinary linear differential equation with constant real coefficients:

$$y^{(k)} + a_1 y^{(k-1)} + \ldots + a_k y = f, \tag{2.1.1}$$

where $f$ is a finite exponentially harmonic sum, i.e., the sum of a finite number of exponentials, sinusoids, and their products with constant coefficients.

The sums of the form

$$\sum_{k=1}^{m} e^{\lambda_k t} \cdot P_k(t)$$

will be called *quasipolynomials*; in general, the $\lambda_k$ and the coefficients of the polynomials $\{P_k\}_1^m$ are complex numbers. The solutions of equation (2.1.1) are precisely the quasipolynomials; they yield a good approximation of many dynamical processes occurring in investigation practice. The quasipolynomials are also solutions of difference equations and of some convolution equations rather common in applied mathematical problems. If the coefficients of equation (2.1.1) and numerical parameters of its right-hand side are unknown, the input information to find the quasipolynomials is its moments, i.e., its values at a finite number of equidistant time instants (or at the multiples of some fixed time unit). Such a situation is quite common in actual practice. For example, the sugar curve is calculated by the glucose concentration in blood taken every half-hour; the data about gallbladder concentration function, about enzyme activity indices are also of discrete nature.

Among the approaches to solving the above problem of identifying equation (2.1.1), we name the Prony algorithm, whose history is about two hundred years old, and its modifications. In this chapter we present in detail a modification of this algorithm and, on the basis of it, develop a method of approximating a discrete function defined on a finite number of nodes of a uniform net by quasipolynomials of fixed order. The theory of entire functions is employed to find conditions under which these algorithms are well posed. The problem of describing the domain of stability of these algorithms arises naturally, because in practice the moments of dynamical indices are taken with some error (for instance, such are readings of instruments that measure relaxation characteristics).

Mathematically, the problem of determining a quasipolynomial $f$ by its moments is equivalent to the problem of determining a rational function $R$ equal to 0 at infinity by the quantities values

$$c_j = \frac{1}{2\pi i} \int\limits_{\Gamma} e^{j\zeta} R(\zeta) d\zeta, \quad j = 0, 1, \ldots, N, \tag{2.1.2}$$

associated with it, where $\Gamma$ is a positively oriented closed Jordan contour in $\mathbb{C}$ enclosing the poles of $R$. The reason is that the Borel transform of a quasipolynomial $f$ is a rational function $R$ whose poles coincide with the quantities $\lambda_k$, and in the above notation the following integral representation of the quasipolynomial (see (1.10.11)) is true:

$$f(z) = \frac{1}{2\pi i} \int\limits_{\Gamma} e^{z\zeta} R(\zeta) d\zeta. \tag{2.1.3}$$

## 2.2. Algorithm of recovering a quasipolinomial by its moments

Since the right-hand side in (2.1.1) is a quasipolynomial, the problem of identifying equation (2.1.1) discussed in 2.1 is equivalent to the problem of recovering a quasipolynomial $g$ by its moments in the case where $g$ is the solution of a homogeneous differential equation

$$y^{(n)} + a_1 y^{(n-1)} + \ldots + a_n y = 0 \tag{2.2.1}$$

with constant coefficients. In 2.2 we consider the existence, uniqueness and stability for solutions of this problem.

**2.2.1 Necessary information about difference equations.** Consider the linear homogeneous difference equation of order $n$ with constant coefficients.

$$y_{k+n} + p_1 y_{k+n-1} + \ldots + p_n y_k = 0, \quad k \in \mathbb{Z}_+, \tag{2.2.2}$$

where $\mathbb{Z}_+ = \{0, 1, 2, \ldots\}$, $p_1, \ldots p_n \in \mathbb{C}$, $p_n \neq 0$.

**Theorem 2.2.1.** *Let $q_i$, $i \in \{1, ..., m\}$, be a root of multiplicity $r_i$ of the characteristic equation*

$$z^n + p_1 z^{n-1} + \ldots + p_n = 0$$

*for (2.2.2) (we have $r_l + ... + r_m = n$). Then*

$$F_i = \{q_i^k,\ kq_i^k,\ \ldots,\ k^{r_i - 1} q_i^k;\ k \in \mathbb{Z}_+\}$$

*is a system of $r_i$ linearly independent solutions of (2.2.2) for every $i \in \{1, ..., m\}$. If $\bigcup_{i=1}^{m} F_i = \{f_1, \ldots, f_n\}$, then:*

*a) every solution of equation (2.2.2) has the form*

$$y_k = \sum_{j=1}^{n} A_j f_j(k), \quad k \in \mathbb{Z}_+,$$

*where $A_l, ..., A_n$ are arbitrary complex constants;*

*b)*

$$\begin{vmatrix} f_1(k) & f_2(k) & \ldots & f_n(k) \\ f_1(k+1) & f_2(k+1) & \ldots & f_n(k+1) \\ \vdots & \vdots & \ddots & \vdots \\ f_1(k+n-1) & f_2(k+n-1) & \ldots & f_n(k+n-1) \end{vmatrix} \neq 0 \quad \forall\, k \in \mathbb{Z}_+.$$

**Theorem 2.2.2.** *Let $R_n$ be the collection of all solutions for various difference equations (2.2.2) of order $n$. A discrete function $y : \mathbb{Z}_+ \to \mathbb{C}$ belongs to $R_n \setminus R_{n-1}$ if and only if $\mathcal{D}_n(k) = 0\ \forall\, k \in \mathbb{Z}_+$, and $\mathcal{D}_{n-1}(k_0) \neq 0$ for at least one integer $k_0 \geqslant 1$, where*

$$\mathcal{D}_\nu(k) = \begin{vmatrix} y_k & y_{k+1} & \ldots & y_{k+\nu} \\ y_{k+1} & y_{k+2} & \ldots & y_{k+\nu+1} \\ \vdots & \vdots & \ddots & \vdots \\ y_{k+\nu} & y_{k+\nu+1} & \ldots & y_{k+2\nu} \end{vmatrix}, \quad \nu = n,\ n-1;$$

*moreover, if $\mathcal{D}_{n-1}(k_0) \neq 0$, then $\mathcal{D}_{n-1}(k) \neq 0\ \forall\, k \in \mathbb{Z}_+ \setminus \{0\}$.*

It is known that there is a close relationship between equations (2.2.1) and (2.2.2): if $e^{\lambda t}$ is a solution of (2.2.1), then $e^{\lambda k}$, $k \in \mathbb{Z}_+$, is a solution of some equation of the form (2.2.2) the characteristic equation of which has the number $q = e^{\lambda}$ as a root. If $\{a_i\}_1^n \subset \mathbb{C}$, the following more general result is true.

**Theorem 2.2.3.** *Let $y = g(t)$, $t \in \mathbb{R}$, be any solution of an equation of the form (2.2.1). Suppose that the characteristic polynomial $P_n(z) = z^n + a_1 z^{n-1} + \cdots + a_n$ of the latter admits the following expansion:*

$$P_n(z) = \prod_{j=1}^{k} (z - \lambda_j)^{\alpha_j}, \quad z \in \mathbb{C}; \qquad \sum_{j=1}^{n} \alpha_j = n.$$

*Then* $y_k = g(k)$, $k \in \mathbb{Z}_+$, *is the solution of an equation of the form* (2.2.2) *such that the corresponding characteristic polynomial* $T_n(z) = z^n + p_1 z^{n-1} + \cdots + p_n$ *expands as follows:*

$$T_n(z) = \prod_{j=1}^{k} (z - e^{\lambda_j})^{\alpha_j}, \quad z \in \mathbb{C}; \quad p_n = (-1)^n exp\{-a_1\}.$$

**2.2.2. Conditions for the existence and uniqueness of a quasipolynomial of order $n$ with given moments.** Fix $d > 0$. Let $H_n(d)$ (respectively, $H_n^-(d)$) be the class of solutions of various linear homogeneous differential equations of order $n$ of the form (2.2.1) such that $\{a_i\}_1^n \subset \mathbb{C}$ and the corresponding characteristic equation has roots only in the strip $\Pi(d) = \{z \in \mathbb{C} : |\mathrm{Im}\, z| < \pi/d\}$ (in the semistrip $\Pi^-(d) = \Pi(d) \cap \{z \in \mathbb{C} : \mathrm{Re}\, z < 0\}$). Denote by $B_n(d)$ (respectively, $B_n^-(d)$) the class of Borel transforms of the functions belonging to $H_n(d)$).

The class $B_n(d)$ coincides with the class of rational functions of the form $R = Q/L$, where $Q, L$ are polynomials without common nonconstant factor, $\deg Q < \deg L \leqslant n$, and the poles of the function $R$ are included in $\Pi(d)$ (in $\Pi^-(d)$ in the case of the subclass $H_n^-(d)$). For these classes the following interpolation theorems are true. In them, $\{c_0, c_1, \ldots, c_N\} \subset \mathbb{C}$, and the number $N \geqslant 2n - 1$ is fixed arbitrarily.

**Theorem 2.2.4** (uniqueness theorem). *In* $H_n(d)$ (*in* $B_n(d)$), *there may exist only one function* $g$ (*respectively,* $R$) *satisfying the conditions*

$$g(kd) = c_k, \quad k = 0, 1, \ldots, N, \tag{2.2.3}$$

*respectively,*

$$\frac{1}{2\pi i} \int_{\Gamma} e^{kd\lambda} \cdot R(\lambda) d\lambda = c_k, \quad k = 0, 1, \ldots, N, \tag{2.2.4}$$

*where* $\Gamma$ *is a closed Jordan contour enclosing the poles of* $R$ (*the Borel transform of* $g$.)

**Theorem 2.2.4′** (existence theorem). *The following assertions are equivalent:*

1) *in* $H_n(d)$ (*respectively, in* $H_n^-(d)$) *there exists a function* $g$ *satisfying condition* (2.2.3);

2) *there is a unique vector* $p = (p_1, \ldots, p_n) \in \mathbb{C}^n$ *such that*

$$c_{k+n} + p_1 c_{k+n-1} + \ldots + p_n c_k = 0, \; k = 0, 1, \ldots, N - n, \tag{2.2.5}$$

*and the equation*

$$T_n(z) := z^n + p_1 z^{n-1} + \ldots + p_n = 0 \tag{2.2.6}$$

*has roots only in* $\mathbb{C}\setminus(-\infty,0]$ (*respectivelyin* $S^- := \{z \in \mathbb{C} : |z| < 1\}\setminus(-1,0]$);

3) *in* $B_n(d)$) *there exists a function* $R$ *satisfying condition* (2.2.4), *and* $R$ *is the Borel transform of a function* $g$ *belonging to* $H_n(d)$ (*respectively, to* $H_n^-(d)$).

By (1.3), the assertions of the theorem concerning the class $B_n(d)$ are equivalent to the respective assertions about the class $H_n(d)$. It suffices to prove the theorems for $d = 1$, and then to change the variables $t \to td^{-1}$.

**Proof of Theorem 2.2.4.** Let functions $g_i \in H_n(1)$, $i = 1, 2$, satisfy condition (2.2.3). Then $\varphi := g_2 - g_1$ is a quasipolynomial in the class $H_m(1)$ where $m \leqslant 2n$. By Theorem 2.2.3, $y_k = \varphi(k)$, $k \in \mathbb{Z}_+$, is a solution of some difference equation

$$y_{k+m} + p_1 y_{k+m+1} + \ldots + p_m y_k = 0, \quad k \in \mathbb{Z}_+,$$

of the form (2.2.2). Since by assumption (see (2.2.3)) $\varphi(k) = 0$, $k = 0, 1, \ldots, N$, where $N \geqslant 2n - 1$, induction shows that $\varphi$ is an entire function of exponential type such that $\varphi(k) = 0$, $k = 0, 1, \ldots$. Its Borel transform is a rational function and has a finite number of poles in $\Pi(1)$. Consequently, the convex hull of these poles (the conjugate diagram) is also included in $\Pi(1)$. Then, by the Polya theorem (see T. $A_1$, the Introduction), the indicator diagram $I$ of $\varphi$ is included in $\Pi(1)$ and its width with respect to the imaginary axis is less than $2\pi$. Now, the Carlson theorem (see T. 1.10.16) implies that $\varphi = 0$, $g_1 = g_2$ ▶

**Proof of Theorem 2.2.4′.** 2) → 1). Let $q_i$ be a root of multiplicity $r_i$ of the polynomial $T_n$ (see (2.2.6)), $i = 1, \ldots, m$ ($r_1 + .... + r_m = n$). By assumotion, $q_i \neq 0$ for $i = 1, \ldots, m$ and, consequently, $p_n \neq 0$. Then the difference equation (2.2.2) whose coefficients are the coordinates of the vector $p$ has order $n$. By Theorem 2.2.1, the general solution of (2.2.2) is represented as follows:

$$y_k = \sum_{i=1}^{m} q_i^k \sum_{j=0}^{r_j - 1} A_{ij} k^j, \quad k = 0, 1, \ldots, \tag{2.2.7}$$

where $\{A_{ij}\} \subset \mathbb{C}$. By (2.2.5), recalling that $N > n$, we conclude that there exists a unique solution $\{\delta_j\}_0^\infty$ of equation (2.2.2) such that

$$\delta_k = c_k, \quad k = 0, 1, \ldots, N. \tag{2.2.8}$$

Let $\gamma_i = \ln q_i$, $i = 1, \ldots, m$, where we mean the principal value of the logarithm. Since, by assumption, $\operatorname{Re} q_i > 0$ if $\operatorname{Im} q_i = 0$, $i = 1, \ldots, m$, it follows that $\{\gamma_i\}_1^m \subset \Pi_n(1)$ and the function

$$g(t) = \sum_{i=1}^{m} e^{\gamma_i t} \sum_{j=0}^{r_j - 1} A_{ij} t^j, \quad t \in \mathbb{C}, \tag{2.2.9}$$

satisfies an equation of the form (2.2.1) such that the root of multiplicity $r_i$ of the characteristic equation of this equation is $\gamma_i$, $i = 1, \ldots, m$, i.e., $g \in H_n(1)$. The "if" part of Theorem 2.2.2 and the uniqueness of a vector $p \in \mathbb{C}^n$ satisfying (2.2.5) imply that the sequence $\{\delta_j\}$ (see (2.2.8)) cannot be a solution of a difference equation of order $n-1$. Therefore, $A_{ij} \neq 0$ in (2.2.7)-(2.2.9) for $j = r_i - 1$, $i = 1, \ldots, m$, and, consequently, $g \in H_n(1) \setminus H_{n-1}(1)$. It is obvious that $g(k) = \delta_k$, $k = 0, 1, \ldots$. Reversing the arguments and applying the "only if" part of Theorem 2.2.2 in the end, we prove the implication 1)→ 2) ▶

**Remark.** In the case of the class $H_n^-(d)$, for describing the set $D_n$ of admissible values of the coefficients of the polynomial $T_n$ (see (2.2.6)) it is possible to consider the rational function $T_n \circ \psi$, where the function $\psi(w) = -\left(\dfrac{1+w}{1-w}\right)^2$ maps $\mathbb{R}^2_- := \{z \in \mathbb{C} : \operatorname{Re} z < 0;\ \operatorname{Im} z < 0\}$ conformally onto $S^-$ (see assertion 2) of T. 2.2.4′). This leads to the problem of describing the coefficients of the polynomials $Q(w) = (1-w)^{2n} \cdot R(w)$ of power $2n$ such that $Q$ has a root in the quadrant $\mathbb{R}^2_-$. For example, if $n = 2$, we have (see 2.2.4)

$$D_2 \cap \{(p_1, p_2) \in \mathbb{C}^2 : \operatorname{Im} p_1 = \operatorname{Im} p_2 = 0\} =$$

$$= \{(x_1, x_2) \in \mathbb{R}^2 : -1 - x_2 < 2\sqrt{x_2},\ 0 < x_2 < 1\}.$$

**2.2.3. Skew-symmetric entire functions of many variables.** To study the stability domain of the above identification problem, we need some auxiliary information.

**Definition 2.2.5.** An entire function $f$ defined on $\mathbb{C}^k\,(k \geqslant 2)$ is said to be *skew-symmetric* if for every $i \in \{1, \ldots, k-1\}$ we have

$$f(z_1, \ldots, z_i, z_{i+1}, \ldots, z_k) \equiv -f(z_1, \ldots, z_{i+1}, z_i, \ldots, z_k). \tag{2.2.10}$$

The class of such functions is denoted by $A(\mathbb{C}^k)$.

**Property 2.2.6.** *Let $\Sigma_k$ be the class of $k$-permutations. Then*

$$f(z_{\sigma(1)}, \ldots, z_{\sigma(k)}) \equiv \varepsilon(\sigma) \cdot f(z_1, \ldots, z_k) \quad \forall f \in A(\mathbb{C}^k),\ \sigma \in \Sigma_k,$$

*where $\varepsilon(\sigma) = \pm 1$ is the signature of the permutation. In particular, if $\sigma$ is a transposition, then $\varepsilon(\sigma) = -1$* (cf. (2.2.10)).

**Property 2.2.7.** *For $z_i = z_j$, $i \neq j$, $i, j \in \{1, \ldots, k\}$, we have $f(z) = 0$.*

We mention several specific properties of functions belonging to $A(\mathbb{C}^k)$.

**Property 2.2.8.** *Every function $f \in A(\mathbb{C}^k)$ admits the following expansion in $\mathbb{C}^k$:*

$$f = W \cdot g_f, \quad \textit{where } W(z) = \prod_{1 \leqslant i < j \leqslant k} (z_j - z_i) \tag{2.2.11}$$

*is the Vandermonde determinant of order $k$ and $g_f$ is a symmetric entire function.*

**Property 2.2.9.** *The partial derivative* $\dfrac{\partial f}{\partial z_j}$ *of a function* $f \in A(\mathbb{C}^k)$ *belongs to the class* $A(\mathbb{C}^{k-1})$ *with respect to the variables* $z_1, ..., z_k$ *except for* $z_j$, *where* $j \in \{1, ..., k\}$.

**Property 2.2.10.** *The mixed derivative* $\dfrac{\partial^2 f}{\partial z_i \partial z_j}$ *of a function* $f \in A(\mathbb{C}^k)$ *belongs to the class* $A(\mathbb{C}^2)$ *with respect to the variables* $z_i, z_j$.

Let $\alpha$ be a point in $\mathbb{C}^k$ all coordinates of which are different, and let

$$\alpha_1 = \ldots = \alpha_{s_1} = \gamma_1, \ldots, \alpha_{s_{m-1}+1} = \ldots = \alpha_k = \gamma_m, \tag{2.2.12}$$

where $1 \leqslant s_1 < s_2 < \ldots < s_{m-1} < s_m = k$ $(m < k)$, $\gamma_i \neq \gamma_j$ if $i \neq j$. Consider the differential operator

$$D(a_1, \ldots, a_k)f = \frac{\partial^{a_1 + \ldots + a_k} f}{\partial z_1^{a_1} \ldots \partial z_k^{a_k}}, \qquad a_i \geqslant 0.$$

**Property 2.2.11.** *If* $a_i = a_j \geqslant 0$ *for* $i \neq j$, $i, j \in (s_l, s_{l+1})$, *and* $l \in \{0, 1, \ldots, m-1\}$ $(s_0 = 0;\ s_m = k)$, *then* $[D(a_1, \ldots, a_k)f](\alpha) = 0$.

◀ Let, for example, $i = 1, j = 2$. By Property 2.2.9, the function

$$g := \frac{\partial^{a_3 + \ldots + a_k} f}{\partial z_3^{a_3} \ldots \partial z_k^{a_k}}$$

belongs to $A(\mathbb{C}^2)$ with respect to the variables $z_1, z_2$. Let $N_g = \dfrac{\partial^2 g}{\partial z_1 \cdot \partial z_2}$, $a_1 = a_2 = \beta$. Then $N_g^\beta = D(a_1, \ldots, a_k)f \in A(\mathbb{C}^2)$ (see Prop. 2.2.10). It remains to use Property 2.2.7 ▶

Thus, the number $[D(a_1, \ldots, a_k)f](\alpha)$ can be different from 0 if the orders of differention in each variable are different in groups with equal coordinates. It is convenient to arrange them in ascending order. Property 2.2.6 implies the following statement.

**Property 2.2.12.** *In the above notation* (see Property 2.2.6 and (2.2.12)), *let* $r_i = s_i - s_{i-1}$, $\sigma_i \in \Sigma_{r_i}$, $s_0 = 0$, $\tau_i(l) = \sigma_i(l) - 1$, $i = 1, \ldots, m$, *where* $i = 1, \ldots, m$, $l = 1, 2, \ldots, r_i$. *Put*

$$Df(\alpha) = [D(0, 1, \ldots, r_1 - 1; \ldots; 0, 1, \ldots, r_m - 1)f](\alpha). \tag{2.2.13}$$

*Then*

$$[D(\tau_1(1), \ldots, \tau_1(r_1); \ldots; \tau_m(1), \ldots, \tau_m(r_m))f](\alpha) = Df(\alpha) \prod_{k=1}^{m} \varepsilon(\sigma_k).$$

**Theorem 2.2.13** (a multidimensional analog of the de l'Hospital rule). *If* $P, Q \in A(\mathbb{C}^k)$ *and* $DQ(\alpha) \neq 0$ (see (2.2.13)), *then the function* $F = P/Q$ *admits analytic continuation to a neighborhood of the point* $\alpha$ *such that*

$$F(\alpha) = \lim_{z \to \alpha;\ z \notin N_Q} \frac{P(z)}{Q(z)} = \frac{DP(\alpha)}{DQ(\alpha)}, \tag{2.2.14}$$

*where* $N_Q$ *is the set of zeros of* $Q$.

◀ By (2.2.11), differentiation rules, and Property 2.2.11, we obtain the formula

$$Df(\alpha) = DW(\alpha) \cdot g_f(\alpha). \tag{2.2.15}$$

Let $g_P$ and $g_Q$ be the functions associated with $P$ and $Q$, respectively (see Prop. 2.2.8). Put $\psi = g_P/g_Q$. Formula (2.2.11) implies the relation

$$N_Q = N_1 \cup N_2, \tag{2.2.16}$$

where $N_1, N_2$ are the zero sets of the functions $W$ and $g_Q$, respectively. Using formula (2.2.15), under the assumptions of the theorem we find that $g_Q(\alpha) \neq 0$ and

$$DP(\alpha)/DQ(\alpha) = \psi(\alpha) = \lim_{z \to \alpha;\ z \notin N_2} \psi(z).$$

Moreover, $\psi$ is a holomorphic function in a neighborhood of $\alpha$. Combining this with (2.2.16) and recalling that $\psi = F$ (see (2.2.11)), we obtain our claim ▶

**Remark.** The order of the operator $D$ is equal to

$$r = \frac{1}{2}\sum_{i=1}^{n}(r_i - 1)r_i = \frac{1}{2}\left(\sum_{i=1}^{m} r_i^2 - k\right).$$

In the notation of Theorem 2.2.14, $D$ can be replaced with any operator of the same order described in Property 2.2.12. For the differential operator $H$ of order $s \neq r$ (see (2.2.15)) the analog of the de l'Hospital rule fails. If $s < r$, then, by Property 2.2.11, $HQ(\alpha) = 0 \ \forall Q \in A(\mathbb{C}^k)$. If $s > r$, $DQ(\alpha) = 0$ and $HQ(\alpha) \neq 0$, then formula (2.2.14) fails if $D$ is replaced with $H$ in it. This is shown by the following example.

**Example.** Let $\alpha = (0, 0)$. We denote $H = \dfrac{\partial^2}{\partial z_1 \partial z_2^2}$,

$$P(z_1, z_2) = z_1^3 - z_2^3, \quad Q(z_1, z_2) = (z_1 - z_2)(z_1 + z_2)^2.$$

Then $r = 1$, $s = 3$, $P, Q \in A(\mathbb{C}^2)$, and $DQ(\alpha) = 0$, where $D := D(0, 1)$ (see (2.2.13)); moreover, $HQ(\alpha) = -2$ and $HP(\alpha) = 0$. On the other hand, direct inspection shows that the limit $L = \lim_{z_i \to \alpha;\ z_1 \neq z_2} P/Q$ does not exist. But if $\alpha = (\beta, \beta)$, where $\beta > 0$, then $L = 3/4$, which is in agreement with formula (2.2.14): $DQ(\alpha) = -4\beta^2 \neq 0$.

**2.2.4. The Prony operator and the domain of its asymptotic stability.** The operator of recovering a function in $H_n(d)$ by its moments (see Subsection 2.2.2) is called the *Prony operator*. We find its general form.

The vector $p$ in condition 2) of Theorem 2.2.4′ is determined by the system $c = (c_0, c_1, \ldots, c_{2n-1})$ of the first $2n$ numbers of the sequence $\{c_i\}_0^N$, i.e., $p = p(c)$. Consequently, this system also determines the roots of the polynomial $T_n$ (see (2.2.6)). We denote by $M_n$, $M_n^-$ the sets of all $c \in \mathbb{C}^{2n}$ that determine the vector $p = p(c)$ uniquely and, such that, moreover, all roots of the associated polynomial $T_n$ lie in $\mathbb{C}\setminus(-\infty, 0)$ (respectively, in $S^-$) (i.e., we consider the case where the above problem of recovering a quasipolynomial is uniquely solvable). These sets are open, and $M_n^- \subset M_n$. The Prony algorithm acts precisely on the set $M_n$.

First, system (2.2.5) for $k = 0, 1, \ldots, n-1$ generates a continuous operator

$$B : M_n \to \mathbb{C}^n, \quad Bc = p, \quad p = (p_1, \ldots, p_n), \tag{2.2.17}$$

and the vector $p$ determines the moments $c_{2n}, \ldots, c_N$ by formula (2.2.5). Therefore, a perturbation (i.e., a small change) of the first $2n$ moments determines the perturbation of other moments. Next, the roots $\{q_i\}_1^n$ of equation (2.2.6) depend continuously on its coefficients (every root is repeated according to its multiplicity, see T. A.1.17, Appendix 1). The numbers

$$\{z_i = d^{-1} \ln q_i\}_1^n, \text{ where } \ln q = \ln|q| + i \arg q, \tag{2.2.18}$$

are contained in $\Pi(d)$ and are the indices of the exponentials of the required function $g \in H_n(d)$. The constant coefficients of $g$ are found from the condition $g(kd) = c_k$, $k = 0, 1, \ldots, n-1$. Eventually, we arrive at the following result.

**Theorem 2.2.14** (an explicit formula for the Prony operator). *Let* $q_i = e^{z_i d}$, $i = 1, \ldots, n$ (see (2.2.18)), *and let* $c = (c_0, c_1, \ldots, c_{2n-1}) \in \mathbb{C}^{2n}$. *Denote*

$$Q_k = Q_k(z) = \begin{vmatrix} q_1^k & \ldots & q_n^k \\ q_1^{k+1} & \ldots & q_n^{k+1} \\ \ldots & \ldots & \ldots \\ q_1^{k+n-1} & \ldots & q_n^{k+n-1} \end{vmatrix}, \qquad z \in \mathbb{C}^n,$$

$$P_k(t) = P_k(t, z) = \begin{vmatrix} 0 & e^{z_1 t} \quad \ldots \quad e^{z_n t} \\ c_k & \\ c_{k+1} & Q_k \\ \vdots & \\ c_{k+n-1} & \end{vmatrix}, \qquad (t, z) \in \mathbb{C} \times \mathbb{C}^n;$$

$k = 0, 1, \ldots, n-1$. *In the notation of formula* (2.2.17), *the operator that recovers the functions of class* $H_n(d)$ *by their moments (the Prony operator)*

$$A : M_n \to H_n(d), \quad A(M_n) = H_n(d), \quad Ac = g,$$

*described in Theorems* 2.2.4, 2.2.4′, *has the following form:*

*a)*

$$g(t) = -P_k(t)/Q_k, \quad t \in \mathbb{C}, \quad k = 0, 1, \ldots, n-1, \tag{2.2.19}$$

*if the* $\{q_i\}_1^n$ *are simple roots of the polynomial* $T_n$ (see (2.2.6)) *and the vector* $z$ *satisfies condition* (2.2.18)*;*

*b)*

$$g(t) = -\frac{DP_k(t;\alpha)}{[DQ_k(\alpha)]}, \ t \in \mathbb{C}, \ k = 0, 1, \ldots, n-1, \tag{2.2.20}$$

*if the roots* $\{q_i\}_1^n$ *of* $T_n$ *are multiple, the coordinates* $\{\alpha_i = d^{-1} \ln q_i\}_1^n$ *of the vector* $\alpha = (\alpha_1, ..., \alpha_n)$ *satisfy condition* (2.2.12), *and* $D$ *is the operator studied in* 2.2.2 *and applied to the functions* $P_k(t; z_1, \ldots, z_n)$, $Q_k(z_1, \ldots, z_n)$ *in the variables* $z_1, ..., z_n$ *at the point* $\alpha$ (see (2.2.13)).

◀ Formula (2.2.19) for $g$ is consistent. Indeed, multiplying the first, respectively, the second line of the determinants $Q_k$, $P_k(t)$ by the coordinate $p_n \neq 0$ of the vector $p = Bc$ (see 2.2.17), adding to them the lines with larger numbers, multiplied, respectively, by $p_{n-1}, \ldots, p_1$, and using (2.2.5), (2.2.6), we find: $P_k/Q_k = P_{k+1}/Q_{k+1}$, $k = 0, 1, \ldots, n-1$. Recalling that $\exp\{z_j d\} = q_j$, $j = 1, \ldots, n$, after elementary transformations of the determinants in the numerator of formula (2.2.19), we obtain:

$$g(kd) = -P_0(kd)/Q_0 = c_k, \ k = 0, \ 1, \ldots, n-1,$$

$$g[(k+n-1)d] = -P_k[(k+n-1)d]/Q_k = c_{k+n-1}, \ k = 1, \ldots, n.$$

But $g \in H_n(d)$, and from Theorem 2.2.3 we deduce that $\mathcal{A}c = g$.

In the case of multiple roots formula (2.2.20) is proved similarly. It suffices to observe that, for example, $DQ_0(\alpha)$ is the determinant obtained from $Q_0$ by differentiation in columns, in accordance with the multiplicity of each root. If the element $\{\alpha_i\}_1^n$ satisfies (2.2.12) and, say, $q_1 = e^{\alpha_1 d}$ is a root of $T_n$ of multiplicity $r_1$, then (see (2.2.13)) the second column of the determinant $Q_0(z)$ is differentiated once with respect to $z_2$, the third is differentiated twice with respect to $z_3$, etc.; the column with the number $r_1 = s_1$ is differentiated $(r_1 - 1)$ times with respect to $z_{s_1}$. Together with Theorem 2.2.1, this implies that $DQ_0(\alpha) \neq 0$. This is valid for every $\alpha$ in the set $N_Q := \{z \in \mathbb{C}^n : Q_0(z) = 0\}$ ▶

We introduce a topology in $H_n(d)$ by indicating a subbase of the system of neighborhoods at every point (i.e., a family of sets such that the finite intersections of its elements form a base of the topology at this point). Let $g \in H_n(d)$. Fix arbitrarily $\varepsilon > 0$ and $k \in \mathbb{N}$. The set $U(g; \varepsilon; k) = \{f \in H_n(d) : ||f - g|| < \varepsilon\}$ is an element of the subbase, where

$$||f - g||_k = \sup \{|f(t) - g(t)| \exp\{-[h_g(\arg t) + 1/k]|t|\} : \ t \in \mathbb{C}\},$$

and $h_g$ is the indicator of $g$ (see (1.10.3)).

**Theorem 2.2.15.** *Prony's operator* $\mathcal{A}$ (see T. 2.2.14) *is continuous in the topology of* $H_n(d)$.

◀ Fix arbitrarily $c \in M_n$, $k \in \mathbb{N}$, and $\tau \in (0,\ 1/k)$.

1. In the above notation, suppose that $T_n$ has different roots $\{q_i(c)\}_1^n$, where $T_N$ is the polynomial the coefficients of which are determined by the vector $p = Bc$ (see (2.2.6), (2.2.17)). Then there is a compact neighborhood $V = V(c) \subset M_n$ of $c$ such that the functions of the set $K(c) := \{\mathcal{A}b,\ b \in V\}$ are linear combinations of $n$ exponentials with nonzero coefficients; moreover, if $f(t) = \sum_{j=1}^{n} B_j e^{w_j t} \in K(c)$, then both the indices $\{w_j\}_1^n$ of the exponentials and the coefficients $\{B_j\}_1^n$ are continuous functions on $V$ (see (2.2.17), (2.2.19) and T. A.1.17). Let $g = \mathcal{A}c$. Then $g(t) = \sum_{j=1}^{n} A_j e^{z_j t}$. It is obvious that the indicator $h_g(\theta)$ of $g$ is equal to $\max\{\operatorname{Re} z_j e^{i\theta} : 1 \leqslant j \leqslant n\}$, $\theta \in \mathbb{R}$. From the foregoing we see that the neighborhood $V$ can be chosen in such a way that every element $f \in K(c)$ satisfies the condition $|w_j - z_j| < \tau$, $j = 1, \dots, n$. Now we estimate the function $f = \mathcal{A}b$, where $b \in V$:

$$|f(t,b)| \leqslant e^{|t| h_g(\arg t)} \sum_{j=1}^{n} |B_j| e^{|t| \operatorname{Re}(w_j - z_j)}, \quad t \in \mathbb{C}.$$

Therefore, there exists a constant $D_\tau = D_\tau(V) > 0$ such that

$$|[\mathcal{A}b](t)| < D_\tau \exp\{|t|[h_g(\arg t) + \tau]\} \quad \forall b \in V. \tag{2.2.21}$$

2. Some of the numbers $\{q_i(c)\}_1^n$ coincide. Then, in the representation of the function $g$, there are powers of the independent variable (see (2.2.9)) that can disappear after small perturbations of the point $c$. This is explained by the following fact: the discriminant set containing the point $p = Bc$ (see (2.2.17)) coincides with the zero set of a certain polynomial on $\mathbb{C}^n$. We consider the function $P_0(t;\ z; b')$ defined on $\mathbb{C} \times \mathbb{C}^n \times \mathbb{C}^n$ that differs from $P_0(t)$ (see (2.2.19)) in the fact that the first column $(c_0, c_1, \dots, c_{n-1})$ is replaced with the column $b' := (b_0, b_1, \dots, b_{n-1})$ of coordinates of a vector $b \in \mathbb{C}^{2n}$. We take also the function $Q_0 = Q_0(z)$, $z \in \mathbb{C}^n$ (see (2.2.19)). Let

$$G(t;\ z;\ b') = -\frac{P_0(t;\ z;\ b')}{Q_0(z)}, \quad t \in \mathbb{C},\ z \in \mathbb{C}^n \setminus N_Q, \tag{2.2.22}$$

where $b' \in \mathbb{C}^n$. In the proof of Theorem 2.2.14 it was shown that $DQ_0(\alpha) \neq 0 \ \forall \alpha \in N_Q$. Since the functions $P_0$, $Q_0$ are skew-symmetric with respect to $z_1, \dots, z_n$ (see Def. 2.2.5), the function $G$ as a function of $z_1, \dots, z_n$ for fixed values of the other variables extends analytically to $\mathbb{C}^n$ (see T. 2.2.13), and, at the same time, is symmetric. Therefore, $G$ determines a function $\tilde{G} = \tilde{G}(t;\ \{z\};\ b')$ defined on the $n$-fold symmetric product $SP^n(\mathbb{C})$ (see Subsection A.1.5), where $\{z\} = \{z_j\}_1^n$ is the class of equivalence in $SP^n(\mathbb{C})$ containing the element $z = (z_1, \dots, z_n) \in \mathbb{C}^n$. By Theorem A.1.17 there exists a homeomorphism $H : \mathbb{C}^n \to SP^n(\mathbb{C})$. For the elements of $SP^n(\mathbb{C})$ all coordinates of which do not belong to the interval $(-\infty, 0]$ of the real axis, the following mapping $L$ to $SP^n(\mathbb{C})$ is well defined:

$$L\{q_j\}_1^n = \{z_j\}_1^n, \quad z_j = d^{-1} \cdot \ln q_j, \quad j = 1, \ldots, n.$$

Therefore, the mapping

$$T = L \circ H \circ B : M_n \to SP^n(\mathbb{C}) \tag{2.2.23}$$

is continuous and one-to-one, and all coordinates of the elements of the set $T(M_n)$ belong to $\Pi(d)$.

We take a compact neighborhood $V = V(c) \subset M_n$ of an arbitrary point $c \in M_n$ and consider (see (2.2.23)) the function $f(t;\, b) := \tilde{G}(t;\, T(b);\, b')$, $t \in \mathbb{C}$. From formulas (2.2.22), (2.2.19), (2.2.20), (2.2.14), we deduce that

$$f(t;\, b) = [\mathcal{A}b](t), \quad (t, b) \in \mathbb{C} \times V. \tag{2.2.24}$$

Let $g = \mathcal{A}c$, and let $\{\alpha\} = T(c)$ (see (2.2.23)). Different numbers occurring among the coordinates of the element $\alpha$ are the indices of the exponentials involved in $g$. At the same time, they are the poles of the Borel transform $\hat{g}$ of $g$, because $g \in H_n(d)$ and $\hat{g} \in B_n(d)$ (see (2.2.24) in 2.2.2). The coordinates of the element $\{z\} = T(b)$ play a similar role for the function $f = \mathcal{A}b$ and its Borel transform $\hat{f}$ for every $b \in V \setminus \{c\}$. We consider the conjugate diagram $I_g$ of $g$ coinciding with the convex hull of the poles of $\hat{g}$ (see the Introduction, T. $A_1$). Put $S_\tau = \{z \in \mathbb{C} : |z| < \tau\}$. Let $O_\tau = I_g + S_\tau$ be the $\tau$-hull of $I_g$, and let $U = \{\{z\} \in SP^n(\mathbb{C}) : \rho(\{z\}, \{\alpha\}) < \tau\}$, where $\rho$ is the distance function in $SP^n(\mathbb{C})$ (see Subsection A.1.5). We fix an initial number $\tau \in (0, 1/k)$. By the structure of the function $\rho$, all coordinates of every element $\{z\} \in U$ belong to $O_\tau$. Since $T$ is continuous at the point $c$, the neighborhood $V$ can be chosen so that $T(V) \subset U$. Now, the preceding arguments show that the functions in the set $\{f = \mathcal{A}b : b \in V\}$ possess the following property: the poles of the Borel transform of each of them belong to a certain compact set $K \subset O_\tau$. Let $\Gamma = \partial O_\tau$ be the positively oriented boundary of $O_\tau$. The following integral representation (see P. 1.10.5) is true:

$$f(t, b) = \frac{1}{2\pi i} \int_\Gamma e^{t\lambda} \cdot \hat{f}(\lambda;\, b)\, d\lambda \quad \forall\, b \in V. \tag{2.2.25}$$

The Borel transform of $G$ (see (2.2.22)) in the variable $t$ has the form

$$\hat{G}(\lambda;\, z;\, b') = -R_0(\lambda;\, z;\, b')/Q_0(z), \quad z \in \mathbb{C}^n \setminus N_Q; \quad b' \in \mathbb{C}^n.$$

Here $R_0$ is the determinant obtained from the determinant $P_0(t;\, z;\, b')$ in (2.2.22) as follows: the exponentials $e^{z_j t}$, $j = 1, \ldots, n$, in the first line are replaced by the rational functions $\dfrac{1}{\lambda - z_j}$, $j = 1, \ldots, n$. Then (see (2.2.25),

(2.2.23)) $\hat{f}(\lambda, b) = \hat{G}(\lambda; T(b), b')$, $b \in V$. If $\lambda \in \Gamma$, $z_j \in K \subset O_\tau$, $j = 1, \dots, n$, and $b' \in \mathbb{C}^n$, then $\hat{G}$ and, with it, $\hat{f}(\lambda, b)$, $(\lambda, b) \in \Gamma \times V$, are continuous functions. This is explained by the following fact: $\hat{G}$ is a holomorphic function in $z_1, \dots, z_n$ in $O_\tau \times \cdots \times O_\tau$ for $\lambda \in \Gamma = \partial O_\tau$ (see the proof of the holomorphy of $G$ in $z_1, \dots, z_n$). Now, (2.2.25) and (2.2.24) imply that inequality (2.2.21) is also valid in the case where some of the numbers $\{q_i(c)\}_1^n$ coincide (see T. $A_1$, the Introduction; Ex. 1.3.20).

3. Let $c$ be any fixed point in $M_n$, and let $V = V(c) \subset M_n$ be its compact neighborhood chosen by the methods discribed in parts 1 and 2 of the proof. Using the continuity of the function $f(t, b)$, $(t, b) \in \mathbb{C} \times V$ proved above, by (2.2.21), where $\tau < 1/k$, and (2.2.24) we see that the function

$$\Phi(t; b) := [\mathcal{A}b](t) \exp\{-[h_g(\arg t) + 1/k]|t|\}$$

is continuous on the compact set $\overline{\mathbb{C}} \times V$ ($\overline{\mathbb{C}}$ is the Riemann sphere), and $\Phi(\infty, b) \equiv 0$, $b \in V$. Consequently, this function is uniformly continuous on $\overline{\mathbb{C}} \times V$ and for every $\varepsilon > 0$ there exists $\delta = \delta(\varepsilon)$ such that

$$|\Phi(t; b) - \Phi(t; c)| < \varepsilon \quad \forall t \in \mathbb{C}, \quad |b - c| < \delta, \ b \in V.$$

Therefore, the Prony operator $\mathcal{A}$ is continuous at the point $c$ ▶

A condition of stability for quasipolynomials on a segment and on the positive semiaxis is provided by the following statement.

**Corollary 2.16.** *Let $T > 0$. For every element $c$ in $M_n$ (in $M_n^-$) and every $\varepsilon > 0$ there exists a number $\delta$ such that for all $t \in [0, T]$ (respectively, $t \in [0, \infty)$) the following inequality is true:*

$$|[\mathcal{A}c](t) - [\mathcal{A}b](t)| < \varepsilon \quad \forall b \in V, \tag{2.2.26}$$

*where $V = V(c)$ is some neighborhood of $c$ included in $M_n$ (respectively, in $M_n^-$). Moreover,*

$$\mathcal{A}(M_n) = H_n(d) \quad (\mathcal{A}(M_n^-) = H_n^-(d)).$$

◀ It suffices to note that the indicator $h_g$ of every function $g \in H_n(d)$ is continuous, and if $g \in H_n^-(d)$, then $h_g(0) < 0$ (see T. $A_1$, the Introduction) ▶

## 2.3. Algorithm of approximation of discrete functions by quasipolinomials (identification algorithm)

**2.3.1. Algorithm of identifying relaxation characteristic that regain the equilibrium.** We dwell on the applied aspect of the Prony algorithm, starting with its modifications that were employed in medico-biological research. Let $f(t)$ be a dynamical characteristic of a homeostasis system. Assume that its behavior is described by a linear differential equation of the form (2.2.1), of order $n$ and with constant real coefficients, and with the boundary condition $y(0) = 0$. If the real part of every root of the characteristic equation for (2.2.1) is negative, then $f(\infty) = y(0) = 0$, and the solution of equation (2.2.1) can be interpreted as a deviation from the steady-state level for a homeostasis parameter in question after an external impact; moreover, this parameter regains its initial level in some time.

Let $F = (f_0, f_1, \dots, f_N) \in \mathbb{R}^{n+1}$, $f_0 = 0$, where $N \geqslant 2n - 1$, be a set of moments of the characteristic $f(t)$ measured in equal lapses of time with step $d > 0$, i.e., at the instances $t_j = jd$, $j = 0, 1, \dots, N$. If the objects under study are homogeneous, the role of the vector $F$ can be played by an appropriate set of mean data of the time characteristic $f(t)$ in question. To approximate it by quasipolynomials of order $n$, we consider the functional

$$\varphi_F(q) = \sum_{j=0}^{N-n} (f_{j+n} + q_1 f_{j+n-1} + \dots + q_n f_j)^2, \quad q \in \mathbb{R}^n,$$

characterizing the discrepancy of the difference equation of order $n$ (see (2.2.2), (2.2.5)) corresponding to equation (2.2.1). To study the approximation algorithm, we need the following properties of the functional $\varphi_F$.

**Theorem 2.3.1.** *Let*

$$\Gamma = \{g = (g_1, \dots, g_N) \in \mathbb{R}^{n-1} : \Delta_j = 0,\ j = 0, 1, \dots, N - 2n + 1\}, \tag{2.3.1}$$

*where*

$$\Delta_0 = \begin{vmatrix} g_{n-1} & g_{n-2} & \cdots & g_0 \\ \vdots & \vdots & \ddots & \vdots \\ g_{2n-3} & g_{2n-4} & \cdots & g_{n-2} \\ g_{2n-2} & g_{2n-3} & \cdots & g_{n-1} \end{vmatrix}, \quad g_0 = 0.$$

*We denote by $\Delta_j$ the determinant obtained from $\Delta_0$ by replacing the last line with the line $(g_k, g_{k-1}, \dots, g_{k-n+1})$ for $k = j + 2n - 2$, $j = 1, 2, \dots, N - 2n + 1$. Suppose*

$$M = \{(0, g, g_N) \in \mathbb{R}^{N+1} : g \in \mathbb{R}^{N-1} \setminus \Gamma,\ g_N \in \mathbb{R}\}.$$

*Then:* 1) *the functional $\varphi_F$ attains the minimum at a unique point $p = p(F)$ only if $F \in M$;* 2) *the function $p(F)$ is continuous on $M$.*

◄ Immediate verification shows that the functional $\varphi_F$ is a strictly convex function on $\mathbb{R}^n$ if and only if there is no vector $g = (\alpha_1, \dots, \alpha_n) \in \mathbb{R}^n \setminus \{0\}$ such that

$$\sum_{k=1}^{n} \alpha_k f_{j+n-k} = 0, \ j = 0, 1, \dots, N - n, \ f_0 = 0. \tag{2.3.2}$$

On the other hand, the existence of such a vector means that the rank of the matrix

$$\begin{Vmatrix} f_{n-1} & f_{n-2} & \cdots & f_0 \\ f_n & f_{n-1} & \cdots & f_1 \\ \vdots & \vdots & \ddots & \vdots \\ f_{N-1} & f_{N-2} & \cdots & f_{N-n} \end{Vmatrix}$$

is less than $n$. This is equivalent to the fact that for $F' = (f_1, ..., f_N)$ in the exceptional set $\Gamma \times \mathbb{R}$ (see (2.3.1)), the functional $\varphi_F$ is not strictly convex and attains its minimum on an affine variety $L$ in $\mathbb{R}^n$ such that $\dim L > 0$. But if $F' \notin \Gamma \times \mathbb{R}$, then (see (2.3.2))

$$\lim_{|q| \to \infty} |q|^{-1} \cdot \varphi_F(q) = \infty. \tag{2.3.3}$$

Therefore, the continuous function $\varphi_F$ attains its minimum at some point $p \in \mathbb{R}^n$, i.e.,

$$\inf\{\varphi_F(q) : q \in \mathbb{R}^n\} = \varphi_F(p). \tag{2.3.4}$$

The horizontal hyperplane $\Pi = \{(u, u_{n+1}) \in \mathbb{R}^n \times \mathbb{R} : u_{n+1} = \varphi_F(p)\}$ is supporting for the epigraph $\operatorname{epi} \varphi_F$ at the point $(p, \varphi_F(p))$. Since $\varphi(F)$ is a strictly convex function (see Cor. 1.5.15)), $p = p(F)$ is the only point where $\varphi_F$ attains the minimum. Condition (2.3.3) means that the asymptotic function $\varphi_F O^+(x)$ is identically equal to $\infty$ for $x \in \mathbb{R}^n \setminus \{0\}$ and, consequently (see T. 1.6.5, (1.6.4)-(1.6.6)), $\operatorname{dom} \varphi_F^* = \mathbb{R}^n$, where $\varphi_F^*$ is the Young transform of $\varphi_F$ (see Def. 1.5.1).

Since the direction vector of $\Pi$ is 0, by Corollary 1.5.15 we obtain

$$p = p(f) = \operatorname{grad} \varphi_F^*(0). \tag{2.3.5}$$

By (2.3.4) and the definition of $\varphi_F^*$, we have $\varphi_F(p) = -\varphi_F^*(0)$. Combining this with (2.3.5) and Theorem 1.4.27, from the continuity of the function $\varphi_F$ we deduce that the claim is true ▶

If $g \in H_n(d) \setminus H_{n-1}(d)$ (see 2.2.2), i.e., $g$ s a solution of equation (2.2.1) and is not a solution of a similar equation of order $n-1$, and, moreover, $g(0) = 0$ , then the vector of moments $(0, g_1, \ldots, g_N)$ belongs to $M$ (see T. 2.3.1 and T. 2.2.4$'$). Therefore, it is natural to expect that the vector of initial data $F := (0, f_1, \ldots, f_N)$ also belongs to $M$. Put $F' = (f_1, ..., f_N)$. Let $T_\varepsilon = \left(\mathbb{R}^{N-1} \setminus O_\varepsilon(\Gamma)\right) \times \mathbb{R}$, where $O_\varepsilon(\Gamma)$ is the $\varepsilon$-neighborhood of the set $\Gamma$ (see (2.3.1)). Assume that $F' \in T_\varepsilon$, where $\varepsilon > 0$ is chosen *a priori* in such a way that errors in measuring the coordinates do not expel $F'$ from the set $T_\varepsilon$. Now; we describe an algorithm of approximating $F$ by quasipolynomials of order $n$.

1. We find the vector $p = (p_1, \ldots, p_n)$ minimizing the functional $\varphi_F$ in $\mathbb{R}^n$.

2. We find the roots $Z_n = \{z_1, \dots, z_n\}$ of the polynomial $z^n + p_1 z^{n-1} + \dots + p_n$. If the required characteristic $f(t)$ is a relaxation charactrristic, i.e., describes a process of homeostasis type and regains the initial level in some time, then, in order that the algorithm can be run further, it is necessary that $Z_n \subset S^-$, where $S^- = \{\in \mathbb{C} : |z| < 1\} \setminus (-1, 0]$ is the unit disk centered at 0 from which we remove the segment $(-1, 0]$ on the negative semiaxis (see T. 2.2.4′).

3. Let $u_0 = 0$. We denote

$$L = \{u \in \mathbb{R}^n : u_{j+n} + p_1 u_{j+n-1} + \dots + p_n u_j = 0, \quad j = 0, 1, N - n\}.$$

Using the method of Lagrange factors, we find the projection $G' = (g_1, \dots, g_n)$ of the vector $F' = (f_1, \dots, f_n)$ to the linear variety $L$ (that is, $|F' - G'|) = \inf\{|F - u| : u \in L\})$. By the formula

$$g_k = -p_1 g_{k-1} - \dots - p_n g_{k-n}, \quad k = N+1, N+2, \dots,$$

we obtain the predicted dynamical characteristic value at any time moment $t_k = kd$, $k = N+1, N+2, \dots$.

4. If $N = 2n-1$, the vector $p$ is found from the system $f_{j+n} + p_1 f_{j+n-1} + \dots + p_n f_g = 0$, $j = 0, 1, \dots, n-1$, and $G' = F'$.

5. By one of formulas (2.2.19) or (2.2.20) corresponding to the cases of simple or multiple roots, we find the quasipolynomial $g$ of order $n$ the vector of moments of which is equal to $(0, g_1, \dots, g_N) = G$.

6. For large $N$ the system of linear equations from which the vector $G'$ is found may happen to be ill-posed. In this case the theoretical prediction curve of relaxation characteristic behavior can be found by a method close to the Prony method. After the vector $p$ is determined (see steps 2 and 4 of the algorithm), it is possible to find the roots of equation

$$z^n + p_1 z^{n-1} + \dots + p_n = 0.$$

If, e.g., the roots $z_i$, $i = 1, 2, \dots, n$, are real, different, and positive (which is expected in accordance with Theorem 2.2.4′, provided the chosen model describes the homeostasis process satisfactorily), then, by the formula

$$\lambda_j = d^{-1} \cdot \ln z_j, \; j = 1, \dots, n,$$

it is possible to find the exponential indices of the quasipolynomial

$$g(t) = A_1 e^{\lambda_1 t} + \dots + A_n e^{\lambda_n t}, \quad A_1 + \dots A_n = 0,$$

approximating the initial vector $F$. The unknown coefficients $A_1, \dots, A_{n-1}$ are found by the least squares method used for minimization of the functional

$$\Omega(A_1, \dots, A_{n-1}) = \sum_{j=1}^{n} [g_j - f_j]^2 = \sum_{j=1}^{N} \left[ \sum_{k=1}^{n-1} A_k (z_k^j - z_n^j) - f_j \right]^2 .$$

7. Standard criteria are used to verify that the theoretical prediction curve of the relaxation characteristic behavior is adequate to the vector of its initial data.

8. The algorithm in question is suitable not only to identify the relaxation characteristics of homeostasis processes, but also to identify dynamical characteristics of a different nature if they are described by equation (2.2.1) with the condition $y(0) = 0$. In this case, recalling condition 2) of Theorem 2.2.4′, it is necessary to require that $Z \subset \mathbb{C} \setminus (-\infty, 0]$.

The function $g(t)$ (see step 5 of the algorithm) is the required approximation of the characteristic $f(t)$. If $Z_n \subset S^-$, then $g(t) \in H_n^-(d) \setminus H_{n-1}^-(d)$ (see 2.2.2), i.e., the real parts of the exponential indices of $g$ are negative.

So, we have constructed the generalized Prony operator

$$\mathcal{A}^* : M \to H_n(d) \setminus H_{n-1}(d), \quad \mathcal{A}^*(F) = g(t), \ t \in \mathbb{C},$$

with the property $\mathcal{A}^*(M^-) \subset H_n^-(d) \subset H_{n-1}^-(d)$, where $M^-$ is a subset of $M$ such that $Z \subset S^-$ (see step 2 of the algorithm). Theorem 2.3.1 and Corollary 2.2.16 imply that all operations of the algorithm are continuous, and, moreover, the algorithm is stable under sufficiently small fluctuations of the initial vector $F$, as is shown by the following theorem.

**Theorem 2.3.2.** *Let $T > 0$. For every vector $F$ in $M$ (in $M^-$) and for every $\varepsilon > 0$ there is a neighborhood $V$ of $F$ included in $M$ (respectively, in $M^-$) such that for all $t \in [0, T]$ (respectively, $t \in [0, \infty)$ ) we have*

$$|[\mathcal{A}^* F](t) - [\mathcal{A}^* G](t)| < \varepsilon \quad \forall G \in V.$$

**2.3.2. Identification algorithm for a relaxation characteristic stabilizing to a different level.** In this case it is natural to assume that the behavior dynamics of such a characteristic is described by a differential equation of the above type which has a nonzero stationary solution. Such is the equation

$$y^{(n)} + a_1 y^{(n-1)} + \cdots + a_n y = a_{n+1}, \tag{2.3.6}$$

where $a_n \neq 0$ if $a_{n+1} \neq 0$. If every root of the characteristic equation for equation (2.3.5) has a negative real part, the solution of equation (2.3.1) describes a relaxation characteristic of a homeostasis process with the finite stabilization level $y(\infty) = a_{n+1}/a_n$. The following equation is the difference analog of equation (2.3.6):

$$y_{k+n} + \sum_{j=1}^{n} p_j y_{k+n-j} + p_{n+1} = 0, \quad k = 0, 1, 2 \ldots, \tag{2.3.7}$$

where $1 + p_1 + \cdots + p_n \neq 0$, i.e., 1 is not a root of the characteristic equation for (2.3.7) (see T. 2.2.1). The constant $p_{n+1}$ in (2.3.7) is related to the stationary solution $B$ of equation (2.3.6) by the formula

$$p_{n+1} = -B(1 + p_1 + \cdots + p_n), \tag{2.3.8}$$

i.e., $B$ is also the constant solution of equation (2.3.7). The problem of recovering a quasipolynomial satisfying (2.3.6) by its moments $c_0, c_1, \ldots, c_N$, where $N \geqslant 2n$, is studied similarly on the basis of the relation

$$c_{k+n} + p_1 c_{k+n-1} + \cdots + p_n c_k + p_{n+1} = 0, \quad k = 0, 1, \ldots, N - n.$$

The approximation algorithm of the initial data vector $F = (f_0, f_1, \ldots, f_N)$, where $N \geqslant 2n$, (the vector of moments of the time characteristic under consideration) is modified accordingly. As opposed to the previous case, we consider the functional

$$\Phi_F(q) = \sum_{k=0}^{N-n} (f_{k+n} + q_1 f_{k+n-1} + \cdots + q_n f_k + q_{n+1})^2, \quad q \in \mathbb{R}^{n+1},$$

reflecting the discrepancy of equation (2.3.7). Starting with this functional and using the outline described above in 2.3.1, it is possible to construct an identification algorithm for the relaxation characteristic of a homeostasis process stabilizing to a new steady-state level. The predicted value of this level is found by formula (2.3.8).

## 2.4. The case of quasipolynomials of order 2 and 3

We consider a situation most common in actual practice, when a model of the dynamical process in question is described by a linear differential equation with constant real coefficients of order 2 or 3. Solutions of these equations and the corresponding difference equations possess some properties that simplify substantially the identification algorithm presented in 2.3.

**2.4.1. The case of quasipolynomials of order 2.** 1. First of all, the structure of the exceptional set $\Gamma$ (see (2.3.1)) outside of which the initial data vector $F$ must lie is easy to study in this case.

**Property 2.4.1.** *In the notation of Theorem 2.3.1, for $n = 2$ the set $\Gamma$ is included in the hyperplane $g_1 = 0$ and the condition $F \in M$ is always fulfilled.*

◀ It suffices to observe that, for a sufficiently small time lapse, the coordinates $f_0$ and $f_1$ of $F$ can be chosen so that $f_0 = 0$ and $f_1 \neq 0$ ▶

2. The verification of the fact that the roots $Z_2 = \{z_1, z_2\}$ of the equation

$$z^2 + p_1 z + p_2 = 0, \tag{2.4.1}$$

belong to the set $S^- = \{z \in \mathbb{C} : |z| < 1\} \setminus (-1, 0]$ or to $\mathbb{C} \setminus (-\infty, 0]$ is easy and can be done without actual finding the roots (see steps 2, 8 of the algorithm in 2.3).

**Property 2.4.2.** *Let* $p = (p_1, p_2)$ *be the vector of real coefficients of equation* (2.4.1). *Then*

*a)* $Z_2 \in S^-$ *if and only if*

$$p \in D_2 = D_{--} \cup \Gamma_- \cup D_-$$

*where*

$$\begin{aligned} D_{--} &= \{q \in \mathbb{R}^2 : -2 < q_1 < 0, \ \max\{0, -1 - q_1\} < q_2 < q_1^2/4\}, \\ \Gamma_- &= \{q \in \mathbb{R}^2 : -2 < q_1 < 0, \ q_1^2 = 4q_2\}, \\ D_- &= \{q \in \mathbb{R}^2 : \ q_1^2/4 < q_2 < 1, \ |q_1| < 2\}; \end{aligned} \tag{2.4.2}$$

*b)* $Z_2 \in \mathbb{C} \setminus (-\infty, 0]$ *if and only if*

$$p \in G_2 = \{q \in \mathbb{R}^2 : \ q_2 > 0, \ q_1 < 2\sqrt{q_2}\}.$$

The proof is elementary and follows from analytic and geometric properties of a square trinomial. The domains $D_2$ and $G_2$ are the domains of asymptotic stability (respectively, stability) on the segment for the algorithm described in 2.3. Fig. 3 shows these domains ($G_2$ is the shaded part of the plane $\mathbb{R}^2$) and the sets they are split into.

Let

$$y'' + a_1 y' + a_2 y = 0 \tag{2.4.3}$$

be a differential equation with constant real coefficients, and let

$$y_{k+2} + p_1 y_{k+1} + p_2 = 0, \quad k = 0, 1, 2, \ldots, \tag{2.4.4}$$

be its difference analog, i.e., if $y = g(t)$ is a solution of equation (2.4.3), then $y_k = g(kd)$, $k = 0, 1, 2, \ldots$ (where $d > 0$ is the value of the uniform grid step) is the solution of equation (2.4.4).

By Theorem 2.2.3, the characteristic equations

$$\lambda^2 + a_1 \lambda + a_2 = 0, \tag{2.4.5}$$

$$z^2 + p_1 z + p_2 \tag{2.4.6}$$

for equations (2.4.2), (2.4.4), respectively, are related as follows: if $\lambda_1, \lambda_2$ are the roots of equation (2.4.5), then $e^{\lambda_1 d}, e^{\lambda_2 d}$ are the roots of equation (2.4.6). The roots of theses equations mutually determine each other if the numbers $a_1, a_2$ are such that the roots of equation (2.4.5) lie in the strip

$\Pi(d) = \{z \in \mathbb{C} : |\mathrm{Im}\, z| < \pi/d\}$ (see 2.2.2). In the following statement we use the notation of Property 2.4.2 and of Fig. 3.

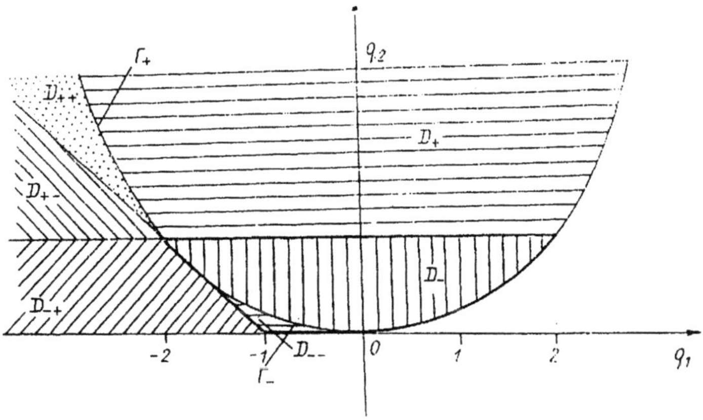


Fig. 3

**Property 2.4.3.** *Let* $p = (p_1, p_2)$ *be the vector of coefficients of equation* (2.4.6), *and let* $\lambda_1, \lambda_2$ *be the roots of equation* (2.4.5). *Then*

*1)* $\lambda_1, \lambda_2 \in \Pi_-(d) := \{z \in \Pi(d) : \mathrm{Re}\, z < 0\}$ *if and only if* $p \in D_2 := D_- \cup \Gamma_- \cup D_{--}^{*)}$ ;

*2)* $\lambda_1, \lambda_2$ *are negative numbers and* $\lambda_1 \neq \lambda_2$ *if and only if* $p \in D_{--}$;

*3)* $\lambda_1 = \alpha + \beta i$, $\lambda_2 = \alpha - \beta i$ *are complex conjugate numbers and* $\alpha < 0$ *if and only if* $p \in D_-$;

*4)* $\lambda_1 = \lambda_2 < 0$ *if and only if* $p \in \Gamma_-$;

*5)* $\lambda_1, \lambda_2 \in \Pi(d)$ *if and only if* $p \in G_2$;

*6)* $\lambda_1, \lambda_2 > 0$ $\lambda_1 \neq \lambda_2$ *if and only if* $p \in D_{++}$,

*7)* $\lambda_1 < 0 < \lambda_2$, $|\lambda_1| > \lambda_2$ *if and only if* $p \in D_{-+}$;

*8)* $\lambda_1 < 0 < \lambda_2$, $|\lambda_1| < \lambda_2$ *if and only if* $p \in D_{|}$ ;

*9)* $\lambda_1 = \alpha + \beta i$, $\lambda_2 = \alpha - \beta i$, $\alpha > 0$ *if and only if* $p \in D_+$;

*10)* $\lambda_1 = \lambda_2 > 0$ *if and only if* $p \in \Gamma_+$.

Formally, the sets mentioned in Property 2.4.3 and shown in Fig. 3 are described as follows: $D_- = \{q \in \mathbb{R}^2 : q_1^2/4 < q_2 < 1\}$; $D_{--} = \{q \in \mathbb{R}^2 : -1 - q_2 < q_1 < -2\sqrt{q_2},\ 0 < q_2 < 1\}$; $D_{|} = \{q \in \mathbb{R}^2 : q_2 > \max\{1, q_1^2/4\}\}$; $D_{++} = \{q \in \mathbb{R}^2 : -1 - q_1 < q_2 < q_1^2/4,\ q_1 < -2\}$; $D_{+-} = \{q \in \mathbb{R}^2 : 1 < q_2 < -1 - q_1,\ q_1 < -2\}$; $D_{-+} = \{Ш \subset \mathbb{R}^F : Ш_\infty < -\infty - Ш_\in,\ / < Ш_\in < \infty\}$; $\Gamma_- = \{q \in \mathbb{R}^2 ; q_2 = q_1^2/4,\ -2 < q_1 < 0\}$; and $\Gamma_+ = \{q \subset \mathbb{R}^2 : q_2 = q_1^2/4,\ q_1 < -2\}$.

So, the characteristic properties of the roots of equation (2.4.5) can be seen immediately by the coefficients of equation (2.4.6).

---

*) The boundary of this domain is indicated in Fig. 3.

3. The system of equations for the coordinates of the projection $G' = \{g_1, \ldots, g_N\}$ of the initial data vector $F'$ (see step 3 of the algorithm in 2.3) to the variety

$$\{u \in \mathbb{R}^N : u_{j+2} + p_1 u_{j+1} + p_2 u_j = 0,\ j = 0, 1, \ldots, N-2,\ u_0 = 0\}$$

in our case looks like this:

$$g_k + \alpha_{k-1} + \alpha_k p_1 + \alpha_{k+1} p_2 = f_k,\ k = 1, 2, \ldots, N, \tag{2.4.7}$$

$\alpha_0 = \alpha_N = \alpha_{N+1} = 0$, and

$$g_{k+2} + p_1 g_{k+1} + p_2 g_k = 0,\ k = 0, 1, \ldots, N-2,$$

where $\alpha_1, \ldots \alpha_{N-1}$ are the Lagrange factors. It suffices to know one coordinate $g_1$ of the projection $G'$ in order to find the solution of equation (2.4.3) satisfying the condition $g(0) = 0$ and approximating the vector $F$.

**Property 2.4.5.** *In the notation of Property* 2.4.2, *suppose that* $p \in G_2$, $g_1 \neq 0$, *and* $d > 0$. *Then there is a unique function* $g(t) \in H_2(d)$ (see 2.2.2) *such that* $g(0) = 0$, $g(d) = g_1$, *and the sequence* $\{g_k = g(kd),\ k = 1, 2, \ldots,\}$ *is a solution of the difference equation* (2.4.4). *The function* $g(t)$ *is determined by the following formula* (see Fig. 3, Property 2.4.3)*:*

*a) if* $p \in D_{--} \cup D_{++} \cup D_{-+} \cup D_{+-}$, $x_1, x_2$ *are the roots of equation* (2.4.6), *and*

$$\lambda_j = d^{-1} \ln x_j,\ j = 1, 2,\ A = g_1 (x_1 - x_2)^{-1},$$

*then* $g(t) = A\left(e^{\lambda_1 t} - e^{\lambda_2 t}\right)$, $t \in \mathbb{R}$;

*b) if* $p \in D_- \cup D_+$, $\alpha = (2d)^{-1} \ln p_2$, *and*

$$\beta = d^{-1} \cdot \arccos \frac{-p_1}{2\sqrt{p_2}},\ A = g_1 \left(p_2 - \frac{p_1^2}{4}\right)^{-1/2}, \tag{2.4.8}$$

*then*

$$g(t) = A e^{\alpha t} \sin \beta t,\ t \in \mathbb{R}; \tag{2.4.9}$$

*c) if* $p \in \Gamma_- \cup \Gamma_+$, $\lambda = d^{-1} \ln(-p_1/2)$, *and* $A = -2g_1 (p_1 d)^{-1}$, *then* $g(t) = A t e^{\lambda t}$, $t \in \mathbb{R}$.

4. For large $N$ system (2.4.7) may happen to be ill-posed. In this case, if, for instance, conditinon b) of Property 2.4.5 is fulfilled, the constant $A$ in (2.4.9) is found by the formula (see (2.4.8))

$$A = \left(\sum_{k=1}^{N} \gamma_k^2\right)^{-1} \cdot \sum_{k=1}^{N} f_k \gamma_k, \qquad \gamma_k = (\sqrt{p_2})^k \sin k\beta d,\ k = 1, \ldots, N.$$

The constant $A$ is the minimum point of the functional

$$\Omega(B) = \sum_{j=1}^{N} (B\gamma_k - f_k)^2, \ B \in \mathbb{R}.$$

**2.4.2. The case of quasipolynomials of order 3.** Application of the algorithm of 2.3 becomes more complicated in comparison with the above considerations.

**Property 2.4.1′.** *In the notation of Theorem* 2.3.1, *for* $n = 3$ *the set* $\Gamma$ *has the form*

$$\Gamma = \left\{g \in \mathbb{R}^{N-1} : g_1^2 g_{k+2} = g_1 g_2 g_{k+1} + (g_1 g_3 - g_2^2) g_k, \ k = 2, \dots, N-3\right\}.$$

Let $Z_3 = \{z_1, z_2, z_3\}$ be the set of all roots of the equation

$$T_3(z) := z^3 + p_1 z^2 + p_2 z + p_3 = 0, \tag{2.4.10}$$

where $p = (p_1, p_2, p_3) \in \mathbb{R}^3$. We put

$$Y = \{q \in \mathbb{R}^3 : \ 1 + q_1 + q_2 + q_3 > 0, \ q_3 < 0\}, \tag{2.4.11}$$

$$Q = A^2 - B^3, \quad A = \frac{q_1^3}{27} - \frac{q_1 q_2}{6} + \frac{q_3}{2}, \quad B = \frac{q_1^2}{9} - \frac{q_2}{3}, \tag{2.4.12}$$

$$\begin{aligned}
D_- &= \{q \in Y : Q < 0, 3 + 2q_1 + q_2 > 0, -3 < q_1 < 0, q_2 > 0\},\\
D_0 &= \{q \in Y : Q = 0, q_1^2 > 3q_2 > 0, 3 + 2q_1 + q_2 > 0, -3 < q_1 < 0,\},\\
D_{00} &= \{q \in \mathbb{R}^3 : \ 3q_2 = q_1^2, \ 27q_3 = q_1^3, \ -3 < q_1 < 0\},\\
D_{rc} &= \{q \in Y : \ Q > 0, \ q_2 - q_3(q_1 - q_3) < 1, \ q_2 < q_1 q_3^{1/3}\},\\
D_{cr} &= \{q \in Y : \ Q > 0, \ q_2 - q_3(q_1 - q_3) < 1, \ q_2 \geqslant q_1 q_3^{1/3}\},\\
D_3 &= D_- \cup D_0 \cup D_{00} \cup D_{rc} \cup D_{cr}.
\end{aligned} \tag{2.4.13}$$

**Property 2.4.2′.** *The set* $Z_3$ *is included in* $S^- = \{z \in \mathbb{C}, \ |z| < 1\} \setminus (-1, 0]$ *if and only if the vector* $p = (p_1, p_2, p_3)$ *of real coefficients of equation* (2.4.10) *belongs to the domain* $D_3$ (see (2.4.13)).

◀ *The "only if" part.* 1. Assume that $Z_3$ is included in $(0, 1)$ and does not contain multiple roots. Let, for definiteness, $0 < z_1 < z_2 < z_3 < 1$. The Viete formulas

$$p_1 = -(z_1 + z_2 + z_3), \quad p_2 = z_1 z_2 + z_1 z_3 + z_2 z_3, \quad p_3 = -z_1 z_2 z_3 \tag{2.4.14}$$

immediately show that $p_3 < 0$, $-3 < p_1 < 0$, $p > 0$. The value $Q$ (2.4.12) is the determinant of equation (2.4.10), and roots of the latter are real and different only if $Q < 0$. By the formula (see (2.4.10))

$$1 + p_1 + p_2 + p_3 = T_3(1) = (1 - z_1)(1 - z_2)(1 - z_3) \tag{2.4.15}$$

and condition $Z_3 \subset (0,1)$, we conclude that $p_1 + p_2 + p_3 + 1 > 0$, i.e., $p \in Y$ (see (2.4.11)).

Consider the function

$$F(z) = 3 - 2(z_1 + z_2 + z_3) + z_1 z_2 + z_1 z_3 + z_2 z_3, \quad z \in \mathbb{R}^3.$$

Let $K = \{z \in \mathbb{R}^3 : \ 0 < z_i < 1, \ i = 1,2,3\}$. Then $\dfrac{\partial F}{\partial z_3} = -2 + z_1 + z_2 < 0 \ \forall \, z \in K$. Consequently, the function $F(z)$ is strictly monotone decreasing in the variable $z_3$ and

$$F(z) = f(z_1, z_2, z_3) > F(z_1, z_2, 1) = (1 - z_1)(1 - z_2) > 0 \ \forall \, z \in K. \qquad (2.4.16)$$

Together with (2.4.14) this implies that $3 + 2p_1 + p_2 > 0$. So, $p \subset D_-$ (see (2.4.13)).

2. Suppose that equation (2.4.10) has a root of multiplicity 2. Then $Q = 0$ (see (2.4.12)) and $Z_3 \subset (0,1)$. The inequality

$$3 + 2p_1 + p_2 = F(z) > 0 \qquad \forall \, z \in K$$

(see (2.4.14) and (2.4.16)) remains valid if the vector $z$ has coinciding coordinates, e.g., $z_2 = z_3 = \alpha$. By formula (2.4.14), we have: $p_1 = -z_1 - 2\alpha$, $p_2 = 2\alpha z_1 + \alpha^2$. Consequently,

$$p_1^2 - 3p_2 = (\alpha - z_1)^2 > 0, \qquad p_1^2 > 3p_2. \qquad (2.4.17)$$

Combining this with (2.4.10), we find that $p \in D_0$ (see (2.4.13)).

3. Suppose that equation (2.4.10) has a single root $\alpha$ of multiplicity 3. Then $\alpha \in (0,1)$, and from (2.4.14) it follows that $p_1^2 = 3p_2$, $p_1^3 = 27p_3$, $-3 < p_1 < 0$. So, $p \in D_{00}$ (see (2.4.13)).

4. It remains to analyse only one possibility: let, for instance, $z_1 \in (0,1)$, and let $z_2, z_3$ be two complex conjugate numbers in $S_-$. We put $r = |z_2| = |z_3|$, $z_2 = \alpha + \beta i$, $z_3 = \alpha - \beta i$. Then (see (2.4.14) and (2.4.15))

$$p_3 = -z_1 r^2 < 0, \quad 1 + p_1 + p_2 + p_3 = (1 - z_1)[(1-\alpha)^2 + \beta^2] > 0, \quad (2.4.18)$$

i.e., $p \in Y$ (see. (2.4.11)). By formula (2.4.14), we find:

$$p_3(p_1 - p_3) - p_2 + 1 = (1 - r^2)[1 - 2\alpha z_1 + (r z_1)^2] > 0, \qquad (2.4.19)$$

because $|\alpha| < r$. After similar transformations, we conclude that

$$p_2 - p_1 p_3^{1/3} = s^2(s^2 - t^2)(s^2 - 2st\cos\varphi + t^2), \qquad (2.4.20)$$

where $\cos\varphi = \alpha/r$, $s = \sqrt[3]{r}$, and $t = \sqrt[3]{z}$. Hence, recalling that $Q > 0$ (see (2.4.12) and (2.4.13)), we see that if $z_1 < r$, then $p \in D_{rc}$, and if $z_1 \geqslant r$, then $p \in D_{cr}$.

So, $p \in D_3$ if $Z_3 \subset S^-$ (see (2.4.13)).

*The "if" part.* 1. Let $p \in D_-$ (see (2.4.13)). Since $Q < 0$, the constant $B$ in (2.4.12) satisfies the condition $B < 0$, i.e., $3p_2 < p_1^2$. Therefore, the derivative $T_3'(z) = 3z^2 + 2p_1 z + p_2$ of the polynomial $T_3(z)$ (see (2.4.10)) has real zeros $x_1, x_2$. Hence, recalling that $T_3'(0) = p_2 > 0$, $T_3'(1) = 3 + 2p_1 + p_2 > 0$, and the function $y = T_3'(z)$, $z \in \mathbb{R}$, attains its minimum at the point $z_0 = -p_1/3 \in (0,1)$ (see (2.4.13)), we see that $\{x_1, x_2\} \subset (0,1)$. Consequently, the function $y = P_3(z)$ increases on the intervals $(-\infty, 0)$, $(1, \infty)$. But $T_3(0) = p_3 < 0$, and $T_3(1) = 1 + p_1 + p_2 + p_3 > 0$. Therefore, $Z_3 \subset (0,1)$ (see Fig. 4).

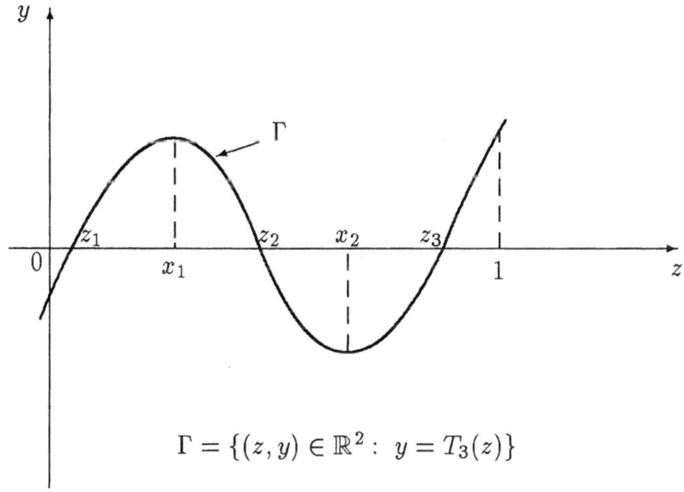


Fig. 4

2. Let $p \in D_0$ (see (2.4.13)). Since $Q = 0$, equation (2.4.10) has a multiple root $\alpha \in \mathbb{R}$. By (2.4.17), we conclude that its multiplicity is equal to 2. If $z_1$ is a simple root of equation (2.4.10) and $z_2 = z_3 = \alpha \in Z_3$, then, the condition $p_3 = -z_1\alpha^2 < 0$ (see (2.4.14), (2.4.13), and (2.4.11)) implies that $z_1 > 0$, and, the condition (see (2.4.20))

$$0 < 1 + p_1 + p_2 + p_3 = (1 - z_1)(1 - \alpha)^2,$$

implies that $z_1 < 1$, i.e., $z_1 \in (0,1)$. The number $\alpha$ is a simple root of the derivative $T_3'(z) = 3z^2 + 2p_1 z + p_2$ of $T_3(z)$ (see (2.4.10)). Let $x$ be another root of $T_3'$. Then, arguing as above in part 1, we conclude that $\{\alpha, x\} \subset (0,1)$.

3. Let $p \in D_{00}$ (see (2.4.13)). Then (2.4.10) implies that $T_3(z) = (z + p_1/3)^3$ for $z \in \mathbb{C}$, i.e., $z_0 = -p_1/3$ is the only root of the polynomial $T_3(z)$, and $z_0 \in (0,1)$. This means that $Z_3 \subset (0,1)$.

4. Let $p \in D_{rc} \cup D_{cr}$ (see (2.4.13)). Since $Q > 0$, the set $Z_3$ consists of one

real number $z_1$ and a pair of complex conjugate numbers $z_2, z_3$. Then formulas (2.4.18) and (2.4.19) imply $Z_3 \subset S^-$.

So, $Z_3 \subset S^-$ if $p \in D_3$ (see (2.4.13)) ▶

The partition of the domain $D_3$ (see (2.4.13)) corresponds to the following possible types of integral curves of the equation

$$y''' + a_1 y'' + a_2 y' + a_3 y = 0, \quad y(0) = y(\infty) = 0 \tag{2.4.21}$$

with constant real coefficients the characteristic equation of which has all roots with negative real part:

$$\begin{aligned}
g(t) &= A(e^{\lambda_1 t} - e^{\lambda_3 t}) + B(e^{\lambda_2 t} - e^{\lambda_3 t});\\
g(t) &= A(e^{\lambda_1 t} - e^{\lambda_2 t}) + Bte^{\lambda_2 t};\\
g(t) &= (A + Bt)te^{\lambda t};\\
g(t) &= Ae^{\alpha t}\sin\beta t + B(e^{\lambda t} - e^{\alpha t}\cos\beta t), \quad \lambda < \alpha;\\
g(t) &= Ae^{\alpha t}\sin\beta t + B(e^{\lambda t} - e^{\alpha t}\cos\beta t), \quad \lambda \geqslant \alpha.
\end{aligned}$$

Here $\lambda_i < 0$ for $i = 1, 2, 3$, $\lambda < 0$, $\alpha < 0$, and $A, B, \beta \in \mathbb{R}$. The theoretical prediction curves of relaxation characteristic behavior, obtained starting with the vectors of discrete laboratory data by the algorithm described in 2.3, coincide with the above integral curves of equation (2.4.21).

## 2.5. The case of wave-type homeostasis processes *)

Up to this point, it has been assumed that in the dynamical equilibrium the numerical characteristics of a homeostasis system are constant. Now we consider the case where such a characteristic $S = S(t)$ oscillates perceptibly about a certain average level. This property is exhibited by physiological indices of a homeostasis system that are associated with certain rhythms of its vital activity. The information about an index under consideration at a fixed time moment is not a number, but a curve that gives an insight into the dynamics of this index. Among such indices we name, for instance, bioelectric potentials related to the activity of the brain, heart, eye muscles, etc.

**2.5.1. Mathematical approach to identifying wave-type homeostasis characteristics.** Let the information signal generated by the system and reflecting the behavior dynamics of some physiological index be determined by an almost-periodic function that is the sum of a finite number of sinusoids:

$$S(t) = \sum_{j=1}^{n} A_j \sin(\omega_j t + \theta_j) + B = \sum_{j=1}^{n} (a_j \cos\omega_j t + b_j \sin\omega_j t) + B, \tag{2.5.1}$$

*) This section was written with support of the Krasnoyarsk Territorial Science Foundation, grant 1F28 for 1993.

where $0 < \omega_1 < \omega_2 < \cdots < \omega_n$. Suppose that at the moment $t = 0$ the system is subject to an external impact under the effect of which the amplitude, frequency, and phase of each harmonic comprising the signal in (2.5.1) change to regain their original value in some time. Suppose that the signal of such a relaxation characteristic admits a representation of the form

$$S(t) = \sum_{j=1}^{n} \{\alpha_j(t) \cos[w_j(t)t] + \beta_j(t) \sin[w_j(t)t]\} + B, \tag{2.5.2}$$

where $\alpha_j, \beta_j$, and $w_j$, $j = 1, \dots, n$, are real quasipolynomials of order $m$ such that the indices of all exponentials of each of them are either negative or have negative real part. The above relaxation condition means that (see (2.5.1), (2.5.2))

$$\alpha_j(0) = \alpha_j(\infty) = a_j, \ \beta_j(0) = \beta_j(\infty) = b_j, \ w_j(0) = w_j(\infty) = \omega_j,$$

where $j = 1, \dots, n$. Suppose we want to identify the parameters of the model (2.5.2) on the basis of a system of the information curves for a certain index of the system in question at the moment $t_k = kd$, $k = 0, 1, \dots, N$, where $d > 0$ is a time lapse. If the interval $T$ when these curves were recorded is sufficiently small, the amplitudes, phases and harmonic frequencies in (2.5.2) can be thought of as constants, i.e.,

$$S(t) \sim \varphi_k(t), \qquad t_k \leqslant t \leqslant t_k + T, \quad k = 0, 1, \dots, N_1, \tag{2.5.3}$$

where

$$\varphi_k(t) = \sum_{j=1}^{n} \{\alpha_j(t_k) \cos[w_j(t_k)t] + \beta_j(t_k) \sin[w_j(t_k)t]\} + B.$$

The basis for identifying the parameters of $\varphi_k$ is its values $f_{kj} = \varphi_k(t_k + jd)$, $j = 0, 1, \dots, N_2$, on a uniform grid with sufficiently small step $d > 0$; here $dN_2 \leqslant T$, $N_2 \geqslant 3n$, and the identification method is a specific version of the Prony method, as will be explained below. The parameters of the $\varphi_k$, $k = 0, 1, \dots, N_1$, are the values of quasipolynomials on a uniform grid; so, the parameters of the function $S$ (see (2.5.2)) can now be found by the method described above in 2.2–2.4.

**2.5.2. The Prony method of harmonic expansion.** We consider the problem of approximating a sequence $s_0, s_1, \dots, s_N$ of real numbers by uniform readings of an almost-periodic signal of the form (2.5.1), where $n$ is a fixed number. We dwell on the conditions under which this algorithm is well posed.

1. Assume $N = 3n$. Then under certain conditions the signal of the form (2.5.1) can be synthesized by the sequence $s_k = S(t_0 + kd)$, $d = 0, 1, \dots, 3n$, of its $3n + 1$ values on the uniform grid, where $t_0 \geqslant 0$.

Let $S_n(d)$ be the class of signals of the form (2.5.1) that satisfy the condition $\omega_n = \max\limits_{1 \leqslant i \leqslant n} \omega_i < \pi/d$. First, from Theorems 2.2.4 and 2.2.4′ we deduce the following statements about the existence and uniqueness of a function belonging to $S_n(d)$ and interpolating the given real sequence $\{s_i\}_0^N$, where $N \geqslant 3n$.

**Theorem 2.5.1.** *In the class $S_n(d)$ there exists at most one function $g$ satisfying the condition*

$$g(t_0 + kd) = s_k, \qquad k = 0, 1, \ldots, N. \tag{2.5.4}$$

**Theorem 2.5.2.** *The class $S_n(d) \setminus S_{n-1}(d)$ contains a function $g$ satisfying (2.5.4) if and only if there is a unique vector $(p_1, \ldots, p_{2n}, p_{2n+1}) \in \mathbb{R}^{2n+1}$ such that*

$$s_{k+2n} + \sum_{i=1}^{2n} p_i s_{k+2n-i} + p_{2n+1} = 0, \quad k = 0, 1, \ldots, N - 2n, \tag{2.5.5}$$

$$p_i = p_{2n-i}, \quad i = 1, \ldots, n-1, \qquad p_{2n} = 1, \tag{2.5.6}$$

*and, moreover, all roots of the equation*

$$T_{2n}(z) := z^{2n} + p_1 z^{2n-1} + \cdots + p_{2n-1} z + p_{2n} = 0 \tag{2.5.7}$$

*are different and lie in $\{z \in \mathbb{C} : \ |z| = 1\} \setminus \{1; -1\}$.*

Condition (2.5.6) arises because if $w$ is a root of equation (2.5.7), then so is $w^{-1}$. Finding the roots of the polynomial $T_{2n}$ can be reduced to the study of a related polynomial of order $n$.

**Proposition 2.5.3.** *In the notation of Theorem 2.5.2, there exists a polynomial $P_n$ of degree $n$ having different roots on the segment $[-1; 1]$ and such that $P_n \circ \lambda(z) \equiv z^{-n} \cdot T_{2n}(z)$, $z \in \mathbb{C}$, where $\lambda(z) = \frac{1}{2}(z + z^{-1})$ is the Zhukovskiĭ function. For $n = 2$ we may take $P_2(w) = 4w^2 + 2p_1 w + p_2 - 2$, where*

$$\max\{2p_1 - 2, -2p_1 - 2\} < p_2 < \frac{p_1^2}{4} + 2, \qquad |p_1| < 4. \tag{2.5.8}$$

The domain in $\mathbb{R}^2$ described by (2.5.8) is shown in Fig. 5.

For the proof it suffices to observe that

$$(z + z^{-1})^m = \sum_{k=0}^{l-1} C_m^k (z^{m-2k} + z^{2k-m}) + \begin{cases} C_m^l, & m = 2l, \\ C_m^l (z + z^{-1}), & m = 2l+1, \end{cases}$$

where $C_m^0 = 1$ and $l = [m/2]$ ($[q]$ is the integral part of a number $q$).

The following modification of the Prony procedure described above provides the way to the synthesis of the signal in $S_n(d)$ starting with a sampling of its uniform readings $s_k = S(t_0 + kd)$, $k = 0, 1, \ldots, 3n$.

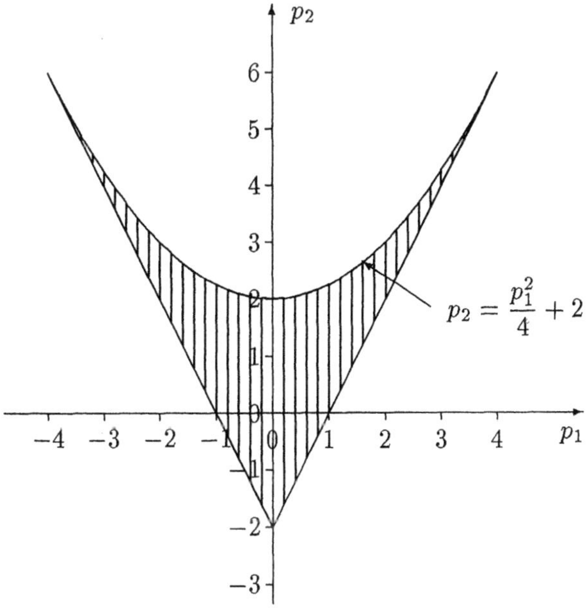


Fig. 5

1) We find the solution $(p_1, \ldots, p_n, p_{2n+1})$ of the system

$$s_{k+2n} + \sum_{j=1}^{2n} p_j s_{k+2n-j} + p_{2n+1} = 0, \ k = 0, 1, \ldots, n, \tag{2.5.9}$$

under condition (2.5.6).

2) We find the roots $Z_{2n} = \{z_i, \overline{z}_i\}_1^n$ of the polynomial $T_{2n}$ (2.5.7) or the roots $W_n = \{w_i = (z_i + \overline{z}_i)/2\}_1^n$ of the polynomial $P_n$ associated with it (see P. 2.5.3). The synthesis is possible if $T_{2n}$, $P_n$ has no multiple roots and $Z_{2n} \subset \{z \subset \mathbb{C} : \ |z| = 1\} \setminus \{1; -1\}$ or $W_n \subset (-1, 1)$. In particular, for $n = 2$ this is valid if (2.5.8) is true. There is no loss of generality in assuming that $\operatorname{Re} z_i > 0$, $i = 1, \ldots, n$, and $\operatorname{Im} z_1 > \operatorname{Im} z_2 > \cdots > \operatorname{Im} z_n$. Suppose $\alpha_i = \arg z_i = \arccos w_i$, $i = 1, \ldots, n$. Let $d > 0$ be a time lapse. We find the angular frequencies $\omega_i = d^{-1} \cdot \alpha_i$, $i = 1, \ldots, n$.

3) The required signal can be constructed by the method described above in 2.2.4.

By the *Prony operator* we mean an operator $\mathcal{A}^*$ of recovering a function of class $S_n(d)$ by its moments (see T. 2.5.1 and T. 2.5.2).

**Theorem 2.5.4** (an explicit formula for the Prony operator). *Let a sequence* $s = \{s_k\}_0^{3n}$ *satisfy the assumptions of Theorem* 2.5.2. *Denote*

$\Delta = t - t_0$. *Then the only function* $S \in S_n(d)$ *satisfying the condition* $S(t_0 + kd) = s_k$, $k = 0, 1, \ldots, 3n$, *has the form*

$$S(t) = -P(t)/Q + B, \quad \mathcal{A}^*(\mathcal{J}) = \mathcal{S}, \tag{2.5.10}$$

*where in the above notation* (see (2.5.9))

$$B = -p_{2n+1} \cdot [p_n + 2(1 + p_1 + \cdots + p_{n-1})]^{-1}, \tag{2.5.11}$$

$$P(t) = \left| \begin{array}{c|cccc} 0 & \varphi_1(t) & \psi_1(t) \ldots & \varphi_n(t) & \psi_n(t) \\ \hline s_0 - B & & & & \\ s_1 - B & & & & \\ s_2 - B & & Q & & \\ \vdots & & & & \\ s_{2n-1} - B & & & & \end{array} \right|,$$

*where*

$$\varphi_k(t) = \cos \omega_k (t - t_0), \ \psi_k(t) = \sin \omega_k (t - t_0), \quad k = 1, \ldots, n,$$

*and*

$$Q = \begin{vmatrix} a_{01} & b_{01} & \cdots & a_{0n} & b_{0n} \\ a_{11} & b_{11} & \cdots & a_{1n} & b_{1n} \\ a_{21} & b_{21} & \cdots & a_{2n} & b_{2n} \\ \vdots & \vdots & \ddots & \vdots & \vdots \\ a_{2n-1,1} & b_{2n-1,1} & \cdots & a_{2n-1,n} & b_{2n-1,n} \end{vmatrix},$$

*where*

$$a_{kr} = \cos k\alpha_r, \ b_{kr} = \sin k\alpha_r, \quad k = 0, 1, ..., 2n-1, \ r = 1, 2, ..., n;$$

*moreover,*

$$Q = 2^{n(n-1)} \cdot \prod_{i=1}^{n} \sin \alpha_i \cdot W^2(cos\alpha_1, \ldots, cos\alpha_n). \tag{2.5.12}$$

*Here* $W(q)$ *is the Vandermonde determinant of order* $n$ *in the variables* $q_1, \ldots, q_n$.

◀ Since 1 is not a root of $T_{2n}$ (see Theorem 2.5.2), the denominator of the fraction in (2.5.11) is not equal to zero. The sequence $c_k = s_k - B$, $k = 0, 1, \ldots, 3n$, satisfies the formula

$$c_{k+2n} + p_1 c_{k+2n+1} + \cdots + p_{2n} c_k = 0, \quad k = 0, 1, \ldots, n. \tag{2.5.13}$$

Since for $\{s_k\}_0^{3n}$ the assumptions of Theorem 2.5.2 are satisfied, we see that $(p_1, p_2, \ldots, p_{2n})$ is the only solution of system (2.5.13), and $-1$ is not a zero of the polynomial $T_{2n}$ (see (2.5.6) and (2.5.7)). Therefore, the sequence $\{c_k\}_0^n$ satisfies the conditions of Theorem 2.4$'$. There exists a unique function $g \in S_n(d)$ such that $g(kd) = c_k$, $k = 0, 1, \ldots, 3n$. Its construction is dictated by formula (2.2.19): $g = -P_0(t)/Q$, where $n$ should be replaced by $2n$, $z_{2j-1} = i\omega_j$, $z_{2j} = -i\omega_j$, $q_{2j-1} = e^{i\alpha_j}$, and $q_{2j} = \exp\{-i\alpha_j\}$, $j = 1, 2 \ldots, n$. We transform the determinants $P_0(t)$ and $Q_0$ simultaneously: 1) we add the columns containing $q_{2j}$ to the columns containing $q_{2j-1}$, 2) we factor out the number 2 from the columns obtained, 3) we subtract the newly obtained columns from the columns containing $q_{2j}$, and 4) we factor out $(-i)$ from the resulting columns. Here $j = 1, 2, \ldots, n$. As a result, we obtain $g(t) = -P(t)/Q$ (see T. 2.5.4) and $S(t) = g(t - t_0) + B$, $t \in \mathbb{R}$, is the required function of class $S_n(d)$.

It only remains to prove (2.5.12). By the factorization formula for the Vandemonde determinant, we have:

$$Q_0 = \prod_{j=1}^{n} (q_{2j} - q_{2j-1}) \prod_{1 \leqslant i < k \leqslant n} |q_{2k-1} - q_{2i-1}|^2 |q_{2k} - q_{2i-1}|^2. \tag{2.5.14}$$

But if $|p| = |q| = 1$, then

$$|p - q|^2 \cdot |\overline{p} - q|^2 = |1 + p^2 - 2p \cdot \operatorname{Re} q|^2 = 4 \left| \frac{p + p^{-1}}{2} - \operatorname{Re} q \right|^2 = 4(\operatorname{Re} p - \operatorname{Re} q)^2.$$

Therefore (see (2.5.14)),

$$Q_0 = (-2i)^n \cdot 4^{\frac{n(n-1)}{2}} \cdot \prod_{j=1}^{n} \sin \alpha_j \cdot W^2(\cos \alpha_1, \ldots, \cos \alpha_n).$$

On the other hand, doing the above transformations for $Q_0$, we find: $Q_0 = (-2i)^n Q$. This implies (2.5.12) ▶

So, the signals of class $S_k(d)$ are synthesized starting with a finite number $3n + 1$ of their uniform readings.

2. Now, let $N > 3n + 1$. To approximate the sequence $s = \{s_i\}_0^N$ by functions of class $S_n(d)$, we consider the functional

$$\varphi_s(q, q_{2n+1}) = \sum_{k=0}^{N-2n} \left[ s_{k+2n} + s_k + \sum_{i=1}^{n-1} c_{ki} q_i + q_n s_{k+n} + q_{2n+1} \right]^2,$$

where $c_{ki} = s_{k+2n-i} + s_{k-i}$, characterizing the discrepancy of the difference equation (2.5.5) of order $2n$. Arguing as in the proof of Theorem 2.3.1, we see that the functional $\varphi_s$ is strictly convex for almost all $s \in \mathbb{R}^{N+1}$. The exceptional set $\Gamma$ is determined by the following condition: the rank of the basic matrix of system (2.5.5) is less than $n+1$. The following theorem is true (see T. 2.3.1).

**Theorem 2.5.5.** *Let* $M = \mathbb{R}^{N+1} \setminus \Gamma$. 1) *The functional* $\varphi_s$ *attains its minimum on* $\mathbb{R}^{N+1}$ *at a unique point* $p = p(s)$ *only if* $s \in M$; 2) *the function* $p = p(s)$ *is continuous on* $M$.

As before, we arrive at the conclusion that it is natural to expect that $s \in M$. Suppose $s \in M_\varepsilon := \mathbb{R}^{N+1} \setminus O_\varepsilon(\Gamma)$, where $O_\varepsilon(\Gamma)$ is the $\varepsilon$-neighborhood of the set $\Gamma$, and $\varepsilon > 0$ is chosen *a priori* so as to the errors in measuring the coordinates of the vector $s$ do not expel it from $M_\varepsilon$. Now, we describe the modified Prony algorithm (cf. 2.3).

1) We find a vector $(p, p_{2n+1})$, where $p = (p_1, \dots, p_n)$, minimizing the functional $\varphi_s$ on $\mathbb{R}^{n+1}$.

2) As at step 2) in the case of $N = 3n+1$, we find the angular frequencies $\{\omega_i\}_1^n$. If the vector $(p, p_{2n+1})$ does not satisfy the assumption of Theorem 2.5.2 (in particular, if for $n = 2$ inequality (2.5.8) fails), it must be admitted that the model is unsatisfactory, and it is necessary to look for approximating signal, e.g., in $S_{n+1}(d)$.

3) We consider the almost-periodic signal

$$\hat{S}(t) = \sum_{i=1}^{n} [a_i \cos \omega_i (t - t_0) + b_i \sin \omega_i (t - t_0)] + B, \qquad t \in \mathbb{R},$$

where the constant $B$ is determined by (2.5.11), and the coefficients are unknown. Their values are found, e.g., by the least squares method in minimizing the functional

$$\Phi(a, b) = \sum_{k=0}^{N} \left[ s_k - \hat{S}(t_0 + kd) \right]^2 .$$

The function $\hat{S}(t)$ obtained in this way is the required signal.

4) If the average $B$ of the almost-periodic signal is known *a priori*, the above algorithm for finding its *harmonic* component $f = \hat{S} - B$ simplifies somewhat. Applying it to the sequence $\{f_i = s_i - B\}_0^N$, $N \geqslant 3n-1$, we take $p_{2n+1} = q_{2n+1} = 0$.

Thus, the generalized Prony operator is constructed (see the notation in Theorem 2.5.5):

$$\mathcal{A}^* : M_N \to S_n(d) \setminus S_{n-1}(d), \quad \mathcal{A}^*(s) = \hat{S}, \quad M_N \subset \mathbb{R}^{N+1}.$$

We equip $S_n(d)$ with the norm usually employed in the study of almost-periodic functions: $\|\hat{S}\| = \sup_{t \in \mathbb{R}} |\hat{S}(t)|$. Theorems 2.2.4′, 2.5.2, 2.5.4, and 2.5.5 imply (cf. T. 2.3.2) the following statement.

**Theorem 2.5.6.** *The operator* $\mathcal{A}^*$ *is continuous in the topology generated by the above norm (the Euclidean topology is used in* $\mathbb{R}^{N+1}$*).*

So, the generalized Prony algorithm is well posed, because it is stable under sufficiently small oscillations of the initial sampling of uniform readings $s = \{s_i\}_0^N$.

## Chapter 3

# Indicator Diagram of an Entire Function of One Variable with Nonnegative Indicator

The idea to construct a plane indicator diagram of an entire function $f$ of order $\rho > 0$, $\rho \neq 1$, of normal type, and with nonnegative indicator $h$ is as follows. By Theorem 1.9.25, the cyclic group of rotations about the origin (the point 0*) in $\mathbb{C}$) with the generator $e^{i2\pi\rho}$ operates on the locally convex curve $\Gamma_h$ (see Def. 1.9.25) associated with $h$. If $h(\varphi) \geqslant 0$, $\varphi \in \mathbb{R}$, then the support straight line $l_\varphi$ to $\Gamma_h$ (see 1.9.16) rotates about 0 as $\varphi$ runs through the interval $(-\infty, \infty)$. If $\rho$ is an integer, then $\Gamma_h$ is a closed curve in $\mathbb{C}$, and in one circuit along it the straight line $l_\varphi$ makes $\rho$ complete turns about 0 in $\mathbb{C}$. If, moreover, $0 < h(\varphi) < \infty$, then $\Gamma_h$ is the image under the mapping $\alpha = \zeta^\rho$ of a closed Jordan curve in $\mathbb{C}$ that bounds a so-called *$\rho$-convex compact set.* This compact set may be viewed as the indicator diagram of $f$. Properties of $\rho$-convex sets are considered in detail in 3.1. The geometry of $\rho$-convex sets is in many ways similar to the geometry of convex sets. For instance, since $\rho$-convex sets are star-like, it is possible to give their analytic discription by an analog of the Minkowski functional.

In 2.2 we consider the Borel $\rho$-transformation determined by the operator $B : [\rho, \infty) \to H_0(\infty)$, where $[\rho, \infty)$ is the space of entire functions of order $\rho$ and of normal type, $H_0(\infty)$ is the space of functions holomorphic in a neighborhood of $\infty$ in $\mathbb{C}$ and equal to 0 at the point $\infty$ (see (1.10.7), (1.10.8)),

$$f(z) = \sum_{k=0}^{\infty} \frac{b_k z^k}{\Gamma[(k+1)/\rho]}, \quad [Bf](\zeta) = \sum_{k=0}^{\infty} \frac{b_k}{\zeta^{k+1}}. \qquad (3.0.1)$$

An analog of the Polya-Martineau-Ehrenpreis theorem in $\mathbb{C}$ (see the Introduction, T. $A_1$ and T. $A_2$) is based on this notion and on the notion of the plane indicator diagram of an entire function with nonnegative indicator. In 3.3 we consider a different form of the analog of the Polya Theorem $A_1$ and apply it to the description of the generalized Borel polygon for a system of integral summation methods for power series.

---

*) This symbol also denotes the number "zero".

## 3.1. Plane $\rho$-convex sets and the indicator diagram

We pass to the geometric properties of certain subsets of $\mathbb{C}$. The class of these sets contains the indicator diagrams of entire functions of order $\rho > 0$ with nonnegative indicator.

### 3.1.1. "Intrinsic" definition of $\rho$-convex sets and their simplest properties.

**Definition 3.1.1.** Let $\rho > 0$, and let $\gamma_\rho = \min\{\pi, \pi/2\rho\}$. We put $\Delta(\theta) = \{re^{i\psi} \in \mathbb{C} : r \geqslant 0,\ |\psi + \theta| \leqslant \gamma_\rho\}$, where $\theta \in \mathbb{R}$. A subset $E$ of $\mathbb{C}$ is said to be *$\rho$-convex* if 1) $0 \in E$, 2) the image of the set $E \cap \Delta(\theta)$ under the mapping $F : \Delta(\theta) \to \mathbb{C},\ F(re^{i\psi}) = r^\rho e^{i\rho\psi}$ is convex for all $\theta \in \mathbb{R}$.

**Remark.** In this definition, it suffices to require that condition 2) be fulfilled for all $\theta$ in some segment of length $2\pi$.

We give an equivalent definition of a $\rho$-convex set. Let $L = (0, \infty) \times \mathbb{R}$ be the Riemann surface of the logarithm, where the complex structure is determined by the local homeomorphism

$$\pi : L \to \mathbb{C}, \quad \pi(r, \theta) = re^{i\theta}. \tag{3.1.1}$$

For every pair of points $Z_j = (r_j, \theta_j) \in L,\ j = 1, 2$, such that

$$0 \leqslant \theta_2 - \theta_1 \leqslant 2\gamma_\rho \text{ and } \gamma_\rho = \min\{\pi, \pi/2\rho\} \tag{3.1.2}$$

we introduce a path connecting these points (cf. (1.9.4)):

$$\Gamma_\rho(Z_1, Z_2) = \{Z \in \Delta :\ Z^\rho = (1 - \lambda)Z_1^\rho + \lambda Z_2^\rho \text{ and } 0 \leqslant \lambda \leqslant 1\}. \tag{3.1.3}$$

Here $\Delta = \Delta(\theta_1, \theta_2) = \{Z \in L : \theta_1 \leqslant \operatorname{Arg} Z \leqslant \theta_2\}$.

**Definition 3.1.1$'$.** A set $E$ in $\mathbb{C}$ is said to be *$\rho$-convex* if the following conditions are fulfilled: 1) $0 \in E$; 2) if $z \in E \setminus \{0\}$, then the segment $[0, z]$ lies in $E$; if $z_j := \pi(Z_j) \in E,\ j = 1, 2$, then $T_\rho := \pi[\Gamma_\rho(Z_1, Z_2)] \subset E$ for every pair of points $Z_1, Z_2 \in L$ satisfying (3.1.2).

**Remarks.** 1. In the notation of Definition 3.1.1$'$, for $\rho \geqslant 1/2$ and $\theta_2 - \theta_1 = \pi/\rho$, the set $T_\rho = [z_1, 0) \cup (0, z_2]$ is a union of half-intervals with common end at the point 0.

2. Let $2(\pi - \gamma_\rho) < \theta_2 - \theta_1 < 2\gamma_\rho$ (cf. (3.1.2)), and let $Z_1' = (r_1, \theta_1 + 2\pi)$. Then the curve $S_\rho := \pi[\Gamma(Z_1', Z_2)]$ is included in $E$. Thus, in this case the set $E$ together with the points $z_1$ and $z_2$ ($z_1 \neq z_2$) also includes the closed curve $S_\rho \cup T_\rho$ passing through these points. For $\rho \leqslant 1/2$ the latter property is always true, and for $\rho \geqslant 1$ it is not valid because $\gamma_\rho \leqslant \pi - \gamma_\rho$ in this case.

3. For $\rho < 1/2$ the curve $\Gamma_\rho(Z_1, Z_2)$ (see (3.1.3)) is defined, in particular, if $\theta_2 - \theta_1 = 2\pi < \pi/\rho$. The assumption of Definition 3.1.1$'$ implies that if $z := \pi(Z_1) = \pi(Z_2) \in E$, then the $\rho$-convex set $E$ together with the point $z \neq 0$ also contains the closed curve $T_\rho$, enclosing 0.

We mention several properties of $\rho$-convex sets.

**Property 3.1.2.** A) *The intersection of a nonempty family of $\rho$-convex sets is $\rho$-convex; b) a $\rho$-convex set is star-like; c) the closure of a $\rho$-convex set is $\rho$-convex; d) the interior of a $\rho$-convex set is nonempty and contains* 0 *for $\rho < 1/2$ if $E \neq \{0\}$; e) if $G$ is an open $\rho$-convex set, then int $\bar{G} = G$; f) if $E$ is a $\rho$-convex set such that $0 \in \overset{\circ}{E}$, then $\bar{\overset{\circ}{E}} = \bar{E}$.*

For $\rho \neq 1$, a $\rho$-convex set can also be described by an analog of the Minkowski functional.

**Definition 3.1.3** (cf. Def. 1.3.13). Let $E$ be a star-like set in $\mathbb{C}$, and let $\rho > 0$. The function

$$p_E(z;\rho) = \inf\{\alpha^\rho > 0 : z\alpha^{-1} \in E\}, \quad z \in E \tag{3.1.4}$$

is called the *Minkowski $\rho$-functional of $E$.*

**Definition 3.1.4.** A function $p : \mathbb{C} \to [0,\infty]$ is said to be *$\rho$-sublinear* if $p(0) = 0$, the function $P \circ \pi$ (see (3.1.1)) is $\rho$-sublinear on $L$ in the sense of Definition 1.9.3, and the values $\theta_1, \theta_2$ occurring in formulas (1.9.4) and (1.9.5) satisfy condition (3.1.2).

Each $\rho$-sublinear function on $\mathbb{C}$ is generated by a nonnegative $\rho$-sublinear function on $L$ admitting infinite values and $2\pi$-periodic in the "angular" coordinate $\theta \in \mathbb{R}$. The following theorem resembles Theorems 1.3.15 and 1.3.16.

**Theorem 3.1.5.** *A function $p\colon \mathbb{C} \to [0,\infty]$ is the Minkowski $\rho$-functional of some closed $\rho$-convex set $E$ if and only if $p$ a lower semicontinuous $\rho$-sublinear function on $\mathbb{C}$ (and $E = \{z \in \mathbb{C} : p(z) \leqslant 1\}$ in the latter case).*

◀ *The "only if" part.* In the notation of Definition 3.1.1, $M_\theta := F[E \cap \Delta(\theta)]$ is a closed convex set in $\mathbb{C}$ containing 0 for every $\theta \in \mathbb{R}$. Let

$$L_\theta = \{te^{i\varphi} : t \geqslant 0;\ |\varphi + \rho\theta| \leqslant \rho\gamma_\rho\},\ q_\theta(te^{i\varphi}) = p(t^{1/\rho}e^{i\varphi/\rho}), te^{i\varphi} \in L_\theta.$$

Since $p$ is the Minkowski $\rho$-functional of the set $E$, we see that $q_\theta$ is the Minkowski functional of the set $M_\theta$. By Theorem 1.3.16, $q_\theta$ is a sublinear function on $L_\theta$. Therefore, $p$ is a $\rho$-sublinear function on $\mathbb{C}$ (see P. 1.9.4). Finally, Theorem 1.3.15 implies that $q_\theta$ is lower semicontinuous on $L_\theta$ and

$$M_\theta = \{w \in L_\theta : q_\theta(w) \leqslant 1\}. \text{ But } p(re^{i\psi}) = q_\theta(r^\rho e^{i\rho\psi}),\ re^{i\psi} \in \Delta(\theta).$$

The "if" part is proved by reversing the arguments ▶

**Theorem 3.1.6.** *Let $\mathfrak{B}_\rho$ be the class of $\rho$-convex sets, where $\rho > 0$ is an arbitrary fixed number, and let $\mathfrak{B} = \bigcap \mathfrak{B}_\rho | \rho > 0$. Then*

1) *if $\rho < \delta$, then $\mathfrak{B}_\rho \subset \mathfrak{B}_\delta$; in particular, if $\rho < 1$ and $E$ is a $\rho$-convex set in $\mathbb{C}$, then $E$ is a convex set in $\mathbb{C}$ containing* 0;

2) *$\mathfrak{B} = \mathfrak{P}$, where $\mathfrak{P} \setminus \{0\}$ is the class of all convex sets $C$ such that the interior of $C$ is a disk centered at* 0.

◀ 1. Let $E \in \mathfrak{B}_\rho$, and let $Z_j = (r_j, \theta_j)$, $j = 1,2$, be any pair of points of $L$ such that $0 < \theta_2 - \theta_1 < \min\{2\pi, \pi/\delta\}$, $z_j := \pi(Z_j) \in E$, $j = 1,2$. By

Definition 3.1.1′, using its notation and taking into account that $E$ is star-like, we conclude that, if $T_\delta \subset E$, then assertion 1) is valid. Since $\rho < \delta$ and $0 < \rho(\theta_2 - \theta_1) < \pi$, there exists a unique number $c \in \mathbb{C}\setminus\{0\}$ with the property Re $cZ_j^\rho = 1$, $j = 1, 2$. Therefore, the curve $\Gamma_\rho(Z_1, Z_2)$ in (3.1.3) admits an equivalent description:

$$\Gamma_\rho(Z_1, Z_2) = \{Z \in \Delta(\theta_1, \theta_2) : \mathrm{Re}\, cZ^\rho = 1\}. \tag{3.1.5}$$

By Definition 3.1.1′, we have (see (3.1.1)):

$$K := \{z = \pi(Z) \in \mathbb{C} : Z \in \Delta(\theta_1, \theta_2);\ \mathrm{Re}\, cZ^\rho \leqslant 1\} \cap \{0\} \subset E. \tag{3.1.6}$$

We calculate the Minkowski $\delta$-functional of the compact set $K$:

$$p_K[\pi(Z); \delta] = (\mathrm{Re}\, cZ^\rho)^{\delta/\rho}, \quad Z \in \Delta(\theta_1, \theta_2), \quad p_K(0; \delta) = 0,$$
$$p_K(z; \delta) = \infty, \quad z \in \mathbb{C} \setminus (\Pi \cup \{0\});\ \Pi = \pi[\Delta(\theta_1, \theta_2)].$$

Let $h_\delta(\theta) := p_K(e^{i\theta}; \delta)$ for $\theta \in [\theta_1, \theta_2]$. Direct calculations yield the formula

$$[\Delta_\delta h_\delta](\theta) := h_\delta''(\theta) + \delta^2 h_\delta(\theta) = \delta(\delta - \rho)|c|^2\ (\mathrm{Re}\, ce^{i\rho\theta})^{\delta/\rho - 2}, \ \theta \in (\theta_1, \theta_2).$$

Since Re $ce^{i\rho\theta} > 0\ \forall\ \theta \in [\theta_1, \theta_2]$ (see (3.1.5)), we have $[\Delta_\delta h_\delta](\theta) > 0\ \forall\ \theta \in (\theta_1, \theta_2)$. By Theorem 1.9.17, $h_\delta$ is a trigonometrically $\delta$-convex function on $(\theta_1, \theta_2)$. Therefore, Proposition 1.9.4 implies that $r^\delta h_\delta(\theta) = p_K(\pi(Z); \delta)$ is a continuous $\delta$-sublinear function on $\Delta(\theta_1, \theta_2)$ and, consequently, $p_K$ is a $\delta$-sublinear function on $\mathbb{C}$ in the sense of Definition 3.1.4. By Theorem 3.1.5, $K$ is a $\delta$-convex set. Hence (see condition 2 of Definition 3.1.1′), $T_\delta \subset K$. But $K \subset E$. Thus, $E \subset \mathfrak{B}_\delta$.

2. We pass to the proof of assertion 2). The inclusion $\mathfrak{P} \subset \mathfrak{B}$ is verified directly (see Definition 3.1.1). Let $E \in \mathfrak{B} \setminus \{0\}$ and let $w \in E \setminus \{0\}$. Assume that $\theta \in \mathrm{Arg}\, w$, and put $r = |w|$. We denote $Z_1 = (r, \theta)$ and $Z_2 = (r, \theta + 2\pi)$. Assertion 1) implies that $\mathfrak{B} = \bigcap \mathfrak{B}_\rho | 0 < \rho < 1/2$. By Definition 3.1.1′, for every $\rho \in (0, 1/2)$ the set $E$ includes the closed curve $T_\rho = \pi[\Gamma_\rho(Z_1, Z_2)]$ and the $\rho$-convex compact set (see (3.1.6), Definition 3.1.1)

$$K_\rho = \{z = te^{i\varphi} \in \mathbb{C} :\ \theta \leqslant \varphi \leqslant \theta + 2\pi,\ \mathrm{Re}\ c_\rho t^\rho e^{i\rho\varphi} \leqslant 1\} \tag{3.1.7}$$

bounded by it. Here the constant

$$c_\rho = [r^\rho \cos\pi\rho]^{-1} \cdot e^{-i\rho(\theta + \pi)} \tag{3.1.8}$$

is determined by the conditions $r^\rho \cdot \mathrm{Re}\ c_\rho e^{i\rho\theta} = 1$, $r^\rho \cdot \mathrm{Re}\ c_\rho e^{i\rho(\theta+\pi)} = 1$. Assertion 1) implies that $K_\rho \subset \mathfrak{B}_\delta$ for $\rho < \delta < 1/2$, i.e., $K_\delta \subset K_\rho$, where $K_\rho$ and $K_\delta$ are compact convex sets. By Theorem 1.3.15, we have $p_\rho \leqslant p_\delta$, where $p_\alpha$ is the Minkowski functional of the set $K_\alpha$ (see (3.1.7)), $\alpha \in (0, 1/2)$. Hence (see (3.1.7), (3.1.8)),

$$p_\rho(z) = p_\rho(te^{i\varphi}) = t(\mathrm{Re}\ c_\rho e^{i\rho\varphi})^{1/\rho} = \frac{t}{r}\left[\frac{\cos\rho(\varphi - \theta - \pi)}{\cos\pi\rho}\right]^{1/\rho}$$

is a monotone nondecreasing function of $\rho$ on $(0, 1/2)$ for an arbitrary fixed $z = te^{i\varphi} \in \mathbb{C}$. By Proposition 1.3.14 (see 4)), we see that the function

$$p(z) = \inf\{p_\rho(z) : \ 0 < \rho < 1/2\} = \lim_{\rho \to 0} p_\rho(z) = |z|/|w|$$

is the Minkowski functional of the convex set $K = \bigcup K_\rho | 0 < \rho < 1/2 \subset E$. Therefore, together with every point $w \neq 0$, the set $E$ also contains the disk $\overset{\circ}{K} = \{z \in \mathbb{C} : |z| < |w|\}$. Thus, $\mathfrak{B} \subset \mathfrak{P}$ ►

The following assertion is entirely similar to the case of convex sets and is based on Theorem 3.1.5 and the method of the proof of Theorem 1.3.17.

**Property 3.1.7.** *Let $E$ be a closed $\rho$-convex set in $\mathbb{C}$ such that $0 \in \overset{\circ}{E}$, $E \neq \mathbb{C}$. We denote by $p_E$ the Minkowski $\rho$-functional of $E$. Then*

$$\overset{\circ}{E} = \{z \in \mathbb{C} : p_E(z) < 1\}, \quad \partial E = \{z \in \mathbb{C} : p_E(z) = 1\}.$$

Not all properties of convex sets can be carried over to the $\rho$-convex sets.

**Example.** Let $E = [-1, 0] \cup F$, where $F = \{re^{i\theta} \in \mathbb{C}\colon\ 0 \leqslant r \leqslant 1, |\theta| \leqslant \pi/(2\rho + 1)\}$, $\rho \geqslant 2$. Then $E$ is a closed $\rho$-convex set, but $\overset{\circ}{E}$ is not $\rho$-convex, because $0 \notin \overset{\circ}{E}$. Moreover, $\overline{\overset{\circ}{E}} \neq E$. So, in the formulation of Property 3.1.2 condition f) $0 \in \overset{\circ}{E}$ is essential.

**3.1.2. Support properties and an "extrinsic" definition of $\rho$-convex sets.** Let $\gamma_\rho = \min\{\pi, \pi/2\rho\}$. We put

$$\Phi_\rho(z) = \begin{cases} |z|^\rho \cos(\rho \operatorname{Arg} z), & |\operatorname{Arg} z| \leqslant \gamma_\rho; \\ 0, \qquad \rho \geqslant 1/2; & \gamma_\rho \leqslant |\operatorname{Arg} z| \leqslant \pi. \end{cases} \tag{3.1.9}$$

The following analog of the Hahn-Banach theorem for $\rho$-sublinear functions is important in the study of support properties of $\rho$-convex sets.

**Theorem 3.1.8.** *Let $p$ be a finite nonnegative $\rho$-sublinear function on $\mathbb{C}$ such that $p(z) \not\equiv 0$. For every point $z_0 \in \mathbb{C}$ there exists a point $w \in \mathbb{C} \setminus \{0\}$ with*

$$\Phi_\rho(z_0 w) = p(z_0), \ \Phi_\rho(zw) \leqslant p(z) \quad \forall\, z \in \mathbb{C}.$$

◄ 1. Let $h(\theta) = p(e^{i\theta})$, $\theta \in \mathbb{R}$. Take $z_0 = te^{i\psi}$ with $p(z_0) > 0$. By Proposition 1.9.4, we conclude that $h$ is a finite function on $P_\rho$ (see Notation). By Theorem 1.9.21, there exists a number $c \in \mathbb{C}$ such that

$$\operatorname{Re} ce^{i\rho\theta} \leqslant h(\theta), \ \theta \in [\psi - \pi/\rho, \ \psi + \pi/\rho], \ \operatorname{Re} ce^{i\rho\psi} = h(\psi). \tag{3.1.10}$$

Since $0 < p(z_0) = t^\rho \cdot h(\psi)$, we have $h(\psi) > 0$ and, consequently, $c \neq 0$. Therefore, for $\rho \geqslant 1/2$ we may assume that $c = |c|e^{i\rho\varphi}$, where $|\varphi + \psi| < \pi/2\rho \leqslant \pi$. We define a function $l_0$ by the conditions

$$l_0(\theta) = \begin{cases} \operatorname{Re} ce^{i\rho\theta}, & |\theta + \varphi| \leqslant \pi/(2\rho) \\ 0, & \pi/(2\rho) \leqslant |\theta + \varphi| \leqslant \pi, \end{cases}$$

and denote by $l$ the $2\pi$-periodic extension of $l_0$ to $\mathbb{R}$. Since $h \geqslant 0$, by (3.1.10) we have $l \leqslant h$.

2. Let $\rho < 1/2$. Since $ce^{i2\pi\rho} \in Dh(\psi - 2\pi)$ (see T. 1.9.25), by analogy with (3.1.10) we find:

$$h(\theta) \geqslant \operatorname{Re} ce^{i(\theta+2\pi)\rho}, \ |\theta - \psi + 2\pi| \leqslant \pi/\rho.$$

Combining this with (3.1.10), for $\rho < 1/2$ we obtain:

$$h(\theta) \geqslant \max\{\operatorname{Re} ce^{i\rho\theta}, \ \operatorname{Re} ce^{i\rho(\theta+2\pi)}\}, \ \theta \in [\psi - 2\pi, \psi]. \tag{3.1.11}$$

In a similar way, we verify the following unequality:

$$h(\theta) \geqslant \max\{\operatorname{Re} ce^{i\rho\theta}, \operatorname{Re} ce^{i\rho(\theta-2\pi)}\}, \ \theta \in [\psi, \psi + 2\pi]. \tag{3.1.12}$$

But $h$ is a $2\pi$-periodic function, therefore, by (3.1.11) and (3.1.12), we find:

$$h(\psi - 2\pi) = h(\psi + 2\pi) = h(\psi) = \operatorname{Re} ce^{i\rho\psi} \geqslant$$
$$\max\{\operatorname{Re} ce^{i\rho(\psi+2\pi)}, \ \operatorname{Re} ce^{i\rho(\psi-2\pi)}\}.$$

Now, elementary transformations yield $c = |c|e^{i\rho\varphi}$, where $|\varphi+\psi| \leqslant \pi$. We note that $[-\varphi - \pi, -\varphi + \pi] \subset [\psi - \pi/, \ \psi + \pi/\rho]$. Combining this with (3.1.10) and recalling that $h$ is a $2\pi$-periodic function, we see that the $2\pi$-periodic extension $l$ to $\mathbb{R}$ of the function $l_0(\theta) := \operatorname{Re} ce^{i\rho\theta}$, $\theta \in [-\varphi - \pi, -\varphi + \pi]$, does not exceed $h$.

3. Taking $w = |c|^{1/\rho} \cdot e^{i\varphi}$, by items 1, 2 of the proof, we verify the theorem in the case where $p(z_0) > 0$. If $p(z_0) = 0$, then, under the assumptions of the theorem, there exists a point $z_1 \in \mathbb{C}$ such that $p(z_1) > 0$. The statements already proved imply the existence of a number $w_1 \in \mathbb{C} \setminus \{0\}$ such that $\Phi_\rho(zw_1) \leqslant p(z)$ $\forall\ z \in \mathbb{C}$. In particular, $0 \leqslant \Phi_\rho(z_0w_1) \leqslant p(z_0) = 0$, i.e., $\Phi_\rho(z_0w_1) = p(z_0)$ ▶

The following assertion describes the class of support curves in the case of $\rho$-convex set.

**Theorem 3.1.9.** *Let $E$ be a closed $\rho$-convex set with $E \neq \mathbb{C}$ and $0 \in \overset{\circ}{E}$. For every point $z_0 \in \partial E$ there exists a point $w \in \mathbb{C} \setminus \{0\}$ such that the curve $\Gamma = \{z \in \mathbb{C} : \Phi_\rho(zw) = 1\}$* (see (3.1.9)) *contains $z_0$ and does not intersect $\overset{\circ}{E}$, moreover, $\Phi_\rho(zw) \leqslant 1$ for all $z \in E$.*

◀ By the assumptions of this theorem and by Theorem 3.1.5, we see that the Minkowski $\rho$-functional $p_E$ of $E$ satisfies the assumptions of Theorem 3.1.8. Applying Property 3.1.7, we verify our claim ▶

The following analog of the strict separation theorem for convex sets is valid for $\rho$-convex sets.

**Theorem 3.1.10.** *Let $E$ be a closed $\rho$-convex set with $E \neq \mathbb{C}$, and let $z_0 \in \mathbb{C} \setminus E$. Then there exists a point $w \in \mathbb{C} \setminus \{0\}$ such that* (see (3.1.9)) $\Phi_\rho(z_0w) > \sup\{\Phi_\rho(zw) : z \in E\}$.

◀ We consider a collection $\{F_\varepsilon,\ \varepsilon > 0\}$ of closed $\rho$-convex sets with the following properties: a) $F_{\varepsilon_1} \subset F_{\varepsilon_2}$ for $0 < \varepsilon_1 < \varepsilon_2$; b) $0 \in \overset{\circ}{F_\varepsilon}\ \ \forall\, \varepsilon > 0$; c) $\cap F_\varepsilon\ |\varepsilon > 0 = E$ (for example, $F_\varepsilon$ may be defined as the closed $\rho$-convex hull of the set $E \cup S(\varepsilon)$, where $S(\varepsilon) = \{z \in \mathbb{C} : |z| \leqslant \varepsilon\}^{*)}$ ). There exists a number $\varepsilon_0 > 0$ and a set $B = F_{\varepsilon_0}$ such that $z_0 \in \mathbb{C} \setminus B$. Since $0 \in \overset{\circ}{B}$ and by Property 3.1.7, for the Minkowski $\rho$-functional $p = p_B$ we obtain $\partial B = \{z \in \mathbb{C} :\ p(z) = 1\}$. Since $z_0 \in \mathbb{C} \setminus F_{\varepsilon_0}$, we conclude that $p(z_0) > 1$ (see T. 3.1.5). Let $\lambda_0 = [p(z_0)]^{-1/\rho}$. Then $\zeta := \lambda_0 z_0 \in \partial B$ and, by Theorem 3.1.9, the support curve $\Gamma = \{z \in \mathbb{C} :\ \Phi_\rho(zw) = 1\}$ passes through $\zeta$, where $w$ is some point in $\mathbb{C} \setminus \{0\}$; moreover, $\Phi_\rho(zw) \leqslant 1\ \forall\, z \in E \subset B$. Therefore,

$$\Phi_\rho(z_0 w) = \Phi_\rho(\zeta w) \cdot \lambda_0^{-\rho} = p(z_0) > 1 \geqslant \sup_{z \in E} \Phi_\rho(zw) \ \blacktriangleright$$

The support properties of $\rho$-convex sets allow us to prove the equivalence of Definition 3.1.1 to the following "extrinsic" definition of a $\rho$-convex set.

**Definition 3.1.11.** Suppose $w \in \mathbb{C}$ and $\nu \in [0, \infty)$. The set $F_\rho(w;\nu) = \{z \in \mathbb{C} : \Phi_\rho(zw) \leqslant \nu\}$ is said to be *elementary $\rho$-convex.* A closed set representable as the intersection of an arbitrary family of elementary $\rho$-convex sets is called a *$\rho$-convex set.* Each of the functions $H(w) := \sup\{\Phi_\rho(zw) : z \in E\}$, $w \in \mathbb{C}$, and $h(\theta) := H(e^{i\theta})$, $\theta \in \mathbb{R}$, is called a *$\rho$-support function* of the given plane set $E$.

The set $F_\rho(w;\nu)$ is $\rho$-convex in the sense of Definition 3.1.1: if $\rho \geqslant 1/2$, $\nu = 0$, then $F_\rho(w;\nu)$ is the exterior of the angle $\{z \in \mathbb{C} :\ |\mathrm{Arg}\ \ w| < \pi/2\rho\}$; if $\rho < 1/2$, $\nu = 0$, then $F_\rho(w;\nu) = \varnothing$; if $\nu > 0$, then the Minkowski $\rho$-functional of the set $F_\rho(w;\nu)$ is equal to $\Phi_\rho(z\zeta)$, where $\zeta = \nu^{-1/\rho} w$, i.e., it is a $\rho$-sublinear function on $\mathbb{C}$ (see T. 3.1.5). The latter is verified directly.

**Proposition 3.1.12.** *Definitions* 3.1.1 *and* 3.1.11 *are equivalent for closed plane sets.*

◀ Let $H$ be the $\rho$-support function of $E$ (see Def. 3.1.11). If $E$ is closed and $\rho$-convex in the sense of Definition 3.1.1, then Theorem 3.1.10 implies the following formula:

$$E = \{z \in \mathbb{C} : \Phi_\rho(zw) \leqslant H(w)\ \forall\, w \in \mathbb{C}\} = \bigcap F_\rho(w; H(w))|w \in \mathbb{C}. \quad (3.1.13)$$

Consequently, $E$ is $\rho$-convex in the sense of Definition 3.1.11. As was mentioned above , the converse is also true for the elementary $\rho$-convex sets. Therefore, by Property 3.1.2, a), it follows that the converse is true in the general case ▶

**3.1.3. Minkowski duality of $\rho$-sublinear functions and $\rho$-convex sets.** Let $X$ be the class of functions $f : \mathbb{C} \to [0, \infty]$, partially ordered in a natural way, i.e., $f_1 \leqslant f_2$ if $f_1(z) \leqslant f_2(z)$ for all $z \in \mathbb{C}$. The lowest upper bound sup $U$ of a set $U \subset X$ is the function $f$ in $\mathbb{C}$ equal to the pointwise supremum of

---

*) The *hull* is defined along the same lines as in the case of convex sets (see Def. 1.3.4).

the functions of the set $U$, i.e., $f(x) = \sup\{g(x) : g \in U\}$, $x \in \mathbb{C}$. Let $P, A$ be subclasses of $X$. The class $P$ is called a *supremal generator of A* if every function $h \in A$ is representable in the form $h = \sup U$, where $U$ is a subset of $P$.

**Definition 3.1.13.** For $\rho > 0$, consider the subset $H_\rho = \{\{\Phi_\rho(zw),\ z \in \mathbb{C}\} : w \in \mathbb{C}\}$ (see (3.1.9)) of $X$ depending on the parameter $w$. A function $h \in X$ is said to be *$H_\rho$-convex* if there exists a set $U \subset H_\rho$ such that $h = \sup U$. We denote the class of $H_\rho$-convex functions by $P(H_\rho, X)$.

**Lemma 3.1.14.** *Let $A_\rho$ be the subclass of $X$ consisting of lower semicontinuous nonnegative ρ-sublinear functions. The class $P(H_\rho, X)$ is a supremal generator for $A_\rho$, and if $h \in A_\rho$ and $U_h = \{p \in P(H_\rho, X) : p \leqslant h\}$, then $h = \sup U_h$.*

◀ 1. It has already been noted that for every fixed $w \in \mathbb{C}$ the function $\Phi_\rho(zw)$ is finite and ρ-sublinear and, consequently, continuous in $z$ on $\mathbb{C}$ (see P. 1.9.4, Prop. 1.9.5). Therefore, the supremum of a family of such functions is a lower semicontinuous ρ-sublinear function (see Ex. 1.2.5), i.e., $P(H_\rho, X) \subset A_\rho$.

2. For every function $h \in A_\rho$ there exists a family $\{h_\varepsilon, \varepsilon > 0\}$ of finite functions in $A_\rho$ such that $h = \sup\{h_\varepsilon : \varepsilon > 0\}$. Indeed, $E = \{z \in \mathbb{C} : h(z) \leqslant 1\}$ is a closed ρ-convex set (see T. 3.1.5). As in the proof of T. 3.1.10, we construct a family of sets $\{F_\varepsilon, \varepsilon > 0\}$ converging to $E$ as $\varepsilon \to 0$. Let $h_\varepsilon$ be the Minkowski ρ-functional of $F_\varepsilon$. By Theorem 3.1.5, $h_\varepsilon \in A_\rho\ \forall\ \varepsilon > 0$. Since $\bigcap F_\varepsilon | \varepsilon > 0 = F$, we see that $h = \sup\{h_\varepsilon : \varepsilon > 0\}$ (see P. 1.3.14, 3)), and since $0 \in \overset{\circ}{F}_\varepsilon$, we see that $h_\varepsilon$ is a finite function.

3. We fix arbitrarily $\delta > 0$ and $z_0 \in C$. Put $M = h(z_0) - \delta$ if $h(z_0) < \infty$, and $M = \delta$ if $h(z_0) = \infty$. There exists $\varepsilon_0 > 0$ such that $M < h_{\varepsilon_0}(z_0) \leqslant h(z_0)$. By Theorem 3.1.8, there exists a function $\Phi_\rho(zw_0) \in H_\rho$ with the properties

$$\Phi_\rho(z_0 w_0) = h_{\varepsilon_0}(z_0),\ \Phi_\rho(zw_0) \leqslant h_{\varepsilon_0}(z) \leqslant h(z)\ \forall\ z \in \mathbb{C}.$$

So, $U_h = \{p \in H_\rho : p \leqslant h\}$ is the required set: $h = \sup U_h$ ▶

**Theorem 3.1.15.** *Let $\mathfrak{B}_\rho$ be the class of closed ρ-convex sets, partially ordered by inclusion. The Minkowski duality $\nu : A_\rho \to \mathfrak{B}_\rho$, $\nu(H) = E_H$, where*

$$E_H = \{z \in \mathbb{C} : \Phi_\rho(zw) \leqslant H(w) \quad \forall\ w \in \mathbb{C}\}, \tag{3.1.14}$$

*establishes isomorphism of the classes $A_\rho$ and $\mathfrak{B}_\rho$* (see 3.1.3, T. 1.3.15, P. 3.1.12, L. 3.1.14). *Moreover, $H := \nu^{-1}(E)$ is the ρ-support function of the set $E \in \mathfrak{B}_\rho$. Furthermore, if $A'_\rho$ is the subclass of finite functions in $A_\rho$, $\mathfrak{B}'_\rho$ is the subclass of compact sets in $\mathfrak{B}_\rho$, and $\nu' = \nu | A'_\rho$, then $\nu' : A'_\rho \to \mathfrak{B}'_\rho$ is an isomorphism.*

◀ First, $E_H \in \mathfrak{B}_\rho\ \forall\ H \in A_\rho$ (see Def. 3.1.11, P. 3.1.12). By Lemma 3.1.14, $H$ is the ρ-support function of $E_H\ \forall\ H \in A_\rho$ (see (3.1.14)). Therefore, $\nu$ is a monomorphism. On the other hand, if $E \in \mathfrak{B}_\rho$, then $E = E_H$, where $H$ is the ρ-support function of $E$ (see 3.1.13)). But $H \in A_\rho$ by Lemma 3.1.14. Consequently, $\nu$ is an epimorphism. So, $\nu$ is an isomorphism.

Now, let $H \in A'_\rho$, and let $h(\theta) = H(e^{i\theta})$ for $\theta \in \mathbb{R}$. Since $h$ is a continuous $2\pi$-periodic function (see P. 1.9.4, Prop. 1.9.5), we see that $m := \sup\{h(\theta) : \theta \in \mathbb{R}\} < \infty$. Consequently, $H(w) \leqslant |w|^\rho \ \forall\ w \in \mathbb{C}$. Therefore, $E_H \subset \{z \in \mathbb{C} : |z|^\rho \leqslant m\}$, i.e., $E_H = \nu'(H)$ is a compact set. On the other hand, if $E \in \mathfrak{B}'_\rho$, then $H := [\nu']^{-1}(E)$ is the $\rho$-support function of the compact set $E$, i.e., $H \in A'_\rho$ ▶

The following assertion is a consequence of the Minkowski duality.

**Proposition 3.1.16.** *Let $S$ be a nonempty set in $\mathbb{C}$, and let $H$ be its $\rho$-support function. Then* (see (3.1.14)) *$E_H$ is the closed $\rho$-convex hull of $S$, i.e., the intersection of all closed $\rho$-convex sets containing $S$.*

◀ In the notation of Theorem 3.1.15, let $E \subset \mathfrak{B}_\rho$, $S \subset E$, and let $H_E$ be the $\rho$-support function of $E$. Then, by Lemma 3.1.14, $H$, $H_E \in A_\rho$. By Theorem 3.1.15, the inequality $H \leqslant H_E$ implies $E_H \subset E$. On the other hand, it is obvious that $S \subset E_H$ ▶

**3.1.4. Plane indicator diagram of an entire function of order $\rho > 0$ and with nonnegative indicator.** Let $f$ be an entire function of order $\rho > 0$ and of normal type with indicator $h_f$ (see (1.10.3) ).

**Definition 3.1.17.** Let $h(\theta) = \max\{0, h_f(\theta)\}$ for $\theta \in \mathbb{R}$, and let $\Phi_\rho$ be the function defined by formula (3.1.9). The plane set

$$X_h = \{z \in \mathbb{C} : \Phi_\rho(ze^{i\theta}) \leqslant h(\theta) \quad \forall\ \theta \in \mathbb{R}\}. \tag{3.1.15}$$

is called the *indicator diagram* of $f$.

Let $H(z) = r^\rho h(\theta)$, where $z = re^{i\theta} \in \mathbb{C}$. Recalling Theorem 1.10.2, Proposition 1.9.4, Definition 3.1.4, and Theorem 3.1.15, in the notation of Theorem 3.1.15 we have: $H \in A'_\rho$, $X_h = E_H$.

Therefore, the following assertion is valid.

**Proposition 3.1.18.** *The indicator diagram $X_h$ is a $\rho$-convex compact set the $\rho$-support function of which coincides with $h$* (see Def. 3.1.11).

**Remark.** If $X_h$ is a closed $\rho$-convex domain (which is true if $h(\theta) > 0 \ \forall\ \theta \in \mathbb{R}$), then the boundary of $X_h$ has the properties mentioned above in the introduction to Ch. 3. If $\rho = 1$ and $h_f(\theta) \geqslant 0$, then Definition 3.1.1 differs from the usual definition of the indicator diagram $I_h$ (see the Introduction and 3.1.10) only by the fact that $X_h$ is the reflection of $I_h$ in the real axis.

## 3.2. Analog of the Polya theorem for an entire function of order $\rho \neq 1$ and with nonnegative indicator

We shall present an analog of the classical version of the G. Polya Theorem (see the Introduction, T. $A_1$) for entire functions of order $\rho > 0$ and of normal type with nonnegative indicator. Let $\rho > 0$, and let $[\rho, \infty) = \{f\}$ be the class of entire functions $f$ on $\mathbb{C}$ that satisfy the inequality $|f(z)| \leqslant M \exp\{m|z|^\rho\} \ \forall\ z \in$

$\mathbb{C}$ for some $M = M(f)$ and $m = m(f)$. We formulate the following analog of Proposition 1.10.4.

**Proposition 3.2.1.** *The operator* $B$ (see (3.0.1), the Borel $\rho$-transformation) *establishes an isomorphism* $B: \; [\rho, \infty) \to H_0(\infty)$, *and if* $f \in [\rho, \infty)$ *and* $\sigma = \varlimsup_{r \to \infty} r^{-\rho} \cdot \ln M_f(r)$, *then* $\{z \in \mathbb{C} : \; |z|^\rho \leqslant \sigma\}$ *is the smallest disk with center at zero outside of which the function* $\hat{f} = Bf$ *is holomorphic.*

The proof of Proposition 3.2.1 is based on Theorem 1.10.3, and follows the outline of the proof of Proposition 1.10.4.

Let $P_\rho^+$ be the class of finite nonnegative $2\pi$-periodic trigonometrically $\rho$-convex functions on $\mathbb{R}$, and let $[\rho, h(\theta)) = \{f\}$ be the class of entire functions $f$ satisfying

$$|f(re^{i\theta})| < A \exp\{[h(\theta) + \delta] r^\rho\}, \quad re^{i\theta} \in \mathbb{C}, \tag{3.2.1}$$

for every $\delta > 0$ and some $A = A(f; \delta) > 0$. Then $[\rho, \infty) = \bigcup [\rho, h(\theta)] | h \in P_\rho^+$.

To formulate the basic result of this section, we need auxiliary notions.

**Definition 3.2.2.** Consider the space $H(\mathbb{C})$ of entire functions endowed with the topology of uniform convergence on compact subsets of $\mathbb{C}$, and denote by $H^*$ its dual space. The elements of $H^*$ are called *analytic functionals.* A compact set $K \subset \mathbb{C}$ is called a *determining set for a functional* $\mu$ if for every neighborhood $\omega$ of $K$ there exists a constant $C_\omega$ such that

$$|\mu(\varphi)| \leqslant C_\omega \sup_{z \in \omega} |\varphi(z)| \quad \forall \; \varphi \in H(\mathbb{C}). \tag{3.2.2}$$

A compact set $K$ is called a *$\rho$-convex support of the functional* $\mu$ if $K$ is a $\rho$-convex determining set for $\mu$ (see Def. 3.1.1), and $K$ is minimal among the sets with this property (i.e., whenever $L$ is a $\rho$-convex determining set for $\mu$ and $L \subset K$, we have $L = K$).

Now we formulate an analog of Theorem $A_1$ (see the Introduction).

**Theorem 3.2.3.** *For* $\rho > 0$, *let* $h \in P_\rho^+$, *and let* $X_h$ *be the $\rho$-convex compact set the $\rho$-support function of which* (see (3.1.15)) *is equal to* $h$. *Suppose* $f$ *is an entire function. The following statements are equivalent:*

1) $f \in [\rho, \infty)$, *and the indicator diagram* (see Def. 3.1.17) *of* $f$ *is equal to* $X_h$ (*or the nonnegative truncation* $h_f^+ = \max\{0, h_f\}$ *of the indicator* $h_f$ *coincides with* $h$);

2) $\hat{f} := Bf \in H_0(\infty)$ (see (3.0.1)), *and the smallest among the $\rho$-convex compact set* $K$ *such that* $\hat{f}$ *admits analytic continuation outside* $K$ (*i.e., the conjugate diagram of* $f$) *coincides with* $X_h$;

3) *there exists an analytic functional* $\mu$ *such that the Laplace $\rho$-transform of* $\mu$ *coincides with* $f$, *i.e.,*

$$\tilde{\mu}(z) := \mu_\omega[E_\rho(zw)] \equiv f(z), \; z \in \mathbb{C}, \tag{3.2.3}$$

*where*

$$E_\rho(u) = \sum_{k=0}^{\infty} \frac{u^k}{\Gamma[(k+1)/\rho]},$$

*and* $X_h$ *is a unique* $\rho$*-convex support of* $\mu$.

First, we prove several lemmas relating the elements of the spaces $[\rho, \infty)$, $H_0(\infty)$, and $H^*$ in various ways.

**Lemma 3.2.4.** *Let* $\psi \in H(\mathbb{C})$, *and let* $\mu \in H^*$. *Then*

$$\mu_w[\psi(zw)] = f(z), \quad z \in \mathbb{C}, \tag{3.2.4}$$

*is an entire function. Moreover, if*

$$\psi^{(k)}(0) \neq 0, \ k = 0, 1, 2 \ldots, \tag{3.2.5}$$

*then* $\mu$ *is a unique element of* $H^*$ *satisfying* (3.2.4) *for given* $\psi$ *and* $f$.

◀ Consider the function

$$B_z(\Delta, w) = \begin{cases} [\psi((z+\Delta)w) - \psi(zw)]\Delta^{-1}, & \Delta \neq 0; \\ \psi'(zw) \cdot w, & \Delta = 0. \end{cases}$$

This is an entire function in each of the variables $\Delta$ and $w$ (for any fixed value of the other variable). Therefore, by the Hartogs fundamental theorem, $B_z(\Delta, w)$ is an entire function of the variables $\Delta$, $w$ and, consequently, it is uniformly continuous on every compact subset of $\mathbb{C}^2$. Hence, $B_z(\Delta, w) \rightrightarrows \psi'(zw) \cdot w$ as $\Delta \to 0$ for $w \in K$, where $K$ is a compact subset of $\mathbb{C}$ *). Since $\mu \in H^*$, there exists a finite limit

$$\lim_{\Delta \to 0} \Delta^{-1} \cdot [f(z+\Delta) - f(z)] \ \forall \ z \in \mathbb{C},$$

i.e., $f(z)$ is an entire function.

If the second assertion of the lemma is false, then there exists a nonzero element $\nu \in H^*$ such that $\nu_w[\psi(z, w)] \equiv 0$, $z \in \mathbb{C}$. We put $\nu(w^k) = c_k$ and $b_k = (k!)^{-1} \cdot \psi^{(k)}(0)$ for $k = 0, 1, \ldots$. Then $\nu_w[\psi(zw)] = \sum_{k=0}^{\infty} c_k b_k z^k \equiv 0$, so that (3.2.5) and the uniqueness theorem show that $\nu = 0 \in H^*$, a contradiction ▶

**Lemma 3.2.5.** *If* $f \in [\rho, h(\theta)]$ (see 3.2.1), *where* $h \in P_\rho^+$, *then the Borel* $\rho$*-transform* $\hat{f} = Bf$ *of* $f$ (see (0.1)) *admits analytic continuation outside the* $\rho$*-convex compact set* $X_h$.

We observe that (see (3.1.13), (3.1.15))

$$\mathbb{C} \setminus X_h = \bigcup \mathcal{D}_\rho(\theta; h(\theta)) \mid \theta \in (-\pi, \pi],$$

$$\mathcal{D}_\rho(\theta; \nu) = \{z \in \mathbb{C} : \Phi_\rho\left(ze^{i\theta}\right) > \nu\}.$$

The lemma is a consequence of the following integral representation of the Borel $\rho$-transform, which is valid for every $\theta \in (-\pi, \pi]$ (see (0.1)):

$$\hat{f}(\zeta) = [Bf](\zeta) = \rho \int_0^{\infty(\arg z = \theta)} \exp\{-(z\zeta)^\rho\} f(z) dz. \tag{3.2.6}$$

*) Here $\rightrightarrows$ indicates uniform convergence on $K$.

The integral in (3.2.6) is a function holomorphic in $\mathcal{D}_\alpha(\theta; \langle(\theta))$. The proof of (3.2.6) can be found in [33, p. 330] (we also refer to the method of the proof of Proposition 1.10.5).

**Lemma 3.2.6.** *If $f \in [\rho, \infty)$, then there exists an unique element $\mu = \mu_f \in H^*$ such that $\tilde{\mu} = f$* (see (3.2. 3)) *and, moreover,*

$$\mu(\varphi) = \frac{1}{2\pi i} \int_{\Gamma_R} \varphi(\zeta) \hat{f}(\zeta) d\zeta, \quad \varphi \in H(\mathbb{C}), \tag{3.2.7}$$

*where $\Gamma_R = \{\zeta \in \mathbb{C} : |\zeta| = R\}$ is a positively oriented circumference in the domain of convergence of the Borel $\rho$-transform $\hat{f}$ of $f$.*

◀ We have the formula

$$f(z) = \frac{1}{2\pi i} \int_{\Gamma_R} E_\rho(z\zeta) \hat{f}(\zeta) d\zeta, \quad z \in \mathbb{C},$$

9(see (3.2.3)), which is verified directly by termwise integration of the Laurent series for the integrand. So, $\tilde{\mu} = f$ (see (3.2.7)) and, by Lemma 3.2.4, $\mu$ is the only element of $H^*$ with this property. ▶

**Lemma 3.2.7.** 1. *If $\mu \in H^*$ and $K$ is a determining set for $\mu$* (see Def. 3.2. 2), *then $f := \tilde{\mu}$* (see (3.2.3)) *is an entire function, moreover, $f \in [\rho, h_K(\theta)]$, where $h_K(\theta) = \sup\limits_{w \in K} \Phi_\rho\left(we^{i\theta}\right)$ is the $\rho$-support function of $K$.*

2. *Conversely, if $K$ is a $\rho$-convex compact set and $f \in [\rho, h(\theta)]$, where $h = h_K$, then $K = X_h$* (see 3.1.15)) *is a determining set for the functional $\mu$* (3.2.7) *associated with $f$* (see lemma 3.2.6).

◀ 1. Let $a_\varepsilon = h_K + \varepsilon$, where $\varepsilon > 0$. We have $\alpha_\varepsilon \in P_\rho^+ \ \forall\ \varepsilon > 0$. By Theorem 3.1.15 and (3.1.15), we see that $F_\varepsilon := X_{\alpha_\varepsilon}$ is a $\rho$-convex compact set for every $\varepsilon > 0$ and, moreover, a) $\bigcap\limits_{\varepsilon > 0} F_\varepsilon = K$, $F_{\varepsilon_1} \subset F_{\varepsilon_2}$ for $0 < \varepsilon_1 < \varepsilon_2$; b) $K \subset \overset{\circ}{F}_\varepsilon \ \forall\ \varepsilon > 0$. For given $\delta > 0$ there exist numbers $\varepsilon > 0$ and $\tau > 0$ such that $\varepsilon + \tau \sup\{|w|^\rho,\ w \in F_\varepsilon\} < \delta$. The function $\Phi_\rho(e^{i\theta})$ is the indicator of the Mittag-Leffler function $E_\rho$ (see (3.2.3)). Therefore, there exists a constant $C_\tau$ with the property

$$|E_\rho(re^{i\theta})| < C_\tau \cdot \exp\{r^\rho[\Phi_\rho(e^{i\theta}) + \tau]\}, \quad re^{i\theta} \in \mathbb{C}.$$

By Lemma 3.2.4, $\tilde{\mu}$ is an entire function. Using the above inequality for $E_\rho$, recalling that $K$ is a determining set for $\mu$, and putting $\omega = F_\varepsilon$ in (3.2.2), we obtain: $\tilde{\mu} \in [\rho, h_K(\theta)]$.

2. Since $f \in [\rho, h_K(\theta)]$, by Lemma 3.2.5 we see that $\hat{f} = Bf$ is a function holomorphic in $\mathbb{C} \setminus X_{h_K} = \mathbb{C} \setminus K$ (see T. 3.1.15). By the Cauchy theorem, the functional $\mu$ in (3.2.7) is transformed to

$$\mu(\varphi) = \frac{1}{2\pi i} \int_{\gamma} \varphi(\zeta) \hat{f}(\zeta) d\zeta,$$

where $\gamma = \partial F_\varepsilon$ (see item 1) above) is a positively oriented closed Jordan curve in $\mathbb{C}$ (see Prop. 3.1.7). After elementary estimates of $|\mu(\varphi)|$, applying the maximum modulus principle for analytic functions, we see that $K$ is a determining set for $\mu$ ▶

In the course of the proof of assertion 2 in Lemma 3.2.7, we have verified the following fact.

**Lemma 3.2.8.** *If the Borel $\rho$-transform $\hat{f}$ of an entire function $f \in [\rho, \infty)$ admits analytic continuation outside a $\rho$-convex compact set $K$, then $K$ is a determining set for the functional $\mu$* (see (3.2.7)) *associated with $f$.*

**Proof of Theorem 3.2.3.** 1)→3). By Theorem 3.1.15 and assertion 2 of Lemma 3.2.7, the compact set $X_h$ is a determining set for the functional $\mu$ (see (3.2.7)) associated with $f$ (see Lemma 3.2.6). Let $L$ be a $\rho$-convex determining set for $\mu$, and let $L \subset X_h$. Denote by $h_L$ the $\rho$-support function of $L$. Then $h_L \leqslant h$ and, by assertion 1 of Lemma 3.2.7, $f = \tilde{\mu} \in [\rho, h_L(\theta)]$. Consequently, $h \leqslant h_L$, where $h(\theta) = \max\{0, h_f(\theta)\}$, and $h_f$ is the indicator of $f$. So, $h = h_L$ and, by Minkowski duality, we have $L = X_h$, i.e., $X_h$ is the $\rho$-convex support of $\mu$. But, by assertion 1 of Lemma 3.2.7, $X_h$ is included in an arbitrary determining set for $\mu$. Therefore, $X_h$ is a unique $\rho$-convex support of $\mu$.

3)→2). By assertion 1 of Lemma 3.2.7, $f = \tilde{\mu} \in [\rho, h(\theta)]$. Suppose that the Borel $\rho$-transform $\hat{f}$ admits analytic continuation outside a $\rho$-convex compact set $L \subset X_h$. Then, by Lemma 3.2.8, $L$ is a determining set for $\mu$. Since $X_h$ is the support of $\mu$, we have $L = X_h$.

2)→1). By Lemma 3.2.8, $X_h$ is a determining set for $\mu$ (see (3.2.7)). Then, by assertion 1) of Lemma 3.2.7, $f = \tilde{\mu} \in [\rho, h(\theta)]$. Consequently, $h_f^+ \leqslant h$. Finally, by Lemma 3.2.5, the function $\hat{f}$ admits analytic continuation outside the compact set $X_g$, where $g = h_f^+$. Therefore, $X_h \subset X_g$ and $h \leqslant g$. So, $h_f^+ = h$ ▶

**Remark.** The topology in $H(\mathbb{C})$ mentioned above is determined, e.g., by the system of seminorms $\left\{ ||f||_K = \sup\limits_{z \in K} |f(z)| \right\}$, where $K$ is an arbitrary compact subset of $\mathbb{C}$. For every $\mu \in H^*$, there exists a compact set $K = K(\mu)$ in $\mathbb{C}$ and a constant $A = A(\mu) > 0$ such that $|\mu(f)| \leqslant A||f||_K \ \forall\, f \in H(\mathbb{C})$ (see Prop. 1.10.10). Therefore, Lemmas 3.2.6, 3.2.7 imply isomorphism of the spaces $H^*$ and $[\rho, \infty)$.

## 3.3. The generalized Borel polygon of a power series

We study a modification of the Borel transform of an entire function, namely, a power series. Let

$$f(z) = \sum_{k=0}^{\infty} \frac{b_k z^k}{\Gamma[(k+1)/\rho]}, \quad \rho > 0,$$

be an entire function in the class $[\rho, \infty)$. With $f$, we associate the series

$$g(z) = [\mathcal{A}f](z) = \sum_{k=0}^{\infty} b_k z^k, \tag{3.3.1}$$

converging in a neighborhood of 0. If $f$ is of order $\rho$, the type $\sigma$ of $f$ and the radius of convergence $R$ of the series $g$ are related by the formula (see P. 3.2.1)

$$\sigma = R^{-\rho}. \tag{3.3.2}$$

Analytic continuation of the transform $\hat{f}$ of $f$ (see (3.0.1)) outside a certain $\rho$-convex compact set $K$ was done with the aid of a system of integrals (see L. 3.2.5). Similarly, in this section we consider a system of integral summation methods for the series $g(z) = \hat{f}(1/z)/z$ (see (3.3.1)) in its so-called generalized Borel polygon coinciding with the "inverted" exterior $\tilde{K}$ of the compact set $K$, namely,

$$\tilde{K} = \{z \in \mathbb{C} : z = \zeta^{-1},\ \zeta \in \overline{\mathbb{C}} \setminus K\}. \tag{3.3.3}$$

If $\rho = 1$ and $0 \in K$, then $K$ is a convex set, $\mathbb{C} \setminus K$ is a union of half-planes (see T. 1.3.19), and $\tilde{K}$ is a union of open disks the boundaries of which contain 0.

**3.3.1. Another form of the analog of the Polya Theorem.** A domain $\Omega$ is said to be *$\rho$-circular* if $\Omega = \tilde{K}$ (see (3.3.3)), where $K$ is a $\rho$-convex compact set.

**Theorem 3.3.1.** *In the notation of Theorem* 3.2.3, *let $h \in P_\rho^+$ and let $\Omega_h = \tilde{X}_h$. Denote by $H(\Omega_h)$ the space of all functions holomorphic in $\Omega_h$. The operator* (see (3.3.1)) *$\mathcal{A} : [\rho, h(\theta)] \to H(\Omega_h)$ establishes an algebraic isomorphism between the spaces $[\rho, h(\theta)]$ and $H(\Omega_h)$* *)*, and for a function $f \in [\rho, h(\theta)]$ the following statements are equivalent:*

1) *the indicator $h_f(\theta) = \varlimsup_{r\to\infty} r^{-\rho} \ln |f(re^{i\theta})|$, $\theta \in \mathbb{R}$, satisfies $h_f^+(\theta) := \max\{0, h_f(\theta)\} = h(\theta)\ \forall\ \theta \in \mathbb{R}$;*

2) *the greatest $\rho$-circular domain $\Omega$ to which the series $g = \mathcal{A}f$ extends analytically coincides with $\Omega_h$.*

Theorem 3.3.1 is a consequence of Theorem 3.2.3; the latter implies, in particular, that there is an algebraic isomorphism between $[\rho, h(\theta)]$ and the space $H_0(\overline{\mathbb{C}} \setminus X_h)$ of functions holomorphic in $\overline{\mathbb{C}} \setminus X_h$ and vanishing at the point $\infty$. Since $0 \in X_h$, the spaces $H_0(\overline{\mathbb{C}} \setminus X_h)$ and $H(\Omega_h)$ are isomorphic: $g(z) = f(1/z)/z$ (see (3.0.1), (3.3.1)).

**3.3.2. The system of integral summation methods for power series.** We pass to a more detailed study of the methods of analytic extension of a power series $g$ to the domain $\Omega_h$ (see T. 3.3.1).

---

*) In Chapter 8 it will be shown in a more general situation that this isomorphism is also topological.

Let $\varphi \in (-\pi/2, \pi/2)$, and let $f = \mathcal{A}^{-1}g$ (see (3.3.1)). We consider the integral

$$G_\varphi(z) = \int_0^{\infty(\arg\ \tau=\varphi)} e^{-\tau} f(\tau^{1/\rho}z)\tau^{1/\rho-1}d\tau. \tag{3.3.4}$$

Let $B_g^\varphi$ be the interior of the set of absolute and uniform convergence for the integral $G_\varphi$. In other words, $B_g^\varphi$ is the greatest open set $M$ in $\mathbb{C}$ where the integral $G_\varphi$ converges absolutely and, moreover, uniformly on each compact subset of $M$. Then $G_\varphi$ is a holomorphic function in $B_g^\varphi$. We formulate a supplement to Theorem 3.3.1.

**Proposition 3.3.2.** *For $h \in P_\rho^+$, let $H(z) = |z|^\rho \cdot h(\arg z)$, $z \in \mathbb{C}$. We denote $E_\varphi = \{z \in \mathbb{C} : H(ze^{i\varphi/\rho}) < \cos\varphi\}$, where $\varphi \in (-\pi/2, \pi/2)$. If $f \in [\rho, h(\theta)]$, then the system of integrals $\{G_\varphi : |\varphi| < \pi/2\}$* (see (3.3.4)) *determines an analytic extension of the series $g = \mathcal{A}\{$* (see (3.3.1)) *to $\Omega_h = \tilde{X}_h$, and*

$$E_\varphi \subset B_g^\varphi,\ \Omega_h = \bigcup E_\varphi |\ |\varphi| < \pi/2. \tag{3.3.5}$$

◀ 1. We prove formula (3.3.5). Take $z_0 \in E_\varphi$, $|\varphi| < \pi/2$, and $0 < \varepsilon < \cos\varphi - H(z_0e^{i\varphi/\rho})$. We put $T_\delta(z) = H(ze^{i\varphi/\rho}) - \cos\varphi + \delta|z|^\rho$. Then $T_\delta(z_0) < -\varepsilon$ for sufficiently small $\delta$. Since $H$ is a continuous function (see Prop. 1.9.5), there exists a compact circular neighborhood $U$ of the point $z_0$ such that $T_\delta(z) \leqslant -\varepsilon\ \forall\ z \in U \subset E_\varphi$. By (3.2.1), we obtain the inequality

$$|f(\tau^{1/\rho}z)| \leqslant A\exp\{|\tau| \cdot [H(ze^{i\varphi/\rho}) + \delta|z|^\rho]\}\ \forall\ z \in U,\ \arg\tau = \varphi.$$

The properties of $U$ imply that in fact

$$|f(\tau^{1/\rho}z)| \leqslant A\exp\{|\tau|(\cos\varphi - \varepsilon)\} \quad \forall\ z \in U,\ \arg\tau = \varphi.$$

Therefore, the integral (3.3.4) converges absolutely and uniformly on the compact set $U \subset E_\varphi$. Consequently, $E_\varphi \subset B_g^\varphi$.

2. The remark to Lemma 3.2.5 implies the following formula:

$$\Omega_h \setminus \{0\} = \bigcup \mathcal{D}'_\rho(\theta; h(\theta)) \mid \theta \in (-\pi, \pi],$$
$$\mathcal{D}'_\rho(\theta, \nu) = \{z \in \mathbb{C} : \Phi_\rho(e^{i\theta}/z) > \nu\} \tag{3.3.6}$$

(see (3.1.9)). Let $z_0 \in \Omega_h \setminus \{0\}$. Then $z_0 \in \mathcal{D}'_\rho(\theta; h(\theta))$ for some $\theta \in (-\pi, \pi)$. By (3.2.6), the integral

$$S_\theta(z) = \frac{\rho}{z}\int_0^{\infty(\arg\ w=\theta)} \exp\{-(w/z)^\rho\}f(w)dw$$

determines an analytic extension of the series $g(z) = \hat{f}(1/z)/z$ (see (3.3.1)) to the domain $\mathcal{D}'_\rho(\theta; h(\theta))$. Put $\psi = \arg z_0$ and $\varphi = \rho(\theta - \psi)$. Since $z_0 \in \mathcal{D}'_\rho(\theta; h(\theta))$ and $h(\theta) \geqslant 0$, we have $|\varphi| < \pi/2$. After the change of variables $\tau = (w/z)^\rho$ in the integral $S_\theta$, we find (see (3.3.4)): $S_\theta(z) = G_\varphi(z)\ \forall\ z \in \mathcal{D}'_\rho(\theta; h(\theta))$, $\arg z = \psi$. Obviously, $z_0 \in E_\varphi$ and, by the uniqueness theorem, $S_\theta(z) = G_\varphi(z)\ \forall\ z \in U$, where $U$ is a circular neighborhood of $z_0$ included in $\mathcal{D}'_\rho(\theta; h(\theta)) \cap E_\varphi$. So, the integral $G_\varphi$ determines an analytic extension of the series $g$ to $E_\varphi$. It only remains to note (see (3.3.6)) that $\Omega_h = \bigcap E_\varphi \mid |\varphi| < \pi/2$ ►

**Definition 3.3.3.** In the notation of formula (3.3.1), the set $B_g := B_g^0$ is called the *Borel polygon of the series g*. We call the set $M_g = \bigcap B_g^\varphi \mid |\varphi| < \pi/2$ the *generalized Borel polygon of g*.

The set $B_g$ can be regarded as a sort of a kernel for the set $M_g$, and it contains the disk of convergence of the series $g$.

**Proposition 3.3.4.** *Put* $K = \{\tau \in \mathbb{C} \setminus \{0\} : \operatorname{Re} 1/\tau \geqslant 1\} \cup \{0\}$ *(this is the disk in C having* $[0,1]$ *as a diameter). Let R be the radius of convergence of the series g* (see (3.1)). *Then* $M_g$ *is a star-like domain,* $\{z \in \mathbb{C} : |z| < R\} \subset B_g \subset M_g$, *and*

$$M_g = \{w = \tau^{1/\rho} \cdot z : z \in B_g,\ \tau \in K\}. \tag{3.3.7}$$

◄ Let $f = \mathcal{A}^{-1}g$. Since $f \in B^{-1}[g(1/\zeta)/\zeta]$, we see that $f \in [\rho, \infty)$ (see P. 3.2.1). There exists a function $h \in P_\rho^+$ such that $f \in [\rho, h(\theta)]$ (see 3.2). Let $\varphi \in (-\pi/2, \pi/2)$. By (3.3.5), $0 \in B_g^\varphi$. If $z \in B_g^\varphi \setminus \{0\}$ and $\lambda \in (0,1)$, then, by (3.3.4),

$$G_\varphi\left(\lambda^{1/\rho} z\right) = \lambda^{-1/\rho} \int_0^{\infty(\arg s = \varphi)} e^{-s/\lambda} f(s^{1/\rho} z) s^{-1+1/\rho}\, ds.$$

The inequality $|\exp\{-s/\lambda\}| \leqslant \exp\{-\operatorname{Re} s\}$, $\operatorname{Re} s > 0$, implies that $\lambda^{1/\rho} z \in B_g^\varphi$. So, $M_g$ is a star-like domain. Elementary transformations show that

$$B_g^\varphi = \{z\,(e^{-i\varphi} \cos\varphi)^{1/\rho} : z \in B_g\}. \tag{3.3.8}$$

But $\{e^{-i\varphi} \cos\varphi : |\varphi| < \pi/2\} = \{\tau \in \mathbb{C} \setminus \{0\} : \operatorname{Re} 1/\tau = 1\}$, and the first part of (3.3.7) follows.

Let $\sigma = \varlimsup_{r \to \infty} r^{-\rho} \ln M_f(r)$. Then $R^{-\rho} = \sigma$ (see P. 3.2.1, (3.3.2)), and $0 \leqslant h(\theta) \leqslant \sigma\ \forall\ \theta \in \mathbb{R}$. If $R = \infty$, then $\sigma = 0$, $h(\theta) \equiv 0$ and, by (3.3.5), $\mathbb{C} = B_g = M_g$. If $R < \infty$, then, again by (3.3.5),

$$\{z \in \mathbb{C} : |z| < R\} = \{z \in \mathbb{C} : |z|^\rho < 1/\sigma\} \subset$$

$$\subset \{z \in \mathbb{C} : H(z) < 1\} \subset B_g \subset M_g \ ►$$

So, if, e.g., $\rho$ is 1 or 1/2, and $z \in B_g$, then the set $M_g$ contains a disk (or the domain bounded by a cardiod) having $[0, z]$ as a diameter. The relationship between the Borel polygon and the generalized Borel polygon of a power series is clearly demonstrated by the following example.

**Example.** Let

$$g(z) = \sum_{k=0}^{\infty}[1 + (-i)^k]z^k = (1 + iz)^{-1} + (1 - z)^{-1}, \ |z| < 1.$$

Then $M = (\mathbb{C} \setminus \mathbb{C}_+) \cup K$, $B = \{x + iy \in \mathbb{C} : \ x, y < 1\}$, where $\mathbb{C}_+ = \{x + iy \in \mathbb{C} : \ x \geqslant 0; y \geqslant 0\}$, and $K = \{z \in \mathbb{C} : \ |z - (1 + i)/2| < 1/\sqrt{2}\}$ (see Fig. 6).

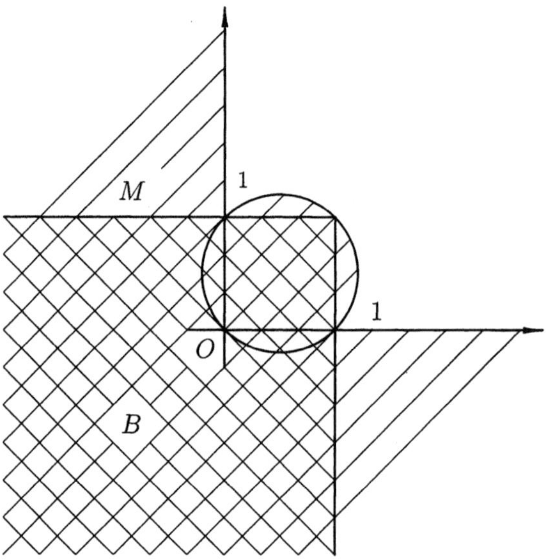


Fig. 6

**Theorem 3.3.5.** *Let $g$ be a convergent power series* (see (3.3.1)), *and let $f = \mathcal{A}^{-1}g$. In the notation of Theorem* 3.3.1, *if $h = h_f^+$, then the set $\Omega_h$ coincides with the generalized Borel polygon $M_g$ of the power series $g$.*

◄ Since $f \in [\rho, h(\theta)]$, Proposition 3.3.2 (see (3.3.5)) implies that $\Omega_h \subset M_g$. Therefore, to prove the theorem, it suffices to show that $M_g \subset \Omega_h$.

Let $Z_g$ be the Mittag-Leffler star of $g$, i.e., the greatest star-like domain in $\mathbb{C}$ with center at 0 where the series $g$ extends analytically. We put $S = \tilde{Z}_g$ (see (3.3.3)) and $T = \{z \in \mathbb{C} : \Phi_\rho(zw) < 1 \ \forall \ w \in S\}$ (see (3.1.9)). We show that the Borel polygon $B_g$ is included in $T$. Let $z_0 \in \mathbb{C} \setminus T$. Then there exists a point $\zeta_0 \in \mathbb{C} \setminus Z_g$ such that $\Phi_\rho(z_0/\zeta_0) \geqslant 1$. Suppose $z_0 \in B_g$. By Proposition 3.3.4, in the notation adopted in it we have

$$\{\zeta = \tau^{1/\rho} z_0 : \tau \in K\} = \{\zeta \in \mathbb{C} : \Phi_\rho(z_0/\zeta) \geqslant 1\} \cup \{0\} \subset M_g \subset Z_g.$$

Therefore, $\zeta_0 \in Z_g$, which is impossible, i.e., $z_0 \notin B_g$. Thus, $B_g \subset T$. The set $S$ is star-like and compact, and the function $\hat{f}(\zeta) = g(1/\zeta)/\zeta \in H_0(\infty)$ admits analytic continuation outside $S$. In particular, it also extends outside the closed $\rho$-convex hull $F$ of $S$. By Theorem 3.2.3, the set $F$ coincides with the indicator diagram $X_h$ of $f$. Let $H(z) = |z|^\rho \cdot h(\arg z)$, $z \in \mathbb{C}$. Recalling that $H$ is the $\rho$-support function of $F$ and of $S$ (see Def. 3.1.11, P. 3.1.16), and using the condition $B_g \subset T$, we find

$$B_g \subset \{z \in \mathbb{C} : \sup_{w \in S} \Phi_\rho(zw) < 1\} = \{z \in \mathbb{C} : H(z) < 1\}.$$

Then, by (3.3.8), $B_g^\varphi \subset \{z \in \mathbb{C} : H(ze^{i\varphi/\rho}) < \cos\varphi\}$ $\forall\ \varphi \in (-\pi/2, \pi/2)$, i.e., $M_g \subset \Omega_h$ (see (3.3.5)) ▶

Chapter 4

# Plane Indicator Diagram of an Entire Function of Order $\rho > 0$ and with Indicator of General Form

This chapter describes the construction of the plane indicator diagram for an entire function $f$ of finite order $\rho > 0$ and with indicator $h$ of variable sign. In particular, it will be shown that in the case of a finite indicator $h$ of general form the locally convex arc $\Gamma_h$ (see Def. 1.9.24) associated with $h$ is the image of a closed Jordan arc $T \subset \mathbb{C}$ under a mapping of the form

$$\alpha(z) = z^{\rho} \cdot \left[1 + \frac{a_1}{z} + \ldots + \frac{a_n}{z^n}\right]. \tag{4.0.1}$$

Here we mean the principal value of the function $z^{\rho}$. The arc $T$ is the boundary of the so-called $(\rho, \alpha)$-convex compact set $X_h$, playing the role of the indicator diagram of $f$. One and the same function $\alpha$ can be chosen for all indicators in the class

$$P_{\rho}(l) = \{h \in P_{\rho} : \ h \geqslant l\}. \tag{4.0.2}$$

Here $P_{\rho}$ is the class of lower semicontinuous $2\pi/$-periodic trigonometrically $\rho$-convex functions defined on $\mathbb{R}$ and taking values in $(-\infty, \infty]$; the class $P_{\rho}$ is partially ordered in the natural manner, i.e., $h \leqslant g$ if $h(\theta) \leqslant g(\theta)$ $\forall\ \theta \in \mathbb{R}$; finally, $l \in M_{\rho}$, where $M_{\rho}$ is the subclass of minimal elements of $P_{\rho}$.

In 4.1 we study the structure of elements of $M_{\rho}$; sections 4.2-4.3 are devoted to the multi-sheeted indicator diagram of an entire function $f$ of order $\rho \neq 1$ (this diagram is situated on the Riemann surface of the function inverse to $\alpha$, see (0.1)). In 4.4 we develop the construction of the plane indicator diagram of $f$. To study the geometry of $(\rho, \alpha)$-convex sets, we apply Minkowski duality (see T. 1.3.15). The construction of the indicator diagram in the case of nonintegral $\rho > 0$ is developed on the basis of the properties of the Riemann surface of the mapping inverse to $\alpha$. These properties are presented systematically in A.1.2. For short, sometimes $\alpha$ is called a polynomial.

## 4.1. Minimal trigonometrically $\rho$-convex functions

**4.1.1. Piecewise-trigonometric functions.** Let $\rho > 0$. A function $q : [\alpha, \beta] \to \mathbb{R}$ is said to be *piecewise-trigonometric* if it is "glued" of a finite number of trigonometric functions of the form

$$\text{Re } pe^{i\rho\theta} = a\cos\rho\theta + b\sin\rho\theta, \quad p = a - bi,$$

with period $2\pi/\rho$. A function $q : \mathbb{R} \to \mathbb{R}$ is said to be *$2\pi$-periodic piecewise-trigonometric* if $q$ is a $2\pi$-periodic extension to $\mathbb{R}$ of a piecewise-trigonometric function $q : [\alpha, \alpha + 2\pi] \to \mathbb{R}$ such that $q(\alpha) = q(\alpha + 2\pi)$ for some $\alpha \in \mathbb{R}$. Let $T_\rho$ be the class of $2\pi$-periodic piecewise-trigonometric functions, and let $l \in T_\rho$. Then $l$ is a continuous function of the form

$$l(\theta) = \text{Re } p_j e^{i\rho\theta}, \quad \theta \in [\theta_{j-1}, \theta_j], \quad j \in \mathbb{Z}. \tag{4.1.1}$$

Here

$$P := \{p_j : j \in \mathbb{Z}\}, \quad \text{and} \quad \Theta := \{\theta_j : j \in \mathbb{Z}\} \tag{4.1.2}$$

are a complex and a real numerical sequence, respectively, such that

$$p_j \neq p_{j+1}, \quad p_j = p_{j+k} e^{i2\pi\rho} \quad \forall\, j \in \mathbb{Z}; \tag{4.1.3}$$

$$\theta_{j+k} = \theta_j + 2\pi \quad \forall\, j \in \mathbb{Z}; \qquad \theta_0 < \theta_1 < \cdots < \theta_k; \qquad k = k(l) \in \mathbb{N}.$$

Let $l : [\theta_0, \theta_k] \to \mathbb{R}$ be a continuous piecewise trigonometric function glued of $k$ trigonometric functions, i.e.,

$$l(\theta) = \text{Re } p_j e^{i\rho\theta}, \quad \theta \in [\theta_{j-1}, \theta_j], \quad j = 1, \ldots, k. \tag{4.1.1$'$}$$

With $l$, we associate two finite sequences

$$P = \{p_j : j = 1, \ldots, k\}, \quad \Theta = \{\theta_j : j = 0, 1, \ldots, k\} \tag{4.1.2$'$}$$

the terms of which satisfy the conditions

$$p_j \neq p_{j+1}, \quad j = 1, 2, \ldots, k-1, \qquad \theta_0 < \theta_1 < \cdots < \theta_k. \tag{4.1.3$'$}$$

For applications it is important to know when a piecewise-trigonometric function $l$ is trigonometrically $\rho$ convex and, in particular, when it belongs to the class $P_\rho$ of $2\pi$-periodic trigonometrically $\rho$-convex functions.

**Proposition 4.1.1.** *Let $P$ and $\Theta$ be sequences satisfying conditions* (4.1.2), (4.1.3), *and let*

$$\mathcal{P}_j = p_{j+1} - p_j, \quad A_j = \text{Re } \mathcal{P}_j e^{i\rho\theta_j}, \quad B_j = \text{Im } \mathcal{P}_j e^{i\rho\theta_j}, \tag{4.1.4}$$

$$l_j(\theta) = \operatorname{Re} p_j e^{i\rho\theta}, \quad \theta \in \mathbb{R}, \tag{4.1.5}$$

*where $j \in \mathbb{Z}$. The following assertions are equivalent:*

1) *The function*

$$l(\theta) = l_j(\theta) \quad \textit{for} \quad \theta \in [\theta_{j-1}, \theta_j], \quad j \in \mathbb{Z},$$

*belongs to the class $P_\rho$;*

2) *we have*

$$A_j = 0, \quad B_j < 0 \quad \forall\, j \in \mathbb{Z}; \tag{4.1.6}$$

3) *for every index $j \in \mathbb{Z}$ there exists a pair of numbers $\varphi_1^{(j)}, \varphi_2^{(j)}$, such that*

$$\begin{aligned} 0 < \varphi_2^{(j)} - \varphi_1^{(j)} < \pi/\rho, \qquad \theta_{j-1} < \varphi_1^{(j)} < \theta_j < \varphi_2^{(j)} < \theta_{j-1}, \\ l_j(\varphi_1^{(j)}) > l_{j+1}(\varphi_1^{(j)}), \qquad l_j(\varphi_2^{(j)}) < l_{j+1}(\varphi_2^{(j)}). \end{aligned} \tag{4.1.7}$$

*A similar statement is true for sequences $P, \Theta$ satisfying* (4.1.2′), (4.1.3′). *In this case $j = 1, 2, \ldots, k-1$ in* (4.1.4), (4.1.6), (4.1.7) *and $j = 1, \ldots, k$ in* (4.1.5).

◀ 1) → 2). The condition $A_j = 0 \;\; \forall\, j \in \mathbb{Z}$ is equivalent to continuity of $l$ at the points of the sequence $\Theta$. The condition $B_j < 0 \quad \forall\, j \in \mathbb{Z}$ is equivalent to the inequality

$$l'_-(\theta_j) < l'_+(\theta_j) \quad \forall\, j \in \mathbb{Z}, \tag{4.1.8}$$

where $l'_-$ and $l'_+$ are the left and the right derivative of $l$, respectively. Condition (4.1.8) is true indeed for the trigonometrically $\rho$-convex functions (see (4.1.1), T. 1.9.18). Equality in (4.1.8) is impossible because $p_j \neq p_{j+1} \quad \forall\, j \in \mathbb{Z}$ (see (4.1.3)).

2) → 3). Let $M_j = l_{j+1} - l_j$, $j \in \mathbb{Z}$ (see (4.1.5)). Formulas (4.1.4), (4.1.5) imply that $M_j(\theta_j) = 0$ and $M_j'(\theta_j) > 0$, i.e., for every fixed $j \in \mathbb{Z}$ the function $M_j$ is strictly monotone increasing in a neighborhood of $\theta_j$. This implies assertion 3).

3) → 1). In the above notation we have

$$M_j(\varphi_1^{(j)}) < 0, \quad M_j(\theta_j) = 0, \quad M_j(\varphi_2^{(j)}) > 0 \quad \forall\, j \in \mathbb{Z}.$$

Hence, using the structure of the trigonometric function $M_j$, for every $j \in \mathbb{Z}$ we find: $l_j(\theta) > l_{j+1}(\theta)$ for $\theta \in (\theta_j - \pi/\rho, \theta_j)$, and $l_{j+1}(\theta) > l_j(\theta)$ for $\theta \in (\theta_j, \theta_j + \pi/\rho)$. Therefore, for every $j \in \mathbb{Z}$ in a sufficiently small neighborhood of $\theta_j$ we have $l(\theta) = \max\{l_j(\theta), l_{j+1}(\theta)\}$. Consequently, $l$ satisfies the assumptions of T. 1.9.14, and $l$ is a trigonometrically $\rho$-convex function ▶

**Remark.** If the sequences $P, \Theta$ satisfy conditions (4.1.3), and formulas (4.1.6), (4.1.7) are valid for $j = 1, \ldots, k$, then they are valid for every $j \in \mathbb{Z}$.

**4.1.2. The structure of minimal elements in $P_\rho$.** The set $M_\rho$ of all minimal elements in $P_\rho$ turns out to be included in the class $T_\rho$ of piecewise-trigonometric functions. To verify this, first we prove a number of auxiliary assertions.

**Lemma 4.1.2.** *Let $h \in P_\rho$, and let $[\varphi_0, \varphi_1] \subset$ int dom $h$. Suppose the function $h$ is differentiable at the points $\varphi_0, \varphi_4.1$. We denote by $Dh(\varphi_j)$ the subdifferential of $h$ at $\varphi_j$, $j = 0, 1$ (see the remark to T. 1.9.21′, where $0 < \varphi_1 - \varphi_0 < \pi/\rho$, $q_j \in Dh(\varphi_j)$, $j = 0, 1$, and, moreover, $q_1 \notin Dh(\varphi_0)$, $q_0 \notin Dh(\varphi_1)$. Let $l_j(\theta) = \operatorname{Re} q_j e^{i\rho\theta}$, $\theta \in \mathbb{R}$, $j = 0, 4.1$. Then there exists $\varphi \in (\varphi_0, \varphi_1)$ such that the function defined by the formulas*

$$l(\theta) = \begin{cases} l_0(\theta), & \theta \in [\varphi_0, \varphi], \\ l_1(\theta), & \theta \in [\varphi, \varphi_1], \end{cases} \tag{4.1.9}$$

*is trigonometrically $\rho$-convex and possesses the following properties:*

$$l(\theta) \leqslant h(\theta), \quad l(\theta) = \max\{l_1(\theta),\ l_2(\theta)\}, \quad \theta \in [\varphi_1, \varphi_2].$$

◀ By the definition of the set $Dh(\varphi_1)$, we have $l_1(\theta) \leqslant h(\theta)$ for $|\theta - \varphi_0| < \delta$, where $\delta$ is a sufficiently small number. Since $q_1 \notin Dh(\varphi_0)$ by assumption, we see that $l_1(\varphi_0) < h(\varphi_0) = l_0(\varphi_0)$. Similarly, $l_0(\varphi_1) < l_1(\varphi_1)$. Consequently, the trigonometric function $\Phi(\theta) := l_1(\theta) - l_0(\theta)$, $\theta \in \mathbb{R}$, satisfies the inequalities $\Phi(\varphi_0) < 0$, $\Phi(\varphi_1) > 0$. Since the period of $\Phi$ is equal to $2\pi/\rho$ and $\varphi_1 - \varphi_0 < \pi/\rho$, it follows that $\Phi$ has a unique zero $\varphi$ on $[\varphi_0, \varphi_1]$. It only remains to use Proposition 4.1.1 (see (4.1.9)) ▶

**Lemma 4.1.3.** *Let $h : (\alpha, \beta) \to \mathbb{R}$ be a trigonometrically $\rho$-convex function, and let $p \in Dh(\gamma)$, where $\gamma \in (\alpha, \beta)$. Then*

$$H_0(\theta) = \begin{cases} h(\theta), & \theta \in (\alpha, \gamma], \\ \operatorname{Re} pe^{i\rho\theta}, & \theta \in [\gamma, \beta) \end{cases}$$

*and*

$$H_1(\theta) = \begin{cases} \operatorname{Re} pe^{i\rho\theta}, & \theta \in (\alpha, \gamma], \\ h(0), & 0 \in [\gamma, \beta] \end{cases}$$

*are trigonometrically $\rho$-convex functions.*

◀ For instance, we consider the function $H_0$. Since $h$ and $\operatorname{Re} pe^{i\rho\theta}$ are trigonometrically $\rho$-convex, it suffices to verify the lemma in a neighborhood of $\gamma$. Let

$$\alpha < \theta_1 < \gamma < \theta_2 < \beta, \quad \theta_1 < \theta < \theta_2, \quad \theta_2 - \theta_1 < \pi/\rho.$$

a) $\theta \geqslant \gamma$. The function $l(\varphi) = \operatorname{Re} pe^{i\rho\varphi}$, $\varphi \in \mathbb{R}$, satisfies formula (1.9.1). Hence, recalling that $l(\theta_1) \leqslant H_0(\theta_1)$, $l_0(\theta) = H_0(\theta)$, and $l(\theta_2) = H_0(\theta_2)$, we conclude that inequality (1.9.2) is true for $H_0$.

b) $\theta_1 < \theta < \gamma$. We consider the function $h_0(\theta) = H_0(\theta) - \text{Re } pe^{i\rho\theta}$, $\theta \in (\alpha, \beta)$. Since $h_0$ is trigonometrically $\rho$-convex on $[\alpha, \gamma]$ and, moreover, $h_0(\theta_1) \geqslant 0$, and the function $T(x) = [\sin\rho(x-\theta_0)]^{-1} \cdot \sin\rho(x-\theta)$, $x \in [\gamma, \theta_2]$, is monotone increasing, from inequality (1.9.2) we deduce that

$$h_0(\theta) \leqslant h_0(\theta_1)\frac{\sin\rho(\gamma-\theta)}{\sin\rho(\gamma-\theta_1)} \leqslant h_0(\theta_1)\frac{\sin\rho(\theta_2-\theta)}{\sin\rho(\theta_2-\theta_1)}.$$

Hence, recalling that $h_0(\theta_2) = 0$, we see that inequality (1.9.2) is true for $h_0$. Therefore, $h_0$ and, with it, $H_0$ are trigonometrically $\rho$-convex ▸

**Lemma 4.1.4.** *Let $\alpha \leqslant \beta < \gamma \leqslant \delta$, and let $\gamma - \beta \geqslant \pi/\rho$. We put $l(\theta) = \text{Re } pe^{i\rho\theta}$, $\theta \in [\alpha, \beta]$, $l(\theta) = \text{Re } qe^{i\rho\theta}$, $\theta \in [\gamma, \delta]$, where $p, q \in \mathbb{C}$. (If $\gamma = \beta + \pi/\rho$, then $l(\gamma) + l(\beta) \geqslant 0$.) Then there exists a trigonometrically $\rho$-convex extension $L$ of $l$ to $[\alpha, \delta]$ that is piecewise trigonometric.*

◂ If $p = q$, then the required function is $L(\theta) = \text{Re } pe^{i\rho\theta}$, $\theta \in [\alpha, \delta]$. Assume that $p \neq q$.

A. Let $K = \{e^{i\rho\theta} : \beta \leqslant \theta \leqslant \gamma\}$. Putting $\Phi(\theta) = \text{Re } (q-p)e^{i\rho\theta}$, $a(\theta) = \text{Re } \Phi(\theta)$, and $b(\theta) = \text{Im } \Phi(\theta)$, $\theta \in [\alpha, \delta]$, we assume that there exists an element $e^{i\rho\psi} \in K$ such that

$$a(\psi) = 0, \quad b(\psi) < 0. \tag{4.1.10}$$

Together with Proposition 4.1.1 (cf. (4.1.10), (4.1.6)), this implies that the required function is given by the formulas

$$L(\theta) = \begin{cases} \text{Re } pe^{i\rho\theta}, & \theta \in [\alpha, \psi], \\ \text{Re } qe^{i\rho\theta}, & \theta \in [\psi, \delta]. \end{cases}$$

We consider the cases in which condition (4.1.10) is fulfilled.

1) $\gamma = \beta + \pi/\rho$. Then, under the assumptions of the lemma, $l(\beta) + l(\gamma) \geqslant 0$, i.e., $a(\beta) \leqslant 0$ and $a(\gamma) \geqslant 0$. Consequently, for sufficiently small $\varepsilon > 0$, on the interval $(\gamma - \varepsilon, \gamma + \varepsilon)$ there exist points $\varphi_1, \varphi_2$ possessing the properties $0 < \varphi_2 - \varphi_1 < \pi/\rho$, $a(\varphi_1) < 0$, and $a(\varphi_2) > 0$. By Proposition 4.1.1, this is the same as condition (4.1.10).

2) $\gamma - \beta \geqslant 2\pi/\rho$. In this case(4.1.10) is established similarly .

B. Suppose the above assumption fails. We have seen that this is possible only if $\pi/\rho < \gamma - \beta < 2\pi/\rho$. In this case

$$a(\beta) > 0, \qquad a(\gamma) < 0. \tag{4.1.11}$$

Then there exists a unique solution of the system

$$\text{Re } se^{i\rho\beta} = l(\beta), \qquad \text{Re } se^{i\rho\gamma} = l(\gamma).$$

Direct calculations show that this solution possesses the properties

$$(s-p)e^{i\rho\beta} = -\frac{ia(\gamma)}{2\sin\rho(\gamma-\beta)}, \qquad (s-q)e^{i\rho\gamma} = -\frac{ia(\beta)}{2\sin\rho(\gamma-\beta)}.$$

Combining this with Proposition 4.1.1 (see (4.1.5)) and recalling (4.1.11), we conclude that the function defined by

$$L(\theta) = \begin{cases} l(\theta), & \theta \in [\alpha, \beta] \cup [\gamma, \delta], \\ \mathrm{Re}\; s e^{i\rho\theta}, & \theta \in [\beta, \gamma] \end{cases}$$

is required ▶

Now, we determine the structure of the class $M_\rho$ of minimal elements in $P_\rho$. First of all, $M_\rho \neq \varnothing$: the function $l_0(\theta) \equiv 0$, $\theta \in \mathbb{R}$, belongs to $M_\rho$, and, moreover, $M_\rho = \{l_0\}$ for $\rho \leqslant 1/2$ (see Prop. 1.9.7, 1.9.9). In the general case the following assertion is valid in the above notation.

**Theorem 4.1.5.** *Let $\rho > 0$. We denote $O_\rho = \{l_0\}$ if $\rho$ is a noninteger, and $O_\rho = \{\mathrm{Re}\; p e^{i\rho\theta} : p \in \mathbb{C}\}$ if $\rho$ is an integer. A function $l \in P_\rho$ belongs to $M_\rho$ if and only if either $l \in O_\rho$, or $\rho > 1/2$ and $l$ belongs to the class $T_\rho$ of $2\pi$-periodic piecewise-trigonometric functions and the sequence $\Theta$* (see (4.1.2)) *associated with $l$ satisfies the additional condition*

$$\theta_j - \theta_{j-1} > \pi/\rho, \quad j = 1, \dots, k. \tag{4.1.12}$$

◀ In view of the remark preceding Theorem 4.1.5, we consider only the case of $\rho > 1/2$. Let $T'_\rho$ be the subclass $T_\rho \cap P_\rho$ distinguished by condition (4.1.2). Property 1.9.8 implies that $T'_\rho \subset M_\rho$.

1. Let $l \in M_\rho$. Then $l(\theta) \not\equiv \infty$ for $\theta \in \mathbb{R}$. We consider the set $F_l$ of all components of $\mathrm{dom}\, l$. Let $\Delta \in F_l$, and let $\mathring{\Delta} \neq \varnothing$. Suppose $\{\psi_0, \psi_1, \psi_2\}$ is an arbitrarily fixed triplet of points in $\mathring{\Delta}$ at which $l$ is differentiable and, moreover, $0 < \psi_j - \psi_{j-1} < \pi/2\rho$, $j = 1, 2$ (see Property 1.9.20) and $Dl(\psi_j) = \{p_j\}$, $j = 0, 1, 2$ (see (1.9.13)). By Lemma 4.1.2, there exist points $\theta_j \in (\psi_j, \psi_{j+1})$, $j = 0, 1$, such that the function

$$r(\theta) = \mathrm{Re}\; p_j e^{i\rho\theta}, \quad \theta \in [\theta_{j-1}, \theta_j], \quad j = 0, 1, 2 \tag{4.1.13}$$

(where $\theta_{-1} = \psi_0$, $\theta_2 = \psi_2$) is trigonometrically $\rho$-convex on $[\psi_0, \psi_2]$. We denote by $L$ the $2\pi$-periodic extension to $\mathbb{R}$ of the function defined as follows

$$l_0(\theta) = \begin{cases} r(\theta), & \theta \in [\psi_0, \psi_2], \\ l(\theta), & \theta \in [\psi_1 - \pi, \psi_0] \cup [\psi_2, \psi_1 + \pi]. \end{cases}$$

By Lemma 4.1.3, the functions

$$r_1(\theta) = \begin{cases} l(\theta), & \theta \in (\psi_1 - \pi, \psi_0], \\ \mathrm{Re}\; p_0 e^{i\rho\theta}, & \theta \in [\psi_0, \theta_0), \end{cases}$$

$$r_2(\theta) = \begin{cases} \mathrm{Re}\; p_2 e^{i\rho\theta}, & \theta \in [\theta_1, \psi_2], \\ l(\theta), & \theta \in [\psi_2, \psi_1 + \pi). \end{cases}$$

are also trigonometrically $\rho$-convex. Therefore, $L \in P_\rho$ (see T. 1.9.14). By Lemma 4.1.2, $r(\theta) \leqslant l(\theta)$ for $\theta \in [\psi_0, \psi_2]$ (see (4.1.13)); therefore, $L \leqslant l$. But $l \in M_\rho$, consequently, $L = l$. Combining this with formula (4.1.13), we conclude that in the set $\{\theta \in \mathbb{R} : Dl(\theta) = \{p_1\}\}$ the component $\Sigma = \Sigma(\psi_1)$ of the point $\psi_1$ contains the interval $(\theta_0, \theta_1)$, which is a neighborhood of $\psi_1$.

The interval $\Sigma(\psi_1)$ is associated with an arbitrarily point $\psi_1$ at which the function $l$ is differentiable. Let $A$ be the set of all points in $\Delta$ with this property. We denote by $\mathfrak{R} = \{\Sigma(\psi) : \psi \in A\}$ the system of intervals associated with the points of $A$ in a similar way. Every two intervals in $\mathfrak{R}$ either do not intersect or coincide and, moreover,

$$l(\theta) = \text{Re } p_\psi e^{i\rho\theta}, \ \theta \in \Sigma(\psi), \quad \psi \in A, \tag{4.1.14}$$

where $\{p_\psi\} = Dl(\psi)$. Since $\bar{A} = \bar{\Delta}$ (see Prop. 1.9.20), $l$ is a piecewise-trigonometric function not differentiable at the ends of every interval in $\mathfrak{R}$.

2. Let $\psi \in A$. We assume that the length $t$ of the interval $\Sigma = \Sigma(\psi)$ is greater than $2\pi$. Then for some $\varphi \in \mathbb{R}$ the interval $\Sigma$ contains the points $\varphi, \varphi + 2\pi$. If $\{p\} = Dl(\varphi)$, then (see (1.9.27))

$$\{pe^{-i2\pi\rho}\} = Dl(\varphi + 2\pi). \tag{4.1.15}$$

But $\varphi + 2\pi \in \Sigma$; consequently, $pe^{-i2\pi\rho} = p$, and $\rho = n$ is an integer if $p \neq 0$. Since the function $l(\theta)$ is $2\pi$-periodic, we see that $l(\theta) = \text{Re } pe^{in\theta}$ for $\theta \in \mathbb{R}$. If $p = 0$, then $l(\theta) = 0$ for $\theta \in \mathbb{R}$. So, in this case $\Sigma = \mathbb{R}$ and $l \in O_\rho$.

3. Let $t < 2\pi$. We put $\Sigma = \Sigma(\psi) = (\varphi_0, \varphi_1)$, $\{p\} = Dl(\psi)$. We show that $t \geqslant \pi/\rho$. Suppose $t < \pi/\rho$. Since $l$ is nondifferentiable at the points $\varphi_0, \varphi_1$ (see item 1), there exist numbers $q_j \in Dl(\varphi_j)$, $j = 0, 1$, different from $p$. This is also valid in the case where $\Sigma$ has a common boundary point with $\Delta$, because $l \in P_\rho$ and

$$l(\theta) = \text{Re } pe^{i\rho\theta}, \qquad \theta \in [\varphi_0, \varphi_1] \tag{4.1.16}$$

(see Property 1.9.21$'$). We consider the functions $\Phi_j(\theta) = \text{Re } q_j e^{i\rho\theta} - l(\theta)$, $j = 0, 1$. Since $\varphi_1 - \varphi_0 \leqslant \pi/\rho$ and $\{q_0, p\} \subset Dl(\varphi_0)$ (see (1.9.13)), we have $\Phi_0(\varphi_0) = 0$ and $\Phi_0(\theta) \leqslant 0 \ \forall \ \theta \in [\varphi_0, \varphi_1]$. Hence, recalling that $\Phi_0$ is a trigonometric function with period $2\pi/\rho$ (see (4.1.16)), we obtain $\Phi_0(\theta) < 0$ $\forall \ \theta \in (\varphi_0, \varphi_1)$. Arguing as above, we find that $\Phi_1(\theta) < 0 \quad \forall \ \theta \in (\varphi_0, \varphi_1)$. We take the function $r$ equal to the $2\pi$-periodic extension of the function

$$r_0(\theta) = \begin{cases} \max\{l_0(\theta), l_1(\theta)\}, & \theta \in [\varphi_0, \varphi_1], \\ \\ l(\theta), \quad \theta \in [\alpha, \varphi_0] \cup [\varphi_1, \alpha + 2\pi], \end{cases} \tag{4.1.17}$$

where $l_j(\theta) = \text{Re } q_j e^{i\rho\theta}$, $j = 0, 1$, $\alpha < \varphi_0 < \varphi_1 < \alpha + 2\pi$. Since $\varphi_1 - \varphi_0 < \pi/\rho$, by Lemma 4.1.2 the representation (4.1.9) is valid for $r_0(\theta)$. Combining this with Lemma 4.1.3 and Theorem 1.9.14, we conclude that $r \in P_\rho$. By the above construction, $r(\theta) \leqslant l(\theta)$ for $\theta \in \mathbb{R}$ and $r(\theta) < l(\theta)$ for $\theta \in [\varphi_0, \varphi_1]$. This contradicts the condition $l \in M_\rho$. So, $t \geqslant \pi/\rho$, and, moreover, $l$ is a piecewise-trigonometric function on each component $\Delta$ with the property $\overset{\circ}{\Delta} \neq \varnothing$. If $\overset{\circ}{\Delta} = \varnothing$, i.e., $\Delta = \{\alpha\}$ is a singleton, then there exists $q \in \mathbb{C}$ such that Re $qe^{i\rho\theta} = l(\theta)$. Combining these arguments with Property 1.9.22 and with Lemma 4.1.4, we conclude that $l$ is a finite piecewise-trigonometric function of class $P_\rho$.

4. To complete the proof, it only remains to exclude the case where $t = \pi/\rho$. Suppose that in the representation (4.1.1) for $l$ there is at least one function $l_j(\theta) = \text{Re}\, p_j e^{i\rho\theta}$ defined on the segment $[\theta_{j-1}, \theta_j]$ with $\theta_j - \theta_{j-1} = \pi/\rho$. Here $j \in \mathbb{Z}$; let, for instance, $j = 1$. Formula (1.9.14), Ch. 1 implies that the subdifferential mapping for $l$ (see Def. 1.9.24) has the form $Dl(\theta) = \{p_j\}$ for $\theta \in (\theta_{j-1}, \theta_j)$ and

$$Dl(\theta_j) = [p_j, p_{j+1}], \ \forall\, j \in \mathbb{Z}. \tag{4.1.18}$$

We denote by $r$ the $2\pi$-periodic extension of the function

$$r_0(\theta) = \begin{cases} \max\{l_0(\theta), l_2(\theta)\}, & \theta \in [\theta_0, \theta_1], \\ l(\theta), \quad \theta \in [\alpha, \theta_0) \cup (\theta_1, \alpha + 2\pi], \end{cases} \tag{4.1.19}$$

(see the above notation in (4.1.5)). From (4.1.1), conditions

$$p_0 \neq p_1 \neq p_2, \quad l(\theta_0) = l_0(\theta_0), \quad l(\theta_1) = l_2(\theta_1), \tag{4.1.20}$$

(see (4.1.3)), and formula (1.9.13), Ch. 1, we deduce that $r$ is a $2\pi$-periodic piecewise-trigonometric function such that $r \leqslant l$, $r \not\equiv l$. Since $\theta_1 - \theta_0 = \pi/\rho$, and $l(\theta) = \text{Re}\, p_1 e^{i\rho\theta}$ for $\theta \in [\theta_0, \theta_1]$, by (4.1.20) we have: $l_0(\theta_1) = l_2(\theta_1)$, $l_0(\theta_0) = l_2(\theta_0)$. Consequently, either $\Phi(\theta) := l_2(\theta) - l_0(\theta) = \text{Re}\, (p_2 - p_0)e^{i\rho\theta}$ is a function of permanent sign on the interval $(\theta_0, \theta_1)$, or $l_0 = l_2$. In the latter case it is obvious that $r \in P_\rho$ (see (4.1.19), (4.1.1)).

Now, let, for instance, $\Phi(\theta) > 0$, $\theta \in (\theta_0, \theta_1)$. Then (see (4.1.19)) $r$ is a piecewise-trigonometric function. As compared to $l$, in the representation of the form (4.1.1), the terms are absent determined by the elements of the set $\{p_1 \exp\{-i2\pi k\rho\} : k \in \mathbb{Z}\}$. Since $\Phi(\theta_0 + \pi/2\rho) > 0$, we have Im $(p_2 - p_0)e^{i\rho\theta_0} < 0$. By Proposition 4.1.1 (see (4.1.6)), $r \in P_\rho$. Since in both cases $r \leqslant l$ and $r \neq l$, we arrive at a contradiction (recall that $l$ is a minimal element of $P_\rho$). Therefore, the assumption $\theta_1 - \theta_0 = \pi/\rho$ is wrong. Thus, $M_\rho \subset T'_\rho$. The theorem is proved ►

**4.1.3. Geometric interpretation of the piecewise-trigonometric functions of class $P_\rho$.** Let $l$ be a piecewise-trigonometric function in $P_\rho$ defined by formula (4.1.1). We study the properties of the plane curve $\Gamma_l$ (see

Def. 1.9.24) associated with $l$. Formula (4.1.18) implies that $\Gamma_l$ is the oriented broken line in $\mathbb{C}$ with the vertices $\{p_j : j \in \mathbb{Z}\}$; we denote this broken line by $P$. Condition (4.1.3) means that, for integral $\rho$, $P$ is a closed broken line with $k$ vertices. For nonintegral $\rho$, the cyclic group $G$ of rotations about 0 with the generator $e^{i2\pi\rho}$ operates on the set of vertices of $P$. If $\rho = m/n$ is a rational number, the group $G$ is finite, and $P$ is a closed broken line with $k \cdot n$ vertices ($m/n$ is an irreducible fraction). If $\rho$ is an irrational number, then $G$ is an infinite group and $P$ is an infinite broken line.

Let $\alpha_j = -\rho\theta_j,\ j \in \mathbb{Z}$. Then $\{\alpha_j : j \in \mathbb{Z}\}$ is a strictly monotone decreasing sequence of numbers such that (see (4.1.3))

$$\alpha_{j-1} - \alpha_j = \alpha_{j-1+k} - \alpha_{j+k} \quad \forall\, j \in \mathbb{Z}. \tag{4.1.21}$$

With it, we associate the sequence $\{a_j = \exp\{i\alpha_j\} : j \in \mathbb{Z}\}$ of unit vectors. Condition (4.1.6) means that for every $j$ the vector $a_j$ is orthogonal to the link $[p_j,\ p_{j+1}]$ of the broken line $P$ and is on the left of $\mathcal{P}_j = p_{j+1} - p_j$. After turning clockwise through the angle $\Delta_j = \alpha_{j-1} - \alpha_j$, the vector $a_{j-1}$ will coincide with $a_j$. The smallest possible angle $\Delta_j$ is shown in Fig. 7. For instance, if $\rho$ is an integer, then the vector $a_j$ returns to its previous position $a_{j+k} = a_j$, having made a turn through an angle of $2\pi\rho$. The tangent to the link with normal vector $a_{j-1}$ also turns clockwise about the vertex $p_j$ through the same angle $\Delta_j$.

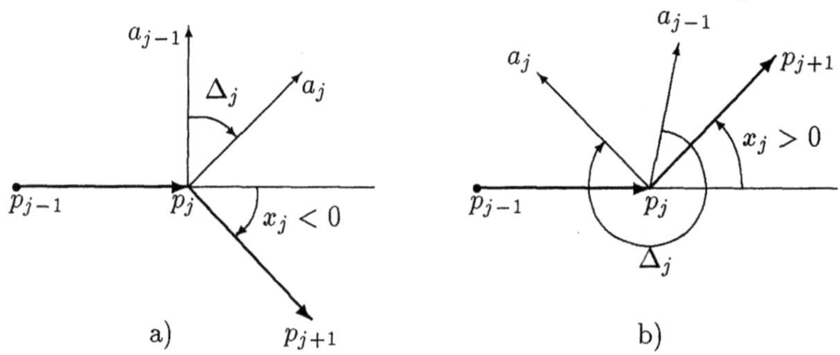


Fig. 7

Thus, each piecewise-trigonometric function belonging to $P_\rho$ gives rise to a broken line with a certain way of rotation of the tangent about the vertices. In the case of minimal trigonometrically $\rho$-convex functions in the class $M_\rho \setminus O_\rho$ (see T. 4.1.5) the rotation about each vertex is through an angle larger than $\pi$. Conversely, each broken line with cyclicity indicated above and with the above way of rotation of the tangent about the vertices can be associated with a piecewise trigonometric function belonging to $P_\rho$ for some $\rho > 0$. To study this relationship, we need to recall a certain notion.

Suppose we are going along a broken line in the direction of its orientation.

We agree that the rotation angle $x_j$ of the tangent about the vertex $p_j$ is *positive* if the tangent turns counterclockwise, and *negative* if it turns clockwise (see Fig. 7).

**Definition 4.1.6.** Let $T$ be an oriented broken line in $\mathbb{C}$ with finite number of vertices. The *angular order* $\delta = \delta(T)$ *of* $T$ is the sum of rotation angles of the tangent about the vertices, divided by $2\pi$ and taken modulo 1; specifically, we agree that this quantity lies in the interval $(1/2, 1/2]^{*)}$.

**Definition 4.1.6′.** Suppose $\gamma \geqslant 0$ and $P = \{p_j : j \in \mathbb{Z}\}$ is a $(k,\gamma)$-cyclic broken line ($k \in \mathbb{N}$), i.e., a broken line the vertices of which satisfy the conditions

$$p_j \neq p_{j+1},\ p_j = p_{j+k}e^{i2\pi\gamma}\ \forall\ j \in \mathbb{Z}.$$

The angular order of the broken line

$$T_j = \{p_j, p_{j+1}, \dots, p_{j+k+1}\}, \tag{4.1.22}$$

where $j \in \mathbb{Z}$ is arbitrary, is called the *angular order of* $P$.

Clearly, Definition 4.1.6′ is consistent: $\delta(T_j) \equiv \text{const}$ for $j \in \mathbb{Z}$, because

$$p_{j+1} - p_j = e^{i2\pi\gamma}(p_{j+k+1} - p_{j+k})\ \forall\ j \in \mathbb{Z}. \tag{4.1.23}$$

**Theorem 4.1.7.** *Let* $P = \{p_j : j \in \mathbb{Z}\}$ *be a* $(k,\gamma)$*-cyclic broken line* (see Def. 4.1.6′) *and* $\delta$ *its angular order. We denote by* $m$ *the number of cusps of the broken line* $T_j$ (see (4.1.22); a cusp is a point at which the rotation angle of the tangent is equal to $\pi$). *Let* $n$ *be the number of points* $T_j$ *where the rotation angle of the tangent is positive. Take an arbitrary number* $\rho > 0$ *the fractional part of which is* $\gamma$. *Suppose* 1) $\rho \geqslant n-\delta$, *or* 2) $\rho \geqslant \varkappa = k-\delta+m$ *and* $\rho > 1/2$. *Then there exists a sequence* $\Theta = \{\theta_j : j \in \mathbb{Z}\}$ *satisfying condition* (4.1.3) (*and in case* 2) *satisfying also inequality* (4.1.12)) *such that the function* $l$ *associated with* $\Theta$ *and with* $P$ *by formula* (4.1.1) *belongs to* $P_\rho \setminus O_\rho$ (*and in case* 2) *we have* $l \in M_\rho \setminus O_\rho$ (see T. 4.1.5)). *The values* $n-\delta$ *and* $\varkappa$ *are the smallest admissible numbers to which* $\rho$ *may be equal in cases* 1) *and* 2), *respectively.*

◄ For the broken line $P$ we consider a sequence of unit vectors $\{a_j\}$ satisfying condition (4.1.6): Re $\mathcal{P}_j\bar{a}_j < 0$, where $\mathcal{P}_j = p_{j+1} - p_j$, $j \in \mathbb{Z}$. Their geometric meaning was explained at the beginning of item $3^0$. By (4.1.23), we find:

$$a_j = a_{j+k}e^{i2\pi\gamma}\ \forall\ j \in \mathbb{Z}. \tag{4.1.24}$$

Let $A = A(P) = \{\alpha_j = \text{Arg}\, a_j : j \in \mathbb{Z}\}$ be the class of strictly monotone decreasing sequences satisfying (4.1.21), and let $\alpha \in A(P)$. Formulas (4.1.24) and (4.1.21) imply that the terms of the sequence $\alpha$ obey the condition

$$\alpha_j - \alpha_{j+k} = 2\pi\rho\quad \forall\ j \in \mathbb{Z}$$

for a certain $\rho > 0$ such that $\{\rho\} = \gamma$. We choose a sequence $\alpha \in A(P)$ such that in either of the cases 1) or 2), $\Delta_j = \alpha_{j-1} - \alpha_j$ is the smallest admissible angle with the property that, after rotation through this angle, the vector $a_{j-1}$ will take the position of the vector $a_j$ $(j \in \mathbb{Z})$. Let $x_j$ be the rotation angle of the tangent to the broken line in question about the vertex $p_j$, $j \in \mathbb{Z}$ (see Fig. 7). Then, in cases 1) and 2), respectively, we have

$$1)\ \Delta_j = \begin{cases} -x_j, & x_j \leqslant 0, \\ 2\pi - x_j, & x_j > 0, \end{cases} \qquad 2)\ \Delta_j = \begin{cases} 2\pi - x_j, & |x_j| < \pi, \\ 3\pi, & x_j = \pi. \end{cases}$$

Recalling that $\alpha_j - \alpha_{j+k} = \sum\limits_{i=j+1}^{j+k} \Delta_i$ and that for $i \in \{1, ..., k\}$ every additional complete turn of the vector $a_{j+i-1}$ to the position of $a_{j+i}$ contributes $2\pi$ to the sum $\alpha_j - \alpha_{j+k}$, we arrive at the following conclusion: the smallest admissible value of $\rho$ in cases 1) and 2) is equal to $n - \delta$ and $\varkappa$, respectively. Let $\theta_j = -\alpha_j/\rho$, $j \in \mathbb{Z}$. By construction, the sequence $\{a_j = e^{-i\rho\theta_j} : j \in \mathbb{Z}\}$ satisfies condition (4.1.6). Therefore, by Proposition 4.1.1 and Theorem 4.1.5, we conclude that $\Theta = \{\theta_j : j \in \mathbb{Z}\}$ is the required sequence ▶

The following example shows that the change of the orientation of the broken line influences the value of $\varkappa$ considered in case 2).

**Example.** Let $M$ be an arbitrary polygon in $\mathbb{C}$ with $k$ vertices the boundary $\partial M$ of which is oriented counterclockwise (clockwise). Suppose $P$ is a $(k, 0)$-cyclic broken line generated by $\partial M$. The angle $\beta$ at a fixed vertex of the polygon $M$ and the rotation angle $x$ for the tangent at the same vertex satisfy either the condition $\beta + x = \pi$ or the condition $\beta - x = \pi$, depending on the chosen orientation of $\partial M$. Hence, since the sum of interior angles of an arbitrary plane polygon with $k$ vertices is equal to $\pi(k-2)$, we see that $\delta(P) = 1$ $(\delta(P) = -1)$. Therefore, $\varkappa(P) = k - 1$ (respectively, $\varkappa(P) = k + 1$).

We apply the results of the above geometric analysis of piecewise-trigonometric functions in order to clarify the structure of minimal trigonometrically $\rho$-convex functions in the simplest cases (see T. 4.1.5).

**Theorem 4.1.8.** 1. *Let $1/2 < \rho < 1$. In the notation of Theorem 4.1.5, a function $l : \mathbb{R} \to \mathbb{R}$ belongs to $M_\rho \setminus O_\rho$ if and only if $l$ is a $2\pi$-periodic extension to $\mathbb{R}$ of a trigonometric function $l : [\alpha, \alpha + 2\pi] \to \mathbb{R}$ for some $\alpha \in \mathbb{R}$ satisfying conditions*

$$l(\alpha) = l(\alpha + 2\pi), \ l'_+(\alpha) > 0, \ \textit{and} \ l'_-(\alpha + 2\pi) < 0.$$

*Moreover, the broken line associated with the function $l$ is formed from $p_0 \in \mathbb{C}$, where* Re $p_0 e^{2\pi i\rho\alpha}(1 - e^{2\pi i\rho}) = 0$, *by turning $p_0$ $n$ times through the angle of $2\pi\rho$ in any direction if $\rho = m/n$ is rational, and infinitely many times through the same angle, consequtively clockwise and anticlockwise if $\rho$ is irrational.*

2. *The broken lines associated with the functions in $M_2 \setminus O_2$ are precisely the boundaries of positively oriented triangles in $\mathbb{C}$.*

3. *The broken lines associated with the functions in* $M_3 \backslash O_3$ *are those named in item* 2, *and also the broken lines the parameters of which are described in the notation of Theorem* 4.1.7 *as follows:* a) $k = 4,\ \delta = 1,\ m = 0$ (*the boundary of a positively oriented quadrangle,* see Example)*;* b) $k = 5,\ \delta = 2,\ m = 0$ (*for instance, a star oriented counterclockwise*); c) $k = 2,\ \delta = 1,\ m = 2$ (*a segment passed twice in different directions).*

◀ Let $\rho > 1/2$, and let $l \in M_\rho \setminus O_\rho$. Denote by $\gamma = \{\rho\}$ the fractional part of $\rho$. Let $P$ and $\Theta$ be the sequences associated with $l$ (see (4.1.2) and (4.1.3)), the definition of which involves the natural number $k = k(l)$. Then $P$ is a $(k, \gamma)$-cyclic broken line (see Def. 4.1.6′). The sequence $\Theta$ defined by formulas (4.1.3) satisfies condition (4.1.12) of Theorem 4.1.5. Hence we obtain the inequality

$$1 \leqslant k < 2\rho. \tag{4.1.25}$$

1. $1/2 < \rho < 1$. By (4.1.25), we have $k = 1$. Part 1 of Theorem 4.1.8 follows now from Theorem 4.1.5 and Proposition 4.1.1.

2. $\rho = 2$. Then $1 \leqslant k \leqslant 3$ (see (4.1.25)). If $k = 1$, then $l \in O_2$, which is impossible. If $k = 2$, then $P$ is a segment passed twice in different directions. Then $\varkappa(P) = 3$ and $\rho \geqslant 3$ (see Theorem 4.1.7), which contradicts the condition $\rho = 2$. It remains to consider the case where $k = 3$. If the vertices of the closed broken line $P$ consisting of three links are on a straight line, then $\varkappa(P) = 4$ and $\rho \geqslant 4$ (see T. 4.1.7). This contradiction shows that $P$ is the boundary of a triangle. By Example to Theorem 4.1.7, we arrive at part 2 of Theorem 4.1.8.

3. $\rho = 3$. Then $1 \leqslant k < 6$ by (4.1.25). We argue as before. For instance, Theorem 4.1.7 proves the possibility of the case where $k = 3,\ \delta = 2,\ m = 0$ (see item 2). The angular order of the star (see case 3 in T. 4.1.8) is equal to 2. It is calculated by parallel translation of rotation angles for the tangent to a fixed point of the plane ▶

**4.1.4. Zorn theorem for the class $P_\rho$.** We study the applicability of the Zorn lemma to the class $P_\rho$.

**Theorem 4.1.9.** *For* $k \in P_\rho$, *the existence of a minimal element* $l \leqslant k$ *is equivalent to the existence of a finite function* $h \in P_\rho$ *with* $h \leqslant k$.

◀ The necessity follows from Theorem 4.1.5: $l$ is a finite function.

*Sufficiency.* Let $P'_\rho$ be the subclass of finite functions in $P_\rho$. Suppose $h \in P'_\rho$ and $h \leqslant k$. We consider the set $K_h = \{g \in P'_\rho : g \leqslant h\}$ and its arbitrary lineraly ordered subset $A$.

Let $H_A(\theta) = \inf\{g(\theta) : g \in A\}$, $\theta \in \mathbb{R}$. For all $a, b > 0$ and $\theta, \psi \in \mathbb{R}$ we have the elementary formula

$$aH_A(\theta) + bH_A(\psi) = \inf\{ag(\theta) + bg(\psi) : g \in A\}. \tag{4.1.26}$$

Since $g \in A \subset P'_\rho$, by (1.9.7) and Def. 1.9.2′ the right-hand side is nonnegative for $a = b = 1$, $\psi = \theta + \pi/\rho$, $\theta \in \mathbb{R}$. Consequently, $H_A$ is a finite function.

Recalling that in formula (1.9.2) the factors at $h(\theta), h(\theta_1)$ and $h(\theta_2)$ are positive for $\theta \in (\theta_1, \theta_2)$, and that $g \in P'_\rho \ \ \forall\, g \in A$, by (4.1.26) we find that $H_A \in P'_\rho$. So, $H_A$ is a lower bound for $A$. By the Zorn lemma, $K_h$ has a minimal element $l \leqslant h$. Let $r \in P_\rho$ satisfy $r \leqslant l$. Then $r \in K_h$ and, therefore, $r = l$. So, $l \in M_\rho$. But $h \leqslant k$, consequently, also $l \leqslant k$ ▶

The following example shows that the assumption of Theorem 4.1.9 may fail.

**Example 4.1.10.** A. Let $\rho > 1$ and let $0 < \epsilon < \min\{1/\sqrt{2}, 2(\rho - 1)\}$. We put $\mu = \rho/\epsilon$. Let $h$ be the $2\pi$-periodic extension to $\mathbb{R}$ of the function defined by $k(\theta) = -\sqrt{\sin \mu\theta}$ for $0 \leqslant \theta \leqslant \pi/\mu$, $k(-\pi/\rho) = 0$, and $k(\theta) = \infty$ for $\theta \in (-\pi/\rho, 0) \cup (\pi/\mu, 2\pi - \pi/\rho)$. Then

$$k''(\theta) + \rho^2 k(\theta) = \frac{(\mu^2 - 4\rho^2)\sin^2 \mu\theta + \mu^2}{4(\sqrt{\sin \mu\theta})^3} \geqslant 0,$$

$\theta \in (0, \pi/\mu)$, i.e., the condition of trigonometric $\rho$-convexity of the function $k$ on the interval $(0, \pi/\mu)$ (see T. 1.9.17) is fulfilled, and, moreover, the function $k$ is continuous on the segment $[0, \pi/\mu]$. Combining this with the unequality $2\pi - \pi/\rho - \pi/\mu > \pi/\rho$ and the relations

$$k(-\pi/\rho) + k(0) = 0, \qquad k(0) = k(-\pi/\rho) = 0, \tag{4.1.27}$$

we see that $k \in P_\rho$ (see Def. 1.9.2′).

Assume that there is a finite function $h$ in $P_\rho$ satisfying $h \leqslant k$. Combining this with formulas (1.9.7), and (4.1.27), we find $h(0) = 0$. By Theorem 1.9.24, Ch. 1, $Dh(0) \neq \varnothing$, i.e., there exists $p \in \mathbb{C}$ such that Re $pe^{i\rho\theta} \leqslant h(\theta)$ for $|\theta| \leqslant \pi/\rho$, and Re $p = h(0)$. But $h \leqslant k$, consequently, $p \in Dk(0)$. On the other hand, the right derivative of the function $k$ at the point 0 is equal to $\infty$ and, by Property 1.9.21′, we have $Dk(0) = \varnothing$. This is a contradiction. Consequently, for this function $k$ the assumption of Theorem 4.1.9 fails.

B. Let $1/2 < \rho < 1$, and let $\rho \leqslant \alpha < 1$. We put $\epsilon = (2\alpha - 1)^2$, and $\mu = \rho/(2\rho - 1)$. Let $k_\alpha$ be the $2\pi$-periodic extension to $\mathbb{R}$ of the function

$$k_\alpha(\theta) = \begin{cases} -(\sin \mu\theta)^\epsilon, & \theta \in [0, \pi(2 - 1/\rho)], \\ \infty, & \theta \in (\pi(2 - 1/\rho), 2\pi). \end{cases}$$

Arguing as above, we prove that 1) $k_\alpha \in P_\rho$ for $\rho \leqslant \alpha, and$ 2) there is no finite functions $h \in P_\rho$ with $h \leqslant k_\alpha$.

For other values of $\rho$ such examples cannot be constructed.

**Proposition 4.1.11.** *For $\rho \in (0, 1/2]$, $\rho = 1$ the functions belonging to $P_\rho$ satisfy the assumption of Theorem* 4.1.9.

◀ For $\rho \leqslant 1/2$ this follows from Property 1.9.22. Let $\rho = 1$, and let $g \in P_\rho$. We denote by $F_g$ a set of all components of dom $g$. The following cases are possible.

1. There is an element $\Delta \in F_g$ with the property $\overset{\circ}{\Delta} \neq \varnothing$. Let $\psi \in \overset{\circ}{\Delta}$, and let $p \in Dg(\psi)$ (see (1.9.13)). Hence, recalling that between $\Delta$ and $\Delta + 2\pi$ there are no other elements of $F_g$ (see Prop. 1.9.22), we conclude that the function $h(\theta) = \text{Re}\, pe^{i\rho\theta}$, $\theta \in \mathbb{R}$, is required .

2. On some half-interval of length $2\pi$ there is one element $\Delta = \{\alpha\} \in F_g$ or two elements $\Delta = \{\alpha\}, \sum = \{\beta = \alpha + \pi\} \in F_g$. Then, there exist numbers $p$, $q \in \mathbb{C}$ such that $g(\alpha) = \text{Re}\, pe^{i\alpha}$, $g(\beta) = \text{Re}\, pe^{i\beta}$. In the first case $h(\theta) = \text{Re}\, pe^{i\theta}$, $\theta \in \mathbb{R}$. In the second case such a function is defined by Lemma 4.1.4 ($g(\alpha) + g(\alpha + \pi) \geqslant 0$, see Def. 1.9.2′) ▶

## 4.2. Many-sheeted diagrams associated with the functions of class $P_\rho$

Let $h$ be a finite trigonometrically $\rho$-convex function of class $P_\rho$. The support straight line $\{p \in \mathbb{C} : \text{Re}\, pe^{i\rho\theta} = h(\theta)\}$ of the locally convex arc $\Gamma_h$ associated with $h$ bounds the half-plane

$$\Pi_h(\theta) = \{p \in \mathbb{C} : \text{Re}\, pe^{i\rho\theta} \geqslant h(\theta)\}. \tag{4.2.0}$$

As the parameter $\theta$ varies, the above straight line moves along $\Gamma_h$, and the half-plane sweeps the many-sheeted surface $E_h$ if $\rho \neq 1$. The surface $E_h$ is located on the polygonal surface $E_l$, where $l$ is an element of $M_\rho$ such that $h \in P_\rho(l)$ (see (4.0.2)). We introduce the notion of the diagram of $h$ $\rho$-sheeted in the vicinity of $\infty$, which is related to the surfaces $E_h$, $E_l$ in a constructive manner. Next, we examine the properties of such surfaces. Moreover, in the case of nonintegral $\rho$ we examine their relationship with the special coverings considered in A.2.4.

**4.2.1. Definition of many-sheeted polygons and polygonal surfaces.** On the Riemann sphere $\overline{\mathbb{C}}$, we consider analogs of well-known notions of a segment, a broken line, and a polygon in $\mathbb{C}$. Let $z, w \in \overline{\mathbb{C}}$, $z \neq w$. By definition, a *segment* on $\overline{\mathbb{C}}$ with vertices (ends) $z$, $w$ is either a usual segment in $\mathbb{C}$ if $z, w \in \mathbb{C}$ or, if one of $z$, $w$ is the point at infinity, a ray closed in the topology of $\overline{\mathbb{C}}$ with vertex at the other point. By a *polygon* on $\overline{\mathbb{C}}$ we mean a closed set in $\mathbb{C}$ bounded by a closed Jordan broken line the links of which are segments in $\overline{\mathbb{C}}$.

**Definition 4.2.1.** A. Let $X$ be a topological space, $T$ a subset of $X$, and $\varphi : T \to \Delta$ a homeomorphism of $T$ onto a certain closed polygon (segment) $\Delta$ on $\overline{\mathbb{C}}$. Then the pair $(\varphi, T)$ (or the set $T$ alone) is called a *topological polygon* (*segment*)*) on $X$. The preimages under $\varphi$ of the vertices and sides of $\Delta$ are called the *vertices* and *sides* of $T$, and the angle of $T$ at its vertex $m$ is called

*) For short, often the word "topological" is omitted.

the *angle* of $\Delta$ at $p = \varphi(m)$. The *orientation of the polygon* $(\varphi, T)$ is the orientation of the image $\varphi(T)$. If $\varphi(m) = \infty$, then the *angle of the polygon* $\Delta$ *at* $\infty$ is the angle at 0 between the images of the sides with a common vertex at $\infty$ under the transformation $1/z$.

B. If $T = \bigcup T_j | j \in J$, where $J = \mathbb{Z}$ or $J = \{1, \dots, k\}$ and each $(\varphi, T_j)$ is a topoloical segment on $X$ with vertices $m_{j-1}, m_j$ $\forall\ j \in J$, then the pair $(\varphi, T)$ (or the set $T$ alone) is called a *topological broken line* on $X$, and the vertices of theses segments are called the *vertices of the broken line* .

We shall need the following modification of the familiar notion of a triangulable manifold.

**Definition 4.2.2.** Let $M$ be a two-dimensional manifold with boundary $\Gamma$ or without boundary ($\Gamma = \varnothing$). Assume that there exists a collection $\Delta = \{M_j : j \in J\}$ (see Def. 4.2.1) of topological polygons on $M$ with the following properties:

1) $\overset{\circ}{M}_i \cap \overset{\circ}{M}_j = \varnothing$, $i \neq j$, $i, j \in J$; $M = \bigcup M_j | j \in J$;

2) if a point $m$ belongs to a side $T$ of a polygon $M_i \in \Delta$, but is not its vertex, then either $T \subset \Gamma$, or there exists only one polygon $M_j \in \Delta$, $j \neq i$, such that $T \subset M_i \cap M_j$ and $M_i \cup M_j$ is a neighborhood of $m$;

3) if $m \in M$ and the set $\Delta_m$ of all polygons in $\Delta$ with vertex $m$ is nonempty, then $\Delta_m$ is finite and can be ordered ($\Delta_m = \{M_{j_1}, \dots, M_{j_k}\}$) in such a manner that each two consequtive polygons have only one side with vertex at $m$ in common and, moreover, the polygons $M_{j_1}$, $M_{j_k}$ have sides $T \subset M_{j_1}$, $S \subset M_{j_k}$ with vertex $m$ such that either $T = S$, or $T \neq S$ and $\{T, S\} \subset \Gamma$*). Furthermore, $M_{j_1} \cup \ldots \cup M_{j_k}$ is a neighborhood of $m$.

Then the system $\Delta$ is called a *triangulation* of $M$, and $M$ is called a *triangulable manifold* or *surface*.

**Definition 4.2.3.** Suppose $M$ is a simply connected two-dimensional manifold with boundary $\partial M$, $\varphi : M \to \overline{\mathbb{C}}$ is a continuous mapping and, moreover, $\varphi : \overset{\circ}{M} \to \overline{\mathbb{C}}$ is a covering (see Definitions A.1.1, A.1.3) Assume (see Definitions 4.2.2, 4.2.1) that:

a) there exists a triangulation $\Lambda = \{(\varphi, M_j) : j \in J\}$ of $M$ consisting of topological polygons on $M$ determined by $\varphi$;

b) the pair $(\varphi, \partial M)$ is a topological Jordan broken line on $M$;

c) in the set $\varphi(M)$ there exists a point $z$ such that the number $N_z$ of elements in the fiber $\varphi^{-1}(z)$ exceeds 1.

If $\varphi^{-1}(\infty) = \varnothing$, and the broken line $(\varphi, \partial M)$ consists of infinitely many links, then the pair $(\varphi, M)$ (or the manifold $M$ alone) is called a *many-sheeted polygonal surface*. In the case where 1) $\Lambda$ is a finite set (consequently, $M$ is compact); 2) $(\varphi, \partial M)$ is a closed broken line; and 3) the set $\varphi^{-1}(\infty)$ contains at most one branching point $m_\infty$ for $\varphi$, the pair $(\varphi, M)$ (or the set $M$ alone) is called a *many-sheeted polygon* .

---

*) The latter condition is fulfilled, in particular, if $k = 1$, i.e., $\Delta_m = \{M_{j_1}\}$, $T \neq S$, and $\{T, S\} \subset M_{j_1}$.

**Remark.** In the notation of Definition 4.2.3, we orient the triangulation polygons $\Lambda$ consistently (see Def. 4.2.1). Since $\varphi$ is open on $\overset{\circ}{M}$ (see Def. A.1.1), we see that $\Lambda$ has the following property: if $M_{ij}$ is a common side of the polygons $(\varphi, M_i)$ and $(\varphi, M_j)$, $i \neq j$, then $\varphi(M_{ij})$ is a common side of the polygons $\varphi(M_i)$, $\varphi(M_j)$, and, moreover, *consistent orientations* of the polygons $(\varphi, M_i)$ and $(\varphi, M_j)$ induce *opposite orientations* on $M_{ij}$. In the case of convex polygons $\varphi(M_i)$ and $\varphi(M_j)$ this means that they are located on different sides of their common edge $\varphi(M_{ij})$. Hence, under the conditions of Definition 4.2.3, $M$ is an orientable manifold.

**Definition 4.2.3′.** A polygonal surface (polygon) $(\varphi, M)$ (see Def. 4.2.3) is said to be *positively oriented* if the orientation induced by $M$ on the boundary $\partial M$ has the property that $\varphi(M)$ remains on the left when bypassed along $\varphi(\partial M)$.

**Definition 4.2.4.** Under the conditions of Definition 4.2.3, let $m \in \partial M$ be a point such that the set $\Lambda_m$ of the triangulation polygons belonging to $\Lambda$ and having a vertex at $m$ is nonempty. We denote by $\Sigma_m$ the sum of the angles of these polygons at $m$. If $\Sigma_m \neq \pi$, then $m$ is called a *vertex of the polygonal surface (of the polygon)* $(\varphi, M)$, and the number $\Sigma_m$ is called *the value of the angle of* $(\varphi, M)$ *at* $m$.

From Definition 4.2.2 and the assertion mentioned in the remark to Definition 4.2.3, it follows that the number $\Sigma_m$ does not depend on the partition $\Lambda$. Also, polygonal surfaces equivalent in the sense of the definition below have identical values of the angles at respective vertices.

**Definition 4.2.5.** Two polygonal surfaces (broken lines) $(\varphi, M)$ and $(\psi, N)$ are said to be *equivalent* if there exists a homeomorphism $\Phi : M \to N$ such that $\varphi = \psi \circ \Phi$.

Equivalent oriented broken lines $(\varphi, M)$ and $(\psi, N)$ are said to be *(anti)isomorphic* if the above homeomorphism $\Phi : M \to N$ is an (anti)-isomorphism of the totally ordered sets $M$ and $N$ (see subsection 1.3.2)*). In what follows, such surfaces (broken lines) will sometimes be identified without special mention.

**Example 4.2.6.** Let $[A, D, B, E, C, A]$ be an oriented closed broken line in $\mathbb{C}$ (see Fig. 8) determining the contour of a star. Consider the triangles $\Delta_1 = \Delta ADB$, $\Delta_2 = \Delta ECA$, and denote by $\Delta_3$ the closure of the exterior of $\Delta BEA$ in the topology of the sphere $\overline{\mathbb{C}}$. Let $i \in \{1, 2, 3\}$. An arbitrary point $m \subset \Delta_i$ is described by the pair $(p, i)$, where $p$ is the complex coordinate of $m$. We glue the triangles $\Delta_1$ and $\Delta_3$ along the common segment $AB$, and the triangles $\Delta_3$ and $\Delta_2$ along the segment $AE$. The resulting manifold $M = \Delta_1 \cup \Delta_2 \cup \Delta_3$ together with the projection $\varphi : M \to \overline{\mathbb{C}}$, $\varphi(p, i) = p \; \forall \; p \in \Delta_i$, $i = 1, 2, 3$, is a positively oriented many-sheeted polygon, one-sheeted in the vicinity of $\infty$, and with the angles $2\pi + \alpha$, $2\pi + \beta$, $\gamma$, $\delta$, and $2\pi + \sigma$ at the vertices $A$, $B$, $C$,

*) An orientation on a topological broken line can be understood as an order relation on it.

$D$, and $E$, respectively; moreover, $\alpha + \beta + \gamma + \delta + \sigma = \pi$.

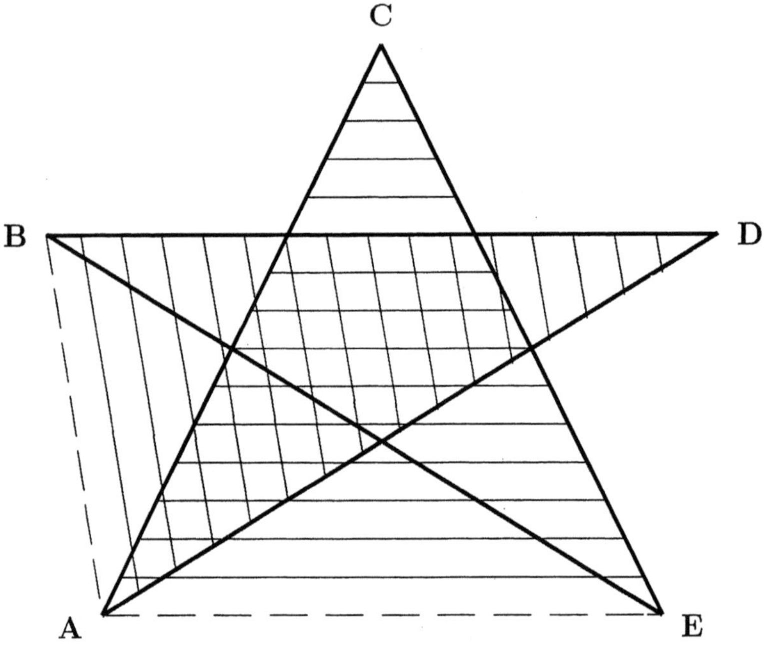


Fig. 8

**4.2.2. The infinitely-sheeted surface $E_h$ associated with a function $h$ in $P_\rho(l)$.** Let $l$ be a piecewise-trigonometric function of class $P_\rho$, and let $h \in P_\rho(l)$ (see (4.0.2)). We formalize the construction of the surfaces $E_l$ and $E_h$ shaped, respectively, by the movement of the half-planes

$$\Pi_l(\theta) = \{(p, \theta) : \ p \in \mathbb{C}; \ \text{Re}\, pe^{i\rho\theta} \geqslant l(\theta)\}, \tag{4.2.1}$$

and $\Pi_h(\theta)$ (see (4.2.0)) as the parametr $\theta$ runs through $\mathbb{R}$. In the set-theoretical union of the half-planes

$$\widetilde{E}_l := \bigcup \Pi_l(\theta) | \theta \in \mathbb{R},$$

we introduce an equivalence relation $F$ as follows:

$$(p, \theta_1) \sim (q, \theta_2) \quad \text{if} \quad p = q, \quad \theta_1, \theta_2 \in \Delta,$$

where $\Delta$ is some connected component of the set

$$X(p, l) := \{\theta \in \mathbb{R} : \ \text{Re}\, pe^{i\rho\theta} \geqslant l(\theta)\}. \tag{4.2.2}$$

We endow the factor-set $E_l = \widetilde{E}_l/F$ with the topology of sum: a set $U \subset E_l$ is open if $\varkappa^{-1}(U) \cap \Pi_l(\theta)$ is open in the relative topology of $\Pi_l(\theta)$ $\forall\, \theta \in \mathbb{R}$, where $\varkappa : \widetilde{E}_l \to E_l$ is the canonical mapping. We denote the elements of $E_l$ by the symbol $(p, \Delta)$, where $p \in \mathbb{C}$ and $\Delta \in X(p, l)$. Next, we define the mapping

$$\varphi_l : \ E_l \to \mathbb{C}; \quad E_l = \widetilde{E}_l/F, \quad \varphi_l[(p, \Delta)] = p. \tag{4.2.3}$$

Since $\Pi_h(\theta) \subset \Pi_l(\theta)$ $\forall\, \theta \in \mathbb{R}$ (see (4.2.0), (4.2.1)), we have:

$$\widetilde{E}_h(\theta) := \bigcup \Pi_h(\theta) | \theta \in \mathbb{R} \subset \widetilde{E}_l; \quad E_h := \varkappa(\widetilde{E}_h) \subset E_l.$$

**Definition 4.2.7.** The set $E_l$ or the pair $(\varphi_l, E_l)$ is called the *surface associated with a piecewise-trigonometric function* $l \in P_\rho$. The set $E_h$, where $h \in P_\rho(l)$ (see (4.0.2)), or the pair $(\varphi_l, E_h)$ is called the *surface associated with* $h$.

So, the concave surface $(\varphi_l, E_h)$ (this is a union of half-planes) is situated on the surface $(\varphi_l, E_l)$. We examine the properties of $(\varphi_l, E_l)$.

Two cases are possible.

1. Put $O_\rho = \{\text{Re}\, p_0 e^{i\rho\theta},\ \theta \in \mathbb{R} :\ p_0 \in \mathbb{C}\}$ for $\rho \in \mathbb{N}$ and $O_\rho = \{l(\theta) \equiv 0,\ \theta \in \mathbb{R}\}$ for $\rho \notin \mathbb{N}$ (see Theorem 4.1.5). If $l \in O_\rho$, then the space $E_l$ is formed by rotation of the half-plane $\Pi_l(\theta)$ about the point $p_0$, i.e., $E_l$ has two logarithmic branching points. One of them is the point $\Lambda_0 = (p_0, \mathbb{R})$ with the property $\varphi_l(\Lambda_0) = p_0$. Let $L = (0, \infty) \times \mathbb{R}$ be the Riemann surface of the logarithm. Consider the mapping

$$\chi_l' : \ L \to E_l; \quad \chi_l'[(r, \varphi)] = (p, \Delta), \quad p = re^{i\varphi} + p_0, \tag{4.2.4}$$

where $\Delta$ is the component of $X(p, l)$ (see (4.2.2)) containing $-\varphi/\rho$. But $X(p, l) = \{\theta \in \mathbb{R} :\ \text{Re}\ (p - p_0)e^{i\rho\theta} \geqslant 0\}$ and, consequently, the length of $\Delta$ does not exceed $\pi/\rho$ for $p \neq p_0$. We have arrived at the following fact.

**Property 4.2.8.** *Suppose* $\rho > 0$, $p_0 \in \mathbb{C}$, *and define* $l(\theta) = \text{Re}\ p_0 e^{i\rho\theta}$ *for* $\theta \in \mathbb{R}$, *where* $p_0 = 0$ *if* $\rho$ *is a noninteger. Let* $L = (0, \infty) \times \mathbb{R}$ *be the Riemann surface of the logarithm. Consider the mapping* $\pi' : L \to \mathbb{C}$, $\pi'(r, \varphi) = p_0 + re^{i\varphi}$. *Then, in the notation of Definition 4.2.7, the surface* $E_l \setminus \{\Lambda_0\}$ *(where* $\Lambda_0 = (p_0, \mathbb{R})$ *is the logarithmic branching point of* $E_l$*) is equivalent to* $L$, *i.e., there exists a homeomorphism* (4.2.4) $\chi_l : L \to E_l \setminus \{\Lambda_0\}$, *and the following diagram is commutative*:

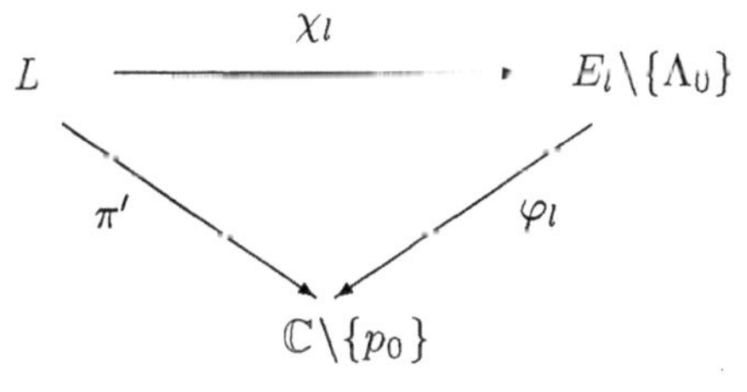


$$\pi' = \varphi_l \circ \chi_l \ ,$$

2. Let $l \in P_\rho \backslash O_\rho$. Then (see Theorem 4.1.5) $l$ is the piecewise-trigonometric function defined by (4.1.1). In this case the space $E_l$ is formed by rotation of the half-plane $\Pi_l(\theta)$ about the point $p_j$ as $\theta$ ranges through $[\theta_{j-1}, \theta_j]$ for every $j \in \mathbb{Z}$. The half-plane $\Pi_l(\theta_j)$ passes through $p_j$ and $p_{j+1}$, and is on the left of the vector $\overline{p_j p_{j+1}}$ (see (4.1.6)). The sequence $P = \{p_j : j \in \mathbb{Z}\}$ determines an oriented $(k, \rho)$-cyclic broken line in $\mathbb{C}$ (see Def. 4.1.6′).

We show that $E_l$ has a logarithmic branching point.

Let $L$ be the Riemann surface of the logarithm; we put

$$\sigma = \max\{l(\theta) : \theta \in \mathbb{R}\}, \quad H_\sigma = \{(r, \varphi) \in L : r > \sigma;\ \varphi \in \mathbb{R}\},$$

$$U_\infty = \{(p, \Delta) \in E_l : |p| > \sigma\}.$$

Consider the operator

$$\chi_l : H_\sigma \to U_\infty;\ \chi_l[(r, \varphi)] = (p, \Delta),\ p = re^{i\varphi}, \tag{4.2.5}$$

where $\Delta$ is the component of $X(p, l)$ (see (4.2.2)) containing $-\varphi/\rho$. Let $(p, \Delta) \in U_\infty$. Then $l(\theta) \not\equiv \operatorname{Re} pe^{i\rho\theta}$, $\theta \in \Delta$. From Property 1.9.8 and the fact that $l \notin O_\rho$, $\rho > 1/2$, it follows that the length of $\Delta$ does not exceed $\pi/\rho$. Therefore, the mapping $\chi_l$ is a homeomorphism. We arrive at the following fact.

**Property 4.2.9.** *In the notation of Theorem* 4.1.5 *and formula* (4.2.3), *for every piecewise-trigonometric function* $l \in P_\rho \setminus O_\rho$ *given by* (4.1.1), *the surface* $(\varphi_l, E_l)$ *associated with* $l$ (see Def. 4.2.7) *has the following properties:*

1. *The oriented* $(k, \rho)$*-cyclic broken line* $P$ *induces on* $\partial E_l$, $E_l$ *an orientation with respect to which* $(\varphi_l, E_l)$ *is a positively oriented* (see Def. 4.2.3′) *polygonal surface with vertices* $\{m_j = (p_j, [\theta_{j-1}, \theta_j]) : j \in \mathbb{Z}\}$, *the angle at* $m_j$ *is equal to* $\rho(\theta_j - \theta_{j-1}) + \pi\ \ \forall\, j \in \mathbb{Z}$, *and* $\varphi_l(m_j) = p_j$, $j \in \mathbb{Z}$.

2. *For* $h \in P_\rho(l)$, *let* $(\varphi_l, E_h)$ *be the surface and* $\Gamma_h$ *the curve in* $\mathbb{C}$ *associated with* $h$ (see Def. 1.9.24) *(if* $h = l$, *the curve* $\Gamma_l = \{p_j : j \in \mathbb{Z}\} = P$ *is a broken line in* $\mathbb{C}$*). Then* $\Gamma_h = \varphi_l(\partial E_h)$. *A similar property is also valid for* $l \in O_\rho$.

3. *The surface* $E_l$ *has a logarithmic branching point the neighborhood* $U_\infty$ *of which* (see (4.2.5)) *is homeomorphically included in the Riemann surface* $L$ *of the logarithm, and in the notation of formula* (4.2.5) *the following diagram is commutative:*

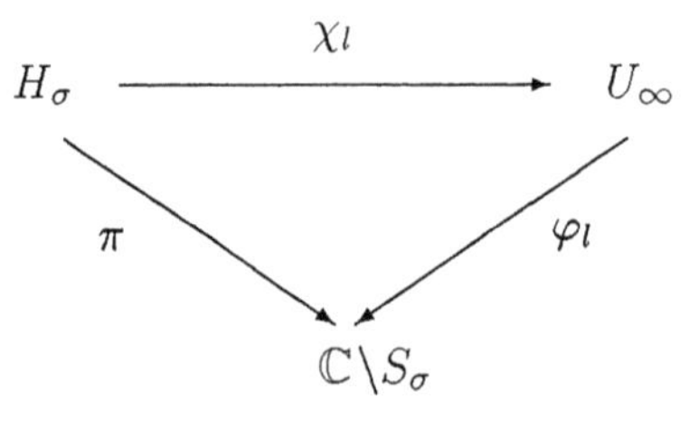


$$\pi = \varphi_l \circ \chi_l\ ,$$

*where* $\pi : L \to \mathbb{C}$, $\pi[(r, \varphi)] = re^{i\varphi}$, *and* $S_\sigma = \{p \in \mathbb{C} : |p| \leqslant \sigma\}$.

Since $l$ is a $2\pi$-periodic function, on the surface $E_l$ the following automorphism is well defined:

$$B: \ E_l \to E_l; \ B[(p, \Delta)] = (q, \Sigma), \ q = pe^{i2\pi\rho} \tag{4.2.6}$$

where $\Sigma = \Delta - 2\pi$. Geometrically, this means that for every $\theta \in \mathbb{R}$ the mapping $B$ is a bijection of the half-plane $\Pi_l(\theta)$ (see (4.2.1)) onto the half-plane $\Pi_l(\theta - 2\pi)$. The mapping $B$ generates a cyclic group $G_E$ of automorphisms acting on $E_l$. Its relationship with the automorphism

$$A_\rho: \ L \to L; \ A_\rho[(r, \varphi)] = (r, \varphi + 2\pi\rho), \ (r, \varphi) \in L \tag{4.2.7}$$

acting on the Riemann surface of the logarithm $L$ is illustrated by the following commutative diagrams (see Properties 4.2.8, 4.2.9):

$$\begin{array}{ccccc} L & \xrightarrow{\chi_l} & E_l\backslash\{\Lambda_0\} & \xrightarrow{\varphi_l} & \mathbb{C}\backslash\{p_0\} \\ \downarrow{\scriptstyle A_\rho} & & \downarrow{\scriptstyle B} & & \downarrow{\scriptstyle e^{i2\pi\rho}} \\ L & \xrightarrow{\chi_l} & E_l\backslash\{\Lambda_0\} & \xrightarrow{\varphi_l} & \mathbb{C}\backslash\{p_0\} \end{array} \qquad \begin{array}{ccccc} H_\sigma & \xrightarrow{\chi_l} & U_\infty & \xrightarrow{\varphi_l} & \mathbb{C}\backslash S_\sigma \\ \downarrow{\scriptstyle A_\rho} & & \downarrow{\scriptstyle B} & & \downarrow{\scriptstyle e^{i2\pi\rho}} \\ H_\sigma & \xrightarrow{\chi_l} & U_\infty & \xrightarrow{\varphi_l} & \mathbb{C}\backslash S_\sigma . \end{array}$$

$$l \in O_\rho \qquad\qquad\qquad\qquad l \in P_\rho\backslash O_\rho$$

From this we deduce that

$$B^k \circ \chi_l = \chi_l \circ A_\rho^k \tag{4.2.8}$$

for every $k \in \mathbb{Z}$ (see (4.2.5)).

Let $l \in P_\rho \backslash O_\rho$. Then on every interval of length $2\pi$ there is a point $\theta'$ such that $l(\theta') > 0$ (see Properties 1.9.7-1.9.8). Without loss of generality we may assume that $\theta' \in [\theta_{-1}, \theta_0]$ (see (4.1.1)). Then $p_0 \neq 0$. The half-plane $\Pi_l(\theta')$ (see (4.2.1)) contains the ray $\widetilde{L}_0 = \{(tp_0, \theta') : t \geqslant 1\}$ with vertex at $p_0$. Let $\varkappa : \widetilde{E}_l \to E_l$ be the canonical mapping (see (4.2.3)); we put

$$\Gamma_0 = \varkappa(\widetilde{L}_0); \quad \Gamma_i = B^{-i}\Gamma_0, \quad i \in \mathbb{Z} \setminus \{0\},$$

where $B^{-i}$ is the element of order $-i$ in the group $G_E$ with generator $B$ (4.2.6). Since $m_0$ is a vertex of $(\varphi_l, \Gamma_0)$ (see Property 4.2.9), the vertex of the ray $(\varphi_l, \Gamma_i)$ coincides with the vertex $m_{ik}$ of the surface $(\varphi_l, E_l)$, where $k = k(l)$ is the number associated with the function $l$ (see (4.1.3)). So,

$$B\Gamma_{i+1} = \Gamma_i, \quad B[m_{\alpha+k}, m_{\alpha+k+1}] = [m_\alpha, m_{\alpha+1}] \ \ \forall \ \alpha \in \mathbb{Z}, \tag{4.2.9}$$

where $[m_\alpha, m_{\alpha+1}] = \{(p, \Delta) \in \partial E_l; \ p \in [p_\alpha, p_{\alpha+1}]; \ -\theta_\alpha/\rho \in \Delta\}$ is the link of the broken line $(\varphi_l, \partial E_l)$ connecting the vertices $m_\alpha$, $m_{\alpha+1}$ of the surface

$(\varphi_l, E_l)$ (see (4.1.3), Property (4.2.9)). By (4.2.9), we conclude that the boundary $\partial E_l$ together with the system of rays $\{(\varphi_l, \Gamma_i) : i \in \mathbb{Z}\}$ breaks the surface $E_l$ into polygonal fundamental domains $\{D_i^+ : i \in \mathbb{Z}\}$ with the property that $\partial D_{i-1}^+ \cap \partial D_i^+ = \Gamma_i$, $i \in \mathbb{Z}$. The closure of each domain $D_i^+$ has a completion $M_i^+$ obtained by adjoining a single point $\Lambda_\infty$ such that $\varphi_l(\Lambda_\infty) = \infty$. Furthermore, $(\varphi_l, M_i^+)$ is a polygon with angle $2\pi\rho$ at the vertex $\Lambda_\infty$. Thus, we have proved the following fact.

**Property 4.2.10.** *On the surface $(\varphi_l, E_l)$, the group $G_E$ of automorphisms with the generator $B$* (see Def. 4.2.7 and (4.2.6)) *acts freely and totally discontinuously. This $B$ satisfies the condition*

$$\varphi_l(Bm) = e^{i2\pi\rho}\varphi_l(m), \quad m \in E_l. \tag{4.2.10}$$

*The group $G_E$ has the following properties:*

1) *Let $G_\rho$ be the cyclic group of automorphisms on the Riemann surface $L$ of the logarithm with the generator $A_\rho$* (see (4.2.7)). *Then the mappings* (see (4.2.4), (4.2.5), (4.2.8))

$$\chi_l : L \to E_l \setminus \{\Lambda_0\} \quad \textit{for } l \in O_\rho; \quad \chi_l : H_\sigma \to U_\infty \quad \textit{for } l \in P_\rho \setminus O_\rho$$

*agree with the action of the groups $G_E$, $G_\rho$.*

2) *For $l \in O_\rho$, the closure of the fundamental domain of the group $G_E$ on $E_l \setminus \{\Lambda_0\}$ coincides with the set $\chi_l(L_\rho)$, where $L_\rho = \{(r, \varphi) \in L;\ r > 0;\ \alpha \leqslant \varphi \leqslant \alpha + 2\pi\rho\}$ is any angle of opening $2\pi\rho$ on $L$. For $l \in P_\rho \setminus O_\rho$, adjoining one point $\Lambda_\infty$ such that $\varphi_l(\Lambda_\infty) = \infty$ yields the completion $M_0^+$ such that $(\varphi_l, M_0^+)$ is a polygon with angle $2\pi\rho$ at the vertex $\Lambda_\infty$.*

Further on we shall need yet another property of the surface $E_l$, characterizing its logarithmic branching point $\Lambda_\infty$.

**Property 4.2.11.** *Let $V_\infty = \{(r, \varphi) \in L : 0 < r < \varepsilon_\infty,\ \varphi \in \mathbb{R}\}$. Consider the automorphism*

$$A : L \to L; \quad A[(r, \varphi)] = (r, \varphi + 2\pi), \quad (r, \varphi) \in L \tag{4.2.11}$$

*acting on $L$. Let $G_L$ be the cyclic group of automorphisms generated by $A$, and let $l \in P_\rho \setminus O_\rho$. There exists a number $\varepsilon_\infty > 0$ and a homeomorphism $\Psi_\infty : U_\infty \to V_\infty$* (see (4.2.5), Definition 1.8 Appendix 1) *that agrees with the action of $G_E$ on $U_\infty$ and of $G_L$ on $V_\infty$ and satisfies the condition*

$$\begin{aligned} A^{-1}\Psi_\infty(m) &= \Psi_\infty \circ B(m), \quad m \in U_\infty, \\ \varphi_l \circ \Psi_\infty^{-1}(T) &\equiv T^{-\rho}, \quad T \in V_\infty, \end{aligned} \tag{4.2.12}$$

*where the symbol $T^{-\rho}$ is understood in the sense of its principal value (i.e., $T^{-\rho} = r^\rho e^{i\rho\varphi}$ if $T = (r, \varphi)$). Let, moreover,*

$$\begin{aligned} T_\theta &= \{p \in \mathbb{C} : \ \mathrm{Re}\ pe^{i\rho\theta} > l(\theta)\}, \\ U_\theta &= \{(p, \Delta) \in E_l : \ p \in T_\theta,\ \theta \in \Delta\}. \end{aligned} \tag{4.2.13}$$

*Then for every sufficiently small* $\varepsilon > 0$ *we have*

$$V(\varepsilon, \theta) \subset \Psi_\infty(U_\theta \cap U_\infty), \tag{4.2.14}$$

*where* $V(\varepsilon,\theta) = \{T \in L : |T| < R_\varepsilon^{-1/\rho};\ \ \rho|\,Arg\,T - \theta| < \pi/2 - \varepsilon\}$, *for some* $R_\varepsilon > 0$ *and all* $\theta \in \mathbb{R}$.

◄ We introduce the mapping $f_\rho : L \to L$ such that $f_\rho[(r,\delta)] = (r^{-1/\rho}, -\delta/\rho)$ and show that the mapping $\Psi_\infty = f_\rho \circ \chi_l^{-1}$ (see (4.2.5)) has the required properties. Let $(p, \Delta) \in U_\infty$. Since $|p| > \sigma$ and $\chi_l$ is a homeomorphism, in the set $Arg\,p$ there exists a unique number $\varphi$ such that $-\varphi/\rho \in \Delta$ and $\chi_l^{-1}[(p,\Delta)] = (|p|, \varphi)$. Therefore,

$$\Psi_\infty[(p,\Delta)] = (|p|^{-1/\rho}, -\varphi/\rho) \quad \forall\ (p,\Delta) \in U_\infty. \tag{4.2.15}$$

Taking $\varepsilon_\infty = \sigma^{-1/\rho}$, we conclude that $\Psi_\infty(U_\infty) = V_\infty$, and $\Psi_\infty$ is a homeomorphism. But $\varphi_l[(p,\Delta)] = p$, and (4.2.12) is fulfilled. Since $B$ acts also on $U_\infty$, from (4.2.6), (4.2.11), and (4.2.15) we see that

$$A^{-1} = \Psi_\infty \circ B \circ \Psi_\infty^{-1}, \quad A^{-k} = \Psi_\infty \circ B^k \circ \Psi_\infty^{-1} \tag{4.2.16}$$

for all $k \in \mathbb{Z}$, i.e., $\Psi_\infty$ agrees with the action of the groups $G_E$ and $G_L$ (see Def. A.1.8).

We take $R_\varepsilon > 0$ in such a manner that $R_\varepsilon \cos\varepsilon \geqslant \sigma$. Then (see (4.2.13))

$$Q_\theta := \{p = re^{i\varphi} \in \mathbb{C} :\ r > R_\varepsilon;\ |\varphi + \rho\theta| < \pi/2 - \varepsilon\} \subset T_\theta$$

and

$$U'_\theta := \{(p,\Delta) \in E_l :\ p \in Q_\theta;\ \theta \in \Delta\} \subset U_\theta \cap U_\infty.$$

Now (4.2.15) implies (4.2.14) ►

**4.2.3. Many-sheeted indicator diagram of an entire function of order $\rho \neq 1$ and with indicator of general form.** Under the action of the group $G_E$ on $E_l$, the links of the broken line $\partial E_l$ are taken to one another, as well as the vertices of the polygonal surface $(\varphi_l, E_l)$ (see (4.2.9)). The angles at the vertices $m_j$ and $m_{j+k}$ of the surface $E_l$ are also identical for every $j \in \mathbb{Z}$ (see Property 4.2.9 and (4.1.3)). Under the action of the generator $B$ of the group $G_E$, the half plane $\Pi_h(0)$, where $h \in F_\rho(l)$, $\theta \in \mathbb{R}$ (see (4.2.0), (4.0.2)), is taken to the half-plane $\Pi_h(\theta - 2\pi)$; any two corresponding points of these half-planes have the same projection to $\mathbb{C}$ for $\rho \in \mathbb{N}$ (see (4.2.10)).

These facts and Properties 4.2.9, 4.2.10 lead to the following construction of the many-sheeted diagram of an entire function of order $\rho \neq 1$.

**Definition 4.2.12.** Suppose $l$ is a piecewise-trigonometric function of class $P_\rho$. For $h \in P_\rho(l)$, let $(\varphi_l, E_h)$ be the surface associated with $h$ (see Def. 4.2.7). We add the point $\Lambda_\infty$ to $\mathcal{A}_h$ and endow the resulting set with the factor-topology of the space of orbits $E_h/G_E$ of the cyclic automorphism group $G_E$

acting on $E_l \supset E_h$ (see Property 4.2.10 and Subsection A.1.2). The space obtained in this way is called the *diagram of $h$ $\rho$-sheeted in the vicinity of $\infty$* (*the indicator diagram of the entire function $f$ of order* $\rho \neq 1$ if $h$ is the indicator of $f$). The set $\mathcal{A}_l$ (containing $\mathcal{A}_h$) is called *the polygon associated with $l$ and $\rho$-sheeted in the vicinity of $\infty$.*

**Remarks.** 1. If $\rho \in \mathbb{N}$, then from (4.2.10) we deduce that the automorphisms of $G_E$ are fiberwise with respect to the mapping $\varphi_l : E_l \to \mathbb{C}$. The latter induces the mapping

$$\hat{\varphi}_l : \mathcal{A}_l \to \overline{\mathbb{C}}; \quad \hat{\varphi}_l(\Lambda_\infty) = \infty; \quad \hat{\varphi}_l(\tilde{m}) := \varphi_l(m), \tag{4.2.17}$$

(see Property 4.2.10), where $\tilde{m}$ is the $G_E$-orbit of an arbitrary point $m \in E_l$. Hence for $l \in M_\rho \setminus O_\rho$ the pair $(\hat{\varphi}_l, \mathcal{A}_l)$ is a poligon $\rho$-sheeted in the vicinity of $\infty$. In the notation of formula (4.1.3), this polygon has $k = k(l)$ vertices, and the angle at each vertex exceeds $\pi$ (see Property 4.2.9). We mention the following facts: 1) $\varphi_l^{-1}(\infty) = \Lambda_\infty \in \overset{\circ}{\mathcal{A}}_l$; 2) there exist neighborhoods $U$ and $W$ of the points $\Lambda_\infty$ in $E_l$ and $\infty$ in $\overline{\mathbb{C}}$, respectively, such that for all $y \in W \setminus \{\infty\}$ the set $\varphi_l^{-1}(y) \cap U$ consists of $\rho$ elements; 3) for $\rho \geqslant 2$ the point $\Lambda_\infty$ is the only branching point in $\overset{\circ}{\mathcal{A}}_l$ of the covering $(\hat{\varphi}_l, \mathcal{A}_l)$.

2. The word "$\rho$-sheeted" in the name of the diagram is explained by Property 4.2.10: the angle at the vertex $\Lambda_\infty$ of the fundamental domain of the group $G_E$ on $E_l$ is equal to $2\pi\rho$.

3. For $\rho \in \mathbb{N}$ and a function $l \in M_\rho \setminus O_\rho$ given by (4.1.1), the topological model of the polygon $(\hat{\varphi}_l, \mathcal{A}_l)$ is produced by clockwise rotation of the half-plane $\Pi_l(\theta)$ (see (4.2.1)) as $\theta$ varies, e.g., on the interval $[\theta_0, \theta_k]$, where $\theta_k = \theta_0 + 2\pi$, followed by gluing $\Pi_l(\theta_0)$ and $\Pi_l(\theta_k)$ at the points with the same projection. Gluing is possible because $\varphi_l[\Pi_l(\theta_0)] = \varphi_l[\Pi_l(\theta_k)]$. So, the polygon $(\hat{\varphi}_l, \mathcal{A}_l)$ has $k = k(l)$ vertices $\{m_1, \ldots, m_k\}$, and $\varphi_l(m_i) = p_i$, $i = 1, \ldots, k$ (see Property 4.2.9). From the construction it follows that the polygon remains on the left when bypassed along $\partial\mathcal{A}_l$ in the direction determined by the order of the vertices $\{m_1, \ldots, m_k\}$. Therefore, $(\hat{\varphi}_l, \mathcal{A}_l)$ is a positively oriented polygon (see Def. 4.2.3$'$). The same orientation is also determined by the order of the vertices of the closed broken line $\mathcal{P} = \{p_1, \ldots, p_k\}$ in $\mathbb{C}$, because $\hat{\varphi}_l(\partial\mathcal{A}_l) = \mathcal{P}$.

For $\rho \notin \mathbb{N}$ and $l \in M_\rho \setminus O_\rho$ such a model $\mathcal{A}_l$ is provided by every polygon $(\varphi_l, M_i^+)$ with sides $\Gamma_i$, $\Gamma_{i+1}$ of the angle at the vertex $\Lambda_\infty$ glued at corresponding points (see Property 4.2.10 and its proof). From (4.2.9) and (4.2.10) it follows that $\varphi_l(\Gamma_i) \neq \varphi_l(\Gamma_{i+1})$. But if $\rho \in \mathbb{N}$, then $\varphi_l(\Gamma_i) = \varphi_l(\Gamma_{i+1})$, and the model is equivalent to the model $(\hat{\varphi}_l, \mathcal{A}_l)$ described earlier.

Now we consider examples of the diagrams of trigonometrically $\rho$-convex functions in the sense of Definition 4.2.12.

**Example 4.2.12$'$.** A. Let $h = h_\varepsilon$ be the function occurring in example 1 b) in 1.9. Then $h_\varepsilon \in P_2$. As the parameter $\theta$ increases from $-\pi$ to $\pi$, the straight line $l_\theta$ (see (1.9.16)) moves clockwise and envelopes the curve $\Gamma_h$ depicted in Fig. 2 b). At the same time, the half-plane $\Pi_h(\theta)$ (see (4.2.0)) bounded by the

Fig. 2 b). At the same time, the half-plane $\Pi_h(\theta)$ (see (4.2.0)) bounded by the straight line $l_\theta$ and located on the left of $l_\theta$ sweeps the diagram $\mathcal{A}_h$ two-sheeted in the vicinity of $\infty$, the half-planes $\Pi_h(-\pi)$ and $\Pi_h(\pi)$ being glued at the points with the same projection.

Let $l = \lim_{\varepsilon \to 0+} h_\varepsilon$. Then $l \in M_2$ and $h_\varepsilon \in P_2(l)$. As $\varepsilon \to 0$, the diagram $\mathcal{A}_h := \mathcal{A}_{h_\varepsilon}$ degenerates into the triangle $\mathcal{A}_l$ associated with $l$ and two-sheeted in the vicinity of the point at infinity. The angles at the vertices of this triangle are equal to $7\pi/3$ (see Def. 4.2.12), and $\mathcal{A}_h \subset \mathcal{A}_l$. We observe that every point of the plane triangle $\Delta$ with vertices $p_1$, $p_2$, $p_3$ is the projection of three points of $\mathcal{A}_l$. On the other hand, the diagram $\mathcal{A}_l$ is formed by clockwise rotation of the half-plane $\Pi_l(\theta)$, $\theta \in [-\pi, \pi]$, through the angle of $4\pi/3$ successively about each of the points $p_1$, $p_2$, $p_3$; this is followed by gluing the half-planes $\Pi_l(-\pi)$and $\Pi_l(\pi)$ as above. The rotation angles are shown in Fig. 9 A).

B. Let $l$ be the $2\pi$-periodic extension to $\mathbb{R}$ of the function $l(\theta) = \operatorname{Re} p_j e^{i4\theta}$, $\theta \in [\theta_{j-1}, \theta_j]$, $j = 1, 2, 3$, where $p_1 = e^{2\pi i/3}$, $p_2 = 1$, $p_3 = e^{-2\pi i/3}$, and $\theta_j = -3\pi/4 + 2\pi j/3$ for $j = 0, 1, 2, 3$. Then $l \in M_4$. As before, the diagram $\mathcal{A}_l$ is formed by rotation of the half-plane $\Pi_l(\theta)$ (as $\theta$ varies on the segment $[\theta_0, \theta_3]$) about the vertices of the same triangle $\Delta$; the half-planes $\Pi_l(\theta_0)$ and $\Pi_l(\theta_3)$ are glued at the end. The difference is that the vertices of $\Delta$ are oriented clockwise and the angle of rotation about each vertex is equal to $8\pi/3$ (see Fig. 9 B). Besides, the triangle $(\hat{\varphi}_l, \mathcal{A}_l)$ is three-sheeted over $\Delta$ and is four-sheeted over $\overline{\mathbb{C}} \setminus \Delta$.

C. Let $l$ be the $2\pi$-periodic extension to $\mathbb{R}$ of the function $l(\theta) = \operatorname{Re} p_j e^{i3\theta}$, $\theta \in [\theta_{j-1}, \theta_j]$, $j = 1, 2$, where $p_1 = -2$, $p_2 = 2$, and $\theta_j = \pi(2j-1)/2$ for $j = 0, 1, 2$. Then $l \in M_3$ (see Theorem 4.1.5). The diagram $(\hat{\varphi}_l, \mathcal{A}_l)$ is a three-sheeted covering of the Riemann sphere $\overline{\mathbb{C}}$ with cut $\Gamma$ connecting the branching points $m_1$ and $m_2$ of multiplicity 2. We have $\varphi_l(m_1) = -2$, $\varphi_l(m_2) = 2$, and $\varphi_l(\Gamma) = [-2, 2]$. This covering has only one additional triple branching point $\Lambda_\infty$. The edges of the cut $\Gamma$ serve as the boundary of the bigon $\mathcal{A}_l$ with the angles of $4\pi$ at the vertices $m_1$ and $m_2$. Fig. 9 C shows the angles of rotation of the half-plane $\Pi_l(\theta)$ (see (4.2.1)) clockwise about the points $p_1$ and $p_2$; these angles give rise to the diagram $\mathcal{A}_l$.

D. Let $\rho \in (1/2, 1)$ and let $l$ be a function of the form (4.1.1). In the notation of formulas (4.1.1)–(4.1.3), suppose $k = 1$, $p_j = \exp\{i\pi\rho\ (1 - 2j\ )\}$, and $\theta_j = 2\pi j$, $j \in \mathbb{Z}$. Then $l \in M_\rho$ (see Proposition 4.1.1, Theorem 4.1.5). The topological model of the diagram $\mathcal{A}_l$ is shown in Fig. 9 D. The rays $\Gamma_0$, $\Gamma_1$ with vertices at the points $p_0$, $p_1$, respectively, are identified at the points equidistant from 0 in $\mathbb{C}$. The diagram $\mathcal{A}_l$ is produced by clockwise rotation of the half-plane $\Pi_l(\theta)$ about the point $p_0$ for $\theta \in [-\alpha, 0]$ and about $p_1$ for $\theta \in [[0, \alpha]$, where $\alpha = \pi(\rho - 1/2)$, followed by gluing the rays $\Gamma_0$ and $\Gamma_1$ ($\Gamma_0 \in \partial\Pi_l(-\alpha)$, $\Gamma_1 \in \partial\Pi_l(\alpha)$, $B\Gamma_1 = \Gamma_0$ (see (4.2.9)).

We describe the topological Jordan broken lines that are the boundaries of polygonal surfaces $E_l$, $\mathcal{A}_l$, where $l \in P_\rho \setminus O_\rho$.

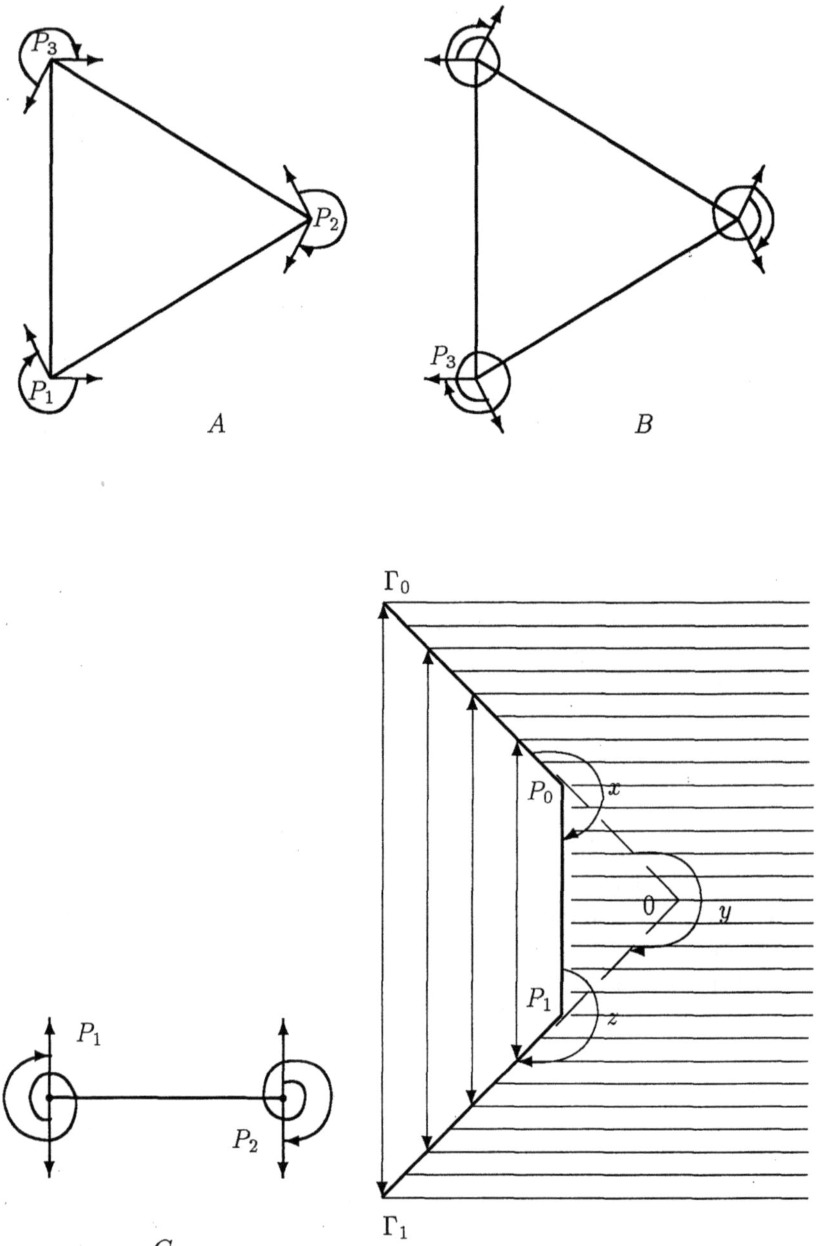


$y = 2\pi\rho;\ x + z = \pi(1 - 2\rho)$

$P_0 = e^{i\pi\rho};\ P_1 = e^{-i\pi\rho}$

D

Fig. 9.

Let $P = \{p_j : j \in \mathbb{Z}\}$ be a $(k,\rho)$-cyclic broken line (see Definition 4.1.6′). We take the set-theoretical union of the segments of the broken line

$$\widetilde{\Gamma} = \bigcup \widetilde{\Gamma}_j | j \in \mathbb{Z}, \quad \widetilde{\Gamma}_j = \{(p,j) : p \in [p_j, p_{j+1}]\} \tag{4.2.18}$$

in $\mathbb{C}$ and glue the neighboring segments at the vertices:

$$(p_j, j-1) \sim (p_j, j) \quad \forall\, j \in \mathbb{Z}. \tag{F}$$

Next, we consider the factor-set $\Gamma_P = \widetilde{\Gamma}/F$ endowed with the factor-topology (see Section A.1) and define the mapping

$$\psi : \Gamma_P \to \mathbb{C}; \quad \psi[(p,j)] = p \quad \forall j \in \mathbb{Z}. \tag{4.2.19}$$

The pair $(\psi, \Gamma_P)$ is called *the infinite broken line associated with the $(k,\rho)$-cyclic broken line $P$ in $\mathbb{C}$*. Since $\psi(\Gamma_P) = P$, the orientation of the broken line $P$ induces an orientation on the topological Jordan broken line $(\psi, \Gamma_P)$. The automorphism (see Def. 4.1.6′)

$$\widetilde{B}_P : (p,j) \to (pe^{i2\pi\rho},\ j-k) \quad \forall\, (p,j) \in \Gamma_P. \tag{4.2.20}$$

acts on $\Gamma_P$. Let $G_P$ be the cyclic group of automorphisms with generator $\widetilde{B}_P$, and let $\gamma_P = \Gamma_P/G_P$ be the space of orbits of the group $G_P$ (see Subsection A.1.2). For $\rho \in \mathbb{N}$, we have

$$\psi(gm) = \psi(m) \quad \forall\, g \in G_P,\ m \in \Gamma_P$$

(see (4.2.19), (4.2.20)). Therefore, for $\rho \in \mathbb{N}$ the following mapping is well defined:

$$\hat{\psi} : \gamma_P \to \mathbb{C}; \quad \hat{\psi}(\tilde{m}) = \psi(m) \quad \forall\, m \in \Gamma_P, \tag{4.2.21}$$

where $\tilde{m}$ is the orbit of the point $m$ on $\Gamma_P$. We call the pair $(\hat{\psi}, \gamma_P)$ *the finite broken line asociated with the $(k,\rho)$-cyclic broken line $P$*. The set $(\bigcup_{j=1}^{k} \widetilde{\Gamma}_j) \setminus \{(p_{k+1}, k)\}$ is a fundamental set for the group $G_P$. For $\rho \in \mathbb{N}$, Definition 4.1.6′ implies that $p_{k+1} = p_1$. Therefore, $\widetilde{B}_P(p_{k+1}, k) = (p_{k+1}, 0) = (p_1, 0) \sim (p_1, 1)$ in accordance with the equivalence relation $(F)$. Hence $(\hat{\psi}, \gamma_P)$ is a closed topological Jordan broken line and $\hat{\psi}(\gamma_P) = \mathcal{P} := \{p_1, \dots, p_k\}$ is a broken line in $\mathbb{C}$ with vertices $p_1, \dots, p_k$. Therefore, the orientation on $\gamma$ is induced by the orientation of the broken line $\mathcal{P}$.

Let $l \in P_\rho \setminus O_\rho$, and let $(\varphi_l, E_l)$ be the surface associated with the function $l$ (see Def. 4.2.7). Using the notation in (4.2.3) and (4.2.19), we consider the mapping

$$\Phi_l : E_l \to \Gamma_P; \quad \Phi_l[(p,\Delta)] = (p,j), \quad p \in [p_j, p_{j+1}], \tag{4.2.22}$$

where $j \in \mathbb{Z}$ and $\Delta = \{\theta_j\}$ for $p \in (p_j, p_{j+1})$. From Definition 4.1.6′ it follows that (see (4.2.6), (4.2.20))

$$\Phi_l \circ B^k = \widetilde{B}_P^k \circ \Phi_l \quad \forall\, k \in \mathbb{Z},$$

i.e., $\Phi_l$ is a homeomorphism compatible with the action of $G_E$ on $\partial E_l$ and of $G_P$ on $\Gamma_P$. Since $\varphi_l = \Phi_l \circ \psi$, the oriented topological broken lines $(\varphi_l, \partial E_l)$ and $(\psi, \Gamma_P)$ are isomorphic (see Def. 4.2.5).

Let $\rho \in \mathbb{N}$. Consider the diagram (see (4.2.17), (4.2.21), (4.2.22))

$$\begin{array}{ccccc} \partial E_l & \xrightarrow{\varkappa_l} & \partial \mathcal{A}_l & \overset{\hat{\varphi}_l}{\searrow} & \\ \Phi_l \downarrow & & \hat{\Phi}_l \downarrow & & \mathcal{P}. \\ \Gamma_P & \xrightarrow{\varkappa_P} & \gamma_P & \underset{\hat{\psi}}{\nearrow} & \end{array} \qquad (4.2.23)$$

Here $\varkappa_l : \partial E_l \to \partial \mathcal{A}_l$, $\varkappa_P : \Gamma_P \to \gamma_P$ are canonical mappings; $\hat{\Phi}_l : \partial \mathcal{A}_l \to \gamma_P$ is a homeomorphism satisfying the condition $\varkappa_P \circ \Phi_l = \hat{\Phi}_l \circ \varkappa_l$ (see Proposition A.1.9). This diagram is commutative because $\partial \mathcal{A}_l = \bigcup \tilde{m} \mid m \in \partial E_l$, where $\tilde{m}$ is the orbit of a point $m \in \partial E_l$, and

$$\begin{aligned} \hat{\psi} \circ \hat{\Phi}_l(\tilde{m}) &= \hat{\psi} \circ \hat{\Phi}_l \circ \varkappa_l(m) = \hat{\psi} \circ \varkappa_P \circ \Phi_l(m) \\ &= \psi \circ \Phi_l(m) = \varphi_l(m) = \hat{\varphi}_l(\tilde{m}) \end{aligned}$$

for every orbit $\tilde{m} \in \partial \mathcal{A}_l$. So, for $\rho \in \mathbb{N}$ the closed oriented topological broken lines $(\hat{\varphi}_l, \partial \mathcal{A}_l)$ and $(\hat{\psi}, \gamma_P)$ are isomorphic.

By Theorem 4.1.7, we arrive at the following statement.

**Theorem 4.2.13.** *Suppose $l$ is a piecewise-trigonometric function of class $P_\rho \setminus O_\rho$, $(\varphi_l, E_l)$ is a positively oriented polygonal surface related to $l$, $(\psi, \Gamma_P)$ is an infinite broken line associated with the oriented $(k,\rho)$-cyclic broken line $P = \{p_j : j \in \mathbb{Z}\}$ in $\mathbb{C}$ the vertices of which are used to define $l$* (see Def. 4.1.6′, (4.1.1)–(4.1.3), (4.2.19)), *$(\hat{\varphi}_l, \mathcal{A}_l)$ is the positively oriented polygon $\rho$-sheeted in the vicinity of $\infty$ and related to $l$ for $\rho \in \mathbb{N}$* (see (4.2.17)), *and $\mathcal{P} = \{p_1, \dots, p_k\}$ is the oriented closed broken line in $\mathbb{C}$ associated with $P$ for $\rho \in \mathbb{N}$. Let $G_E$ and $G_P$ be the cyclic groups of automorphisms with the generators $B$ and $\widetilde{B}_P$* (see (4.2.6), (4.2.20)) *acting freely and totally discontinuously on $E_l$ and on $\Gamma_P$, respectively. Let $\gamma_P = \Gamma_P / G_P$, and let $(\hat{\psi}, \gamma_P)$ be the closed finite broken line related to $P$ for $\rho \in \mathbb{N}$. Then*

1) *the oriented topological broken lines $(\varphi_l, \partial E_l)$ and $(\psi_P, \Gamma)$ are isomorphic and* (see (4.2.22)) *the following diagram is commutative:*

$$
\begin{array}{ccccc}
\partial E_l & \xrightarrow{B^k} & E_l & \searrow^{\varphi_l} & \\
\Phi_l \Big\downarrow & & \Big\downarrow \Phi_l & & P; \\
\Gamma_P & \xrightarrow{(\widetilde{B}_P)^k} & \Gamma_P & \nearrow_{\psi} &
\end{array}
$$

2) *for* $\rho \in \mathbb{N}$ *the closed oriented topological broken lines* $(\hat{\varphi}_l, \partial\mathcal{A}_l)$ *and* $(\hat{\psi}, \gamma_P)$ *are isomorphic, and the diagram* (4.2.23) *is commutative;*

3) *if* $\alpha \in [0, 1)$ *and* $P$ *is an oriented* $(k, \alpha)$*-cyclic broken line, then there exists a number* $\rho$ *the fractional part* $\{\rho\}$ *of which is* $\alpha$*, and a piecewise-trigonometric function* $l \in P_\rho \setminus O_\rho$ *satisfying assertion* 1) *of the theorem, and for* $\alpha = 0$ *satisfying also asserton* 2).

Now we pass to describing the complex structure on $\mathcal{A}_l$. In the case where $l \in P_\rho \setminus O_\rho$, for a coordinate neighborhood of the point $\Lambda_\infty$ we take the set

$$\omega_\infty = \{\Lambda_\infty\} \cup (U_\infty / G_E),$$

where $U_\infty/G_E$ is the canonical image of $U_\infty \subset E_l$ in the space of orbits $E_l/G_E$ (see (4.2.5)). From Property 4.2.11 and Proposition A.1.9 we deduce the existence of a homeomorphism

$$\psi_\infty : \omega_\infty \to S, \quad \psi_\infty(\Lambda_\infty) = 0, \tag{4.2.24}$$

where $S = \{z \in \mathbb{C} : \ |z| < \sigma^{-1/\rho}\}$ , playing the role of a local coordinate in the vicinity of $\Lambda_\infty$ on $\mathcal{A}_l$. Next, in the notation of formula (4.2.13) we have $\overset{\circ}{E}_l = \bigcup U_\theta \mid \theta \in \mathbb{R}$. Since $\rho > 1/2$ and (see (4.2.6))

$$BU_\theta = U_{\theta - 2\pi}, \quad U_\theta \cap U_{\theta-2\pi} = \varnothing \quad \forall\, \theta \in \mathbb{R}, \tag{4.2.25}$$

we see that the canonical mappings

$$\gamma_\theta : U_\theta \to U_\theta / G_E, \quad \theta \in \mathbb{R}, \tag{4.2.26}$$

are homeomorphisms. Therefore,

$$\overset{\circ}{\mathcal{A}}_l \setminus \{\Lambda_\infty\} = \bigcup U_\theta / G_E | \theta \in [-\pi, \pi]$$

and $\psi_\theta := \varphi_l \circ \gamma_\theta^{-1}$ is a local coordinate in $U_\theta/G_E$ (see (4.2.3)) for all $\theta \in (-\pi, \pi]$. Put

$$V_\theta = \Psi_\infty(U_\theta \cup U_\infty), \quad \omega_\theta = \omega_\infty \cap (U_\theta/G_E), \quad \theta \in (-\pi, \pi].$$

From Property 4.2.11 and formulas (4.2.16), (4.2.24) we deduce that $V_\theta \cap V_{\theta-2\pi} = \varnothing \ \ \forall\, \theta \in (-\pi, \pi]$. Therefore, in the following diagram all mappings are homeomorphisms (see (4.2.24), (4.2.26)) for every fixed $\theta \in (-\pi, \pi]$:

$$\begin{array}{ccccc} \partial E_l & \xrightarrow{\quad B^k \quad} & E_l & \searrow^{\varphi_l} & \\ \Phi_l \big\downarrow & & \big\downarrow \Phi_l & & P. \\ \Gamma_P & \xrightarrow{\quad (\widetilde{B}_P)^k \quad} & \Gamma_P & \nearrow_{\psi} & \end{array} \tag{4.2.27}$$

From the diagram we find that $\psi_\theta \circ \psi_\infty^{-1} = \varphi_l \circ \Psi_\infty^{-1} \circ \pi^{-1}$. Hence (see (4.2.12)),

$$\psi_\theta \circ \psi_\infty^{-1}(\tau) \equiv \tau^{-\rho}, \quad \tau \in \pi(V_\theta), \quad \theta \in (-\pi, \pi], \tag{4.2.28}$$

i.e., the charts $(\omega_\infty, \psi_\infty)$, $(U_\theta / G_E, \psi_\theta)$ of the atlas on $\mathcal{A}_l$ are biholomorphically compatible for every $\theta \in (-\pi, \pi]$.

Let $l \in O_\rho$. The mapping

$$F_\rho : L \to L, \quad F_\rho[(r, \varphi)] = (r^{1/\rho}, \varphi/\rho), \quad (r, \varphi) \in L,$$

is compatible with the action on $L$ of the cyclic groups $G_\rho$ and $G_L$ generated by $A_\rho$ and $A$, respectively (see (4.2.7) and (4.2.11)):

$$F_\rho \circ A_\rho = A \circ F_\rho, \quad F_\rho \circ A_\rho^k = A^k \circ F_\rho \quad \forall\, k \in \mathbb{Z}.$$

Formula (4.2.8) and Property 4.2.10 imply that the same is true for the mappings $\chi_l$ and $\chi_l^{-1}$ relative to the groups $G_E$ on $E_l \setminus \Lambda_0$ and $G_\rho$ on $L$. Therefore, $E_l \setminus \{\Lambda_0\} \xrightarrow{\chi_l^{-1}} L \xrightarrow{F_\rho} L$ is a composition of conformal homeomorphisms satisfying the conditions of Proposition 4.1.9 in Appendix 1. Consequently, the mapping

$$(E_l \setminus \{\Lambda_0\})/G_E \xrightarrow{\hat{\chi}_l^{-1}} L/G_\rho \xrightarrow{\hat{F}_\rho} L/G_L = \mathbb{C} \setminus \{0\}$$

is also a composition of conformal homeomorphisms. Here $\hat{\chi}_l^{-1}$ and $\hat{F}_\rho$ are the factor-mappings the domains of which can be completed (see Def. 4.2.12) up to $\mathcal{A}_l$ and $\widehat{L}_\rho$, respectively; moreover, the mappings $\hat{\chi}_l^{-1}$, $\hat{F}_\rho$ and their composition $\Phi = \hat{F}_\rho \circ \hat{\chi}_l^{-1}$ extend up to homeomorphisms

$$\mathcal{A}_l \xrightarrow{\hat{\chi}_l^{-1}} \widehat{L}_\rho \xrightarrow{F_\rho} \overline{\mathbb{C}}; \quad \mathcal{A}_l \xrightarrow{\Phi} \overline{\mathbb{C}}. \tag{4.2.29}$$

In the notation of Definition 4.2.12, we have $\Phi(\Lambda_\infty) = \infty$ and $\Phi(\Lambda_0) = 0$. So, the diagram $\mathcal{A}_l$ is conformally equivalent to the Riemann sphere $\overline{\mathbb{C}}$. In particular, for $l \in O_\rho$, $\rho \in \mathbb{N}$, in the notation of Remark 1 to Definition 4.2.12, the pair $(\hat{\varphi}_l, \mathcal{A}_l)$ is the Riemann surface of the inverse to the function $\alpha(z) = p_0 + z^\rho$, where $p_0$ is the complex number occurring in the definition of $l$: $l(\theta) = \operatorname{Re} p_0 e^{i\rho\theta}$, $\theta \in \mathbb{R}$. We arrive at the following fact.

**Property 4.2.14.** *For every piecewise-trigonometric function* $l \in P_\rho \setminus O_\rho$, *the diagram* $\mathcal{A}_l$ *associated with* $l$ (see Def. 4.2.12) *is a simply connected analytic manifold with boundary of complex dimension* 1. *If* $l \in O_\rho$, *then there exists a conformal homeomorphism* $\Phi : \mathcal{A}_l \to \overline{\mathbb{C}}$ *such that* $\Phi(\Lambda_0) = 0$, $\Phi(\Lambda_\infty) = \infty$. *In particular, if* $\rho \in \mathbb{N}$, $l(\theta) = \operatorname{Re} p_0 e^{i\rho\theta}$ *for* $\theta \in \mathbb{R}$ *(where* $p_0 \in \mathbb{C}$*), and* $\alpha(z) = p_0 + z^\rho$, *then* $\alpha \circ \Phi = \hat{\varphi}_l$, *where* $\hat{\varphi}_l : \mathcal{A}_l \to \overline{\mathbb{C}}$ *is the mapping considered in Remark* 1 *to Definition* 4.2.12.

**Remark.** The set $\widehat{L}_\rho$ (see (4.2.29)) is a topological model of the diagram $\mathcal{A}_l$. For $\rho \in \mathbb{N}$ it is natural to call it *the Riemann surface of the inverse function to* $p = z^\rho$, where the symbol $z^\rho$ is understood in the principal value sense (see A.2.2).

**4.2.4. Inclusion of the many-sheeted diagram of an entire function of integral order $\rho > 1$ in a covering of the Riemann sphere.** We show that for $l \in P_\rho \setminus O_\rho$, $\rho \in \mathbb{N} \setminus \{1\}$, the diagram $(\hat{\varphi}_l, \mathcal{A}_l)$ can be included in a finitely-sheeted covering of the Riemann sphere $\overline{\mathbb{C}}$.

We illustrate the idea of the inclusion method by examples A–C in item 2°. Take the triangle $(\varphi_l, A_l)$ occurring in example $A$ and attach to it the exterior of the plane triangle $\Delta$ with vertices $p_1$, $p_2$, $p_3$. We shall obtain a three-sheeted covering of the Riemann sphere $\overline{\mathbb{C}}$ with four double branching points. Gluing the triangle $\Delta$ and the triangle $(\varphi_l, \mathcal{A}_l)$ in Example $B$ along the common boundary, we obtain a four-sheeted covering of the Riemann sphere with three quadruple branching points and such that $\varphi_l(\Lambda_\infty) = \infty$. A three-sheeted covering of $\overline{\mathbb{C}}$ with three branching points is formed if the sides of the cut $\Gamma$ in the diagram $(\hat{\varphi}_l, \mathcal{A}_l)$ in example C are glued at the points that have the same projection (see Fig. 9 C).

**Theorem 4.2.15.** *Suppose* $\rho \in \mathbb{N} \setminus \{1\}$, $l$ *is a function of class* $P_\rho \setminus O_\rho$ *defined by formula* (4.1.1), $(\hat{\varphi}_l, \mathcal{A}_l)$ *is the polygon associated with the function* $l$ *and* $\rho$*-sheeted in the vicinity of* $\infty$, *and* $\mathfrak{M} = \{m_1, \dots, m_k\}$ *is the ordered sequence of the vertices of this polygon with* $\hat{\varphi}_l(m_j) = p_j$, $j = 1, \dots, k$ (see (4.1.1), (4.1.3)). *Then there exists a finitely-sheeted covering* $\varphi : M \to \overline{\mathbb{C}}$ *of the Riemann sphere* $\overline{\mathbb{C}}$ *such that* 1) $\mathcal{A}_l$ *is homeomorphically included in* $M$ *and* $\varphi\big|_{\mathcal{A}_l} = \hat{\varphi}_l$; 2) $M$ *is a simply connected compact Riemann surface;* 3) *the set* $\mathfrak{M}$ *coincides with the set of all branching points of the covering that project to finite points in* $\mathbb{C}$ *if* $l \in M_\rho$ (see Theorem 4.1.5).

◄ 1. Consider the set $\mathfrak{N}(\partial\mathcal{A}_l) = \{(\psi, B)\}$ of positively oriented polygons finitely sheeted in the vicinity of the point at infinity and such that a) $\psi\big|_{\partial B} = \hat{\varphi}_l\big|_{\partial\mathcal{A}_l}$; b) $\partial B = \partial\mathcal{A}_l$; c) the polygons $(\hat{\varphi}_l, \mathcal{A}_l)$ and $(\psi, B)$ induce opposite orientations on their common boundary.

We show that $\mathfrak{N}(\partial\mathcal{A}_l) \neq \varnothing$. Indeed, $\varphi_l(\partial\mathcal{A}_l) = \mathcal{P} = \{p_1, \dots, p_k\}$ is a closed broken line in $\mathbb{C}$ the orientation of which makes $(\varphi_l, \mathcal{A}_l)$ a positively oriented polygon (see Remark 3 to Definition 4.2.12)

Consider the closed broken line

$$\mathcal{Q} = \{q_j := p_{k+1-j} \ : \ j = 1, \dots, k\} \tag{4.2.30}$$

differing from the broken line $\mathcal{P}$ by orientation. By Theorem 4.2.13, for the $(k,0)$-cyclic broken line $Q = \{q_j := p_{k+1-j} \; : \; j \in \mathbb{Z}\}$ (see Def. 4.1.6′) there exists a number $\delta \in \mathbb{N}$ and a piecewise-trigonometric function $g(\varphi) = \mathrm{Re}\; q_j e^{i\delta\varphi}$, $\varphi \in [\varphi_{j-1}, \varphi_j]$, $j \in \mathbb{Z}$ in $P_\delta \setminus O_\delta$ (the definition of $g$ involves the broken line $Q$) such that the closed oriented topological broken lines $(\hat{\varphi}_g, \partial\mathcal{A}_g)$ and $(\hat{\psi}, \gamma_Q)$ are isomorphic (see Def. 4.2.5). Here $(\hat{\varphi}_g, \partial\mathcal{A}_g)$ is the positively oriented polygon $\delta$-sheeted in the vicinity of $\infty$ and associated with the function $g$, and $(\hat{\psi}, \gamma_Q)$ is the broken line related to $Q$. On the other hand, by Theorem 4.2.13 the oriented broken lines $(\hat{\varphi}_l, \partial\mathcal{A}_l)$ and $(\hat{\psi}, \gamma_P)$ are also isomorphic. If the orientation is neglected, then $P = Q$ and $\gamma_P = \gamma_Q$. Therefore, the broken lines $(\hat{\varphi}_l, \partial\mathcal{A}_l)$ and $(\hat{\varphi}_g, \partial\mathcal{A}_g)$ are equivalent, but have opposite orientations, because so are $\mathcal{P}$ and $\mathcal{Q}$ (see (4.2.30)). Thus, there exists a homeomorphism

$$\Phi : \partial\mathcal{A}_l \to \partial\mathcal{A}_g; \quad \hat{\varphi}_l = \hat{\varphi}_g \circ \Phi.$$

2. Let $A = \mathcal{A}_l$, $B = \mathcal{A}_g$, $M = A \cup B$. We glue $A$ and $B$ with the aid of the mapping $\Phi$, i.e., for $m \in \partial A$ and $n \in \partial B$ we agree that $m \sim n$, if $n = \Phi(m)$, and introduce the factor-topology in $M$. We claim that the mapping

$$\varphi : M \to \overline{\mathbb{C}}, \quad \varphi(m) = \begin{cases} \hat{\varphi}_l(m), & m \in A \\ \hat{\varphi}_g(m), & m \in B \end{cases}, \tag{4.2.31}$$

has the properties indicated in Theorem 4.2.15.

The construction of polygons associated with piecewise-trigonometric functions shows (see Properties 4.2.8, 4.2.9, 4.2.14) that $(\varphi, A)$ and $(\varphi, B)$ are triangulable manifolds with boundary, and there exists a triangulation $\Lambda_A$, of, say, $A$ consisting of topological triangles $\{T_i \; : \; i = 1, \dots, N\}$ such that each $\Delta_i := \varphi(T_i)$ is a triangle in $\overline{\mathbb{C}}$ (i.e., one of its vertices may be $\infty$) and $\varphi : T_i \to \Delta$ is a homeomorphism for every $i = 1, \dots, N$. We may choose triangulations $\Lambda_A$ and $\Lambda_B$ of $A$ and $B$ in such a manner that every point $m$ of the common boundary of $A$, $B$ belongs to the common side of the triangles $T = T(m)$ and $S = S(m)$, where $T \in \Lambda_A$, $S \in \Lambda_B$. Therefore, $M$ is a triangulable topological space. By the so-called addition theorem (see Theorem A.1.18), since $A$ and $B$ are simply connected and are separated by the connected Jordan broken line $A \cap B = \partial A = \partial B$ (see Proposition 4.2.14), we conclude that $M$ is also simply connected.

3. Let $\Gamma = \partial A = \partial B$. We define a complex structure on $M$. On $M \setminus \Gamma = \overset{\circ}{A} \cup \overset{\circ}{B}$ it is introduced as in the proof of Property 4.2.14.

Suppose $m \in \Gamma$ and $m \in \mathrm{int}(T \cap S)$, where $T \in \Lambda_A$ and $S \in \Lambda_B$ are topological triangles considered in item 2°. By construction (see (4.2.31)), $(\varphi, A)$ and $(\varphi, B)$ are positively oriented polygons inducing opposite orientations on their common boundary. Therefore, the triangles $\varphi(T)$ and $\varphi(S)$ in $\mathbb{C}$ are positioned on opposite sides of the common face $\varphi(T \cap S)$ and $\varphi : T \cup S \to \varphi(T) \cup \varphi(S)$ is a homeomorphism. Since $m \in \mathrm{int}(T \cup S)$, $\varphi$ is a local coordinate in the vicinity of $m$.

Let $m \in \Gamma$, and let $\Lambda_m$ be the set of all triangles in the triangulation $\Lambda = \Lambda_A \cup \Lambda_B$ of $M$ that have a vertex at $m$. Suppose that $\Lambda_m$ is nonempty. Then $\Lambda_m = \{T_1, \dots, T_r\}$, where $r \geqslant 4$, and for the same reason each pair $\varphi(T_j)$, $\varphi(T_{j+1})$ of neighboring triangles intersects in $\overline{\mathbb{C}}$ along their common side $\varphi(T_j \cap T_{j+1})$. Here $j = 1, \dots, r$, $T_{r+1} := T_1$. Therefore, the set $\bigcup_{j=1}^{r} \varphi(T_j)$ is a neighborhood of the point $\varphi(m)$ in $\mathbb{C}$ and the sum $\Sigma_m$ of the angles of the triangles $\{\varphi(T_j)\}_1^r$ at the vertex $\varphi(m)$ is a multiple of $2\pi$. If $\Sigma_m = 2\pi$, i.e., the multiplicity of $\varphi$ at $m$ is equal to 1, then, again, $\varphi$ is a local coordinate in the vicinity of $m$. Only in the case where $m = m_j$ $(j = 1, \dots, k)$ is one of the common vertices of the polygons $(\varphi, A)$ and $(\varphi, B)$, the multiplicity $\mu_j$ of $\varphi$ at $m_j$ can exceed 1. From (4.1.12) and Property 4.2.9 it follows that $\mu_j > 1$ if $l \in M_\rho$. In this case $m_j$ is a branching point of $\varphi$. Since $\varphi(m_j) = p_j$, we conclude that there exists a neighborhood $U_j$ of $m_j$ and a local parameter $\psi_j : U_j \to S_\varepsilon := \{z \in \mathbb{C} : |z| < \varepsilon\}$, $\varepsilon > 0$, such that

$$\psi_j(m_j) = 0, \quad \varphi \circ \psi_j^{-1}(t) \equiv p_j + t^{\mu_j}, \quad t \in S_\varepsilon. \tag{4.2.32}$$

So, $M$ is a Riemann surface and $\varphi : M \to \overline{\mathbb{C}}$ is a holomorphic covering of the Riemann sphere $\overline{\mathbb{C}}$. Since $M$ is a compact surface, $\varphi$ is a proper holomorphic mapping. Such a mapping takes each value in $\overline{\mathbb{C}}$ at precisely $n$ points, counting the multiplicities (see A.1.15). Namely, $n = \rho + \delta$ is the total number of sheets in the vicinity of $\infty$ in the polygons $(\varphi, A)$ and $(\varphi, B)$. Thus, $\varphi : M \to \overline{\mathbb{C}}$ is an $n$-sheeted covering ▶

**Remark.** In item 1 of the proof we showed the existence of a concave polygon in $\mathfrak{N}(\partial A_l)$ (see Property 4.2.9). The examples of inclusion considered prior to the proof show that

1) the set $\mathfrak{N}(\partial A_l)$ may also include other polygons (example B);

2) inclusion in a covering of $\overline{\mathbb{C}}$ can be achieved by other methods (in example C by gluing the cut $\Gamma$ at the points with the same projection to $\overline{\mathbb{C}}$). Then the inclusion $A_l$ in $M$ is continuous, but not homeomorphic.

**4.2.5. The polygonal surface associated with the piecewise-trigonometric indicator of an entire function of nonintegral order: inclusion in a covering of $\mathbb{C}$.** The inclusion property described in item 4.2.4 fails for the indicator diagram of an entire function of nonintegral order. Formula (4.2.10) shows that on the surface $(\varphi_l, E_l)$ associated with a trigonometrically $\rho$-convex function $l \in M_\rho$ (see Def. 4.2.7), for $\rho \notin \mathbb{N}$ the automorphism $B$ is not fiberwise, i.e., the condition $\varphi_l(Bm) = \varphi_l(m)$ $\forall\, m \in E_l$ is invalid. However, the surface $(\varphi_l, E_l)$ can be included in a certain infinitely-sheeted covering $\varphi : M \to \mathbb{C}$ of the plane, and the group property of the surface $E_l$ remains true for the manifold $M$. We illustrate this fact by an example.

For $\rho \in (1/2, 1)$, we put $p_j = \exp\{i\pi\rho(1 - 2j)\}$ and $\theta_j = 2\pi j$, $j \in \mathbb{Z}$. We denote by

$$l(\theta) = \text{Re } p_j e^{i\rho\theta} = \cos\rho(\theta + \pi - 2\pi j), \quad \theta \in (\theta_{j-1}, \theta_j], \tag{4.2.33}$$

where $j \in \mathbb{Z}$, a piecewise-trigonometric function occuring in Example D of subsection 4.2.3. Consider the covering $(\varphi, M)$ of $\mathbb{C}$ introduced in Example A.2.10, and suppose that the sequence $\{p_j : j \in \mathbb{Z}\}$ occurring in its definition is chosen as indicated above. We take the mapping $\Phi : E_l \to M$, where (see (4.2.3)) $\Phi[(p, \Lambda)] = (p, j)$ if

$$\tau = \pi(1/\rho - 1), \quad \Delta \cap [\theta_j - \tau, \theta_j + \tau] \neq \varnothing \text{ for } j \in \mathbb{Z}.$$

For $\theta \in [\theta_{j-1}+\tau, \theta_j - \tau]$ the half-plane $\Pi_l(\theta)$ rotates about the point $p_j$ through the angle of $\pi$ for every $j \in \mathbb{Z}$ (see (4.2.1), (4.2.33)). Therefore,

$$\bigcup \Pi_l(\theta) | \theta \in [\theta_{j-1} + \tau, \theta_j - \tau] = \Pi_l(\theta_{j-1} + \tau) \cup \Pi_l(\theta_j - \tau), \ j \in \mathbb{Z},$$

and the mapping $\Phi$ is defined everywhere on $E_l$ and embeds $E_l$ into $M$ homeomorphically (so, $E_l \subset M$). In the notation of formula (4.2.3) and Example A.2.10, we have:

$$\Phi(E_l) = \bigcup M_j^+ | j \in \mathbb{Z}, \quad \varphi_l = \varphi \circ \Phi.$$

Moreover, immediate verification shows that

$$\Phi \circ g = g' \circ \Phi \quad \forall\, g \in G_E, \quad g' \in G_M$$

(see (4.2.6), Property 4.2.10, and Example A.2.10), i.e., the mapping $\Phi$ agrees with the action of $G_E$ on $E_l$ and of $G_M$ on $M$. Fig. 10 shows the closure of the fundamental domain for the group $G_M$. So, the group $G_M$ extends the domain of action of $G_E$ from $E_l$ to $M$.

In the general case, we have the following statement.

**Theorem 4.2.16.** *Suppose $\rho > 0$ is a noninteger, $l \in M_\rho \setminus O_\rho$ is a minimal trigonometrically $\rho$-convex function, $P = \{p_j : j \in \mathbb{Z}\}$ is the $(k, \rho)$-cyclic broken line related to the definition of $l$* (see Def. 4.1.6′, (4.1.3)), *$(\varphi_l, E_l)$ is the positively oriented polygonal surface associated with $l$* (see Def. 4.2.7), *and $\mathfrak{M} = \{m_j : j \in \mathbb{Z}\}$ is the ordered set of all vertices of the surface $(\varphi_l, E_l)$ in the notation of Property* 4.2.9. *Then there exists a $(\rho, n)$-covering* (see Def. A.2.7), *and $\varphi : M \to \mathbb{C}$ such that*

1) *the surface $(\varphi_l, E_l)$ admits fiberwise homeomorphic inclusion in the covering $(\varphi, M)$, i.e., there exists a homeomorphism*

$$\Psi : E_l \to \Psi(E_l) \subset M; \quad \varphi \circ \Psi = \varphi_l; \tag{4.2.34}$$

2) *the set $\mathfrak{M}$ coincides with the set of all branching points of the covering, and $\varphi(m_j) = p_j \ \ \forall\, j \in \mathbb{Z}$;*

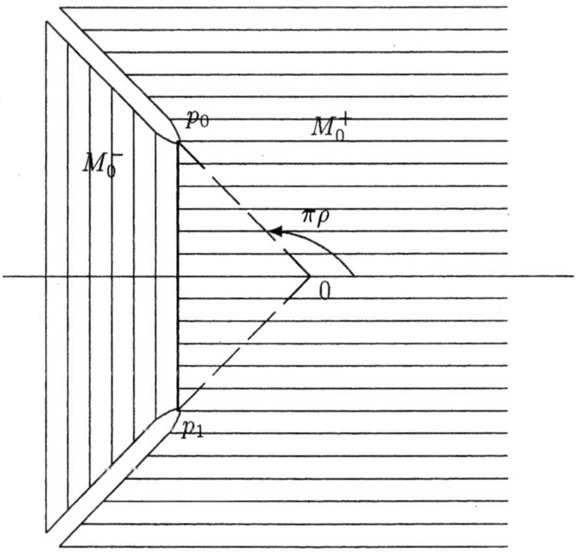


$p_0 = e^{i\pi\rho}$, $p_1 = e^{-i\pi\rho}$; $\quad M_0 = M_0^+ \cup M_0^-$.

Fig. 10

3) *the cyclic automorphism group* $G_M$ *with generator* $B_M$ *acts on* $M$ *freely and totally discontinuously, and* (see (4.2.6) and Property 4.2.10) *the mapping* $\Psi$ *agrees with the action of* $G_E$ *on* $E_l$ *and of* $G_M$ *on* $M$, *i.e.,*

$$\Psi \circ B^r = B_M^r \circ \Psi \quad \forall\, r \in \mathbb{Z}, \tag{4.2.35}$$

*where* $B$ *is the generator of the group* $G_E$ *such that*

$$Bm_{j+k} = m_j \quad \forall\, j \in \mathbb{Z}; \tag{4.2.36}$$

*moreover, if* $\mu_j$ *is the multiplicity of branching at the point* $m_j$ *of the covering* $(\varphi, M)$, *then* $\mu_{j+k} = \mu_j \ \ \forall\, j \in \mathbb{Z}$.

◀ We orient the broken line $(\varphi_l, \partial E_l)$ in the opposite direction. For this, it suffices to alter the orientation of the broken line $P$ because by Theorem 4.2.13 the broken lines $(\varphi_l, \partial E_l)$ and $(\psi, \Gamma_P)$ are isomorphic (see (4.2.19) and Def. 4.2.5). We obtain the broken line

$$Q = \{q_j := p_{-j} : \ j \in \mathbb{Z}\}. \tag{4.2.37}$$

Let $\gamma = \{\rho\}$ be the fractional part of $\rho$. Since $\gamma \in (0,1)$ and $P$ is a $(k,\gamma)$-cyclic broken line, we have

$$q_j \neq q_{j+1}, \quad q_j = q_{j+k} e^{i2\pi(1-\gamma)} \quad \forall\, j \in \mathbb{Z},$$

i.e., $Q$ is a $(k, 1-\gamma)$-cyclic broken line.

By Theorem 4.2.13, for the oriented Jordan topological broken line $(\psi, \Gamma_Q)$ associated with $Q$ (see (4.2.19)) there exists a number $\delta > 0$ with

$$\{\delta\} = 1 - \gamma \tag{4.2.38}$$

and a piecewise-trigonometric function $g \in P_\delta \setminus O_\delta$ such that the boundary $(\varphi_g, \partial E_g)$ of the positively oriented surface $(\varphi_g, E_g)$ associated with $g$ is isomorphic to $(\psi, \Gamma_Q)$. Since the broken lines $(\psi, \Gamma_P)$ and $(\psi, \Gamma_Q)$ differ in orientation only, the following diagram is commutative:

$$\begin{array}{ccccccc} \partial E_l & \xrightarrow{\Phi_l} & \Gamma_P & \xrightarrow{F} & \Gamma_Q & \xrightarrow{\Phi_g^{-1}} & \partial E_g \\ & \searrow^{\varphi_l} & \searrow^{\psi} & & \swarrow^{\psi} & \swarrow^{\varphi_g} & \\ & & & P & & & \end{array} \tag{4.2.39}$$

Here (see (4.2.22)) $\Phi_l,\ \Phi_g : \partial E_g \to \Gamma_Q$ are homeomorphisms, each of which yields isomorphism for the corresponding pair of topological broken lines. Next, $F : \Gamma_P \to \Gamma_Q$ is the identical transformation of the broken line $\Gamma_P = \Gamma_Q$, where (see (4.2.19) and (4.2.37))

$$\Gamma_Q = \{(p, r) : p \in [q_r, q_{r+1}],\ r \in \mathbb{Z}\}; \quad F[(p, j)] = \\ (p, -j-1), \quad p \in [p_j, p_{j+1}], \quad j \in \mathbb{Z}. \tag{4.2.40}$$

Let

$$\Phi = \Phi_g^{-1} \circ F \circ \Phi_l\ . \tag{4.2.41}$$

Then (see (4.2.39)) $\Phi : \partial E_l \to \partial E_g$ is a homeomorphism and

$$\varphi_l = \varphi_g \circ \Phi\ . \tag{4.2.42}$$

We glue the equivalent broken lines $(\varphi_l, \partial E_l)$ and $(\varphi_g, \partial E_g)$ using the mapping $\Phi$:

$$m \sim \Phi(m) \quad \forall\, m \in \partial E_l\ . \tag{4.2.43}$$

Introducing the factor-topology in the set-theoretical union $M = \partial E_l \cup \partial E_k$, we make $M$ a topological space. The mapping (see (4.2.42) and (4.2.43))

$$\varphi : M \to \mathbb{C}, \quad \varphi\,|_{E_l} = \varphi_l; \quad \varphi\,|_{E_g} = \varphi_g \tag{4.2.44}$$

satisfies condition 1) of the theorem. As in the proof of Theorem 4.2.15, we show that $M$ is simply connected, and $(\varphi, M)$ is a covering. Since $l \in M_\rho$, from

Theorem 4.1.5 and Property 4.2.9 we deduce that the angle at every vertex $m_j$ of the polygonal surface $(\varphi_l, E_l)$ is equal to

$$\rho(\theta_j - \theta_{j-1}) + \pi > 2\pi. \tag{4.2.45}$$

Hence, the sum of the angles of the glued surfaces $(\varphi_l, E_l)$ and $(\varphi_k, E_k)$ at the common vertex $m_j$ is $2\pi\mu_j > 2\pi \ \forall\ j \in \mathbb{Z}$; moreover, $\mu_j \in \mathbb{N}$. So, at the point $m_j \in M$ the multiplicity of the mapping $\varphi$ is equal to $\mu_j > 1 \ \forall\ j \in \mathbb{Z}$, i.e., $m_j$ is a branching point of the covering $(\varphi, M)$. Furthermore, $\varphi(m_j) = p_j$ $\forall\ j \in \mathbb{Z}$ (see Property 4.2.9). Since $l$, $g$ are $2\pi$-periodic piecewise-trigonometric functions, we have $\mu_j = \mu_{j+k} \ \forall\ j \in \mathbb{Z}$ (see (4.2.45)). From the structure of polygonal surfaces it follows that the covering $(\varphi, M)$ has no other branching points.

We introduce a complex structure on $M$. For every $j \in \mathbb{Z}$, in some neighborhood $U_j$ of the point $m_j \in \mathfrak{M}$ there exists a local coordinate $\psi_j$ : $U_j \to S_\varepsilon$ for which (4.2.32) is valid, and near the other points the role of a local coordinate is played by $\varphi$. So, $M$ is a Riemann surface.

2. For the function $g$, by Property 4.2.10 we see that there exists a generator $B_1$ of a cyclic group of automorphisms acting freely and totally discontinuously on $E_g$ such that (see (4.2.44), (4.2.38), and (4.2.19))

$$\varphi(B_1 m) = \varphi(m)e^{i2\pi\delta} = \varphi(m)e^{i2\pi(1-\gamma)} \quad \forall\ m \in E_g.$$

Hence

$$\varphi(B_1^{-1} m) = \varphi(m)e^{i2\pi\gamma} = \varphi(m)e^{i2\pi\rho} \quad \forall\ m \in E_g. \tag{4.2.46}$$

On the other hand (see (4.2.41) and (4.2.10))

$$\varphi(Bm) = \varphi(m)e^{i2\pi\rho} \quad \forall\ m \in E_l, \tag{4.2.47}$$

where $B$ is the generator of a group of automorphisms acting on $E_l$. Consider the automorphism $B_M : M \to M$ defined by the formula

$$B_M \big|_{E_l} = B, \quad B_M \big|_{E_g} = B_1^{-1} \tag{4.2.48}$$

and satisfying the condition (see (4.2.46) and (4.2.47))

$$\varphi(B_M m) = \varphi(m)e^{i2\pi\rho} \quad \forall\ m \in M = E_l \cup E_g. \tag{4.2.49}$$

Taking (4.2.43) into account, we see that, to show that $\varphi$ is well defined, it suffices to prove that

$$B \big|_{\partial E_l} = B_1^{-1} \big|_{\partial E_g},$$

i.e. (see (4.2.41), (4.2.43)), $B(m) \sim B_1^{-1} \circ \Phi(m) \ \forall\ m \in \partial E_l$, or

$$\Phi \circ B(m) = B_1^{-1} \circ \Phi(m) \quad \forall\ m \in \partial E_l. \tag{4.2.50}$$

Consider the diagram (see (4.2.39) and (4.2.20))

$$\begin{array}{ccccccc}
\partial E_l & \xrightarrow{\Phi_l} & \Gamma_P & \xrightarrow{F} & \Gamma_Q & \xrightarrow{\Phi_g^{-1}} & \partial E_g \\
{\scriptstyle B}\downarrow & & \downarrow{\scriptstyle \widetilde{B}_P} & & \downarrow{\scriptstyle \widetilde{B}_Q^{-1}} & & \downarrow{\scriptstyle B_1^{-1}} \\
\partial E_l & \xrightarrow{\Phi_l} & \Gamma_P & \xrightarrow{F} & \Gamma_Q & \xrightarrow{\Phi_g^{-1}} & \partial E_g ,
\end{array}$$

where $\widetilde{B}_P$ and $\widetilde{B}_Q$ are the generators of automorphism groups $G_P$ and $G_Q$, acting on $\Gamma_P$ and $\Gamma_Q$, respectively. By Theorem 4.2.13 we have:

$$\Phi_l \circ B = \widetilde{B}_P \circ \Phi_l; \quad \Phi_g \circ B_1 = \widetilde{B}_Q \circ \Phi_g.$$

The second formula shows that

$$B_1^{-1} \circ \Phi_g^{-1} = \Phi_g^{-1} \circ \widetilde{B}_Q^{-1}.$$

Therefore, formula (4.2.50) (which means that the diagram is commutative) will be proved if we show that

$$F \circ \widetilde{B}_P = \widetilde{B}_Q^{-1} \circ F. \tag{4.2.51}$$

But (see (4.2.20)) $\widetilde{B}_Q : \Gamma_Q \to \Gamma_Q$, where

$$\widetilde{B}_Q[(q, r)] = (qe^{i2\pi\delta}, r - k), \quad q \in [q_r, q_{r+1}], \quad r \in \mathbb{Z},$$

and $qe^{i2\pi\delta} \in [q_{r-k}, q_{r-k+1}]$ $\forall$ $r \in \mathbb{Z}$, because $Q$ is a $(k, \delta)$-cyclic broken line. Therefore, the mapping $\widetilde{B}_Q^{-1} : \Gamma_Q \to \Gamma_Q$ is determined by the relation (see (4.2.38))

$$\widetilde{B}_Q^{-1}[(p, \alpha)] = (pe^{i2\pi\rho}, \alpha + k), \quad p \in [q_\alpha, q_{\alpha+1}], \quad \alpha \in \mathbb{Z}.$$

Now formula (4.2.51) is verified directly on the basis of (4.2.40) and (4.2.20). Thus, formulas (4.2.50), (4.2.35), and (4.2.49) are valid. Identity (4.2.36) has already been proved (see (4.2.9)).

3. To prove that $(\varphi, M)$ is a $(\rho, n)$-covering, it remains to verify condition 2) of Definition A.2.7. By Property 4.2.11, there exists an open set $U_0 \subset E \subset M$ on which the automorphism $B_1$ still acts, and there exists a homeomorphism $\Psi_0 : U_0 \to V_0$, where $V_0 = (0, \varepsilon_0) \times \mathbb{R}$, $\varepsilon_0 > 0$, such that

$$\Psi_0(B_1 m) = A^{-1}\Psi_0(m), \quad m \in U_0; \quad \varphi_g \circ \Psi_0^{-1}(T) \equiv T^{-\delta}, \quad T \in V_0.$$

Since $n := \rho + \delta \in \mathbb{N}$ (see (4.2.38), (4.2.41), and (4.2.45)), we find:

$$\Psi_0(B_M m) = A\Psi_0(m), \quad m \in U_0; \quad \varphi \circ \Psi_0^{-1}(T) \equiv T^{\rho - n}, \quad T \in V_0.$$

Now, by (4.2.12), we conclude that the mapping $\varphi : M \to \mathbb{C}$ is a $(\rho, n)$-covering ▶

## 4.3. Relationship of the Polynomials of the Form $\alpha(z) = z^\rho + a_1 z^{\rho-1} + \ldots + a_n z^{\rho-n}$ with Functions of Class $M_\rho$

Now we show how the polygonal surface $(\varphi_l, E_l)$ considered in 4.2 and associated with a minimal trigonometric $\rho$-convex function $l$ is related to the Riemann surface of the inverse of a polynomial of the form

$$\alpha(z) = z^\rho + a_1 z^{\rho-1} + \ldots + a_n z^{\rho-n}, \tag{4.0.1}$$

where $n \geqslant 0$ and $a_n \neq 0$ $(a_0 = 1)$; the case of $n > \rho$ is not excluded (see the beginning of Ch. 4).

**4.3.1. The function class $\mathfrak{P}_\rho(l)$ for $\rho \in \mathbb{N}$.** Consider the following class of polynomials of the form (4.0.1).

**Definition 4.3.1.** Suppose $\rho > 1$ is an integer. For $l \in M_\rho \setminus O_\rho$, consider the systems $\{p_j\}_1^k$, $\{\theta_j\}_0^k$ of complex and real numbers, respectively, occurring in the definition of the function $l$ (see Theorem 4.1.5). Let $\alpha$ be a rational function of the form (4.0.1). We put $D_\alpha = \alpha^{-1}(\mathbb{C})$, i.e., $D_\alpha = \mathbb{C}$ for $n \leqslant \rho$ and $D_\alpha = \mathbb{C}^*$ for $n > \rho$. Finally, let $\mathcal{D}_\alpha(\theta)$ be the component of the set $\{\xi \in D_\alpha : \operatorname{Re} \alpha(\xi) e^{i\rho\theta} > l(\theta)\}$ containing the ray $\{te^{-i\theta} : t \geqslant t_0(\theta) > 0\}$, where $\theta \in \mathbb{R}$. The function $\alpha$ is said to *belong to the class* $\mathfrak{P}_\rho(l)$ if

1) equation $\{z \in \mathbb{C} : \alpha'(z) = 0\}$ has only $k$ roots $X = \{z_1, \ldots, z_k\}$ such that $\alpha(z_j) = p_j$, $j = 1, \ldots, k$;

2) $X \cap \mathcal{D}_\alpha(\theta) = \varnothing \ \ \forall\, \theta \in \mathbb{R}$;
$X \cap \overline{\mathcal{D}_\alpha(\theta)} = \{z_j\} \ \ \forall\, \theta \in (\theta_{j-1}, \theta_j)$, $j = 1, \ldots, k$;
$X \cap \overline{\mathcal{D}_\alpha(\theta_j)} = \{z_j, z_{j+1}\}$, $j = 1, \ldots, k$; $z_{k+1} = z_1$.

If $l \in O_\rho$ and $l(\theta) = \operatorname{Re} pe^{i\rho\theta}$, $\theta \in \mathbb{R}$, then by definition we agree that $\mathfrak{P}_\rho(l) = \{\alpha(z) = z^\rho + p\}$ (see Theorem 4.1.5).

**Example 4.3.2.** A. Let $l$ be a $2\pi$-periodic extension to $\mathbb{R}$ of the function

$$l(\theta) = \operatorname{Re} p_j e^{i2\theta} \quad \text{for} \quad \theta \in [\theta_{j-1}, \theta_j], \quad j = 1, 2, 3,$$

where $p_j = \exp\{\frac{2}{3}\pi i(j-2)\}$, $\theta_j = -\pi + \frac{2}{3}\pi j$, $\theta_0 = -\pi$. Then $l = M_2$ (see Theorem 4.1.5, Proposition 4.1.1, and Theorem 4.1.8).

1. We show that $\alpha(z) := z^2 + 2(3\sqrt{3}z)^{-1} \in \mathfrak{P}_2(l)$. Since $\alpha(z_j) = p_j$, $\alpha'(z_j) = 0$, and $\alpha''(z_j) \neq 0$ for $j = 1, 2, 3$, we find: $z_j = (\sqrt{3})^{-1} \exp\{i2\pi(2 - j)/3\}$ for $j = 1, 2, 3$. The period of $l$ is $2\pi/3$; consequently,

$$\mathcal{D}_\alpha(\theta - 2\pi/3) = \{\xi e^{i2\pi/3} : \xi \in \mathcal{D}_\alpha(\theta)\} \quad \forall\, \theta \in \mathbb{R}. \tag{4.3.1}$$

From the relation $z_j = z_{j+1} e^{i2\pi/3}$, $j = 1, 2, 3$ $(z_4 := z_1)$, we see that, for the given function $\alpha$, it suffices to verify condition 2) of Definition 4.3.1, say, for $\theta \in [\theta_1, \theta_2]$.

2. We show that the set $\mathcal{D}_\alpha(\theta)$ contains the ray $L = \{1/\sqrt{3}+te^{-i\theta} : t > 0\}$ for $\theta \in [\theta_1, \theta_2]$. Let $T(z;\theta) = \text{Re}\ \alpha(z)e^{i2\theta} - l(\theta)$, $\theta \in \mathbb{R}$. Then

$$\begin{aligned} T(z;\theta) &= \text{Re}\ (z - \frac{1}{\sqrt{3}})^2 e^{i2\theta}(1+\frac{2}{\sqrt{3}z}) = \\ &= \text{Re}\ (z - \frac{1}{\sqrt{3}})^2 e^{i2\theta}(1+\frac{2\bar{z}}{\sqrt{3}|z|^2}). \end{aligned} \tag{4.3.2}$$

Hence, $T(1/\sqrt{3}+te^{-i\theta};0) > 0\ \forall\, t>0,\ |\theta| \leqslant \pi/3$. So, there exists a component $G_\alpha(\theta)$ of the set $\{z \in \mathbb{C}^* : T(z,\theta) > 0\}$ containing the ray $L$ for $\theta \in [\theta_1, \theta_2]$; moreover, $z_2 = 1/\sqrt{3} \in \overline{G_\alpha(\theta)} \setminus G_\alpha(\theta)$, $\theta \in [\theta_1,\theta_2]$. Since $\alpha(\xi) \approx \xi^2$ as $\xi \to \infty$, for every $\varepsilon \in (0, \pi/4)$ there exists $r_0(\varepsilon) > 0$ such that

$$\{z = re^{i\lambda} \in \mathbb{C} :\ r > r_0(\varepsilon);\ |\theta+\lambda| < \pi/4 - \varepsilon\} \subset G_\alpha(\theta).$$

Hence, $\{te^{-i\theta} : t > r_0(\varepsilon)\} \subset G_\alpha(\theta)$. Therefore (see Def. 4.3.1), $G_\alpha(\theta) = \mathcal{D}_\alpha(\theta)$ and $\{z_2\} \subset \overline{\mathcal{D}_\alpha(\theta)} \setminus \mathcal{D}_\alpha(\theta)\ \ \forall\, \theta \in [\theta_1, \theta_2]$. In particular, $z_2 \in \partial\mathcal{D}_\alpha(\pi/3)$. Hence (see (4.3.1)), $\{z_1\} \subset \partial\mathcal{D}_\alpha(-\pi/3)$, i.e. $\{z_1, z_2\} \subset \partial\mathcal{D}_\alpha(-\pi/3)$.

3. It only remains to prove that $\{z_1, z_3\} \cap \mathcal{D}_\alpha(\theta) = \varnothing$, $\theta \in [\theta_1, \theta_2)$. Since $\alpha(z_j) = p_j$, $j = 1, 3$, we deduce that

$$T(z_1,\theta) = \sqrt{3}\sin(2\theta - \pi/3); \quad T(z_3,\theta) = -\sqrt{3}\sin(2\theta+\pi/3).$$

Consequently, $\{z_1, z_3\} \cap \mathcal{D}_\alpha(\theta) = \varnothing$, $|\theta| \leqslant \pi/6$. Taking, for instance, $\theta \in (-\frac{\pi}{3}, -\frac{\pi}{6})$, we see that $T(z_1,\theta) < 0$, i.e., $z_1 \notin \mathcal{D}_\alpha(\theta)$, but $T(z_3, \theta) > 0$.

We show that $z_3 \notin \mathcal{D}_\alpha(\theta)$. Take the continuous transformation $w = H(z) := (z - 1/\sqrt{3})e^{i\theta}$, $z \in \mathbb{C}$. Let $\Pi = \{w \in \mathbb{C} : \text{Re}\ w > 0\}$. Then $H(z_3) = (e^{-i2\pi/3} - 1)e^{i\theta}/\sqrt{3}$, $\text{Re}\ H(z_3) = \sin(\theta - \pi/3)$, i.e., $H(z_3) \cap \Pi = \varnothing$ $\forall\, \theta \in (-\pi/3, -\pi/6)$. On the other hand, the image $H(L)$ of the ray $L$ under the mapping $H$ (see 2)) is included in $\Pi$.

We show that the set $H[\mathcal{D}_\alpha(\theta)]$ is also included in $\Pi$. (Since $H$ is one-to-one, this will imply that $z_3 \notin \mathcal{D}_\alpha(\theta)$.) For this, it suffices to prove that $(\partial\Pi) \cap H[\mathcal{D}_\alpha(\theta)] = \varnothing$, because $H[\mathcal{D}_\alpha(\theta)]$ is connected. We have: $\partial\Pi = \{w = i\tau : \tau \in \mathbb{R}\}$. Using the formula $z = H^{-1}(w) = we^{-i\theta} + 1/\sqrt{3}$, by (4.3.2) we find

$$\begin{gathered} T_\alpha(1/\sqrt{3} + i\tau e^{-i\theta};\theta) \equiv -\tau^2(3\tau^2 + 4\sqrt{3}\tau\sin\theta + 3)\times \\ \times(1 + 2\sqrt{3}\sin\theta + 3\tau^2)^{-1}, \quad \tau \in \mathbb{R}. \end{gathered}$$

Hence, $T_\alpha(1/\sqrt{3}+i\tau e^{-i\theta};\theta) < 0$ for $\theta \in (-\pi/3, -\pi/6)$, i.e., $H[\mathcal{D}_\alpha(\theta)] \cap \partial\Pi = \varnothing$ and $z_3 \notin \mathcal{D}_\alpha(\theta)\ \ \forall\, \theta \in (-\pi/3, -\pi/6)$. Since $T_\alpha$ is continuous, we see that $T_\alpha(z_3; -\pi/3) \leqslant 0$, i.e., $z_3 \notin \mathcal{D}_\alpha(\pi/3)$.

If $\theta \in (\pi/6, \pi/3)$, we proceed similarly. As a result, we obtain: $\{z_1, z_2, z_3\} \cap \mathcal{D}_\alpha(\theta) = \varnothing\ \ \forall\, \theta \in [-\pi/3, \pi/3]$. So, all conditions of Definition 4.3.1 are fulfilled for $\alpha$.

B. Let $l \in M_4$ be the function considered in Example 4.2.12′, B. Arguing as in example A with slight modifications, we see that the class $\mathfrak{P}_\rho(l)$ contains the polynomial

$$\alpha(z) = z^4 + az, \quad z \in \mathbb{C}; \quad a = e^{i\pi/4} \cdot 4/3^{3/4}.$$

C. Similarly, we prove that in the case of a function $l$ as in Example 4.2.12′, C the class $\mathfrak{P}_\rho(l)$ contains the polynomial

$$\alpha(z) = z^3 - 3z, \quad z \in \mathbb{C}.$$

Below we shall show that the Riemann surface for the inverse of each function $\alpha \in \mathfrak{P}_\rho(l)$ considered in Example 4.3.2 is equivalent to the corresponding finitely-sheeted covering of the Riemann sphere. This covering includes the polygon $(\varphi_l, \mathcal{A}_l)$ associated with the function $l$ (see the arguments preceding Theorem 4.2.15).

**4.3.2. Embedding the polygon $(\varphi_l, \mathcal{A}_l)$ in the Riemann surface $\mathbf{R}(\alpha^{-1})$ for a polynomial $\alpha \in \mathfrak{P}_\rho(l)$.** We recall the definition of the set $R(\alpha^{-1})$: $R(\alpha^{-1}) = H \cup K$, where $H$, $K$ are the sets of the so-called regular and degenerate function elements, respectively. Let $N = \{z \in \mathbb{C} : \alpha'(z) = 0\}$. Then $H = \{(p, f_z) : p \in \mathbb{C};\ z \in \mathbb{C} \setminus N;\ \alpha(z) = p\}$, where $f_z$ is the germ at $p$ of the holomorphic function $f_z$ defined in a circular neighborhood $V_z$ of $p$ and satisfying

$$f_z(p) = z; \quad \alpha \circ f_z(q) = q; \quad f_z'(q) \neq 0, \quad q \in V_z. \tag{4.3.3}$$

The implicit function theorem implies that, for fixed $z \in \mathbb{C} \setminus N$, the function $f_z(q)$ is the only holomorphic local inverse for $\alpha$ such that the graph $f_z(q)$ passes through the point $(\alpha(z), z)$. Therefore, if $z, w \in \mathbb{C} \setminus N$, $V := V_z \cap V_w \neq \varnothing$, $q_0 \in V$, and $f_z(q_0) = f_w(q_0)$, then $f_z(q) = f_w(q)$ for $q \in V$.

Consider the set $K = \{(p, f_z) : z \in \alpha^{-1}(\infty) \cup N,\ p = \alpha(z)\}$ of "degenerate" elements $R(\alpha^{-1})$. If $n \leqslant \rho$ in (4.0.1), then $\alpha^{-1}(\infty) = \infty$, and for $n > \rho$ we have $\alpha^{-1}(\infty) = \{0, \infty\}$. The second element $f_\infty$ of the pair $(\infty, f_\infty)$ is a germ at 0 of the meromorphic function $f_\infty(t)$, which is holomorphic in the punctured circular neighborhood $S_\infty$ of 0 and has a pole at 0; moreover,

$$\alpha \circ f_\infty(t) \equiv t^{-\rho}, \quad t \in S_\infty; \quad \operatorname*{Res}_{t=0} f_\infty(t) = 1. \tag{4.3.4}$$

The second element $f_z$ of every other pair in the set $K$ is a germ at 0 of the holomorphic function $f_z(t)$ defined in some circular neighborhood $S_z$ of 0 in $\mathbb{C}$ and possessing the following properties for $z = 0$ (in the case when $n > \rho$) and for $z \in N$, respectively (see (4.0.1)):

$$\begin{aligned} &f_0(0) = 0; && \alpha \circ f_0(t) \equiv a_n t^{\rho - n}, && t \in S_0; && f_0'(0) = 1; \\ &f_z(0) = z; && \alpha \circ f_z(t) \equiv p + t^{\mu}, && t \in S_z; && f_z'(0) \neq 0, \end{aligned} \tag{4.3.4′}$$

where $\mu > 1$ is the multiplicity of the mapping $\alpha$ at the point $z \in N$ and $p = \alpha(z)$. The formula $w = f_z(\varepsilon t)$, $q = p + t^\mu$, $t \in S_z$, where $\varepsilon$ is an arbitrary $\mu$-root of unity, describes all possible ways of local uniformization of the set $\{(w, q) \in \mathbb{C}^2 : \alpha(w) - q = 0\}$ in the vicinity of the point $(z, p)$, where $z \in N$ and $p = \alpha(z)$. We agree to say that these uniformizations are *equivalent* to one another. Define the inversion operator

$$\gamma_\alpha : \overline{\mathbb{C}} \to R(\alpha^{-1}), \quad \gamma_\alpha(z) = (\alpha(z), f_z), \tag{4.3.5}$$

and the projection

$$\varphi_\alpha : R(\alpha^{-1}) \to \overline{\mathbb{C}}; \quad \varphi_\alpha[(p, f_z)] = p; \tag{4.3.6}$$

they are related by the formula $\varphi_\alpha \circ \gamma_\alpha = \alpha$.

In a usual way, we introduce a topology in $R(\alpha^{-1})$. In the notation accepted above, let $z$ be an arbitrary point in $\mathbb{C} \setminus N$. We denote by $U_z$ an open set in $\mathbb{C}$ such that $\alpha(z) \in U_z \subset V_z$ (see (4.3.3)), and by $\rho_t(\varphi)$ a germ of the holomorphic function $\varphi = f_z$ at the point $t \in U_z$. Then the system of all sets of the form $[U_z, f_z] = \{(t, \rho_t(\varphi)) : t \in U_z\}$ constitutes a basis of the topology in question at the point $(\alpha(z), f_z) \in H$. With respect to this topology, the mapping $\varphi_\alpha|_H$ is a local homeomorphism, and $\gamma_\alpha \big|_{\mathbb{C}\setminus N}$ is a homeomorphism. The topology in $R(\alpha^{-1})$ is induced from $\overline{\mathbb{C}}$: if $\{G_i : i \in I\}$ is a basis of the system of neighborhoods of the point $z$ in $\overline{\mathbb{C}}$, then $\{\gamma_\alpha(G_i) : i \in I\}$ is a basis of the system of neighborhoods of the element $(\alpha(z), f_z)$ in $R(\alpha^{-1})$. In this way, $H$ is endowed with the same topology as before. Since $\alpha$ is holomorphic, we see that $\varphi_\alpha : R(\alpha^{-1}) \to \overline{\mathbb{C}}$ is a covering (see Def. A.1.1, item A.1.2).

We introduce a complex structure on $R(\alpha^{-1})$. Near any point of $H$, the role of a local coordinate is played by the mapping $\varphi_\alpha$ (see (4.3.6)). Let $O = \alpha^{-1}(\infty) \cup N$. If $w \in O$, then for a coordinate neighborhood of the point $(\alpha(w), f_w)$ in $R(\alpha^{-1})$ we take the set $G_w = \{(\alpha(z), f_z) \in R(\alpha^{-1}) : z \in f_w(S_w)\}$ (see the notation in (4.3.3), (4.3.4), (4.3.4′)). We agree that the system of disks $\{S_w : w \in O\}$ is such that the elements of the set $\{G_w : w \in O\}$ are mutually disjoint and for every $w \in O$ the function $f_w : S_w \to f_w(S_w)$ has an inverse $f_w^{-1}$ (see (4.3.4), (4.3.4′)). For a local coordinate in $G_w$, we take the mapping $\psi_w : G_w \to f_w(S_w)$, where $\psi_w[(\alpha(z), f_z)] = f_w^{-1}(z)$, $w \in O$. From the property of the functions $\{f_w : w \in \overline{\mathbb{C}}\}$ we conclude that the transition mappings are holomorphic.

For instance, let $w \in N$, and for $u \in \mathbb{C} \setminus N$ let $G_u = \{(\alpha(z), f_z) : z \in f_u(V_u)\}$ (see (4.3.3)). If $\alpha(w) = \alpha(u) = p$, then the disk $V_u$ with center at $p$ can be assumed to satisfy $G_u \cap G_w = \varnothing$. Let $\alpha(w) \neq \alpha(u)$. We put $G := G_w \cap G_u \neq \varnothing$. From (4.3.3) it follows that $\alpha(w) \notin V_u$. Since $f_w(S_w) \cap f_u(V_u) \neq \varnothing$, from (4.3.4′) we deduce that $g(t) := \varphi_\alpha \circ \psi_w^{-1}(t) = \alpha(w) + t^\mu$, $t \in \psi_w(G)$. Therefore, the inverse function $g^{-1} = \psi_w \circ (\varphi_\alpha|_G)^{-1}$ is also holomorphic, and $g^{-1}$ is a branch of the multivalued function $\sqrt[\mu]{q - \alpha(w)}$ defined on the open subset $\varphi_\alpha(G)$ of the disk $V_u$ ($V_u$ does not contain the point $\alpha(w)$).

So, $R(\alpha^{-1})$ is a Riemann surface homeomorphic to $\overline{\mathbb{C}}$.

We show that the covering mentioned in Theorem 4.2.15 is equivalent to the covering $\varphi_\alpha : R(\alpha^{-1}) \to \overline{\mathbb{C}}$, where $R(\alpha^{-1})$ is the Riemann surface for the inverse of a certain polynomial $\alpha \in \mathfrak{P}_\rho(l)$.

**Theorem 4.3.3.** *For every function $l \in M_\rho \setminus O_\rho$ there exists a function $\alpha \in \mathfrak{P}_\rho(l)$ such that the polygon $(\widehat{\varphi}_l, \mathcal{A}_l)$* associated with $l$ (see Def. 4.3.1, 4.2.12) *is included homeomorfically and fiberwise in the covering $\alpha : \overline{\mathbb{C}} \to \mathbb{C}$ equivalent to the covering $\varphi_\alpha : R(\alpha^{-1}) \to \overline{\mathbb{C}}$* (see (4.3.6)), *and the inclusion map $\Phi_\alpha : \mathcal{A}_l \to \overline{\mathbb{C}}$ is holomorphic in $\overset{\circ}{\mathcal{A}}_l$ and satisfies the condition $\Phi_\alpha(\Lambda_\infty) = \infty$.*

◀ Let $\varphi : M \to \overline{\mathbb{C}}$ be the holomorpohic covering mentioned in Theorem 4.2.15. Since $M$ is a compact manifold, $\varphi$ is a proper holomorphic mapping. In accordance with Proposition 4.1.16 of Appendix 1, it takes each value in $\overline{\mathbb{C}}$ precisely $n$ times (counting the multiplicity), where $n = \rho + \delta$, and $\delta$ is the number of sheets of the polygon $(\varphi, B)$ in the vicinity of the point at infinity (see the proof of Theorem 4.2.15 and (4.2.31)).

1. By the Riemann theorem, for the simply connected compact Riemann surface $M$ there exists a one-to-one conformal mapping $\Phi^* : M \to \overline{\mathbb{C}}$. Since the automorphisms of the sphere $\overline{\mathbb{C}}$ are linear-fractional mappings, without loss of generality we may assume that

$$\Phi^*(\Lambda_\infty) = \infty; \quad \Phi^*(\Lambda_0) = 0. \tag{4.3.7}$$

Here $\Lambda_\infty \in \mathcal{A}_l$, $\Lambda_0 \in B$, and $\varphi(\Lambda_\infty) = \varphi(\Lambda_0) = \infty$ (the case of $\varphi^{-1}(\infty) \cap B = \varnothing$ is discussed below). The holomorphic mapping $\beta := \varphi \circ (\Phi^*)^{-1}$ is a function meromorphic in $\overline{\mathbb{C}}$. Therefore, $\beta$ is a rational function of power $n$ ($\varphi : M \to \overline{\mathbb{C}}$ is an $n$-sheeted covering). Since $\varphi^{-1}(\infty) = \{\Lambda_\infty, \Lambda_0\}$ and $\delta = n - \rho$, from (4.3.7) we deduce that

$$\beta(z) = b_0 z^\rho + b_1 z^{\rho-1} + \ldots b_n z^{\rho-n}, \tag{4.3.8}$$

where $b_0, b_n \neq 0$. Consider the mapping $\Phi_1 = b_0^{1/\rho} \cdot \Phi^*$, where $b_0^{1/\rho}$ is a $\rho$-root of the coefficient $b_0$ of the polynomial (4.3.8). The function $\alpha_1 := \varphi \circ \Phi_1^{-1}$ is of the form (4.0.1). We take a chart $(\omega_\infty, \psi_\infty)$ of the atlas on $M$ in the vicinity of the point $\Lambda_\infty$. Then there exists a disk $S$ centered at $0 \in \mathbb{C}$ and such that $\varphi \circ \psi_\infty^{-1}(\tau) \equiv \tau^{-\rho}$ for $\tau \in S$ (see (4.2.28) and the proof of Property 4.2.14). Therefore, the holomorphic function $\varkappa := \Phi_1 \circ \psi_\infty^{-1}$ satisfies the identity $\alpha_1 \circ \varkappa(\tau) \equiv \tau^\rho$, $\tau \in S$. There exists a $\rho$-root of unity $\varepsilon$ such that $\varkappa(\tau) = \varepsilon\tau^{-1}(1 + O(\tau))$, $\tau \to 0$.

2. Let $\Phi_\alpha = \varepsilon^{-1}\Phi_1$. We show that $\alpha := \varphi \circ \Phi_\alpha^{-1} \in \mathfrak{P}_\rho(l)$. The function $\alpha$ has the form (4.0.1), and by construction for $f_\infty := \Phi_\alpha \circ \psi_\infty^{-1}$ we have (cf. (4.3.4)):

$$f_\infty(\tau) = \tau^{-1}(1 + O(\tau)), \quad \tau \to 0. \tag{4.3.9}$$

Let $z_j = \Phi_\alpha(m_j)$ for $j = 1, \dots, k$, where $m_j$ is a branching point of the covering $\varphi : M \to \overline{\mathbb{C}}$. Since $\varphi(m_j) = p_j$ (see Property 4.2.9), we have $\alpha(z_j) = p_j$, $j = 1, \dots, k$, and the multiplicty of the mapping $\alpha$ at the point $z_j$ is equal to the number $\mu_j > 1$ occurring in (4.2.15). Therefore, $\alpha$ satisfies condition 1 of Definition 4.3.1 (the covering $\varphi : M \to \overline{\mathbb{C}}$ has no other branching points except $\Lambda_\infty$, $\Lambda_0$ ).

Let $V_\theta = U_\theta/G_E$ be the set considered in the proof of Property 4.2.14 (see (4.2.28)): $T_\theta = \{p \in \mathbb{C} : \text{Re } pe^{i\rho\theta} > l(\theta)\}$. We show that in the notation of Definition 4.3.1 we have

$$\Phi_\alpha(V_\theta) = \mathcal{D}_\alpha(\theta) \quad \forall\, \theta \in \mathbb{R}. \tag{4.3.10}$$

Clearly,

$$\alpha \circ \Phi_\alpha(V_\theta) = \varphi(V_\theta) = T_\theta \quad \forall\, \theta \in \mathbb{R}. \tag{4.3.11}$$

Next, let $\omega_\infty$ be a coordinate neighborhood of the point $\Lambda_\infty$ on $\mathcal{A}_l$ (see (4.2.28)). We observe that in the notation of Property 4.2.11 and formula (4.3.9) we have

$$\Phi_\alpha(V_\theta \cap \omega_\infty) = f_\infty \circ \pi \circ \Psi_\infty(U_\theta \cap U_\infty), \quad \theta \in \mathbb{R}, \tag{4.3.12}$$

where $\pi : L \to \mathbb{C}$; $\pi[(r,\theta)] = re^{i\theta}$, $(r,\theta) \in L$. The mapping $\Psi_\infty$ satisfies condition (4.2.14) for every sufficiently small $\varepsilon > 0$. Therefore, from (4.3.12), (4.3.9) we deduce that the ray $\{re^{-i\theta} : r > r_0(\theta)\}$ is included in $\Phi_\alpha(V_\theta)$ $\forall\, \theta \in \mathbb{R}$. By (4.3.11) we conclude that condition (4.3.10) is satisfied.

Finally, from the structure of the system of sets $\{\Phi_\alpha(V_\theta) : \theta \in \mathbb{R}\}$ we deduce condition 2) of Definition 4.3.1 for the polynomial $\alpha$, because $\mathcal{D}_\alpha(\theta) = \mathcal{D}_\alpha(\theta + 2\pi)$ $\forall\, \theta \in \mathbb{R}$.

3. Now, let $\delta = 0$. Then $\rho = n$, and condition (4.3.7) is replaced by the condition $\Phi^*(m_\infty) = \infty$. We repeat the arguments of items 1, 2. In this case the required function $\alpha$ is a usual polynomial of degree $\rho$.

4. Since $\alpha = \varphi_\alpha \circ \gamma_\alpha = \varphi \circ \Phi_\alpha^{-1}$ (see (4.3.5), (4.3.6)), we conclude that the mapping $H_\alpha := \gamma_\alpha \circ \Phi_\alpha : M \to R(\alpha^{-1})$ embeds the set $\mathcal{A}_l$ in $R(\alpha^{-1})$ homeomorphically, and the covering $\varphi : M \to \overline{\mathbb{C}}$ is equivalent to the covering $\varphi_\alpha : R(\alpha^{-1}) \to \mathbb{C}$, i.e., $\varphi = \varphi_\alpha \circ H_\alpha$.

**Remark.** For a map $\alpha$ of the form (4.0.1), the Riemann surface $R(\alpha^{-1})$ has $r = \max\{\rho, n\}$ sheets. Theorem 4.3.3 provides a lower bound of for $\rho$ if $\alpha \in \mathfrak{P}_\rho(l)$. Since $R(\alpha^{-1})$ has genus 0, the Riemann- Hurwitz formula yields

$$2r = 2 + (\rho - 1) + \sum_{j=1}^{k} (\mu_j - 1), \tag{4.3.13}$$

where $\mu_j$ is the multiplicity of $\alpha$ at the point $z_j$, $j = 1, \dots, k$. Since the polygon $(\widehat{\varphi}, \mathcal{A}_l)$ is included in $R(\alpha^{-1})$ homeomorphically and fiberwise, from Property 4.2.9 we have $2\pi\mu_j > \pi + \rho(\theta_j - \theta_{j-1})$, $j = 1, \dots, k$. By (4.3.13) we find: $r \geqslant \rho + 1/2 - k/4$, where $k$ is the number of portions of a suitable segment of length $2\pi$ such that the function $l$ is trigonometric (see (4.1.1) – (4.1.3)) on each portion.

Let $l \in M_\rho \setminus O_\rho$. Immediate verification shows that the polygon $(\widehat{\varphi}_s, \mathcal{A}_s)$ associated with the function

$$l_s(\varphi) := l(\varphi - 2\pi s/\rho), \quad \varphi \in \mathbb{R}; \quad s \in \{1, \dots, \rho - 1\} \tag{4.3.14}$$

is equivalent to the polygon $(\widehat{\varphi}_l, \mathcal{A}_l)$, i.e., there exists a homeomorphism $\Phi : \mathcal{A}_s \to \mathcal{A}_l$ such that $\widehat{\varphi}_s = \widehat{\varphi}_l \circ \Phi$. We examine the relationship between the elements of the sets $\mathfrak{P}_\rho(l_s)$ and $\mathfrak{P}_\rho(l)$ (see (4.3.14) and Def. 4.3.1).

**Property 4.3.4.** *If $\alpha \in \mathfrak{P}_\rho(l)$, then for every $s \in \{1, ..., \rho - 1\}$ the polynomial*

$$\beta(w) := \alpha(we^{2\pi i s/\rho}), \quad w \in \mathbb{C}, \tag{4.3.15}$$

*belongs to $\mathfrak{P}_\rho(l_s)$.*

◀ If $l$ is defined by formula (4.1.1), then the function $l_s$ (see (4.3.14)) is the $2\pi$-periodic extension of the function

$$l_s(\varphi) = \operatorname{Re} p_j e^{i\rho\varphi}, \ \varphi \in [\theta_{j-1} + \Delta_s, \theta_j + \Delta_s], \tag{4.3.16}$$

where $\Delta_s = 2\pi s/\rho$, $j = 1, \dots, k$. Consider the operator

$$H : \mathbb{C} \to \mathbb{C}; \quad w = H(z) = ze^{-2\pi i s/\rho}, \quad z \in \mathbb{C}. \tag{4.3.17}$$

In the notation of Definition 4.3.1, let $z_1, \dots, z_k$ be the roots of the derivative $\alpha'$ of the polynomial $\alpha \in \mathfrak{P}_\rho(l)$, and let

$$w_j = H(z_j), \quad j = 1, \dots, k. \tag{4.3.18}$$

Then $w_1, \dots, w_k$ are the roots of the derivative of the polynomial $\beta$ (see (4.3.15)), and $\beta(w_j) = p_j$ for $j = 1, \dots, k$, i.e., $\beta$ satisfies condition 1) of Definition 4.3.1.

In the set $\{w : \operatorname{Re} \beta(w)e^{i\rho\varphi} > l_s(\varphi)\}$, consider the component $\mathcal{D}_\beta(\varphi; l_s)$ containing the ray $\{te^{-i\varphi} : t > t_0(\varphi)\}$, where (see (4.3.16)) $\varphi - 2\pi s/\rho \in [\theta_{j-1}, \theta_j]$ for $j \in \{1, \dots, k\}$. In the notation of formula (4.3.17) and Definition 4.3.1, we have

$$\mathcal{D}_\beta(\varphi, l_s) = H[\mathcal{D}_\alpha(\varphi - 2\pi s/\rho)].$$

By (4.3.16), (4.3.18) we see that condition 2) of Definition 4.3.1 is valid ▶

**4.3.3. The class $\mathfrak{P}_\rho(l)$ with nonintegral $\rho > 1/2$.** In the case of nonintegral $\rho > 1/2$ we have a statement similar to Theorem 4.3.3 about the embedding of the polygon $\mathcal{A}_l$ (see Def. 4.2.11) in the Riemann surface for the inverse of a polynomial of the form (4.0.1).

If $\rho$ is a noninteger, it is convenient to treat the polynomial (4.0.1) as a function on $L$:

$$\alpha(Z) = Z^\rho + a_1 Z^{\rho-1} + \ldots + a_n Z^{\rho-n}, \quad Z \in L, \tag{4.3.19}$$

where $n > \rho$. In this case the class $\mathfrak{P}_\rho(l)$ is defined in the following way. Consider an arbitrary polynomial $\alpha$ of the form (4.3.19). Let $N_\alpha$ be the set of all zeros of the derivative $\alpha'$ in $L$, and let $X = \pi(N_\alpha)$. Then $X$ is the set of different roots in $\mathbb{C}$ of the polynomial

$$\rho z^n + (\rho - 1)a_1 z^{n-1} + \ldots + (\rho - n)a_n.$$

For $\theta \in \mathbb{R}$, in the set $\{Z \in L : \operatorname{Re} \alpha(Z)e^{i\rho\theta} > l(\theta)\}$ we consider the component $\mathcal{D}'_\alpha(\theta)$ containing the ray $\{(t, -\theta) \in L,\ t > t_0(\theta) > 0\}$ and put

$$\mathcal{D}_\alpha(\theta) = \pi[\mathcal{D}'_\alpha(\theta)], \tag{4.3.20}$$

where $l$ is a fixed function in $M_\rho$ having the form (4.1.1).

**Definition 4.3.5.** Suppose $\rho \notin \mathbb{N}$ and $l \in M_\rho \setminus O_\rho$. Let $P = \{p_j : j \in \mathbb{Z}\}$, $\Theta = \{\theta_j : j \in \mathbb{Z}\}$ be the sequences of complex and real numbers, respectively, occurring in the definition of the function $l$ (see (4.1.2)). *A mapping $\alpha$ of the form* (4.3.19) is said to *belong to the class* $\mathfrak{P}_\rho(l)$ if the following conditions are satisfied:

1) the elements of the set $N_\alpha$ form a sequence $N_\alpha = \{Z_j : j \in \mathbb{Z}\}$ such that $\alpha(Z_j) = p_j\ \forall\, j \in \mathbb{Z}$;

2) the number $k$ related to the function $l$ coincides with the number of elements in the set $X = \{z_1, \ldots, z_k\}$, and (see item 1), Property 4.2.11, and (4.1.3)) $z_i = \pi(Z_{i+jk})\ \forall\, j \in \mathbb{Z},\ i = 1, \ldots, k;\ Z_{j-k} = AZ_j\ \forall\, j \in \mathbb{Z}$.

3) condition 2) of Definition 4.3.1 is fullfilled with $X$, $\mathcal{D}_\alpha(\theta)$, and $z_j$ replaced by $N_\alpha$, $\mathcal{D}'_\alpha(\theta)$, and $Z_j$, $j \in \mathbb{Z}$.

If $l \in O_\rho$, i.e., $l(\theta) \equiv 0$ (see Theorem 4.1.5), then $\mathfrak{P}_\rho(l) = \{Z^\rho\}$.

**Remark.** Condition 3) can be verified on any other interval of length $2\pi$, because

$$\mathcal{D}'_\alpha(\theta + 2\pi) = \{(\tau, \psi - 2\pi) \in L : (\tau, \psi) \in \mathcal{D}'_\alpha(\theta)\} \quad \forall\, \theta \in \mathbb{R},$$

and, consequently, $\mathcal{D}_\alpha(\theta + 2\pi) = \mathcal{D}_\alpha(\theta)\ \forall\, \theta \in \mathbb{R}$ ( $l$ is a $2\pi$-periodic function).

**Example 4.3.6.** Let $\rho \in (1/2, 1)$, and let

$$l(\theta) = \cos \rho(\theta + \pi - 2\pi j), \quad \theta \in [2\pi(j-1), 2\pi j], \quad j \in \mathbb{Z}.$$

The function $l$ is determined by the $(1, \rho)$-cyclic broken line $P = \{p_j : j \in \mathbb{Z}\}$, where $p_j = \exp\{-i\pi\rho(2j-1)\}$ for $j \in \mathbb{Z}$ (see Def. 4.1.6′). We show that the function

$$\alpha(Z) = Z^\rho - \rho(1-\rho)^{1/\rho - 1} Z^{\rho-1}, \quad Z \in L$$

belongs to $\mathfrak{P}_\rho(l)$.

The sequence $\{Z_j = ((1-\rho)^{1/\rho}, \pi - 2\pi j) \in L : j \in \mathbb{Z}\}$ is the set of zeros of the derivative $\alpha'$. Direct inspection shows that $\alpha(Z_j) = p_j$, $j \in \mathbb{Z}$. We prove that the function $\alpha$ maps the plane with a cut along the negative semiaxis $\Pi_0 = \{Z = (r, \psi) \in L : r > 0; |\psi| < \pi\}$ bijectively and conformally onto the complex plane $M_0$ with cuts along the rays $L_0$ and $L_1$, where $L_j = \{p_j t : t \geqslant 1\}$ for $j = 0, 1$.

For instance, let $j = 0$. We put $\Pi_0^+ = \{z \in \mathbb{C} : |z| > 0; 0 < \arg z < \pi\}$ and denote by $M_0^+$ the upper half-plane in $\mathbb{C}$ with a cut along the ray $L_0$ ($M_0^+$ can be viewed as a triangle with a vertex at $p_0$ and two vertices at $\infty$, and with angles equal to $2\pi$, $(1-\rho)\pi$, and $\rho\pi$, respectively). By the Christoffel-Schwarz formula, the general form of a one-sheeted conformal mapping $f$ of the half-plane $\Pi_0^+$ onto $M_0^+$ satisfying the conditions $f(-\lambda) = p_0$ and $f(\infty) = f(0) = \infty$ is as follows:

$$f(z) = c_1 \int_{z_0}^{z} (t+\lambda) t^{\rho-2}\, dt + c_2 = c_1 \Big(\frac{z^\rho}{\rho} - \frac{\lambda z^{\rho-1}}{1-\rho}\Big) + c_2, \quad z \in \Pi_0^+,$$

where the symbol $z^\nu$ is taken in the principal value sense ($f$ sends 0 to $\infty$, and the angle at the latter point is taken with the minus sign). The function $f$ has the form (4.3.19) for $c_1 = \rho$, $c_2 = 0$. The condition $f(-\lambda) = p_0$ is fulfilled with $\lambda = (1-\rho)^{1/\rho}$. Then $f = \alpha$. Finally, since $f[(0,\infty)] = (-\infty, \infty)$, from the symmetry principle we deduce that the mapping $\alpha$ possesses the required property.

If the edges of the cuts mentioned above are not identified, then $\alpha$ maps the closure of $\Pi_0$ onto the closure of $M_0$ homemorphically, but conformity is violated at the points $Z_0$ and $Z_1$. In the notation of Definition 4.3.5, $k = 1$ and $X = \{-(1-\rho)^{1/\rho}\}$, i.e., condition 2) of Definition 4.3.5 is valid. Condition 3) (saying that $\alpha \in \mathfrak{P}_\rho(l)$) can be verified for $\theta \in (-\pi, \pi]$ (see the remark to Def. 4.3.5).

Let $T_\theta = \{p \in \mathbb{C} : \operatorname{Re} pe^{i\rho\theta} > l(\theta),\ |\theta| \leqslant \psi\}$, where $\psi = \pi(1 - 1/2\rho)$. Then $T_\theta \subset M_0$. The property of $\alpha$ described above implies the existence of a unique domain $G_\alpha(\theta) \subset \Pi_0$ such that $\alpha[G_\alpha(\theta)] = T_\theta$. Let $\Gamma_0 = \{(t, -\theta) \in L : t \geqslant R\}$. For sufficiently large $R$ we have $\alpha(\Gamma_\theta) \subset T_\theta$. Hence, $\Gamma_\theta \subset G_\alpha(\theta)$, $G_\alpha(\theta) = \mathcal{D}'_\alpha(\theta)$. Since $\alpha : \overline{\mathcal{D}'_\alpha(\theta)} \to \overline{T}_\theta$ is a homeomorphism and since $p_0 \in T_\theta$ for $\theta \subset [-\psi, 0]$ and $p_1 \in T_\theta$ for $\theta \in [0, \psi]$, it follows that condition 3) of Definition 4.3.5 is fulfilled for $|\theta| \leqslant \psi$. Now, let $\theta \in (\psi, \pi]$ and let $\varphi = \pi(1 + 1/2\rho)$. Then the ray $L_1 \subset T_\theta$ and the half-plane $T_\theta$ lie in the plane $T_\psi \cup L_1 \cup T_\varphi$ with a cut along the ray that is the continuation of $L_1$. Since $\alpha : \overline{\mathcal{D}'_\alpha(\psi)} \to \overline{T}_\psi$ is a conformal homeomorphism and, in particular, the function $\alpha$ maps the ray $\Gamma = \{(t, -\pi) \in L : t \geqslant (1-\rho)^{1/\rho}\}$ onto $L_1$ homeomorphically, by the symmetry principle the mappings

$$\alpha : \overline{\mathcal{D}'_\alpha(\varphi)} \to \overline{T}_\varphi; \quad \alpha : \overline{\mathcal{D}'_\alpha(\theta)} \to \overline{T}_\theta \quad \forall\, \theta \in (\psi, \pi]$$

are also conformal homeomorphisms. Since $p_1 \in \overline{T}_\theta \setminus T_0$, we see that $Z_1 \in \overline{\mathcal{D}'_\alpha(\theta)} \setminus \mathcal{D}'_\alpha(\theta)$ $\forall\, \theta \in (\psi, \pi]$, i.e., condition 3) of Definition 4.3.5 is valid for the mapping $\alpha$ and all $\theta \in (\psi, \pi]$. We argue similarly for $\theta \in (-\pi, -\psi)$. Thus, $\alpha \in \mathfrak{P}_\rho(l)$.

Starting with the system of sets $\{M_j : j \in \mathbb{Z}\}$, we construct a $(\rho, 1)$-covering $\varphi : M \to \mathbb{C}$ (see Example A.2.10). As we shall see below, this covering is equivalent to the covering $\varphi_\alpha : H_\alpha \to \mathbb{C}$, where $H_\alpha$ is the Riemann surface for $\alpha^{-1}$ (see (A.2.12)).

For every noninteger $\rho > 1$, Example 4.3.6 admits generalization to the case of a function trigonometric on intervals of length $2\pi$.

Suppose $n \in \mathbb{N} \setminus \{1\}$, $n-1 < \rho < n$, and $p_j = (-1)^{n-1}\exp\{\rho(\pi - 2\pi j)i\}$ for $j \in \mathbb{Z}$. We put

$$l_n(\theta) = \operatorname{Re}\ p_j e^{i\rho\theta} = (-1)^{n-1}\cos\rho(\theta + \pi - 2\pi j), \quad \theta \in [\theta_{j-1}, \theta_j], \quad j \in \mathbb{Z}, \tag{4.3.21}$$

where $\theta_j = 2\pi j$. Then, in the notation of formula (4.1.4), for every $j \in \mathbb{Z}$ we have $A_j = 0$,

$$B_j = \operatorname{Im}\ (p_{j+1} - p_j)e^{i\rho\theta_j} = (-1)^n 2\sin\pi\rho = 2\sin\pi(n - \rho - 1) < 0,$$

i.e., condition (4.1.6) is fulfilled. By Proposition 4.1.1 and Theorem 4.1.6, $l_n \in M_\rho \setminus O_\rho$. For $\lambda > 0$, let

$$\lambda^\rho = (\rho-1)(\rho-2)\cdots(\rho-n+1)(n-\rho)/n!. \tag{4.3.22}$$

Consider the function

$$\alpha_n(Z) = \sum_{j=0}^{n} Z^{\rho-j}\binom{n}{j}\frac{\rho\lambda^j}{\rho-j}, \quad Z \in L, \tag{4.3.23}$$

where $\binom{n}{j}$ are the binomial coefficients. For $n = 1$, the functions $l_n$, $\alpha_n$ coincide with the functions considered in Example 4.3.6. We compute the derivative of $\alpha_n$:

$$\alpha_n'(Z) = \rho\sum_{j=0}^{n}\binom{n}{j}\lambda^j Z^{\rho-j-1} = \rho Z^{\rho-1-n}(\pi(Z)+\lambda)^n, \quad Z \in L; \quad \pi(Z) = |Z|e^{i\,Arg\,Z}.$$

Then $\{Z_j = (\lambda, \pi - 2\pi j) \in L : j \in \mathbb{Z}\}$ is the sequence of zeros of $\alpha_n'$. In order to verify condition 1) of Definition 4.3.4 for $\alpha_n$, we write:

$$\alpha_n(Z_j) = \rho\lambda^\rho e^{i\pi\rho(1-2j)}\sum_{s=0}^{n}\binom{n}{s}(-1)^s/(\rho-s), \quad j \in \mathbb{Z}.$$

The formula $\alpha_n(Z_j) = p_j$, $j \in \mathbb{Z}$, now follows from (4.3.21), (4.3.22) and the lemma below.

**Lemma 4.3.7.** *We have*

$$T(\rho) := \sum_{s=0}^{n} \binom{n}{s} \frac{(-1)^s}{\rho - s} = \frac{(-1)^n n!}{\rho(\rho-1)\cdots(\rho-n)}. \qquad (4.3.24)$$

◀ By the uniqueness theorem for meromorphic functions, it suffices to prove the identity for $\rho > 1$. Using the method of integral representations with $\rho > n$, we find:

$$T(\rho) = \sum_{s=0}^{n} (-1)^s \binom{n}{s} \int_0^1 x^{\rho-s-1}\, dx = \int_0^1 x^{\rho-1} \sum_{s=0}^{n} (-1)^s \binom{n}{s} x^{-s}\, dx.$$

So,

$$T(\rho) = (-1)^n \int_0^1 (1-x)^n x^{\rho-n-1}\, dx = (-1)^n B(n+1, \rho-n),$$

where $B$ is the beta-function. By the Euler formula, now we have

$$T(\rho) = (-1)^n \frac{\Gamma(n+1)\Gamma(\rho-n)}{\Gamma(\rho+1)} = \frac{(-1)^n n!}{\rho(\rho-1)\cdots(\rho-n)} \qquad ▶$$

Condition 2) (saying that $\alpha_n \in \mathfrak{P}_\rho(l_n)$, see (4.3.23)) is verified by the method similar to that used in Example 4.3.5 for $n = 1$.

If $\rho \notin \mathbb{N}$, the functions of class $\mathfrak{P}_\rho(l)$ have geometric properties similar to the properties of the functions of class $\mathfrak{P}_\rho(l)$ for $\rho \in \mathbb{N}$ (see Theorem 4.3.3).

**Theorem 4.3.8.** *Let $\rho \notin \mathbb{N}$. For every function $l \in M_\rho \setminus O_\rho$ there exists a polynomial $\alpha \in \mathfrak{P}_\rho(l)$* (see Def. 4.3.5) *such that the polygonal surface $(\varphi_l, E_l)$ associated with $l$ admits homeomorphic inclusion $F$ into the covering $\alpha : L \to \mathbb{C}$ related to $\alpha$. Moreover, $F$ has the following properties:*

1) *$F : E_l \to L$ is a fiberwise mapping, i.e., $\alpha \circ F = \varphi_l$;*

2) *$F : \overset{\circ}{E_l} \to F(\overset{\circ}{E_l})$ is a holomorphic homeomorphism ;*

3) *the mapping $F$ agrees with the actions of the cyclic groups $G_E$ on $E_l$ and $G_L$ on $L$* (see Properties 4.2.10, 4.2.11), *i.e.,*

$$F \circ B^s = A^s \circ F \quad \forall\, s \in \mathbb{Z},$$

*where $B$ and $A$ are generators of groups $G_E$ and $G_L$, respectively.*

◀ Let $(\varphi, M)$ be the $(\rho, n)$-covering considered in the proof of Theorem 4.2.16 and having the following properties: $M = E_l \cup E_g$, where $l$ is the given function in $M_\rho \setminus O_\rho$ and $g \in P_\delta \setminus O_\delta$ with $\delta = n - \rho$; here $E_l$ and $E_g$ stand for the polygonal surfaces associated with the functions $l$ and $g$, respectively (see Def. 4.2.7) and glued in the due way at boundary points. Therefore, if $\mathcal{A}_l$ and $\mathcal{A}_g$ are polygons associated with the functions $l$ and $g$ ($\mathcal{A}_l$ is $\rho$-sheeted near $\infty$ and $\mathcal{B}_g$ is $\delta$-sheeted near $\infty$, see Def. 4.2.12), and $\mathfrak{M}$ is the topological extension of the space of orbits $\mathfrak{M}'$ of the group $G_M$ on $M$ with the aid of two points $\Lambda_\infty$, $\Lambda_0$, then $\mathfrak{M} = \mathcal{A}_l \cup \mathcal{A}_g$, and $\Lambda_\infty \in \mathcal{A}_l$, $\Lambda_0 \in \mathcal{A}_g$. From Property 4.2.12 and Theorem A.2.8 we deduce that $\mathfrak{M}$ is a compact Riemann surface.

On the other hand, $\mathcal{A}_l$, $\mathcal{A}_g$, and $\mathfrak{M}$ are simplicial spaces (see subsection A.1.5). We verify this, say, for $\mathcal{A}_l$. By Property 4.2.10, a model for $\mathcal{A}_l$ is the polygon $(\varphi_l, M_0^+)$ the sides $\Gamma_0$ and $\Gamma_1$ of which are glued at appropriate points (see the proof of Property 4.2.10). We triangulate $M_0^+$ to ensure the following condition: if $m_1$, $m_2$ are two different points in $M_0^+$ belonging to one and the same orbit of the group $G_M$ on $E_l$ and, say, $m_1$ is a vertex of one of the triangulation simplexes, then so is $m_2$. Gluing $\Gamma_0$ and $\Gamma_1$ at the points belonging to one and the same orbit of the group $G_M$, we get a triangulation on $\mathcal{A}_l$. But $\mathcal{A}_l$ and $\mathcal{A}_g$ are simply connected sets, and $\mathcal{A}_l \cap \mathcal{A}_g = \partial E_l / G_M$ is a closed path (see Properties 4.2.10 and 4.2.14) in $\mathfrak{M}$. Hence, $\mathfrak{M}$ is a simply connected set (see Theorem A.1.18). Moreover, $M$ is a simply connected surface (see the proof of Theorem 4.2.16). Therefore, the covering $\varphi : M \to \mathbb{C}$ satisfies all conditions of Theorem A.2.12.

Let $\{m_j : j \in \mathbb{Z}\}$ be the branching points of the covering $\varphi : M \to \mathbb{C}$, let $F = \Phi^{-1}$, where $\Phi : L \to M$ is a homeomorpohism, and let $\alpha = \varphi \circ \Phi$ be the polynomial of the form (4.3.19) mentioned in Theorem A.2.12. We put $Z_j \doteq F(m_j)$, $j \in \mathbb{Z}$. It only remains to show that $\alpha \in \mathfrak{P}_\rho(l)$.

Using the notation of Theorem 4.2.16, from that theorem we deduce that $\alpha(Z_j) = \varphi(m_j) = p_j$, $j \in \mathbb{Z}$, and the multiplicity of the mapping $\alpha$ at the point $Z_j$ coincides with $\mu_j > 1 \ \forall\, j \in \mathbb{Z}$. From formula (4.2.36) and formula (A.2.46) we find:

$$AZ_{j+k} = Z_j \quad \forall\, j \in \mathbb{Z} \quad (Z_j = F(m_j)). \tag{4.3.25}$$

Therefore, conditions 1), 2) of Definition 4.3.5 are satisfied: $\alpha'(Z_j) = 0 \ \forall\, j \in \mathbb{Z}$.

For $\theta \in \mathbb{R}$, we put $T_\theta = \{p \in \mathbb{C} : \operatorname{Re} pe^{i\rho\theta} > l(\theta)\}$ and $U_\theta = \{(p, \Delta) \in E_l : p \in T_\theta;\ \theta \in \Delta\}$ (see (4.2.3)). Since $\varphi_l = \varphi \mid_{\partial E_l}$ (see the proof of Theorem 4.2.16), we have

$$\alpha \circ F(U_0) = \varphi_l(U_\theta) = T_\theta \quad \forall\, \theta \in \mathbb{R}. \tag{4.3.26}$$

The homeomorphism $\Psi_\infty : U_\infty \to V_\infty$ associated with the $(\rho, n)$-covering $\varphi : M \to \mathbb{C}$ is chosen as indicated in Property 4.2.11 (see Def. A.2.7, and the proof of Theorem 4.2.16). Therefore,

$$W_\theta := \left\{T \in L :\ |T| < R_\varepsilon^{-1/\rho};\ \rho |\, Arg\ T - \theta| < \frac{\pi}{2} - \varepsilon \right\} \subset V_\infty$$

(see (4.2.14)) and, consequently, $\Psi_\infty^{-1}(W_\theta) \subset U_\theta$. By property 3) of the mapping $\Phi$ in Theorem A.2.12, the ray $\{(r, -\theta) : r > r_0(\theta)\}$ is included in $F(U_0)$. So, from (4.3.26) we deduce that

$$F(U_\theta) = \mathcal{D}'_\alpha(\theta) \quad \forall\, \theta \in \mathbb{R}. \tag{4.3.27}$$

Taking into account the geometic structure of the surface $(\varphi_l, E_l)$, from (4.3.25), (4.3.27) we deduce condition 3) in Definition 4.3.5. Thus, $\alpha \in \mathfrak{P}_\rho(l)$ ▶

Suppose $\rho \notin \mathbb{N}$, $s \in \mathbb{Z}\backslash\{0\}$, and $l \in M_\rho \backslash O_\rho$. Let $l_s$ be the trigonometrically $\rho$-convex function defined by formula (4.3.16) and associated with $l$ (in the notation of that formula, $j \in \mathbb{Z}$, and $(\varphi_s, l_s)$ is the polygonal surface associated with $l_s$ (see (4.2.2) and (4.2.3)). We have

$$F: E_l \to E_s; \quad F[(p, \Delta)] = (p, \Delta_s), \quad (p, \Delta) \in E_l,$$

where $\Delta$ and $\Delta_s$ are components of the sets $X(p; l)$ and $X(p; l_s)$, respectively, and $\Delta_s = \Delta + 2\pi s/\rho = \{\varphi + 2\pi s/\rho : \varphi \in \Delta\})$. Since $\varphi_l = \varphi_s \circ F$, the surfaces $(\varphi_l, E_l)$ and $(\varphi_s, E_s)$ are equivalent. The following assertion is a modification of Property 4.3.4 and is proved similarly.

**Property 4.3.4′.** *In the previous notation, if* $\alpha \in \mathfrak{P}_\rho(l)$ (see (4.3.20) and Def. 4.3.5), *then for every* $s \in \mathbb{Z} \setminus \{0\}$ *the polynomial* $\beta(W) = \alpha(A_{s/\rho}W)$, $W \in L$, *belongs to the class* $\mathfrak{P}_\rho(l_s)$, *where*

$$A_\nu : L \to L; \quad A_\nu[(r, \varphi)] = (r, \varphi + 2\pi\nu), \quad (r, \varphi) \in L.$$

## 4.4. Plane $(\rho, \alpha)$-convex sets and the indicator diagram of an entire function of order $\rho > 0$

We introduce the notion of the plane indicator diagram of an entire function of order $\rho > 0$, $\rho \neq 1$, and examine the properties of such diagrams.

**4.4.1. The plane indicator diagram of an entire function of order $\rho > 0$.** The geometric image in $\mathbb{C}$ of the indicator $h$ of an entire function can be found with the help of the mapping $\alpha$ of the form (4.0.1): the locally convex curve $\Gamma_h$ (see Def. 1.9.24) associated with $h$ is the image under such a mapping of the closed curve in $\mathbb{C}$ that bounds the "indicator diagram". If, for instance, $\rho \in \mathbb{N}$, and the index of $\Gamma_h$ relative to $p_0$ is $\rho$, then $\Gamma_h$ is the image under the mapping $\alpha(z) = z^\rho + p_0$ of a certain closed Jordan curve in $\mathbb{C}$ that encloses the point $z = 0$ and bounds the $\rho$-convex compact set $X_h$ (see Subsection 3.1.4). Now we pass to precise definitions.

Suppose $l \in M_\rho$, $h \in P_\rho(l)$, and $\alpha \in \mathfrak{P}_\rho(l)$ (see Definitions 4.3.1 and 4.3.5). With the polynomial $\alpha$ and the function $h$, we associate a system of sets $\{\mathcal{D}_\alpha(\theta; h) : \theta \in \operatorname{dom} h\}$; for $h = l$ this system turns into the system $\{\mathcal{D}_\alpha(\theta) : \theta \in \mathbb{R}\}$ considered in Definitions 4.3.1 and 4.3.5. If $\rho \in \mathbb{N}$, then $\mathcal{D}_\alpha(\theta, h)$ is the component of the set $\{z \in D_\alpha : \operatorname{Re} \alpha(z)e^{i\rho\theta} > h(\theta)\}$ containing the ray $\{te^{-i\theta} : t > t_0(R)\}$. Here $D_\alpha = \alpha^{-1}(\mathbb{C})$ (see Def. 4.3.1). If $\rho \notin \mathbb{N}$, $\rho > 1/2$, then (see Def. 4.3.5)

$$\mathcal{D}_\alpha(\theta; h) = \pi(\mathcal{D}'_\alpha(\theta; h)), \quad \theta \in \operatorname{dom} h,$$

where $\mathcal{D}'_\alpha(\theta; h)$ is the component of the set $\{Z \in L : \text{Re}\, \alpha(Z)e^{i\rho\theta} > h(\theta)\}$ containing the ray $\{(t, -\theta) \in L : t > t_0(R)\}$. In particular, if $l = l_0$, where $l_0(\theta) \equiv 0$ for $\theta \in \mathbb{R}$, $h \in P_\rho(l_0)$, $\alpha(z) = z^\rho$, and $\gamma_\rho = \min\{\pi, \pi/2\rho\}$, then for all $\theta \in \text{dom}\, h$ we find:

$$\mathcal{D}_\alpha(\theta; h) = \{z = re^{i\varphi} \in \mathbb{C} : r^\rho \cos\rho(\theta + \varphi) > h(\theta),\ |\varphi + \theta| \leqslant \gamma_\rho\}. \quad (4.4.1)$$

For $l \in M_\rho$, $h \in P_\rho(l)$, and $\alpha \in \mathfrak{P}_\rho(l)$ we put

$$X_h = \mathbb{C} \setminus \bigcup \mathcal{D}_\alpha(\theta; h) | \theta \in \text{dom}\, h, \quad (4.4.2)$$

$$\mathcal{D}_\alpha(\theta) := \mathcal{D}_\alpha(\theta; l); \quad G_\alpha = \bigcup \mathcal{D}_\alpha(\theta) | \theta \in \mathbb{R}. \quad (4.4.3)$$

In particular, if $l \in O_\rho$, i.e., $l(\theta) = \text{Re}\, p_0 e^{i\rho\theta}$ for $\theta \in \mathbb{R}$ and $\alpha(z) = z^\rho + p_0$, where $p_0 \in \mathbb{C}$ and $p_0 = 0$ if $\rho \notin \mathbb{N}$ (see Theorem 4.1.5, Definition 4.3.1), then in the notation of formulas (4.4.1)–(4.4.3) we have:

$$\mathcal{D}_\alpha(\theta) = \{z = re^{i\varphi} : |\varphi + \theta| < \gamma_\rho\}, \quad \theta \in \mathbb{R}, (4.4.3')$$

and, moreover, $\mathcal{D}_\alpha(\theta) = \mathbb{C} \setminus \{0\}$, $\theta \in \mathbb{R}$, for $\rho < 1/2$; consequently, for every $\rho > 0$ we have

$$G_\alpha = \mathbb{C} \setminus \{0\}, \quad X_l = \{0\}.$$

The role of the set $X_h$ is clarified by the following supplement to Theorems 4.3.3 and 4.3.8. We formulate this supplement in the notation of Definition 4.2.12 and Remark 1 to it.

**Theorem 4.4.1.** *For $\rho > 0$, $l \in M_\rho$, and $h \in P_\rho(l)$, let $\mathcal{A}_l$ be the polygon $\rho$-sheeted near $\infty$ and associated with the function $l$; let $\mathcal{A}_h$ be the diagram of $h$ $\rho$-sheeted near $\infty$, and let $\alpha \in \mathfrak{P}_\rho(l)$. There exists a homeomorphic mapping* (see (4.4.2), (4.4.3))

$$\Phi_\alpha : \mathcal{A}_l \to \overline{G}_\alpha \cup \{\infty\} \quad (4.4.4)$$

*holomorphic and conformal in $\overset{\circ}{\mathcal{A}}_l$ and possessing following the properties:*

$$\Phi_\alpha(\Lambda_\infty) = \infty, \quad \Phi_\alpha(\overset{\circ}{\mathcal{A}}_h) = \overline{\mathbb{C}} \setminus X_h; \quad \Phi_\alpha(\partial\mathcal{A}_h) = \partial X_h;$$

*moreover, for $\rho \in \mathbb{N}$ the following diagrams are commutative:*

*where $\Gamma_h$ is the curve in $\mathbb{C}$ associated with $h$* (see Def. 1.9.24).

◀ 1. For $\rho \in \mathbb{N}$, we take the polynomial $\alpha$ in $\mathfrak{P}_\rho(l)$ and the homeomorphic and conformal inclusion $\Phi_\alpha$ of $\mathcal{A}_l$ in $\overline{\mathbb{C}}$ considered in the proof of Theorem 4.3.3. Since $(\widehat{\varphi}_l, \mathcal{A}_l)$ is a topological polygon (see Def. 4.2.3 and the remark to Def. 4.2.12), we have $\overline{\overset{\circ}{\mathcal{A}}_l} = \mathcal{A}_l$. From the formula $\overset{\circ}{\mathcal{A}}_l \setminus \{\Lambda_\infty\} = \bigcup V_\theta | \theta \in (-\pi, \pi]$ (see the proof of Theorem 4.3.3), we find that

$$\Phi_\alpha(\mathcal{A}_l) = \overline{\Phi_\alpha(\overset{\circ}{\mathcal{A}}_l)} = \overline{G}_\alpha \cup \{\infty\} ,$$

i.e., (4.4.4) is valid. The automorphisms in the group $G_E$ still act on $E_h$, because $h$ is a $2\pi$-periodic function (see (4.2.6)). The inclusions

$$E_h \subset E_l, \quad \mathcal{D}_\alpha(\theta; h) \subset \mathcal{D}_\alpha(\theta) \quad \forall\, \theta \in \mathbb{R}$$

imply that $\Phi_\alpha(\overset{\circ}{\mathcal{A}}_h) = \overline{\mathbb{C}} \setminus X_h$ (see (4.4.2)).

2. For $\rho \notin \mathbb{N}$, we take the polynomial $\alpha$ belonging to $\mathfrak{P}_\rho(l)$ and the holomorphic inclusion $F : E_l \to L$ of $E_l$ in $L$ that were considered in the proof of Theorem 4.3.8. Since the mapping $F$ agrees with the action of the groups $G_E$ on $E_l$ and $G_L$ on $L$, from Proposition A.1.9 we deduce that the factor-mapping $\widehat{F}$ for $F$ is a holomorphic homeomorphism. It only remains to show that $\Phi_\alpha := \widehat{F}$ is the required mapping. This is done by the same procedure as in the case of $\rho \in \mathbb{N}$, on the basis of formulas (4.3.28) and (4.3.21).

So, the set $X_h$ is a geometric image in $\mathbb{C}$ of the trigonometrically $\rho$-convex function $h \in P_\rho(l)$ (see (4.0.2)). It is natural to use this fact in the definition of the plane indicator diagram of an entire function.

**Definition 4.4.2.** Suppose $\rho > 0$, $l \in M_\rho$, $\alpha \in \mathfrak{P}_\rho(l)$, and $h \in P_\rho(l)$. In the notation of formula (4.4.3), the set $X_h$ is called the *$(\rho,\alpha)$-diagram of the function* $h$ or the *$(\rho,\alpha)$-indicator diagram of the entire function* $f$ of order $\rho$ and normal type if $h = h_f$ is the indicator of $f$.

**Remarks.** 1. For $l = l_0$, $h \in P_\rho(l_0)$, dom $h = \mathbb{R}$, $\alpha(z) = z^\rho$ (see (4.4.1)) Definition 4.4.2 turns into Definition 3.1.17.

2. For $\rho = 1$ we have $l(\theta) = \operatorname{Re} p_0 e^{i\theta}$ for $\theta \in \mathbb{R}$, where $p_0 \in \mathbb{C}$, and $\alpha(z) = z + p_0$. Therefore, if $h \in P_1(l)$, then $X_h + p_0 = K_h$, where

$$K_h = \{z \in \mathbb{C} : \operatorname{Re} z e^{i\theta} \leqslant h(\theta) \quad \forall\, \theta \in \mathbb{R}\}$$

is the "conjugate diagram", and $p_0 \in K_h$. Since $I_h = \{z \in \mathbb{C} : \bar{z} \in K_h\}$ (see the Introduction), we see that $I_h$ coincides with $X_h$ up to translation and complex conjugation.

We pass to describing a certain class of sets in $\mathbb{C}$ that contains the $(\rho,\alpha)$-diagrams of functions in $P_\rho(l)$, where $l \in M_\rho$. This class is a natural generalization of the class of $\rho$-convex sets (see Section 3.1).

**4.4.2. Plane $(\rho,\alpha)$-convex sets with integral $\rho \geqslant 2$.** We give an intrinsic definition of a $(\rho,\alpha)$ -convex set for $\rho \in \mathbb{N}$.

**Definition 4.4.3.** For $l \in M_\rho$, let $\alpha \in \mathfrak{P}_\rho(l)$, and let $X_l$ be the $(\rho, \alpha)$-diagram of $l$ (see Def. 4.4.2). A set $\Omega \subset \mathbb{C}$ is said to be $(\rho, \alpha)$-*convex* if: 1) $X_l \subset \Omega$; 2) for every $\theta \in \mathbb{R}$ the image of any nonempty set $\Omega \cap \mathcal{D}_\alpha(\theta)$ under $\alpha$ is convex.

**Remark.** From formula (4.4.3′) and Theorem 1.3.12 it follows that for $l \in O_\rho$ (i.e., $l(\theta) = \text{Re}\ p_0 e^{i\rho\theta}$, $\theta \in \mathbb{R}$, see Theorem 4.1.5) and $\alpha(z) = z^\rho + p_0$, this definition of a $(\rho, \alpha)$-convex set is equivalent to the definition of a $\rho$-convex set (see Def. 3.1.1). Then $X_l = \{0\}$. In the general case the set $X_l$ does not reduce to a point as we shall see below.

We examine a number of properties of the sets $\{\mathcal{D}_\alpha(\theta) : \theta \in \mathbb{R}\}$, $X_l$, and $G_\alpha$ related to the definition of a $(\rho, \alpha)$-convex set for $\rho \in \mathbb{N}$, where $\alpha \in \mathfrak{P}_\rho(l)$ and $l \in M_\rho \setminus O_\rho$.

**Property 4.4.4.** *The set $X_l$ is compact in $\mathbb{C}$.*

◀ We put $\sigma = \max\{l(\theta) : \theta \in \mathbb{R}\}$ and

$$N(R; \theta, \delta) = \{te^{i\varphi} \in \mathbb{C} : t > \mathbb{R};\ |\varphi + \theta| < \delta\}. \tag{4.4.5}$$

For every sufficiently small $\varepsilon > 0$ there exists $R > 0$ such that in the notation of formula (4.0.1) for all $\theta \in \mathbb{R}$ and $z \in N(R; \theta; \varepsilon)$ we have

$$\text{Re}\ \ \alpha(z)e^{i\rho\theta} = \text{Re}\ \ (ze^{i\theta})^\rho \left[1 + \frac{a_1}{z} + \ldots + \frac{a_n}{z^n}\right] \geqslant R^\rho(1-\varepsilon)\cos\varepsilon\rho > \sigma.$$

Hence, $N(R; \theta; \varepsilon) \subset \mathcal{D}_\alpha(\theta)\ \ \forall\, \theta \in \mathbb{R}$. Therefore, $\{z \in \mathbb{C} :\ |z| > R\} \subset G_\alpha$ and $F_\alpha \subset S_R$. It remains to note that $X_l$ is a closed set ▶

**Property 4.4.5.** *For $\theta \in \mathbb{R}$, let $T_\theta = \{p \in \mathbb{C} :\ \text{Re}\ pe^{i\rho\theta} > l(\theta)\}$. The function $\alpha$ is a conformal homeomorphism of $\mathcal{D}_\alpha(\theta)$ onto $T_\theta$; moreover, $\alpha : \overline{\mathcal{D}_\alpha(\theta)} \to \overline{T}_\theta$ is a homeomorphism.*

To prove Property 4.4.5, we need the following well-known corollary to the monodromy theorem.

**Lemma 4.4.6.** *Suppose $\mathcal{D}$ and $X$ are Riemann surfaces, $\alpha : \mathcal{D} \to X$ is a conformal mapping, and $T := \alpha(\mathcal{D})$ is a simply connected domain in $X$. Then $\alpha$ is a univalent.*

◀ Suppose there exist two points $z_0$, $z_1$ in $\mathcal{D}$ such that $z_0 \neq z_1$ and $p := \alpha(z_1) = \alpha(z_2)$. Consider a path $f : [0,1] \to \mathcal{D}$ connecting them. Then $\alpha \circ f : [0,1] \to T$ is a closed path in $T$. Since $\alpha$ is locally univalent, in the vicinity of $p$ there exist germs $\varphi_i$, $i = 0, 1$, of the inverse to $\alpha$ such that $\varphi_i(p) = z_i$, $i = 0, 1$. Along the curve $\{\alpha \circ f(t) \in T : t \in [0,1]\}$ the germ $\varphi_0$ continues analytically, and turns into $\varphi_1 \neq \varphi_0$. This contradicts the monodromy theorem ▶

◀ For $\varepsilon > 0$, consider the half-plane $T_\theta^\varepsilon = \{p \in \mathbb{C} : \text{Re}\ pe^{i\rho\theta} > l(\theta) - \varepsilon\}$ cut along $\Gamma_j(\theta) = \{p_j - te^{-i\rho\theta},\ 0 \leqslant t < \varepsilon\}$ for $\theta \in (\theta_{j-1}, \theta_j)$, $j = 1, \ldots, k$; if $\theta = \theta_j$, this plane is cut along $\Gamma_j(\theta_j)$ and $\{p_{j+1} - te^{-i\rho\theta_j} : 0 \leqslant t < \varepsilon\}$, $j = 1, \ldots, k$, where $p_{k+1} = p_1$; here $\{p_j\}_1^k$ and $\{\theta_j\}_0^k$ are the systems of numbers occuring in the definition small $\varepsilon > 0$ the mapping $\alpha : \mathcal{D}^{\varepsilon}_{\alpha}(\theta) \to T^{\varepsilon}_{\theta}$ satisfies the conditions of Lemma 4.4.6. Therefore, there exists a conformal homeomorphism $g_{\theta} : T^{\varepsilon}_{\theta} \to \mathcal{D}^{\varepsilon}_{\alpha}(\theta)$ such that

$$\alpha[g_{\theta}(p)] = p \quad \forall\, p \in T^{\varepsilon}_{\theta}. \tag{4.4.6}$$

From the properties of algebraic functions, in the notation of Definition 4.3.1 we deduce that (see (4.3.5))

$$\lim_{p\to p_j} g_{\theta}(p) = z_j; \quad p \in T^{\varepsilon}_{\theta}, \quad \theta \in (\theta_{j-1}, \theta_j];$$

$$\lim_{p\to p_{j+1}} g_{\theta_j}(p) = z_{j+1}, \quad p \in T^{\varepsilon}_{\theta_j}; \quad p_{k+1} = p_1; \quad z_{k+1} = z_1,$$

for $j = 1, \ldots, k$; moreover, for $p \in T^{\varepsilon}_{\theta}$ we have

$$\lim_{|p|\to\infty} p^{-1}[g_{\theta}(p)]^{\rho} = 1 \tag{4.4.7}$$

(see (4.0.1), (4.4.6)). Therefore, Property 4.4.5 is valid ▶

**Corollary 4.4.7.** *Suppose $\theta \in \mathbb{R}$, $\zeta_0 \neq \zeta_1$ ; $\zeta_j \in \overline{D_{\alpha}(\theta)}$, and $q_j = \alpha(\zeta_j)$, $j = 0, 1$. There is a unique path*

$$\gamma_{\alpha}(\zeta_0; \zeta_1; \theta) = \{f : [0,1] \to \overline{D_{\alpha}(\theta)}\} \tag{4.4.8}$$

*such that $f(j) = \zeta_j$, $j = 0, 1$, and $\alpha \circ f(\lambda) \equiv \lambda q_1 + (1-\lambda) q_0$ for $\lambda \in [0,1]$.*

The path $\gamma_{\alpha}(\zeta_0; \zeta_1; \theta)$ is the analytic Jordan arc determined by the function $g_{\theta}$ in question (see (4.4.7)).

Let $(\widehat{\varphi}_l, \mathcal{A}_l)$ be the polygon associated with a function $l \in M_{\rho} \setminus O_{\rho}$ (see Def. 4.2.1 and Remark 2 to it). As a supplement to Theorem 4.4.1, we study some more properties of the inclusion of $(\widehat{\varphi}_l, \mathcal{A}_l)$ in the covering $\alpha : \overline{\mathbb{C}} \to \overline{\mathbb{C}}$, where $\alpha$ is an *arbitrarily fixed function* in $\mathfrak{P}_{\rho}(l)$. It turns out that this inclusion is not always a homeomorphism (see the remark to Theorem 4.2.15).

Let

$$S^1 - \{e^{i\theta} : \theta \in \mathbb{R}\}; \quad \widetilde{l}(\omega) = l(Arg\,\omega), \quad \omega \subset S^1.$$

Then the function $\widetilde{l}$ is well-defined, because $l$ is $2\pi$-periodic. From the structure of the group $G_E$ that acts on the polygonal surface $(\varphi_l, E_l)$ associated with $l$, it follows that for $\rho \in \mathbb{N}$ and $l \in M_{\rho} \setminus O_{\rho}$ every element of $A_l \setminus \{\Lambda_{\infty}\}$ is represented by a pair $(p, \delta)$, where $p \in \mathbb{C}$ and $\delta = \delta(p)$ is a nonempty component of the set $\{\omega \in S^1 : \mathrm{Re}\, p\omega^{\rho} \gtrless \widetilde{l}(\omega)\}$ (see Def. 4.2.1, (4.2.3), and (4.2.6)).

**Property 4.4.8.** *For $l \in M_{\rho} \setminus O_{\rho}$, let $\alpha \in \mathfrak{P}_{\rho}(l)$, and let $(\widehat{\varphi}_l, \mathcal{A}_l)$ be the polygon associated with $l$. There exists a continuous mapping $\Phi_{\alpha} : \mathcal{A}_l \to \overline{G}_{\alpha} \cup \{\infty\}$* (see (4.4.3)) *with the following properties* : a) *if $\Lambda_{\infty} = \varphi_l^{-1}(\infty)$, then $\Phi_{\alpha}(\Lambda_{\infty}) = \infty$* ; b) *$\Phi_{\alpha} : \mathring{\mathcal{A}}_l \to G_{\alpha} \cup \{\infty\}$ is a holomorphic homeomorphism, and*

$$\varphi_l \circ \Phi_{\alpha}^{-1}(z) \equiv \alpha(z), \quad z \in G_{\alpha}; \quad \text{c)}\ \Phi_{\alpha}(\partial \mathcal{A}_l) = \partial G_{\alpha}.$$

◀ Let $\psi < \theta < \psi + \pi/\rho$. In the notation of Property 4.4.5 we have $T_\psi \cap T_\theta \neq \varnothing$ and $\mathcal{D} := \mathcal{D}_\alpha(\theta) \cap \mathcal{D}_\alpha(\psi) \neq \varnothing$ (see (4.4.3)). For every point $z \in \mathcal{D}$ there exists a germ at $p = \alpha(z)$ of a unique function holomorphic near $p$ and satisfying (4.3.3). We observe that $p \in T_\psi \cap T_\theta$ (see Property 4.4.5); so, denoting by $g_\varphi : \overline{T}_\varphi \to \overline{\mathcal{D}_\alpha(\varphi)}$, $\varphi \in \mathbb{R}$, the inverse of the polynomial $\alpha$ (see the proof of Property 4.4.5), from the uniqueness of analytic continuation we obtain:

$$g_\theta(p) \equiv g_\psi(p), \quad p \in \overline{T}_\theta \cap \overline{T}_\psi. \tag{4.4.9}$$

Let $V_\theta = \{(p, \delta) \in \mathcal{A}_l : p \in T_\theta;\ e^{i\theta} \in \delta\}$. Then $\varphi_l : \overline{V}_\theta \setminus \{\Lambda_\infty\} \to \overline{T}_\theta$ is a homeomorphism and $l$ is a $2\pi$-periodic function. Therefore,

$$\mathcal{A}'_l := \overset{\circ}{\mathcal{A}}_l \setminus \{\Lambda_\infty\} = \bigcup V_\theta | \theta \in (-\pi, \pi]. \tag{4.4.10}$$

On $\mathcal{A}_l$, we consider the function

$$\Phi_\alpha(m) := [g_\theta \circ \varphi_l](m), \quad m \in \overline{V}_\theta, \quad \theta \in (-\pi, \pi]; \quad \Phi_\alpha(\Lambda_\infty) = \infty.$$

We show that this definition is consistent. Suppose $m_0 \in \mathcal{A}_l$, $m_0 = (p, \delta) \in \overline{V}_\theta \cap \overline{V}_\psi$, and $|\theta - \psi| < \pi/\rho$. From (4.4.9) and Property 4.4.5 we deduce that $g_\theta(p) = g_\psi(p)$. Let $|\theta - \psi| \geqslant \pi/\rho$, and let, for instance, $\theta < \psi$. Then there exists a sequence $\{\varphi_j\}_0^{s+1}$ such that $\varphi_0 = \theta < \varphi_1 < \ldots < \varphi_s < \varphi_{s+1} = \psi$; $\varphi_j - \varphi_{j-1} < \pi/\rho$, $j = 1, \ldots, s+1$. Since $e^{i\varphi} \in \delta\ \forall\, \varphi \in [\theta, \psi]$, we have $m_0 \in \overline{V}_\varphi$ $\forall\, \varphi \in [\theta, \psi]$. Therefore, (4.4.9) implies that $g_\theta(p) = g_{\varphi_1}(p) = \ldots = g_{\varphi_s}(p) = g_\psi(p)$. By (4.4.7),

$$\lim_{|p| \to \infty; p \in T_\theta} g_\theta(p) = \infty\ .$$

The continuity of the mapping $\varphi_l : \mathcal{A}_l \to \overline{\mathbb{C}}$ and Property 4.4.5 imply that $\Phi_\alpha$ is a continuous mapping conformal in $\mathcal{A}'_l$. Moreover,

$$\Phi_\alpha(V_\theta) = \mathcal{D}_\alpha(\theta) \quad \forall\, \theta \in \mathbb{R} \quad \text{and} \quad \Phi_\alpha(\mathcal{A}'_l) = G_\alpha \tag{4.4.11}$$

(see (4.4.3), (4.4.10)). Since $\mathcal{A}_l = \overline{\mathcal{A}'_l}$, and $\Phi_\alpha$ is a continuous mapping, (4.4.11) shows that $\Phi_\alpha(\mathcal{A}_l) = \overline{G}_\alpha \cup \{\infty\}$ and $\Phi_\alpha(\partial \mathcal{A}_l) = \partial G_\alpha$.

So, $\Phi_\alpha : \overset{\circ}{\mathcal{A}}_l \to G_\alpha \cup \{\infty\}$ is a conformal mapping with pole of order 1 at $\Lambda_\infty$ and continuous up to the boundary. But $\partial \mathcal{A}_l$ is a connected subset of $\mathcal{A}_l$. Therefore, so is the set $\partial G_\alpha = \Phi_\alpha(\partial \mathcal{A}_l)$. Hence $G_\alpha \cup \{\infty\}$ is a simply connected domain in $\overline{\mathbb{C}}$. Now from Lemma 4.4.6 we deduce property b) ▶

**Property 4.4.9.** *In the notation of Property* 4.4.8 *and formula* (4.4.3), *the set* $K(z) = \{e^{i\theta} : z \in \overline{\mathcal{D}_\alpha(\theta)}\}$ *consists of one component for all* $z \in G_\alpha$, *and* 1) *if* $\Phi_\alpha$ *is the mapping dealt with in Property* 4.4.8, *then* $\Phi_\alpha^{-1}(z) = (\alpha(z), K(z))$ $\forall\, z \in G_\alpha$; 2) *if* $K(z) = \{e^{i\varphi} : \varphi \in [\psi_1, \psi_2]\}$, *then* $\psi_2 - \psi_1 \leqslant \pi/\rho\ \ \forall\, z \in G_\alpha$.

◀ Let $z \in \overline{G}_\alpha$. From Property 4.4.5 we deduce that every component of the set $K(z)$ is a component of the set $\{e^{i\theta} : p \in \overline{T}_\theta\}$, where $p = \alpha(z)$. Suppose

$K(z)$ contains two components $\delta_1$, $\delta_2$. Then $m_j := (p, \delta_j) \in \mathcal{A}_l$, $j = 1, 2$, and if $z \in G_\alpha$, then $m_j \in \overset{\circ}{\mathcal{A}}_l$, $j = 1, 2$. The structure of the mapping $\Phi_\alpha$ shows (see the proof of Property 4.4.8) that $\Phi_\alpha(m_j) = z$, $j = 1, 2$. This is impossible for $z \in G_\alpha$, because $\Phi_\alpha : \overset{\circ}{\mathcal{A}}_l \to G_\alpha$ is one-to-one (see Property 4.4.8). This ensures Property 4.4.9, with the exception of assertion 2). But that assertion follows from 1) and the structure of every element $(p, \delta)$ in $\mathcal{A}'_l$ (see (4.4.10) and the proof of Property 4.2.9) ▶

Property 4.4.8 (see b)) immediately implies the following statement.

**Property 4.4.10.** *In the notation of Definitions* 4.3.1, 4.4.2, 4.4.3 *and formulas* (4.4.3), (4.4.8), *let* $\Gamma_j = \gamma_\alpha(z_j; z_{j+1}; \theta_j)$ *for* $j = 1, \ldots, k$, *and let* $z_{k+1} = z_1$. *Then*

$$\partial G_\alpha = \partial X_l = \bigcup_{j=1}^{k} \Gamma_j.$$

So, the boundary of the $(\rho, \alpha)$-diagram of the function $l$ is a closed curved broken line with vertices at the roots of $\alpha'$, and under the mapping $\alpha$ its links become straight line segments.

Property 4.4.10 yields a method of constructing $(\rho, \alpha)$-diagrams for the functions of class $M_\rho \setminus O_\rho$.

**Example 4.4.11.** A. Let $l$ be the $2\pi$-periodic extension to $\mathbb{R}$ of the function $l(\theta) = -2\cos 3\theta$ for $\theta \in [-\pi/2, \pi/2]$ and $l(\theta) = 2\cos 3\theta$ for $\theta \in [\pi/2, 3\pi/2]$, and let $\alpha(z) = z^3 - 3z$. In the notation of Theorem 4.1.5 we have: $l \in M_3$, $p_1 = -2$, $p_2 = 2$, $\theta_0 = -\pi/2$, $\theta_1 = \pi/2$, and $\theta_2 = 3\pi/2$. From the conditions $\alpha(z_j) = p_j$ and $\alpha'(z_j) = 0$, $j = 1, 2$, we find: $z_1 = 1$, $z_2 = -1$. Let

$$T_\alpha(z; \theta) = \mathrm{Re}\ [\alpha(z) - l(\theta)]e^{i3\theta}, \quad \theta \in \mathbb{R}. \tag{4.4.12}$$

Since the period of $l$ is $\pi$ and $T_\alpha(z;\theta) = \mathrm{Re}\ (z-1)^2(z+2)e^{i3\theta}$, $|\theta| \leqslant \pi/2$, by the method described in Example 4.3.2 we show that $\alpha \in \mathfrak{P}_3(l)$. The function $\alpha$ decreases on the segment $[-1, 1]$ and maps its bijectively onto the segment $[-2, 2] \subset \overline{T}_{\pi/2} \cap \overline{T}_{-\pi/2}$ (see Property 4.4.5). Now Property 4.4.10 yields $\Gamma_1 = \Gamma_2 = X_l = [-1, 1]$.

B. Let $l$ be the same as above; we put $z_0 = \sqrt[3]{1/4}$ and $\alpha(z) = z^3 - 6z_0^2 z - 3z_0^4/z$. The derivative $\alpha'$ has two double roots $\pm z_0$, and $\alpha(\pm z_0) = \mp 2$. We use the formula

$$T_\alpha(z; \theta) = \mathrm{Re}\ (z - z_0)^3(z + 3z_0)e^{i3\theta}/z, \quad |\theta| \leqslant \pi/2$$

(see (4.4.12)) and, as in Examples 4.3.2 and 4.4.11 A, obtain the relation $\alpha \in \mathfrak{P}_3(l)$. By Property 4.4.10, $\partial G_\alpha = \partial X_l = \Gamma_1 \cup \Gamma_2$, where $\Gamma_1$ and $\Gamma_2$ are arcs connecting the points $-z_0$ and $z_0$; moreover (see Property 4.4.5), $\Gamma_1 \subset \partial\mathcal{D}_\alpha(-\pi/2)$ and $\Gamma_2 \subset \partial\mathcal{D}_\alpha(\pi/2)$. Therefore, the points of the set $\partial X_l$ satisfy the condition $T_\alpha(z; \pm\pi/2) = 0$ (see (4.4.12)). In the Cartesian coordinate system $(z = x + iy)$, this condition is equivalent to the formula $y^4 - 2y^2(x^2 - 3a) -$

$3(x^2-a)^2=0$, where $a=z_0^2$. Since $\psi(t)=t-3a+\sqrt{(t-3a)^2+3(t-a)^2}$ is a monotone decreasing function on the segment $[0,a]$ and $\psi(t)\geqslant 0\ \forall\, t\in[0,a]$, we arrive at the following equation for the set $\partial G_\alpha=\partial X_l : y^2=\psi(x^2)$, $|x|\leqslant z_0$. So, $X_l$ is a compact set in $\mathbb{C}$ with Jordan boundary. Moreover, $X_l$ is centrally symmetric and symmetric relative to both coordinate axes.

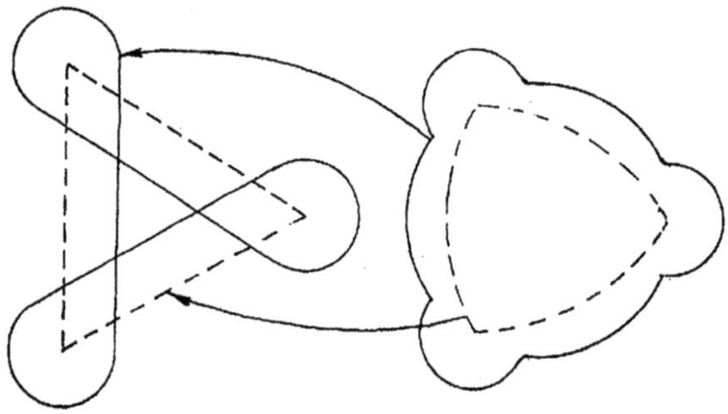

Fig. 11

C. Let $l\in M_2$, and let $\alpha\in\mathfrak{P}_2(l)$ be the functions occurring in Example 4.3.2. By Property 4.4.10, $\partial X_l=\Gamma_1\cup\Gamma_2\cup\Gamma_3$, where the arc $\Gamma_3$ connects the points $z_1$, $z_3$, and the arcs $\Gamma_1$, $\Gamma_2$ are obtained by rotation of the arc $\Gamma_3$ about 0 through the angles of $\pm 2\pi/3$. Since $\Gamma_3\subset\partial\mathcal{D}_\alpha(\pi)$ (see Property 4.4.5), the points of the set $\Gamma_3$ satisfy the equation Re $\alpha(z)=-1/2$, equivalent in the Cartesian coordinate system $(z=x+iy)$ to the equation $y^4-y^2/2-\psi(x)=0$, where $\psi(x)=x^4+x^2/2+2x/(3\sqrt{3})$. The function $\gamma(x)=\psi(x)/x$ is monotone increasing for $x\in(-\infty,\infty)$, and $\gamma(-1/\sqrt{3})=-1/6\sqrt{3}<0$; $\gamma(-1/2\sqrt{3})=\sqrt{3}/8>0$. By the Cauchy theorem, on the interval $(-1/\sqrt{3},-1/2\sqrt{3})$ there exists a unique zero $x_0$ of the function $\gamma$, and on the segment $[x_0,-1/2\sqrt{3}]$ we have $0\leqslant\gamma(x)\leqslant\sqrt{3}/8$ and $-1/16\leqslant x\gamma(x)=\psi(x)\leqslant 0$. We conclude that the equation of the arc $\Gamma_3$ is as follows:

$$y^2=1/4-\sqrt{1/16+\psi(x)},\quad x\in[x_0,-1/2\sqrt{3}].$$

Hence the $(\rho,\alpha)$-diagram $X_l$ of the function $l$ contains the triangle with vertices $(1/\sqrt{3})\exp\{i2\pi k/3\}$, $k=-1,0,1$, and lies in the disk of radius $1/\sqrt{3}$ and with center at 0. The set $X_l$ is shown in Fig. 11 together with the $(\rho,\alpha)$-diagram of the function $h(\theta)=l(\theta)+\varepsilon$, $\theta\in\mathbb{R}$, and the curves $\Gamma_l$, $\Gamma_h$ associated with $l$ and $h$, respectively (see Def. 1.9.24, Fig. 2 B, and Example 4.2.12′, A).

We indicate some conditions under which the set $\mathring{\mathcal{A}}_l$ can be replaced by $\mathcal{A}_l$ in Property 4.4.8.

**Property 4.4.12.** *Let $(\widehat{\varphi}_l,\mathcal{A}_l)$ be the polygon associated with the function $l$, and let $\alpha\in\mathfrak{P}_\rho(l)$. The following conditions are equivalent:*

1) *the set* $K(z)=\{e^{i\theta} : z\in\overline{\mathcal{D}_\alpha(\theta)}\}$ *has one component for every* $z\in\overline{G}_\alpha$, *and the mapping* (see Property 4.4.8) $\Phi_\alpha^{-1} : \overline{G}_\alpha\to\mathcal{A}_l$, $\Phi_\alpha^{-1}(z)=(\alpha(z),K(z))$ *is a homeomorphism* ;

2) *there exists a homeomorphic inclusion* $\Phi_\alpha$ *of the polygon* $\mathcal{A}_l$ *into the covering* $\alpha:\overline{\mathbb{C}}\to\overline{\mathbb{C}}$ *such that* $\alpha\circ\Phi_\alpha=\widehat{\varphi}_l$;

3) *the common boundary of the sets* $G_\alpha$ *and* $X_l$ *is a closed Jordan curve.*

Property 4.4.12 is a consequence of Properties 4.4.5, 4.4.8, and 4.4.10. For the functions occurring in Example 4.4.11, conditions 1)–3) are fulfilled in cases B and C, but not in case A. The proof of Theorem 4.4.1 shows that the subclass $\mathfrak{P}'_\rho(l)$ of $\mathfrak{P}_\rho(l)$ distinguished by the conditions in Property 4.4.12 is not empty.

In the general case the Riemann surface of the inverse to the function $\alpha$ of the form (4.3.1) can be obtained from the polygon $(\widehat{\varphi}_l,\mathcal{A}_l)$ by gluing some edges of the polygon $(\widehat{\varphi}_l,\mathcal{A}_l)$ at the points with the same projection to $\overline{\mathbb{C}}$ (see the remark to Theorem 4.2.15).

We mention several properties of $(\rho,\alpha)$-convex sets proved in the same way as the corresponding properties of $\rho$-convex sets (see Property 3.1.2).

**Property 4.4.13.** a) *The intersection of a nonempty family of* $(\rho,\alpha)$-*convex sets is a* $(\rho,\alpha)$-*convex set*; b) *the closure of a* $(\rho,\alpha)$-*convex set is a* $(\rho,\alpha)$-*convex set*; c) *if* $G$ *is an open* $(\rho,\alpha)$-*convex set, then* $\operatorname{int}\overline{G}=G$; d) *if* $E$ *is a* $(\rho,\alpha)$-*convex set such that* $X_l\subset\mathring{E}$, *then* $\overline{\mathring{E}}=\overline{E}$.

**4.4.3. Minkowski duality for $(\rho,\alpha)$-convex sets and functions of class $P_\rho(l)=\{h\in P_\rho : h\geqslant l\}$ for integral $\rho\geqslant 2$.** We establish isomorphism between some natural classes of $(\rho,\alpha)$-convex sets and appropriate subclasses of $P_\rho(l)$, or even of the smaller class $Q_\rho(l)=\{h\in P_\rho : h>l\}$. Here $h>l$ means that $h(\theta)>l(\theta)\ \ \forall\,\theta\in\mathbb{R}$.

For $\alpha\in\mathfrak{P}_\rho(l)$, we put $H_\rho(l)=\{r_\alpha(\theta;\zeta) : \zeta\in\mathbb{C}\}\subset P_\rho(l)$, where (see Def. 4.4.3)

$$r_\alpha(\theta;\zeta)=\begin{cases}\operatorname{Re}\alpha(\zeta)e^{i\rho\theta}, & \zeta\in\mathcal{D}_\alpha(\theta),\quad \theta\in\mathbb{R},\\ l(\theta), & \zeta\in\mathbb{C}\setminus\mathcal{D}_\alpha(\theta),\quad \theta\in\mathbb{R}.\end{cases}\tag{4.4.13}$$

If $l\in O_\rho$, then (see (3.1.9)) $r_\alpha(\theta,\zeta)=\Phi_\rho(\zeta e^{i\theta})+l(\theta)$. With any function $h$ in $P_\rho(l)$ or in $Q_\rho(l)$, we associate the sets

$$\begin{aligned}X_h&=\{\zeta\in\mathbb{C} : r_\alpha(\,\cdot\,;\zeta)\leqslant h\};\\ \Omega_h&=\{\zeta\in\mathbb{C} : r_\alpha(\,\cdot\,;\zeta)<h\}.\end{aligned}\tag{4.4.14}$$

Sometimes we denote these sets by $X(h)$, $\Omega(h)$ or $X(h;\alpha)$, $\Omega(h;\alpha)$. To examine Minkowski duality, we need several auxiliary statements.

**Property 4.4.14.** *For* $h\in P_\rho(l)$ $(Q_\rho(l))$, *the set* $X_h$ $(\Omega_h)$ *is a closed (open)* $(\rho,\alpha)$-*convex set, and for* $h\not\equiv\infty$ we have

$$X_h=\overline{\Omega}_h,\quad \Omega_h=\mathring{X}_h,\quad h\in Q_\rho(l).\tag{4.4.15}$$

◀ Let $X_h(\theta) = \{\zeta \in \mathbb{C} : r_\alpha(\theta, \zeta) \leqslant h(\theta)\}$, $\theta \in \mathbb{R}$. Since $X_h = \cap X_h(\theta) \,|\theta \in \mathbb{R}$ $\forall\, h \in P_\rho(l)$, we have $X_l \subset X_h(\theta)$ $\forall\, \theta \in \mathbb{R}$ (see Def. 4.4.3). Consider the set $Z(\theta, \psi) := X_h(\theta) \cap \overline{\mathfrak{D}_\alpha(\psi)}$ for $\theta \in \mathbb{R}, \psi \in [\theta, \theta + 2\pi]$. In the notation of Property 4.4.5, for $\psi \in [\theta, \theta + \pi/\rho] \cup (\theta + 2\pi - \pi/\rho, \theta + 2\pi]$ we have:

$$\alpha[Z(\theta, \psi)] = \{p \in \overline{T}_\psi : \text{Re } pe^{i\rho\theta} \leqslant h(\theta)\}.$$

If $\psi \in [\theta + \pi/\rho, \theta + 2\pi - \pi/\rho]$ $(\rho \geqslant 2)$, then $\alpha[Z(\theta, \psi)] = \overline{T}_\psi$. So, $X_h$ is a $(\rho, \alpha)$-convex set. In a similar way we show that $\Omega_h$ is also $(\rho, \alpha)$-convex (see (4.4.14)).

For every $\varepsilon \in (0, 2\pi)$ and some $R = R(\epsilon)$ (see (4.4.5), the proof of Property 4.4.4) we have $N(R; \theta; \pi/2 - \varepsilon) \subset \mathcal{D}_\alpha(\theta)$ $\forall\, \theta \in \mathbb{R}$. Moreover, $\max\{l(\theta), \text{Re } \alpha(\zeta)e^{i\rho\theta}\}$ is a continuous function on $\mathbb{R} \times \mathbb{C}$. Consequently, the function $r_\alpha$ is also continuous. On the other hand, $-h$ is an upper semicontinuous function on $\mathbb{R}$ continuous on dom $h$ (see Ch. 1). Hence, using Exercise 1.2.7, we conclude that

$$Y(\zeta) := \sup\{r_\alpha(\theta; \zeta) - h(\theta) : \theta \in [-\pi, \pi]\}$$

is a continuous function. But $r_\alpha, h$ are $2\pi$-periodic functions in the variable $\theta$. Therefore,

$$X_h = \{\zeta \in \mathbb{C} : Y(\zeta) \leqslant 0\},\ \Omega_h = \{\zeta \in \mathbb{C} : Y(\zeta) < 0\},$$

and (4.4.15) follows.

We show that, as in the case where $l(\theta) = l_0(\theta) \equiv 0$ for $\theta \in \mathbb{R}$, and $\alpha(z) = z^\rho$, the class $H_\rho(l)$ is important in the structure of the class $P_\rho(l)$ for all $l \in M_\rho \setminus \{l_0\}$.

In analogy with the notion considered in Section 3.1 (see Def. 3.1.11), we introduce the following definition.

**Definition 4.4.15.** For $M \subset \mathbb{C}$, the function $h_M(\theta) = \sup\{r_\alpha(\theta; \zeta) : \zeta \in M\}$ (see (4.4.13)) is called *the $(\rho, \alpha)$-support function for $M$.*

**Property 4.4.16.** *The class $P_\rho(l)$ contains $(\rho, \alpha)$-support functions of plane sets, in particular, the elements of the set $H_\rho(l)$.*

◀ If $l \in O_\rho$, then (see (3.1.9))

$$r_\alpha(\theta; \zeta) = \Phi_\rho(\zeta e^{i\theta}) + l(\theta),\ \ (\theta; \zeta) \in \mathbb{R} \times \mathbb{C}.$$

Together with Lemma 3.1.14 this implies Property 4.4.16 for the case of $l \in O_\rho$. Now, let $l \in M_\rho \setminus O_\rho$.

We show that $H_\rho(l) \subset P_\rho(l)$. Fix $\zeta \in \mathbb{C}$. If $\zeta \in X_l$ (see Def. 4.4.3), then $r_\alpha(\cdot; \zeta) = l$. Let $\zeta \in G_\alpha$ (see (4.4.3)). Then by Property 4.4.9

$$K(\zeta) = \{e^{i\theta} : \zeta \in \overline{\mathcal{D}_\alpha(\theta)}\} = \{e^{i\theta} : \theta \in [\varphi_1, \varphi_2]\}, \tag{4.4.16}$$

where $0 < \varphi_2 - \varphi_1 \leq \pi/\rho$. For $\zeta \in G_\alpha$, formula (4.4.13) is equivalent to the formula

$$r_\alpha(\theta;\zeta) = \begin{cases} \operatorname{Re} \alpha(\zeta)e^{i\rho\theta}, & e^{i\theta} \in K(\zeta); \\ l(\theta), & e^{i\theta} \in S^1 \setminus K(\zeta), \end{cases} \tag{4.4.17}$$

where $S^1 = \{e^{i\theta} : \theta \in \mathbb{R}\}$. Let $p = \alpha(\zeta)$. Then $\operatorname{Re} pe^{i\rho\theta} > l(\theta)$ for some $\theta \in (\varphi_1, \varphi_2)$. Since $l$ satisfies (4.1.10), from (4.4.17) we conclude that for $l \in M_\rho \setminus O_\rho$ two cases are possible.

1. In the notation of formula (4.1.3), there exists an index $j \in \{1, \dots, k\}$ such that $[\varphi_1, \varphi_2] \subset (\theta_{j-1}, \theta_j)$. Then for fixed $\zeta \in G_\alpha$ the piecewise trigonometric function $r_\alpha(\theta;\zeta)$ is determined by the sequences

$$\{p_1, \dots, p_j, p, p_j, \dots, p_k\} \quad \text{and} \quad \{\theta_0, \theta_1, \dots, \theta_{j-1}, \varphi_1, \varphi_2, \theta_j, \dots, \theta_k\}.$$

From the definition of the set $K$ we obtain: $\operatorname{Re} pe^{i\rho\varphi_j} = l(\varphi_j)$, $j = 1, 2$, and from (4.4.13) we obtain: $p \neq p_j$. By (4.4.13) and the relation $l \in P_\rho$, we see that $r_\alpha(\cdot, \zeta)$ is a continuous $2\pi$-periodic piecewise trigonometric function satisfying condition 3) of Proposition 4.1.1. Consequently, $r_\alpha(\cdot, \zeta) \in P_\rho(l)$ for all $\zeta \in \mathbb{C}$.

2. There exists an index $j \in \{1, \dots, k\}$ such that $\theta_j \in [\varphi_1, \varphi_2]$. If $\theta_j \in (\varphi_1, \varphi_2)$, then the function $r_\alpha(\theta, \zeta)$ is determined by the sequences

$$\{p_1, \dots, p_j, p, p_{j+1}, \dots, p_k\} \quad \text{and} \quad \{\theta_0, \theta_1, \dots, \theta_{j-1}, \varphi_1, \varphi_2, \theta_{j+1}, \dots, \theta_k\}.$$

As in case 1, we see that $r_\alpha(\cdot, \zeta) \in P_\rho(l)$. For the same reason, this fact is also valid when one of the ends of the segment $[\varphi_1, \varphi_2]$ belongs to the sequence $\Theta$. From formulas (4.4.16) and (4.1.12) it follows that there are no other possible locations of the segment $[\varphi_1, \varphi_2]$ with respect to the elements of $\Theta$. It remains only to note that $P_\rho(l)$ contains the supremum of every its subset (see Exercise 1.2.5) ▶

**Property 4.4.17.** *The class $H_\rho(l)$ is a supremal generator for the classes $P_\rho(l)$ and $Q_\rho(l)$. Moreover, the $(\rho, \alpha)$ support function of the set $X_h$ or $\Omega_h$ coincides with $h$ for all $h$ in $P_\rho(l)$ or $Q_\rho(l)$* (see (4.4.13), (4.4.14) and Subsection 3.1.1).

◀ Let $h \in P_\rho(l)$. We show that $h$ coincides with the $(\rho, \alpha)$-support function $g$ of the set $X_h$ (see (4.4.14)). Obviously, $g \leqslant h$. If $h(\theta) = l(\theta)$, then $g(\theta) = l(\theta)$. Let $\Delta$ be any fixed component of the set dom $h$; take $\theta \in \Delta$ with $h(\theta) > l(\theta)$. Assume that $\mathring{\Delta} \neq \varnothing$ and $\theta \in \mathring{\Delta}$. By Theorem 1.9.21′ there exists a number $p \in \mathbb{C}$ with the property

$$h(\theta) = \operatorname{Re} pe^{i\rho\theta};\ h(\varphi) \geqslant \operatorname{Re} pe^{i\rho\varphi},\ |\varphi - \theta| \leqslant \pi/\rho. \tag{4.4.18}$$

Therefore, from Property 4.4.5 we deduce the existence of a point $\zeta$ in $\overline{\mathcal{D}_\alpha(\theta)}$ such that $\alpha(\zeta) = p$. For the function $r_\alpha(\cdot;\zeta)$ corresponding to $\zeta$ (see 4.4.13) formula (4.4.16) is valid in which $e^{i\theta} \in K(\zeta)$. Now, (4.4.18) implies that $h(\theta) = r_\alpha(\theta;\zeta)$ and $h(\varphi) \geqslant r_\alpha(\varphi;\zeta)$ for $\varphi \in \mathbb{R}$. Since $\zeta \in X_h$ (see (4.4.14), we have $g(\theta) \geqslant h(\theta)$ and, consequently, $g(\theta) = h(\theta)$.

Let $\theta \in \partial\overset{\circ}{\Delta}$. Since $h, g \in P_\rho$, the functions $h$ and $g$ are continuous in $\Delta$ (see Property 4.4.16, Section 1.9). Therefore, since $h(\varphi) = g(\varphi)$ for $\varphi \in \overset{\circ}{\Delta}$, we have $h(\theta) = g(\theta)$ if $\theta \in$ dom $h$ or $h = \infty$.

Let $\overset{\circ}{\Delta} = \varnothing$, i.e., $\Delta = \{\theta\}$. Then (see (4.4.13), (4.4.14) and Def. 4.4.15)

$$g(\theta) = \sup\{r_\alpha(\theta;\zeta) : \ \zeta \in X_h \cap \overline{\mathcal{D}_\alpha(\theta)}\}. \tag{4.4.19}$$

Consider the set

$$N_\theta(h) = \{p \in \mathbb{C} : \text{ Re } pe^{i\rho\theta} \geqslant l(\theta); \text{ Re } pe^{i\rho\varphi} \leqslant h(\varphi), |\varphi - \theta| \leqslant \pi/\rho\}. \tag{4.4.20}$$

By (4.4.19),

$$g(\theta) = \sup\{\text{Re } pe^{i\rho\theta} : \ p \in N_\theta(h)\}, \ \ \theta \in \mathbb{R}. \tag{4.4.21}$$

Since the length of every component of $R \setminus$ dom $h$ is at least $\pi/\rho$ (see Property 1.9.22), from the inequality

$$-l(\theta) \leqslant l(\theta \pm \pi/\rho) \leqslant h(\theta \pm \pi/\rho), \ \ \theta \in \mathbb{R}$$

(see (1.9.7)) we see that

$$N_\theta(h) = \{p \in \mathbb{C} : \ l(\theta) \leqslant \text{Re } pe^{i\rho\theta} \leqslant h(\theta)\}.$$

Hence (see (4.4.21)), $g(\theta) = h(\theta)$.

Now suppose $\theta \in \mathbb{R} \setminus \overline{\text{dom } h}$, $\Delta_1$ is the component of $\mathbb{R} \setminus$ dom $h$ containing $\theta$, and $d$ is the length of $\Delta_1$. Let $I(\theta) = (\theta - \pi/2\rho, \theta + \pi/2\rho)$. If $I(\theta) \subset \Delta_1$, then $\{re^{-i\rho\theta} : \ r \geqslant r_0\} \subset N_\theta(h)$ and $g(\theta) = \infty$. If there exists a point $\psi \in I(\theta)$ such that $\psi \in \partial\Delta_1$ and, say, $\theta < \psi$, $(\theta, \psi) \subset \Delta_1$, then, since $d \geqslant \pi/\rho$, we have: dom $h \cap I(\theta) \subset (\psi, \theta + \pi/2\rho)$. There exists a number $r_0 = r_0(\psi)$ such that $\{r \exp\{-i(\rho\psi - \pi/2)\} : \ r \geqslant r_0\} \subset N_\theta(h)$. Therefore, $g(\theta) = h(\theta) = \infty$. Thus, $g = h$.

If $h \in Q_\rho(l)$, then, by (4.4.15), from the above we deduce that $h$ is the $(\rho, \alpha)$-support function of $\Omega_h$ ▶

**Property 4.4.18.** *Let $\mathfrak{B}_i(\rho, \alpha)$, $i = 1, 2$, be the classes of closed (respectively, open) $(\rho, \alpha)$-convex sets in $\mathbb{C}$. Then* (see (4.4.14))

$$F = X(h_F), \ F \in \mathfrak{B}_1(\rho, \alpha); \ \ F = \Omega(h_F), \ \ F \in \mathfrak{B}_2(\rho, \alpha), \tag{4.4.22}$$

*where $h_F$ is the $(\rho, \alpha)$-support function of the set $F$.*

◀ If $F = \mathbb{C}$, then (4.4.22) is obvious. Suppose $F \neq \mathbb{C}$. Consider two situations.

1. Let $F \in \mathfrak{B}_1(\rho, \alpha)$. It suffices to prove that for every $\zeta \in \mathbb{C} \setminus F$ there is $\varphi \in \mathbb{R}$ such that (see 4.4.13)

$$r_\alpha(\varphi; \zeta) > h_F(\varphi). \tag{4.4.23}$$

Let $p = \alpha(\zeta)$. Two cases are possible (they were considered in the proof of Property 4.4.16); we use the notation accepted in that statement).

(a) Let $[\varphi_1, \varphi_2] \subset (\theta_{j-1}, \theta_j)$ for some $j \in \{1, \dots, k\}$. Then $\varphi_2 = \varphi_1 + \pi/\rho$ and $K(\zeta) = \{e^{i\theta} : |\theta - \psi| \leqslant \pi/2\rho\}$, where $\psi = \varphi_1 + \pi/2\rho$. Consider the closed convex set $B := N_\psi(h_F)$ defined by formula (4.4.20). If $\overset{\circ}{B} = \varnothing$, then $\operatorname{Re} pe^{i\rho\psi} > l(\psi) = h_F(\psi)$ (see (4.4.21)), i.e., $r_\alpha(\psi; \zeta) > h_F(\zeta)$. Suppose $\overset{\circ}{B} \neq \varnothing$, and put $T_\psi = \{q \in \mathbb{C} : \operatorname{Re} qe^{i\rho\psi} > l(\psi)\}$. If $u \in \overset{\circ}{B}$, then $[u, p] \cap B = [w, p]$ is a segment one of the ends of which (say, $w$) is in $T_\psi \cap \partial B$. We draw a support straight line $\Gamma$ to $B$ at the point $w \in \partial B$. Then there exist numbers $\varphi, c \in \mathbb{R}$ with the properties

$$\begin{gathered} |\varphi - \psi| < \pi/2\rho; \; l(\varphi) < c < \operatorname{Re} pe^{i\rho\varphi}, \\ \Gamma = \{q \in \mathbb{C} : \operatorname{Re} qe^{i\rho\varphi} = c\}. \end{gathered} \tag{4.4.24}$$

Let $A = N_\varphi(h_F)$ (see (4.4.20)). In particular, $\Gamma$ is a support straight line to the set $B \cap A$. Since $A$ is convex, $\Gamma$ is a support straight line to $A$ at the point $w \in \partial A$. From (4.4.24) and (4.4.21) by the separation theorem for convex sets we obtain:

$$c = \sup\{\operatorname{Re} qe^{i\rho\varphi} : q \in A\}; \; r_\alpha(\varphi; \zeta) = \operatorname{Re} pe^{i\rho\varphi} > h_F(\varphi),$$

i.e., inequality (4.4.23) is valid.

We argue similarly if $\theta_j$ coincides, for instance, with $\varphi_1$.

(b) Let $\theta_j \in (\varphi_1, \varphi_2)$ for some $j \in \{1, \dots, k\}$. Inequality (4.4.23) is proved in the same way as above. Instead of $B$ we consider the set $N_{\theta_j}(h_F)$, which contains the segment $[p_j, p_{j+1}]$ (see the proof of Property 4.4.16).

2. Let $F \in \mathfrak{B}_2(\rho, \alpha)$. Then $\overline{F} \in \mathfrak{B}_1(\rho, \alpha)$ (see Property 4.4.13). Therefore, from (4.4.14) we have: $\overline{F} = X(h_{\overline{F}}) = X(h_F)$. Properties 4.4.13 and 4.4.14 imply that $F = \overset{\circ}{\overline{F}} = \operatorname{int} X(h_F) = \Omega(h_F)$ ►

**Theorem 4.4.19.** *Minkowski duality*

$$\begin{aligned} \nu_1 : \quad & P_\rho(l) \to \mathfrak{B}_1(\rho, \alpha), \quad && \nu_1(h) = X_h; \\ \nu_2 : \quad & Q_\rho(l) \to \mathfrak{B}_2(\rho, \alpha), \quad && \nu_2(h) = \Omega_h, \end{aligned}$$

*is an isomorphism of partially ordered sets and $\nu_i^{-1}(F) = h_F$ is the $(\rho, \alpha)$-support function of the set $F \in \mathfrak{B}_i(\rho, \alpha)$, $i = 1, 2$. Moreover, the subclass $P'_\rho(l)$ of finite functions belonging to $P_\rho(l)$ is isomorphic to the class $\mathfrak{B}'_1(\rho, \alpha)$ of compact $(\rho, \alpha)$-convex sets.*

◄ First, $X_h \in \mathfrak{B}_1(\rho, \alpha)$ (see Property 4.4.14). By Property 4.4.17, $h$ is the $(\rho, \alpha)$-support function of $X_h$, $\forall\, h \in P_\rho(l)$. Therefore, $\nu_1$ is a monomorphism.

On the other hand, if $F \in \mathfrak{B}_1(\rho, \alpha)$, then $F = X_h$, where $h$ is the $(\rho, \alpha)$-support function of $F$ (see Property 4.4.18). But $h \in P_\rho(l)$ by Property 4.4.16. Hence $\nu_1$ is an epimorphism. Thus, $\nu_1$ is an isomorphism. For $\nu_2$ this property is verified similarly.

We prove the second statement. It suffices to show that $\nu_1[P'_\rho(l)] = \mathfrak{B}'_1(\rho, \alpha)$. Since $r_\alpha$ is continuous (see (4.4.13)), we see that the $(\rho, \alpha)$-support function $h_F$ of every compact set $F \subset \mathbb{C}$ is finite. Therefore, $\mathfrak{B}'_1(\rho, \alpha) \subset \nu_1[P'_\rho(l)]$.

Let $h \in P'_\rho(l)$. Suppose $X_h = \nu_1(h)$ is unbounded. Then in $X_h$ there exists a sequence $\{z_n\}_1^\infty$ such that $\lim\limits_{n\to\infty} z_n = \infty$. Let $e_n = z_n/|z_n|$, $n \in \mathbb{N}$. Since $S^1 = \{c \in \mathbb{C} : \ |c| = 1\}$ is compact and includes $\{c_n\}_1^\infty$, there exists an element $c_0 = e^{i\psi_0}$ and a subsequence $\{c_{n_k}\}_1^\infty$ such that $\lim\limits_{k\to\infty} c_{n_k} = c_0$. By (4.4.3), we see that $z_{n_k} \in \mathcal{D}_\alpha(-\psi_0)$ for $k \geqslant k_0$, and $r_\alpha(-\psi_0; z_{n_k}) \to \infty$ as $k \to \infty$. On the other hand, $\{z_{n_k}\} \subset X_h \cap \mathcal{D}_\alpha(-\psi_0)$, and, consequently,

$$r_\alpha(-\psi_0; z_{n_k}) \leqslant h(-\psi_0) < \infty, \ k \geqslant k_0,$$

a contradiction. So, $X_h$ is compact in $\mathbb{C}$, consequently, $\nu_1[P'_\rho(l)] \subset \mathfrak{B}'_1(\rho, \alpha)$ ▶

Consider several consequences of Theorem 4.4.19.

**Property 4.4.20.** *Suppose $G$ is an open $(\rho, \alpha)$-convex set in $\mathbb{C}$ including a compact set $F$, $h_F$ is the $(\rho, \alpha)$-support function of $F$ and $E$ is the closed $(\rho, \alpha)$-convex hull of $F$. Then the compact set $X(h_F)$ associated with $h_F$ is $E$* (see (4.4.14), Theorem 4.4.19), *and $X(h_F) \subset G$.*

◀ Let $E \in \mathfrak{B}_1(\rho, \alpha)$ (see Property 4.4.18), and let $h_E$ be the $(\rho, \alpha)$-support function of $E$. By Property 4.4.16, $\{h_F, h_E\} \subset P_\rho(l)$. Since $h_F \leqslant h_E$, by Theorem 4.4.19 we have $X(h_F) \subset E$. On the other hand, it is obvious that $F \subset X(h_F)$, i.e., the first assertion is proved. By Property 4.4.18, $G = \Omega(h_G)$, where $h_G$ is the $(\rho, \alpha)$-support function of $G$ (see (4.4.14)). Since $r_\alpha(\theta; \zeta)$ is continuous in $\zeta$ on $\mathbb{C}$ for every fixed $\theta \in \mathbb{R}$, and since $F$ is compact, we have $h_F < h_G$. Hence, $X(h_F) \subset \Omega(h_G) = G$ ▶

**Property 4.4.21.** *Let $h \in Q_\rho(l)$. Then there exists a sequence $\{h_k\}_1^\infty \subset P'_\rho(l)$ such that*

$$h_k < h_{k+1} < h, \ \forall\ k \in \mathbb{N}; \ \lim_{k\to\infty} h_k(\theta) = h(\theta) \ \forall\ \theta \in \mathbb{R}.$$

◀ For the open set $\Omega_h$ in $\mathbb{C}$ there exists an expanding sequence of compact sets $\{K_n\}_1^\infty$ satisfying

$$K_n \subset \overset{\circ}{K}_{n+1}, \ n \in \mathbb{N}; \ \bigcup_{n=1}^{\infty} K_n = \Omega_h. \tag{4.4.25}$$

Starting with this sequence, we construct a sequence $\{L_n\}_1^\infty$ of $(\rho, \alpha)$-convex compact sets with similar properties. We put $L_1 := X_l$ (see Def. 4.4.3). There exists a compact set $K_{n_1}$ such that $L_1 \subset K_{n_1}$. Let $L_2$ be the closed

$(\rho, \alpha)$-convex hull of $K_{1+n_1}$. From (4.4.25) and Property 4.4.20 we deduce that $L_1 \subset \overset{\circ}{L}_2 \subset L_2 \subset \Omega_h$. In the same way, starting with $L_2$ we obtain a $(\rho, \alpha)$-convex compact set $L_3$ such that $L_2 \subset \overset{\circ}{L}_3 \subset \Omega_h$, and so on. Let $h_n$ be the $(\rho, \alpha)$-support function of $L_n$. Since $L_n \subset \overset{\circ}{L}_{n+1} \subset L_{n+1} \subset \Omega_h$, we arrive at the required inequality (see (4.4.15)). The limit relation is a consequence of the formula $\bigcup_{n=1}^{\infty} L_n = \Omega_h$ (see (4.4.25)): the limit exists and is the $(\rho, \alpha)$-support function of $\Omega_h$, i.e., this limit coincides with $h$ (see Theorem 4.4.19) ▶

**Property 4.4.22.** *The boundary $\partial\Omega$ of every bounded $(\rho, \alpha)$-convex domain $\Omega$ is a closed rectifiable Jordan curve.*

◀ Let $h$ be the $(\rho, \alpha)$-support function of $\Omega$, and let $\mathcal{A}_h$ be the surface lying in $\mathcal{A}_l$ and considered in Theorem 4.4.1. Let $\Phi_\alpha$ be the mapping occurring in Property 4.4.8. Since $X_l \subset \Omega$, we have $\mathbb{C} \setminus \Omega \subset G_\alpha$ (see (4.4.3), Definition 4.4.3). By Property 4.4.8, $\Phi_\alpha \to \overline{\mathbb{C}} \setminus \Omega$ is a homeomorphism. Consequently, it suffices to prove that $\partial\mathcal{A}_h$ is a rectifiable Jordan curve.

We observe that $\partial\mathcal{A}_h = \partial(E_h \setminus G_E)$ (in the notation of Theorem 4.4.1), and $\partial E_h = \{(p, \Delta) \in E_l : p \in \Gamma_h\}$, where $\Gamma_h$ is the curve in $\mathbb{C}$ associated with $h$ as dictated by formula (1.9.26). Since $h$ is finite, Theorem 1.9.25 and Property 1.9.26 imply that $\partial\mathcal{A}_h$ is a closed rectifiable curve. For $\theta \in \mathbb{R}$, we put

$$M(\theta) = \{(p, \delta) \in \mathcal{A}_l : p \in Dh(\theta);\ e^{i\theta} \in \delta\}$$

(see (1.9.25), (1.9.13)). The set $M(\theta)$ is either a singleton, or a topological segment in $\mathcal{A}_l$ (see Theorem 1.9.21). Moreover,

$$\partial\mathcal{A}_h = \bigcup M(\theta) | \theta \in \mathbb{R}. \tag{4.4.26}$$

We want to know when $M(\theta_1) \cap M(\theta_2) = \varnothing$ for some $\theta_1 < \theta_2 < \theta_1 + 2\pi$. Let $(p_0, \delta_0) \in M(\theta_1) \cap M(\theta_2)$. Then $\delta_0 = \{e^{i\theta} : \theta \in \Delta_0\}$, where $\Delta_0$ is the component of $X(p, l)$ (see (4.2.2)) containing either $\theta_1$ or $\theta_2$. Since $h > l$ and the length of $\Delta_0$ does not exceed $\pi/\rho$ (see the proof of Property 4.2.7), we may have, for instance, $\theta_1, \theta_2 \in \overset{\circ}{\Delta}_0$ and $|\theta_1 - \theta_2| < \pi/\rho$. The formulas Re $p_0 e^{i\rho\theta_j} = h(\theta_j)$, $j = 1, 2$ (see 1.9.13)) and $h \in P_\rho$ imply that $h(\theta) \leqslant$ Re $p_0 e^{i\rho\theta}$, $\theta \in [\theta_1, \theta_2]$. On the other hand, from (1.9.13) we have: Re $p_0 e^{i\rho\theta} \leqslant h(\theta)$. So, $h(\theta) =$ Re $p_0 e^{i\rho\theta}$ for $\theta \subset [\theta_1, \theta_2]$, and $Dh(\theta) = \{p_0\}$ for $\theta \in (\theta_1, \theta_2)$ (see (1.9.14)). If the sets $Dh(\theta_i)$, $i = 1, 2$, are segments in $\mathbb{C}$, then $\pi + \rho(\theta_1 - \theta_2)$ is the angle between these segments and $p_0$ is their common end (see Theorem 1.9.20); if, e.g., $Dh(\theta_1)$ is a segment and $Dh(\theta_0)$ is a point, then $p_0$ is one of the ends of $Dh(\theta_1)$. In either case the part $\bigcup Dh(\theta) | \theta \in [\theta_1, \theta_2]$ of the curve $\Gamma_h$ is a Jordan curve in $\mathbb{C}$. A similar statement is true about the part $\bigcup M(\theta) | \theta \in [\theta_1, \theta_2]$ of the boundary $\partial\mathcal{A}_h$ (see (4.4.26)). This argument is also valid in the case where $\theta_2, \theta_1 + 2\pi \in \overset{\circ}{\Delta}_0$. So, $\partial\mathcal{A}_h$ is a Jordan curve ▶

The following property is a consequence of (4.4.14) and Property 4.4.18.

**Property 4.4.23.** *Suppose $F$ is a closed $(\rho,\alpha)$-convex set and $h$ is its $(\rho,\alpha)$-support function. Then* (see (4.4.14))

$$\mathbb{C}\setminus F = \mathbb{C}\setminus X_h = \bigcup \mathcal{D}_\alpha(\theta; h(\theta))|\theta \in \mathbb{R},$$

*i.e., in formulas* (4.4.2), (4.4.14) *the symbol $X_h$ denotes one and the same set.*

Note also the following fact, which will be useful below.

**Property 4.4.24.** *Suppose $g, h \in P'_\rho(l)$ and $g > h$* (see Theorem 4.4.19). *Then there exists a monotone increasing sequence of numbers $\{\varphi_j\}_0^N$ such that $\varphi_N = \varphi_0 + 2\pi$, $\varphi_{j+1} - \varphi_j < \pi/2\rho$ for $j = 0, 1, \ldots, N-1$, and $\partial\Omega_g \subset \bigcup \mathcal{D}(\varphi_j; h)|j = 1, ..., N$* (see (4.4.2), (4.4.14)).

◀ 1. Let $\{\varphi_j\}_0^N$ be an arbitrary sequence with the described properties; we introduce the functions $l_j(\theta) = a_j \cos\rho\theta + b_j \sin\rho\theta$, $\theta \in \mathbb{R}$, where the coefficients are determined by the conditions $l_j(\varphi_k) = h(\varphi_k)$ for $k = j-1, j$ and $j = 1, \ldots, N$ (see the remark to Def. 1.9.1). Consider the $2\pi$-periodic extension to $\mathbb{R}$ of the function

$$f(\theta) = l_j(\theta),\ \theta \in [\varphi_{j-1}, \varphi_j],\ j = 1, \ldots, N. \tag{4.4.27}$$

We have $f'_-(\varphi_j) = l'_j(\varphi_j) = \rho A(\varphi_{j-1}, \varphi_j)$ and $f'_+(\varphi_j) = l'_{j+1}(\varphi_j) = \rho A(\varphi_{j+1}, \varphi_j)$, $j = 1, \ldots, n$, where $A(\varphi, \theta) := A(\varphi)$ is the function defined by formula (1.9.10) and $\varphi_{N+1} = \varphi_1 + 2\pi$. Since $\varphi_{j+1} - \varphi_{j-1} < \pi/\rho$ and $A(\varphi, \varphi_j)$ is a monotone increasing function on the set $[\varphi_{j-1}, \varphi_{j+1}]\setminus\{\varphi_j\}$, $j = 1, \ldots, N$, we have $f'_-(\varphi_j) < f'_+(\varphi_j)$, $j = 1, \ldots, N$, and $f \in P_\rho$ (see the proof of Theorem 1.9.18, P. 4.1.1). From Definition 1.9.1 we obtain $h \leqslant f$ and $f \in P'_\rho(l)$.

2. Now, we take a sequence $\{\varphi_j\}_0^N$ in such a manner that $f < g$. Choosing $0 < \delta < \pi/\rho$, we put

$$\begin{aligned} m(\theta;\delta) &= \max\{h(\theta+\varphi) : \varphi \in [-\delta,\delta]\}; \\ M(\theta;\delta) &= \min\{g(\theta+\varphi) : \varphi \in [-\delta,\delta]\}; \\ T(\theta,\delta) &= m(\theta,\delta) - M(\theta,\delta)\cdot\cos(\rho\delta/2). \end{aligned}$$

The functions $h, g$ are continuous and $2\pi$-periodic (see Property 1.9.5) and, consequently, uniformly continuous. Therefore, $m, M$, and $T$ are continuous functions on $\mathbb{R}\times[0,\pi/\rho]$. The function $T$ increases in the variable $\delta$ as $\delta \to +0$ for every fixed $\theta \in \mathbb{R}$. Next, $\lim\limits_{\delta\to 0+} T(\theta,\delta) = h(\theta) - g(\theta)$, and by the Dini test the convergence is uniform with respect to $\theta \in \mathbb{R}$. Therefore, if $\varepsilon_0 := \inf\{[g(\theta) - h(\theta)] : \theta \in \mathbb{R}\}$, then for every $\varepsilon \in (0, \varepsilon_0)$ there exists a number $\delta_0 = \delta_0(\varepsilon)$ such that $T(\theta,\delta) < h(\theta) - g(\theta) + \varepsilon < 0$ for all $\theta \in \mathbb{R}$ and all $\delta \in [0, \delta_0]$. So,

$$m(\theta,\delta) < M(\theta,\delta)\cos(\rho\delta/2),\ \theta \in \mathbb{R};\ \delta \in [0,\delta_0]. \tag{4.4.28}$$

We require additionally that $|\varphi_j - \varphi_{j-1}| \leqslant \delta_0$, $j = 1, 2, \ldots, N$, and estimate $f$ from above (see (4.4.27)). By (1.9.1), (4.4.28), we see that

$$f(\theta) \leqslant m(\theta,\delta_0)\cdot[\cos(\rho\delta_0/2)]^{-1} < M(\theta,\delta_0) \leqslant g(\theta)\ \forall\ \theta \in \mathbb{R},$$

i.e., $f < g$, and $f \in P'_\rho(l)$ (see item 1).

3. In the notation of formula (4.4.14) we have: $\mathbb{C} \setminus X_f \supset \mathbb{C} \setminus \Omega_g$. Together with Property 4.4.23, this implies that

$$\partial\Omega_g \subset \bigcup \mathcal{D}_\alpha(\theta; f(\theta)) | \theta \in \mathbb{R}. \tag{4.4.29}$$

Let $\Pi_f(\theta) = \{p \in \mathbb{C};\ \operatorname{Re} pe^{i\rho\theta} > f(\theta)\}$ for $\theta \in \mathbb{R}$. For $j = 1, \ldots, N$ the function $l_j$ in (4.4.27) satisfies $l_j(\theta) = \operatorname{Re} p_j e^{i\rho\theta}$, $\theta \in [\varphi_{j-1}, \varphi_j]$, where $p_j \in \mathbb{C}$. Since $|\varphi_j - \varphi_{j-1}| < \pi/\rho$, from (1.9.1) we conclude that

$$\Pi_f(\theta) \subset \Pi_f(\varphi_{j-1}) \cup \Pi_f(\varphi_j),\ \theta \in (\varphi_{j-1}, \varphi_j),\ j = 1, \ldots, N.$$

Therefore (see Property 4.4.5 and (4.4.9)) for $j = 1, \ldots, N$ we have

$$\mathcal{D}_\alpha(\theta; f(\theta)) \subset \mathcal{D}_\alpha(\theta; f(\varphi_{j-1})) \cup \mathcal{D}_\alpha(\theta; f(\varphi_j)),\ \theta \in (\varphi_{j-1}, \varphi_j).$$

But $f(\varphi_j) = h(\varphi_j)$, $j = 0, 1, \ldots, N$. Together with (4.4.29), this yields Property 4.4.24 ▶

**4.4.4. A property of the indicator diagram.** From Proposition 4.4.23 we conclude that the construction of the indicator diagram of an entire function of order $\rho \in \mathbb{N}$ (see (4.4.2)) has the properties dealt with at the beginning of Chapter 4.

**Corollary 4.4.25.** *Consider a function $h \in P_\rho$ that is not identically equal to $\infty$ and coincides with the indicator of an entire function $f$ of order $\rho > 1/2$. Let $l \in M_\rho$. Suppose that $h \in P_\rho(l)$ and $\alpha \in \mathfrak{P}_\rho(l)$. Then the $(\rho,\alpha)$-indicator diagram $X_h$* (see (4.4.2)) *of $f$ is a closed $(\rho,\alpha)$-convex set. If $h > l$ and $h$ is a finite function, then the boundary $\partial X_h$ of the set $X_h$ is a closed Jordan curve.*

**4.4.5. The case of $\rho \notin \mathbb{N}$.** The results of items 4.4.2-4.4.4 have a complete analog for $\rho \notin \mathbb{N}$. The statements and proofs become more complicated because the Riemann surface $L$ of the logarithm must be involved. Mostly, results and definitions given in $2^0$-$4^0$ persist for $\rho \notin \mathbb{N}$. In this case we do not reproduce them below. When making references, we tacitly assume that these results are valid for every $\rho > 0$.

Suppose $\rho \notin \mathbb{N}$ and $\rho > 1/2$. Take $l \in M_\rho \setminus O_\rho$ and $\alpha \in \mathfrak{P}_\rho(l)$ (see Def. 4.3.5).

**Property 4.4.5′.** *For every function $\alpha$ Property 4.4.5 is valid with $\mathcal{D}_\alpha(\theta)$ replaced by $\mathcal{D}'_\alpha(\theta)$.*

The proof of Property 4.4.5′ is based on Lemma 4.4.6 and is similar to the proof of Property 4.4.5.

**Property 4.4.8′.** *Let $(\varphi_l, E_l)$ be the polygonal surface and $\mathcal{A}_l$ the polygon associated with a function $l \in M_\rho \setminus O_\rho$* (see Def. 4.2.12), *and let $\alpha \in \mathfrak{P}_\rho(l)$. Then there exists a continuous inclusion $\Phi : E_l \to L$ with the following properties:*

1. *$\Phi$ agrees with the action of the group $G_E$ on $E_l$ and of the group $G_L$ on $L$ and, in the notation of Properties 4.2.10 and 4.2.11, satisfies $\alpha \circ \Phi = \varphi_l$, $\Phi \circ B = A \circ \Phi$.*

2. *Let $\Phi_\alpha$ denote the extension to $\mathcal{A}_l$ by continuity of the factor-mapping corresponding to $\Phi$. Then $\Phi_\alpha : \overset{\circ}{\mathcal{A}}_l \to \{\infty\} \cup G_\alpha$ is a conformal homeomorphism (see (4.4.3)) and*

$$\Phi_\alpha(\mathcal{A}_l) = \overline{G}_\alpha \cup \{\infty\}; \ \ \Phi_\alpha(\Lambda_\infty) = \infty; \ \ \Phi_\alpha(\partial \mathcal{A}_l) = \partial G_\alpha.$$

◀ 1. By Property 4.4.5′, for every $\theta \in \mathbb{R}$ there is a homeomorphism $g_\theta : \overline{T}_\theta \to \overline{\mathcal{D}'_\alpha(\theta)}$ that is holomorphic in $T_\theta$ and coincides with the inverse of $\alpha$. Let $U_\theta = \{(p, \Delta) \in E_l : \ p \in T_\theta; \theta \in \Delta\}$. Then

$$E_l = \bigcup \overline{U_\theta} | \theta \in \mathbb{R}; \quad \overset{\circ}{E_l} = \bigcup U_\theta | \theta \in \mathbb{R},$$

and $\varphi_l : \ \overline{U}_\theta \to \overline{T}_\theta$ is a homeomorphism for every $\theta \in \mathbb{R}$. On $E_l$, we consider the mapping $\Phi : \ E_l \to L$, where

$$\Phi(m) = [g_\theta \circ \varphi_l](m), \ \ m \in \overline{U}_\theta, \ \theta \in \mathbb{R}.$$

This mapping is well defined (see the proof of Property 4.4.8). From the properties of the mappings $\{g_\theta : \ \theta \in \mathbb{R}\}$ it follows that $\Phi$ is a continuous embedding of $E_l$ in $L$ conformal on $E$, and

$$\Phi(\overline{U}_\theta) = \overline{\mathcal{D}'_\alpha(\theta)}, \ \Phi(\overline{U}_\theta) = \mathcal{D}'_\alpha(\theta) \ \forall \ \theta \in \mathbb{R}.$$

By (4.2.6) and the remark to Definition 4.3.5, we conclude that the following diagram is commutative for every $\theta \in \mathbb{R}$:

$$\begin{array}{ccc} \overline{U}_\theta & \xrightarrow{\Phi} & \overline{\mathcal{D}'_\alpha(\theta)} \\ \downarrow B & & \downarrow A \\ \overline{U}_{\theta-2\pi} & \xrightarrow{\Phi} & \overline{\mathcal{D}'_\alpha(\theta - 2\pi)} \end{array}$$

Thus, the mapping $\Phi$ has property 1.

2. Let $G'_\alpha = \bigcup \mathcal{D}'_\alpha(\theta) | \theta \in \mathbb{R}$. From item 1 we see that $G'_\alpha = \Phi(\overset{\circ}{E_l})$, and $\Phi : \overset{\circ}{E_l} \to G'_\alpha$ is a conformal local homeomorphism that agrees with the action of $G_E$ on $E_l$ and of $G_L$ on $L$. Using Lemma 4.4.6, we see that $\Phi$ is a conformal homeomorphism (see the proof of Property 4.4.8). Then this property is shared by the factor-mapping for $\hat{\Phi} : \overset{\circ}{\mathcal{A}_l} \setminus \{\Lambda_\infty\} \to G_\alpha$ (see item A.1.9). As in the case of an integer $\rho > 0$, we have

$$\lim_{p \to \infty; p \in T_\theta} |g_\theta(p)| = \infty;$$

consequently, $\Phi$ possesses property 2 ▶

In the statement of the following supplement to Properties 4.4.5′, 4.4.8′, we use the notation accepted there and in (4.4.3).

**Corollary 4.4.26.** *Suppose $\theta, \psi \in \mathbb{R}$ satisfy*

$$0 \leqslant \theta - \psi < \pi \min\{1/\rho, 2 - 1/\rho\}.$$

*Taking $\alpha \in \mathfrak{P}_\rho(l)$, we put $\mathcal{D} = \mathcal{D}_\alpha(\theta) \cup \mathcal{D}_\alpha(\psi)$, $\mathcal{D}' = \mathcal{D}'_\alpha(\theta) \cup \mathcal{D}'_\alpha(\psi)$. The operator $\pi : L \to \mathbb{C}$ on the set $\mathcal{D}$ admits an inverse $\pi^{-1}(\theta, \psi)$ such that the mappings*

$$\pi^{-1}(\theta, \psi) : \mathcal{D} \to \mathcal{D}' \text{ and } \alpha \circ \pi^{-1}(\theta, \psi) : \mathcal{D} \to T_\theta \cup T_\psi$$

*are homeomorphisms. This assertion remains valid for the set $\overline{\mathcal{D}}$ for all $l \in M_\rho \setminus O_\rho$, except for the case where* (see (4.1.2) and (4.1.3) for $k = 1$)

$$l(\varphi) = \operatorname{Re} p_j e^{i\rho\varphi}, \ \varphi \in [\theta_{j-1}, \theta_j], \ j \in \mathbb{Z}, \tag{4.4.30}$$

*and $p_{j+1} = p_1 e^{-i2\pi\rho j}$, $\theta_j = \theta_0 + 2\pi j$, $j \in \mathbb{Z}$; here the parameters $\theta$ and $\psi$ satisfy the conditions $\psi \leqslant \theta_0 + 2\pi s \leqslant \theta$ for some $s \in \mathbb{Z}$. In the exceptional case indicated, these mappings are defined and homeomorphic on the set $\overline{\mathcal{D}} \setminus \{z_1\}$, where $z_1$ is the only element of the set $X$ in Definition* 4.3.5.

◀ The set $\overline{T}_\theta \cup \overline{T}_\psi$ is an angle in $\mathbb{C}$, and $T_\theta \cap T_\psi \neq \varnothing$. From Property 4.4.5′ we conclude that $\alpha : \overline{\mathcal{D}'} \to \overline{T}_\theta \cup \overline{T}_\psi$ is a homeomorphism (see the proof of Property 4.4.8′). Therefore, it suffices to prove that, under the above restrictions, the set $\overline{\mathcal{D}'}$ is in a fundamental domain of the group $G_L$ on $L$. By Property 4.4.8′, this is equivalent to the same statement for the set $\overline{U} = \Phi^{-1}(\overline{\mathcal{D}'})$, (where $U = U_\theta \cup U_\psi$) and the group $G_E$ on $E_l$ (see the proof of Property 4.4.8′).

Assume that there exists a point $m$ in $E_l$ and a number $r \in \mathbb{Z}$ such that $\{m, B^r m\} \subset \overline{U}$, where $B$ is the generator of $G_E$ and $B^r$ is an element of $G_E$ of order $r$ (see (4.2.6)). Let $m = (p, \Delta)$ (see item 4.2.2), and let $\mu$ be the length of $\Delta$. Suppose, for instance, that $m \in \overline{U}_\theta$. If $B^r m \in \overline{U}_\theta$, then $\theta \in \Delta \cap (\Delta - 2\pi r)$. This is possible only if $r = 1$ *) and $\mu \geqslant 2\pi$. If $m \in \overset{\circ}{E}_l$, then $\mu \leqslant \min\{\pi/\rho, \pi(2 - 1/\rho)\} < 2\pi$ (the distance between the neighboring segments in the system $X(p, l)$ does not exceed $\pi/\rho$, see (4.2.2) and the proof of Property 4.2.9). But $\mu \geqslant 2\pi$, therefore, $m \in \partial E_l$. Now, from Property 4.2.9 we deduce that $\mu = 2\pi$, which is possible only if $m$ and $Bm$ are neighboring vertices of the surface $(\varphi_l, E_l)$ associated with the function $l$ of the form (4.4.30). If $B^r m \in U_\psi$ with $\psi < \theta$, then $\psi \in \Delta - 2\pi r$ and $\theta \in \Delta$. Similar arguments show that, under the restrictions imposed on $\theta, \psi$, the latter is possible only if $r = 1$ in the exceptional case described in the corollary. So, the assumption is true only if $l$ is of the form (4.4.30) with some $m = m_{s+1} = (p_{s+1}, [\theta_s, \theta_{s+1}])$ and $Bm = m_s = (p_s, [\theta_{s-1}, \theta_s])$, $\psi \leqslant \theta_s \leqslant \theta$, $s \in \mathbb{Z}$ (see 4.2.6)). Since $\Phi \circ B = A \circ \Phi$

*) The case of $r = -1$ reduces to this one, because $m = B(B^{-1} m)$.

(see Property 4.4.8′), by Property 4.2.9 and Definition 4.3.5 we see that the points $Z_j := \Phi(m_j)$, $j = s, s+1$, where

$$\alpha(Z_j) = p_j, \ \ j = s, s = 1; \ AZ_{s+1} = Z_s, \tag{4.4.31}$$

are roots of $\alpha'$, and $\{Z_s, Z_{s+1}\} \subset \overline{\mathcal{D}'}$. The final assertion of the corollary concerns the point $z_1 := \pi(Z_s) = \pi(Z_{s+1}) \in X \cap \overline{\mathcal{D}}$ ▶

**Remark.** In particular, the statement is valid for $\theta = \psi$ and for the operator $\pi_\theta^{-1} := \pi^{-1}(\theta, \theta)$. For simplicity, in what follows the mappings $\alpha \circ \pi_\theta^{-1}$ and $\alpha \circ \pi^{-1}(\theta, \psi)$ are denoted by $\alpha$, and their inverses by $\alpha^{-1}$. In particular, $\alpha^{-1}(p) := \pi[g_\theta(p)]$, $p \in T_\theta$ (in the notation of the proof of Property 4.4.8′).

For instance, the above remark shows that Definition 4.4.3 of a $(\rho, \alpha)$-convex set and formula (4.4.13) for the function $r_\alpha(\theta; \zeta)$ persist for $\rho \notin \mathbb{N}$. From Corollary 4.4.26 it follows that the function $r_\alpha(\theta, \zeta)$ is continuous in $\zeta$ on $\mathbb{C}$. In particular, this is true in the exceptional case described in Corollary 4.4.26. To verify this, it suffics to use formula (4.4.31), where $\{Z_s, Z_{s+1}\} \subset \overline{\mathcal{D}'_\alpha(\theta_s)}$, and to refer to the gluing conditions for piecewise trigonometric functions: $l(\theta_s) := \text{Re}\, p_s e^{i\rho\theta_s} = \text{Re}\, p_{s+1} e^{i\rho\theta_s}$. If $l \in O_\rho$ with $\rho \notin \mathbb{N}$, then $l(\theta) \equiv 0$ and $\alpha(Z) = Z^\rho$. Then we put (see (3.1.9)) $r_\alpha(\theta; \zeta) := \Phi_\rho(\zeta e^{i\theta})$,

$$\mathcal{D}'_\alpha(\theta) := \{Z \in L : \ |Arg\ Z + \theta| < \min\{\pi/2\rho, \pi\}\}, \ \mathcal{D}_\alpha(\theta) = \pi(\mathcal{D}'_\alpha(\theta)).$$

We observe some specific features of the structure of the common boundary of the sets $X_l$ and $G_\alpha$ for $\rho \notin \mathbb{N}$. In the notation of Definition 4.3.5, let $\{Z_j, Z_{j+1}\} \subset \overline{\mathcal{D}'_\alpha(\theta_j)} \cap N_\alpha$, $j \in \mathbb{Z}$, and let $\{p_j, p_{j+1}\} \subset \overline{T}_{\theta_j}$. By Property 4.4.5, there exists a unique path $f_j : [0,1] \to \overline{\mathcal{D}'_\alpha(\theta_j)}$ such that $f_j(0) = Z_j$, $f_j(1) = Z_{j+1}$, and $\alpha \circ f_j(t) = (1-t)p_j + tp_{j+1}$. Let $S_j$ be the support of the path $f_j$ in $L$. Then $\gamma(z_1, z_2; \theta_j) := \pi(S_j)$ is a curve in $\mathbb{C}$ connecting the points $z_1 := \pi(Z_j)$ and $z_2 := \pi(Z_{j+1})$ of the set $X$. Since $Z_{j-k} = AZ_j \ \forall\, j \in \mathbb{Z}$ and $\{Z_{j-k}, Z_{j-k+1}\} \subset \overline{\mathcal{D}'_\alpha(\theta_j - 2\pi)}$, we see that $\pi(S_j) = \pi(S_{j-k}) \ \forall\, j \in \mathbb{Z}$. Therefore, in $\mathbb{C}$ there exist only $k$ different curves $\{\Gamma_j\}_1^k$ connecting, respectively, the points $z_j$ and $z_{j+1}$, $j = 1, \ldots, k$ of the set $X$, where $z_{k+1} := z_1$. If $k = 1$, then $\Gamma_1$ is a closed curve in $\mathbb{C}$ passing through $z_1$. With this distinction, Property 4.4.10 is also valid for $\rho \notin \mathbb{N}$. For instance, for the function $\alpha$ in Example 4.3.6 with $\rho \in (1/2, 1)$, the set $X_l$ in Definition 4.4.3 is a compact subset of $\mathbb{C}$ that contains 0 and is bounded by an analytic arc $\Gamma_1$ passing through the point $z_1 = -(1-\rho)^{1/\rho}$ and coinciding with the projection to $\mathbb{C}$ of the inverse image of the segment $\{(1-t)e^{i\pi\rho} + te^{-i\pi\rho} : \ t \in [0,1]\}$ under the mapping $\alpha$ when $|Arg\ Z| \leqslant \pi$. Other assertions and definitions of Section 4.4, together with proofs, are extended without considerable changes to the case where $\rho \notin \mathbb{N}$.

Chapter 5

# Spaces of Entire Functions of Order $\rho > 0$ with Restrictions on the Indicator

In this chapter the Polya theorem about the relationship between the indicator diagram and the conjugate diagram of an entire function of finite power (see the Introduction, Theorems $A_1$, $A_2$; Theorem 1.10.8) is extended in various ways to entire functions of order $\rho > 0$, $\rho \neq 1$. We also consider applications of these extensions. The key notion is that of the generalized Borel transformation associated with the polynomial

$$\alpha(z) = z^\rho \left[1 + \frac{a_1}{z} + \ldots + \frac{a_n}{z^n}\right],$$

various properties of which were examined in Chapter 4. We consider the operator $B_\alpha$ that takes an entire function

$$f(z) = \sum_{k=0}^{\infty} \frac{b_k z^k}{\Gamma\left(\dfrac{k+1}{\rho}\right)} \tag{5.0.1}$$

of order $\rho$ and of finite type to the function

$$\widehat{f}(\zeta) = [B_\alpha f](\zeta) = \sum_{k=0}^{\infty} \frac{b_k}{[u(\zeta)]^{k+1}}, \tag{5.0.2}$$

holomorphic in a certain neighborhood $G$ of the point $\infty$. Here $u(\zeta)$, $\zeta \in G$, is a one-sheeted and nonvanishing branch of the multivalued function $[\alpha(\zeta)]^{1/\rho}$ such that

$$\lim_{\zeta\to\infty} \zeta^{-1} u(\zeta) = 1. \tag{5.0.3}$$

Section 5.1 is of auxiliary nature. In it we examine asymptotic properties of a certain entire function $K_\alpha$ of two variables (namely, the kernel of a certain integral representation of $f$) in terms of its generalized Borel transform $\widehat{f} = B_\alpha f$. Section 5.2 is devoted to various forms of the analog in question of the

Polya theorem. The spaces $[\rho, h(\theta))$, $[\rho, h(\theta)]$ of entire functions are realized as spaces of analytic functionals on $(\rho, \alpha)$-convex sets; the structure of these sets is determined by the structure of the plane indicator diagram examined in Chapter 4 (see Def. 4.4.2). Applications of the analogs of the Polya theorem are illustrated in Section 5.3. There the author wanted only to indicate the lines of application, without any attempt of making radical progress in any of these directions.

## 5.1. Entire function of two variables associated with a polynomial in $\mathfrak{P}_\rho(l)$

Suppose $\rho > 0$, $l \in M_\rho$, and $\alpha \in \mathfrak{P}_\rho(l)$ (see Definitions 4.3.1, 4.3.5). With the polynomial $\alpha$ we shall associate a certain entire function $K_\alpha$ of two variables related to the generalized Borel transformation (5.0.1), (5.0.2). Now we pass to some well-known results of asymptotic analysis required for the study of $K_\alpha$.

**5.1.1. Information about the asymptotics of functions.** We start with some results on the asymptotics of functions representable by integrals.

**Watson lemma.** *Suppose $a > 0$, $\alpha > 0$, $\beta > 0$, $f(x)$ is a continuous function on the segment $[0, a]$, and $f(0) \neq 0$. Put $S_\varepsilon = \{\lambda \in \mathbb{C} : |\arg \lambda| \leqslant \pi/2 - \varepsilon\}$, where $\varepsilon \in (0, \pi/2)$. Then for $\lambda \in S_\varepsilon$ we have*

$$\int_0^a x^{\beta-1} f(x) \exp\{-\lambda x^\alpha\}\, dx = \alpha^{-1}\Gamma(\beta/\alpha)[f(0) + o(1)]\lambda^{-\beta/\alpha}$$

*as $|\lambda| \to \infty$.*

*The saddle-point method.* We describe a method of estimating the asymptotic behavior as $t \to \infty$ of integrals of the form

$$\Phi(t) = \int_\gamma e^{tg(\zeta)} f(\zeta)\, d\zeta, \quad t > 0, \tag{5.1.1}$$

called the *saddle-point method.* Here $g$ and $f$ are functions analytic in a domain $D \subset \mathbb{C}$, and $\gamma$ is an oriented piecewise-smooth contour in $D$. A point $c \in D$ is called a *saddle point* of $g$ if at $c$ the derivative $g'$ is equal to 0. Every component of the set

$$G := \{\zeta \in D : \operatorname{Re} g(\zeta) < \operatorname{Re} g(c)\}$$

is called a *valley* of the point $c$. Every saddle point $c$ has an open neighborhood $U$ such that the number of components of the set $G \cap U$ is equal to the *multiplicity* $k$ of the zero of the function

$$g_c(\zeta) = g(c) - g(\zeta), \quad \zeta \in D,$$

at $c$. The latter means that $k$ is a natural number and

$$g_c(\zeta) = (\zeta - c)^k h(\zeta), \quad \zeta \in U, \tag{5.1.2}$$

where $h$ is analytic in $U$ and $h(c) \neq 0$. In this case $c$ is called a *saddle point of order* $k-1$, where $k \geqslant 2$. In the neighborhood $U$ of the saddle point $c$ of $g$, in every valley $V \subset G \cap U$ (see (5.1.2)) there is only one *steepest descent line*, determined by the requirement Im $g(z) = \text{Im } g(c)$, $z \in \gamma \cap V$. A contour $\gamma$ is said to be *saddle* if it can be deformed in such a way that the value of the integral (5.1.1) does not change and the resulting contour $\gamma_1$ satisfies the following conditions:

1) the contour $\gamma_1$ passes through one or several saddle points in such a manner that its remaining part lies in valleys of these points;

2) if $\gamma_1$ passes through a saddle point $c$, then the set $(U \cap \gamma) \setminus \{c\}$ consists of two steepest descent lines located in different valleys of $c$ ($U$ is a sufficiently small neighborhood of $c$);

3) if $\gamma_1$ is an unbounded contour, then

$$\lim_{|\zeta| \to \infty; \zeta \in \gamma} \text{Re } g(\zeta) = -\infty.$$

So, if $\gamma$ is a saddle contour, there is no loss of generality in assuming that this contour satisfies conditions 1)–3). In what follows we adopt this.

Let $\gamma$ be a saddle contour. The *contribution of the saddle point* $c$ (into the integral $\Phi(t)$, see (5.1.1)) is defined to be the integral

$$\Phi_c(t) = \int\limits_{\gamma \cap U} e^{tg(\zeta)} f(\zeta)\, d\zeta, \quad t > 0, \tag{5.1.3}$$

over a sufficiently small neighborhood $\gamma \cap U$ of the point $c$ on $\gamma$.

**Theorem A.** *In formula* (5.1.1), *let* $\gamma$ *be a saddle contour. Then, as* $t \to \infty$, *the value of the integral* $\Phi(t)$ (see (5.1.1)) *is asymptotically equivalent to the sum* $S(t) := \sum_{i=1}^{k} \Phi_{c_i}(t)$, $t > 0$, *of the contributions of all saddle points* $\{c_1, \ldots, c_k\}$ *through which the contour* $\gamma$ *passes, i.e.*,

$$\lim_{t \to \infty} \Phi(t)/S(t) = 1.$$

The Watson lemma immediately implies the following asymptotic formula for the saddle point contribution.

**Theorem B.** *In the notation of formulas* (5.1.2), (5.1.3), *suppose that the following conditions are fulfilled*:

1) $c$ *is a saddle point of* $g$ *of order* $k-1$, *where* $k \geqslant 2$;

2) *the contour $\gamma$ is oriented so that in a sufficiently small neighborhood $U$ of $c$ it goes (through $c$) from some valley to the adjacent valley that lies in the contour's positive direction relative to the original valley* *);

3) *a branch $\omega = M(\zeta)$ of the multivalued function $[g_c(\zeta)]^{1/k}$ is fixed* (see (5.1.2)) *such that $M$ maps the contour $\gamma \cap U$, consisting of two steepest descent lines, onto the oriented curve*

$$\omega(\tau) = \begin{cases} (1-2\tau)a\exp\{-2\pi i/k\}, & \tau \in (0, 1/2], \\ (2\tau - 1)b, & \tau \in [1/2, 1) \end{cases}$$

*which is a part of the boundary of the angle of opening* $2\pi/k$ (with vertex at 0); *here $a, b > 0$*);

4) *$m \geqslant 0$ is the multiplicity of zero of the function $f$ at $c$, i.e.,*

$$f(\zeta) = (\zeta - c)^m \varphi(\zeta), \quad \zeta \in U,$$

*where $\varphi$ is a function holomorphic in $U$, and $\varphi(c) \neq 0$.*

*Then we have the following formula for the asymptotic behavior as $t \to \infty$ of the contribution of the saddle point $c$* (see (5.1.3)):

$$V_c(t) = e^{tg(c)} \frac{\Gamma((m+1)/k)}{kt^{(m+1)/k}} \left[ \frac{\left(1 - e^{-2\pi i(m+1)/k}\right) f^{(m)}(c)}{m! A^{(m+1)/k}} + o(1) \right],$$

*where $A = -g^{(k)}(c)/k!$.*

**Remarks.** 1. Since $M'(c) = A^{1/k}$, the choice of the branch indicated in the theorem is equivalent to a choice of an angle the opening of which lies among the numbers $\{(\arg A + 2\pi l)/k : l = 0, 1, \dots, k-1\}$; specifically, after rotation through this angle, the tangent at $c$ to the line of the steepest descent to the second valley becomes directed as the real axis (we have used the terminology of condition 2) of the theorem).

2. The Watson lemma implies that Theorems A and B are also valid as $t \to \infty$, $t \in \{z \in \mathbb{C} : |\arg z| \leqslant \alpha\}$, where $\alpha \in (0, \pi/2)$, if the contour $\gamma$ in (5.1.1) is saddle relative to the system of functions $\{e^{i\theta} g(\cdot) : |\theta| \leqslant \alpha\}$, where the function $g$ may depend on additional parameters.

*Asymptotic and integral properties of the Mittag-Leffler function.*

**Theorem C.** *Let $\rho > 1/2$. For the Mittag-Leffler function $E_\rho(z) = \sum_{k=0}^{\infty} \frac{z^k}{\Gamma[(k+1)/\rho]}$, $z \in \mathbb{C}$, and for every $\alpha \in (\pi/2\rho, \pi)$ we have the following asymptotic formula as $z \to \infty$:*

$$E_\rho(z) = \begin{cases} \rho z^{\rho-1} \exp\{z^\rho\} + O(1/z), & |\arg z| \leqslant \alpha; \\ O(1/z), & \alpha \leqslant |\arg z| \leqslant \pi. \end{cases}$$

*) I.e., the second valley must occur immediately after the first in the by-pass of $c$ counterclockwise (for instance, along the boundary of a disk in $U$ centered at $c$).

**Theorem D.** *Suppose $\rho > 1/2$, $\pi/2\rho < \beta \leqslant \min\{\pi, \pi/\rho\}$, and $a > 0$. Let $\gamma(a;\beta)$ be the contour passed in the direction of increase of the argument and consisting of the rays $\{te^{\pm i\beta} : t \geqslant a\}$ and the circular arc $\{ae^{i\varphi} : |\varphi| \leqslant \beta\}$. Then the Mittag-Leffler function admits the integral representation*

$$E_\rho(z) = \frac{\rho}{2\pi i} \int\limits_{\gamma(a;\beta)} \frac{\omega^{\rho-1} \exp\{\omega^\rho\}\, d\omega}{\omega - z}, \qquad |z| < a.$$

**5.1.2. The function $K_\alpha$: the definition and simplest properties.** Take $R_0 > 0$ in such a way that for $R > R_0$ the circle $C_R = \{\zeta \in \mathbb{C} : |\zeta| = R\}$ is included in $G$, and put $S_R = \{\zeta \in \mathbb{C} : |\zeta| < R\}$. We orient the boundary $C_R$ of the disk $S_R$ positively and consider the function

$$K_\alpha(\lambda, z) = \frac{1}{2\pi i} \int\limits_{C_R} \frac{E_\rho[\lambda u(\zeta)]u'(\zeta)\, d\zeta}{\zeta - z}, \qquad (\lambda, z) \in \mathbb{C} \times S_R\,, \tag{5.1.4}$$

where

$$E_\rho(w) = \sum_{k=0}^{\infty} \frac{w^k}{\Gamma\left(\dfrac{k+1}{\rho}\right)}, \qquad w \in \mathbb{C}\,, \tag{5.1.5}$$

is the Mittag-Leffler function. Obviously, the right-hand side in (5.1.4) does not depend on $R$ for $R > R_0$, and $K_\alpha$ is an entire function of two complex variables. Moreover, the condition $\alpha(z) = z^\rho$ (which is valid, e.g., if $0 < \rho \leqslant 1/2$) implies by (5.0.3) that

$$K_\alpha(\lambda, z) \equiv E_\rho(\lambda z), \quad (\lambda, z) \in \mathbb{C}^2. \tag{5.1.6}$$

Substituting (5.1.5) in (5.1.4) and integrating the series termwise, we obtain the following expansion for $K_\alpha$:

$$K_\alpha(\lambda, z) = \sum_{k=0}^{\infty} \lambda^k q_k(z; R)/\Gamma\left(\frac{k+1}{\rho}\right), \qquad z \in S_R\,,$$

where

$$q_k(z; R) = \frac{1}{2\pi i} \int\limits_{C_R} \frac{[u(\zeta)]^k u'(\zeta)}{\zeta - z}, \qquad z \in S_R\,. \tag{5.1.7}$$

The latter function admits an analytic extension $q_k(z)$ to $|z| \geqslant R$. Let $R' > |z| > R$. By (5.1.7), we have

$$q_k(z; R') = [u(z)]^k u'(z) + q_k(z; R). \tag{5.1.8}$$

Formula (5.1.7) does not depend on $R$ and yields the trace on $S_R$ of the entire function $q_k$. From (5.1.7), (5.1.8) we conclude that

$$q_k(z) = [u(z)]^k u'(z) + O(1/z), \quad z \to \infty. \tag{5.1.9}$$

By the structure of $u$ (see (5.0.3)), since $u'(z) \to 1$ as $z \to \infty$, we see that $q_k$ is a monic polynomial of power $k$. We arrive at the following statement.

**Theorem 5.1.1.** *For $\rho > 1/2$, formula* (5.1.4) *defines an entire function $K_\alpha$ of two complex variables, which expands in the series*

$$K_\alpha(\lambda, z) = \sum_{k=0}^{\infty} \lambda^k Q_k(z).$$

*This series converges absolutely and uniformly on every compact subset of $\mathbb{C}^2$; here $Q_k$ is a polynomial of power $k$ with coefficient of $z^k$ equal to $\left[\Gamma\left(\frac{k+1}{\rho}\right)\right]^{-1}$*

We derive another integral representation for $K_\alpha$. This representation is convenient for the study of the asymptotic properties of $K_\alpha$ for $\rho > 1/2$, $\alpha(z) \neq z^\rho$, where formula (5.1.6) fails.

Suppose $0 < \psi_2 - \psi_1 < \pi \min\{1/\rho, 2 - 1/\rho\}$ and $c_j > l(\varphi_j)$, $j = 1, 2$. We denote $D_\alpha(\theta, c) = \{z \in D_\alpha(\theta) : \operatorname{Re} \alpha(z)e^{i\rho\theta} > c\}$, $I = D_\alpha(\psi_1; c_1) \cup D_\alpha(\psi_2; c_2)$, and $L(\psi_1, \psi_2) = \{\lambda \in \mathbb{C} : \psi_1 < Arg\,\lambda < \psi_2\}$. Consider the integral

$$K_\alpha(\lambda, z; \psi_1, \psi_2) = \frac{\lambda^{\rho-1}}{2\pi i} \int_T \frac{\exp\{p\lambda^\rho\}\,dp}{\alpha^{-1}(p) - z}, \tag{5.1.10}$$

where $\lambda \in L(\psi_1, \psi_2)$, $z \in \mathbb{C} \setminus \overline{I}$, $\alpha^{-1}$ is the inverse of the restriction of $\alpha$ to $\overline{I}$, and $T = T(\psi_1, \psi_2; c_1, c_2) := \partial\alpha(I)$ is the negatively oriented boundary of the domain $\alpha(I)$. In the integral (5.1.10), we change the variables $p = qe^{i2\pi\rho}$. Since $\alpha^{-1}(qe^{i2\pi\rho} = \alpha^{-1}(q)$ (see the remark to Corollary 4.26 in Ch. 4), and $l$ is a $2\pi$-periodic function, we deduce that $K_\alpha(\,\cdot\,; \psi_1, \psi_2) = K_\alpha(\,\cdot\,; \psi_1 + 2\pi, \psi_2 + 2\pi)$ for $c_i > |z|$, $i = 1, 2$. Therefore, after a suitable choice of the integration contour, the integral in (5.1.10) can be viewed as a function defined for $\lambda$ in an angle on the plane $\mathbb{C}$, rather than on the Riemann surface of the logarithm.

**Theorem 5.1.2.** *The integral $K_\alpha(\lambda, z; \psi_1, \psi_2)$ in* (5.1.10) *extends analytically to $\mathbb{C}^2$ up to the function $K_\alpha$ considered in Theorem* 5.1.1.

◀ Let $\theta = (\psi_1 + \psi_2)/2$, and let $\gamma = \{w \in \mathbb{C} \setminus \{0\} : Arg\,w = \theta\}$. By the uniqueness theorem for holomorphic functions, it suffices to prove Theorem 5.1.2 for $\lambda \in \gamma$. Using standard tools of complex analysis, we see that in (5.1.10) the contour $T$ can be replaced by the contour $T_\lambda = T_1 \cup T_\smile \cup T_2$, where

$$T_j = \{te^{i\varphi_j} : t \geqslant a^\rho/|\lambda|^\rho\}, \quad j = 1, 2;$$
$$T_\smile = \{a^\rho e^{i\varphi}/|\lambda|^\rho : \varphi_1 \leqslant \varphi \leqslant \varphi_2\}.$$

Here $\varphi_1 = -\rho\psi_2 - \pi/2 - \rho\varepsilon$ and $\varphi_2 = -\rho\psi_1 + \pi/2 + \rho\varepsilon$, where $0 < \varepsilon < (\psi_2 - \psi_1)/2$. The number $a > 0$ is taken so that $T_\lambda \subset \alpha(I)$,

$$\{\alpha^{-1}(p) : p \in T_\lambda\} \subset \{\zeta \in \mathbb{C} : |\zeta| > R\},$$

where $I = D_\alpha(\psi_1; c_1) \cup D_\alpha(\psi_2; c_2)$ (see (5.1.10)), and $R > 0$ is the number occurring in formula (5.1.4). Let $\Pi_\lambda = \Pi_1 \cup \Pi_\smile \cup \Pi_2$, where $\Pi_j = \{re^{i\theta_j} : r \geqslant a/|\lambda|\}$, $j = 1, 2$; $\theta_1 = -(\psi_2 + \varepsilon + \pi/2\rho)$; $\theta_2 = \varepsilon - \psi_1 + \pi/2\rho$; $\Pi_\smile = \{ae^{i\varphi}/|\lambda| : \theta_1 \leqslant \varphi \leqslant \theta_2\}$. The function $u$ in (5.0.3) associated with $\alpha$ has the property

$$\{u^{-1}(v) : v \in \Pi_\lambda\} \subset \{\zeta \in \mathbb{C} : |\zeta| > R\}, \quad R < a/|\lambda|,$$

because $\alpha^{-1}(v^\rho) \equiv u^{-1}(v)$, $v \in \Pi_\lambda$. Therefore (see (5.1.10), (5.1.1)),

$$K_\alpha(\lambda, z; \psi_1, \psi_2) = \frac{\rho\lambda^{\rho-1}}{2\pi i} \int\limits_{\Pi_\lambda} \frac{v^{\rho-1}\exp\{(v\lambda)^\rho\}\, dv}{u^{-1}(v) - z}, \quad |z| < R.$$

Now, $u^{-1}(\infty) = \infty$, and for the function $\varphi(\tau) = 1/[u^{-1}(\tau) - z]$ we have $\varphi(v) = \dfrac{1}{2\pi i}\displaystyle\int\limits_{\gamma_R} \frac{\varphi(\tau)\, d\tau}{v - \tau}$, where $\gamma_R = \{u(Re^{i\theta}) : |\theta| \leqslant \pi\}$. Interchanging the order of integration, we find:

$$K_\alpha(\lambda, z; \psi_1, \psi_2) = \frac{\rho\lambda^{\rho-1}}{(2\pi i)^2} \int\limits_{\gamma_R} \frac{d\tau}{u^{-1}(\tau) - z} \int\limits_{\Pi_\lambda} \frac{v^{\rho-1}\exp\{(v\lambda)^\rho\}\, dv}{v - \tau}. \tag{5.1.11}$$

Putting $\beta = \dfrac{\pi}{2\rho} + \dfrac{\psi_2 - \psi_1}{2} - \varepsilon$, we denote by $\gamma(a; \beta)$ the contour described in Theorem D (see Subsection 5.1.1). Substituting $w = v\lambda$ in the inner integral in (5.1.11), we obtain:

$$K_\alpha(\lambda, z; \psi_1, \psi_2) = \frac{\rho}{(2\pi i)^2} \int\limits_{\gamma_R} \frac{d\tau}{u^{-1}(\tau) - z} \int\limits_{\gamma(a;\beta)} \frac{w^{\rho-1}\exp\{w^\rho\}\, dw}{w - \tau\lambda}. \tag{5.1.12}$$

From formula (5.1.12) and Theorem D we deduce Theorem 5.1.2 ▶

After elementary transformations, formula (5.1.10) and Theorem C imply the following statement.

**Corollary 5.1.3.** *If* $\rho \in \mathbb{N}$, $l(\theta) = \operatorname{Re} p_0 e^{i\rho\theta}$ *for* $\theta \in \mathbb{R}$, $p_0 \in \mathbb{C}$, *and* $\alpha(z) = z^\rho + p_0$ (see Definition 3.1), *then*

$$K_\alpha(\lambda, z) = E_\rho(\lambda z)e^{p_0\lambda^\rho}, \quad (\lambda, z) \in \mathbb{C}^2.$$

**Remark.** Corollary 5.1.3 together with formula (5.1.6) yield a description of the function $K_\alpha$ associated with an arbitrary polynomial in $\mathfrak{P}_\rho(l)$ in the case where $l \in O_\rho$ (see Theorem 4.1.5).

**5.1.3. Asymptotic properties of the function $K_\alpha(\lambda, z)$.** Suppose $\rho > 1/2$, $l \in M_\rho \setminus O_\rho$, and $\alpha \in \mathfrak{P}_\rho(l)$. We start with examining the asymptotic behavior of $K_\alpha$ as $|z| \to \infty$.

Fix $\lambda \in \mathbb{C}$. If $\lambda = 0$, we have $K_\alpha(0; z) \equiv [\Gamma(1/\rho)]^{-1}$ (see Theorem 5.1.1). Let $\lambda \neq 0$. By analogy with formula (5.1.9), from (5.1.4) we find:

$$K_\alpha(\lambda, z) = E_\rho[\lambda u(z)]u'(z) + O(1/z), \quad z \to \infty. \tag{5.1.13}$$

Now we apply the asymptotic formula for $E_\rho$ with $\rho > 1/2$ (see Theorem C in 1°) and formula (5.1.13). Using (5.0.3) and the relation $\lim_{z\to\infty} u'(z) = 1$, we find the indicator (in $z$) of the function $K_\alpha(\lambda, z)$:

$$h_K(\varphi; \lambda) = \begin{cases} \operatorname{Re}\,(e^{i\varphi}\lambda)^\rho, & |\varphi + \arg\lambda| < \pi/2\rho \\ 0, & \pi/2\rho \leqslant |\varphi + \arg\lambda| \leqslant \pi \end{cases}; \tag{5.1.14}$$

moreover, the limit involved in the definition of the indicator exists. By Theorem 1.10.12, we conclude that the function $K_\alpha(\lambda, z)$ has completly regular growth in $z$.

We arrive at the following statement.

**Theorem 5.1.4.** *Suppose $\rho > 1/2$, $l \in M_\rho \setminus O_\rho$, and $\alpha \in \mathfrak{P}_\rho(l)$. The function $K_\alpha(\lambda, z)$ defined by formula* (5.1.4) *has order $\rho$ and completly regular growth in the variable $z$ for all $\lambda \in \mathbb{C} \setminus \{0\}$; the indicator is determined by formula* (5.1.14).

Now, we examine the asymptotics of $K_\alpha$ as $|\lambda| \to \infty$. We shall need the following formula (obtained by changing the variables $p = \alpha(\zeta)$ in the integral (5.1.10) with subsequent integration by parts):

$$K_\alpha(\lambda, z; \psi_1, \psi_2) = \frac{1}{2\pi i\lambda} \int_{\partial I} \frac{\exp\{\alpha(\zeta)\lambda^\rho\}d\zeta}{(\zeta - z)^2}. \tag{5.1.15}$$

Let $0 < \Delta < \pi/2\rho$, and let $\{\theta_i\}_0^k$ be the sequence of numbers occurring in the definition of $l$ (see Theorem 4.1.5). We fix arbitrarily a number $\theta \in [\theta_0, \theta_0 + 2\pi]$. Moreover, let $F$ be a compact subset of $\mathbb{C} \setminus \{\zeta_1, \dots, \zeta_k\}$, where $\{\zeta_i\}_1^k$ is the set of roots of the equation $\alpha'(\zeta) = 0$ (see Def. 4.3.1).

1. First, we consider the case where $\theta \in (\theta_{j-1}, \theta_j)$ for some $j \in \{1, 2, \dots, k\}$.

A. Suppose $\Delta \in (0, \pi/2\rho)$ is such that $[\theta - \Delta, \theta + \Delta] \subset (\theta_{j-1}, \theta_j)$. For $\delta \in (0, \Delta)$, we put $F_+ = F \cap [\mathcal{D}_\alpha(\theta - \delta) \cup \mathcal{D}_\alpha(\theta + \delta)]$. Next, we choose $c_1$ and $c_2$ such that $c_1 > \operatorname{Re}\, z e^{i\rho(\theta-\Delta)}$; $c_2 > \operatorname{Re}\, z e^{i\rho(\theta+\Delta)}$ $\forall\ z \in F_+$, and define $Y_\Delta = \partial D$, where

$$D = \mathcal{D}_\alpha[\theta - \Delta; l(\theta - \Delta)] \cup \mathcal{D}_\alpha[\theta + \Delta; l(\theta + \Delta)].$$

By (5.1.15), for $\lambda \in L(\theta - \Delta; \theta + \Delta)$, $z \in F_+$ we find:

$$K_\alpha(\lambda, z) = \lambda^{\rho-1}\alpha'(z)\exp\{\alpha(z)\lambda^\rho\} + N_\alpha(\lambda, z), \tag{5.1.16}$$

where

$$N_\alpha(\lambda, z) = \frac{1}{2\pi i\lambda} \int_{Y_\Delta} \frac{\exp\{\alpha(\zeta)\lambda^\rho\}}{(\zeta - z)^2}\, d\zeta.$$

The contour $Y_\Delta$ is saddle relative to the function $g(\zeta) = \alpha(\zeta)e^{i\rho\theta}$: this contour goes in the positive (counterclocckwise) direction from one valley to an adjacent valley; next, $Y_\Delta$ passes through the saddle point $\zeta_j$ such that $\alpha'(\zeta_j) = 0,\ \alpha(\zeta_j) = p_j$, and

$$\text{Re } g(\zeta) < \text{Re } g(\zeta_j) \quad \forall\, \zeta \in Y_\Delta \setminus \{z_j\}.$$

Let $k_j$ be the multiplicity of the root $\zeta_j$ of the equation $\alpha(\zeta) - p_j = 0$. Since $\alpha(\zeta)\lambda^\rho = g(\zeta)|\lambda|^\rho \exp\{i\rho(Arg\,\lambda - \theta)\}$, applying Theorem B for $m = 0$ and $k = k_j$ to the integral $N_\alpha(\lambda, z)$ with $\lambda \in L(\theta - \delta, \theta + \delta)$, we obtain

$$\begin{aligned} N_\alpha(\lambda, z) &= \frac{\exp\{\alpha(\zeta_j)\lambda^\rho\}}{2\pi i\lambda^{1+\rho/k_j}} \left\{ \frac{[1 - \exp(-i2\pi/k_j)]}{k_j(\zeta_j - z)^2 \cdot e^{-iq_j}} \times \right. \\ &\left. \times \Gamma\left(\frac{1}{k_j}\right) \left[\frac{k_j!}{\left|\alpha^{(k_j)}(\zeta_j)\right|}\right]^{1/k_j} + o(1) \right\} \end{aligned} \tag{5.1.17}$$

as $\lambda \to \infty$, where $\varphi_j$ is the angle between the positive direction of the $x$-axis and the descent direction from the saddle point $\zeta_j$ along the contour $Y_\Delta$, and where $z \in F_+$. The asymptotic estimate in (5.1.17) is uniform in $z \in F_+$ because the integration contour remains saddle (see Remark 2 to Theorem B).

B. If $z \in F \setminus F_+$, then for $\lambda \in (\theta - \varepsilon, \theta + \varepsilon)$ we have (see (5.1.16)

$$K_\alpha(\lambda, z) = K_\alpha(\lambda, z; \theta - \varepsilon, \theta + \varepsilon) = N_\alpha(\lambda, z), \tag{5.1.18}$$

and in the definition of $N_a$ the contour $Y_\Delta$ is replaced by the analogous contour $Y_\varepsilon$ corresponding to the parameter $\varepsilon < \delta < \Delta$. Again, the asymptotic expansion for $K_\alpha$ as $|\lambda| \to \infty$ is given by a formula like (5.1.17).

2. Now suppose $\theta \in \{\theta_i\}_1^k$, say, $\theta = \theta_j$; we put

$$F_+ = F \cap [D_\alpha(\theta_j - \delta) \cup D_\alpha(\theta_j) \cup D_\alpha(\theta_j + \delta)];$$

and choose $c_1$ and $c_2$ such that

$$c_1 > \text{Re } z e^{i\rho(\theta_j - \Delta)}; \quad c_2 > \text{Re } z \cdot \exp[i\rho(\theta_j + \Delta)] \quad \forall\, z \in F_+.$$

Here $0 < \delta < \Delta < \pi/2\rho$ is the contour resembling the outline of the letter W and consisting of four links:

$$T_1 = \{p_j + t\exp\{i(-\rho\theta + \rho\Delta + \pi/2)\} : t \geqslant 0\},$$

$$T_2 = \{p_{j+1} + t\exp\{-i(\rho\theta + \rho\Delta + \pi/2)\} : t \geqslant 0\},$$

and the segments $[p_j, p_0]$, $[p_0, p_{j+1}]$, where $p_0$ is the intersection point of the straight lines $\{p \in \mathbb{C} : \text{Re } (p - p_j)\exp\{i\rho(\theta + \delta)\} = 0\}$ and $\{p \in \mathbb{C} : 0 = \text{Re } (p - p_{j+1}) \times \exp\{i\rho(\theta - \delta)\}\}$, and $p_j, p_{j+1}$ are the complex numbers occurring in the definition of the function $l$. The contour $T_\Delta$ is oriented from $T_2$ towards $T_1$. The parameter $\Delta$ is assumed to be so small that the branch $\alpha^{-1}$ of the inverse to $\alpha$ dealt with in the definition of the function $K_\alpha(\lambda, z; \theta - \Delta, \theta + \Delta)$ admits analytic continuation into the domain $E_\Delta$ bounded by the contour $Y_\Delta$. As above, we verify (5.1.16) with $z \in F_+$ and $\lambda \in L(\theta - \Delta, \theta + \Delta)$. The contour $\alpha^{-1}(T_\Delta)$ is saddle and passes through two saddle points $\zeta_j, \zeta_{j+1}$ of the function $g(\zeta) = \alpha(\zeta)e^{i\rho\theta}$ ($\alpha(\zeta_i) = p_i$ and $\alpha'(\zeta_i) = 0$ for $i = j-1, j$). These points give two contributions into the asymptotic expansion of $N_\alpha(\lambda, z)$ as $|\lambda| \to \infty$ (see (5.1.17)) corresponding to $\zeta_j, \zeta_{j+1}$. If $z \in F \setminus F_+$ and $\lambda \in L(\theta - \varepsilon, \theta + \varepsilon)$ with $\varepsilon < \delta < \Delta$, then the asymptotic expansion for $K_\alpha$ as $\lambda \to \infty$ is similar to that treated above (see (5.1.18)).

The expansions obtained are valid for every fixed $z \neq \zeta_j$ in case 1 and for every fixed $z \neq \zeta_j, \zeta_{j+1}$ in case 2. If, for instance, $z = \zeta_j$ in case 1, then one of the factors in the integrand in (5.1.10) admits the following representation after changing the variables $p = \alpha(\zeta)$:

$$\alpha'(\zeta)(\zeta - \zeta_j)^{-1} = (\zeta - \zeta_j)^{k_j - 2}\varphi(\zeta),$$

where $\varphi$ is a function holomorphic near $\zeta_j$, and $\varphi(\zeta_j) \neq 0$. Therefore, using Theorems 5.1.2 and B, we find a constant $A_j \neq 0$ such that

$$K_\alpha(\lambda, \zeta_j) = \lambda^{-1+\rho/k_j} \exp\{p_j \lambda^\rho\}[A_j + o(1)], \ \lambda \in L(\theta - \varepsilon, \theta + \varepsilon),$$

as $|\lambda| \to \infty$. In case 2, the expansion of $K_\alpha$ is modified similarly, for $z = \zeta_{j+1}$ or $z = \zeta_j$, So, if $K_\alpha^+(\lambda, z)$ is the main term of the asymptotic expansions deduced above for the function $K_\alpha(\lambda, z)$ as $|\lambda| \to \infty$, then we can find $\varepsilon > 0$ such that for every $\theta \in \mathbb{R}$, $z \in \mathbb{C}$, and $\lambda \in L(\theta - \varepsilon, \theta + \varepsilon)$ we have:

$$\lim_{|\lambda| \to \infty} \left[K_\alpha^+(\lambda, z)\right]^{-1} \cdot K_\alpha(\lambda, z) = 1,$$

and the convergence is uniform in the variable $z$ on every compact set $F \subset \mathbb{C} \setminus \{\zeta_i\}_1^k$. We arrive at the following statement (see Def. 1.10.11).

**Theorem 5.1.5.** *For every fixed $z \in \mathbb{C}$ the function $K_\alpha(\lambda, z)$ has completly regular growth in the variable $\lambda$ and its indicator at the order $\rho$ coincides with the function $r_\alpha(\theta; z) \in H_\rho(l)$ (see (4.4.13)), and, as $t \to \infty$, we have $t^{-\rho} \times \ln |K_\alpha(te^{i\theta}; z)| \to r_\alpha(\theta; z)$ uniformly with respect to $\theta \in \mathbb{R}$ and $z \in F$, where $F$ is an arbitrary compact subset of $\mathbb{C} \setminus \{\zeta_i\}_1^k$.*

## 5.2. The analog of Borel transformation and realization of the spaces $[\rho, h(\theta))$, $[\rho, h(\theta)]$

**5.2.1. Generalized Borel's transformation and associated integral representations.** For $\rho > 1/2$ and $h \in P_\rho$, let [Re $h(\theta)$) and $[\rho, h(\theta)]$ be spaces of entire functions satisfying

$$\overline{\lim_{t\to\infty}}\, t^{-\rho}\ln\left|f\left(te^{i\theta}\right)\right| < h(\theta) \quad \forall\,\theta\in\mathbb{R}$$

(respectively, $\leqslant h(\theta)$). In the first case it is assumed that $\mathfrak{M}(h) := \{g \in P_\rho : g < h\} \neq \varnothing$, and in the second case that $h(\theta) < \infty\ \forall\ \theta \in \mathbb{R}$. We can find a function $l \in M_\rho$ such that $l \leqslant h$, and if $\mathfrak{M}(h) \neq \varnothing$, then also $l < h$ (see Theorem 4.1.9). In $[\rho, h(\theta))$ we introduce the topology of inductive limit of the sequence $\{\mathcal{A}_k\}_1^\infty$ of $B$-spaces with the norms

$$\|f\|_k = \sup_{z\in\mathbb{C}} |f(z)|\exp\{-h_k(\arg z)|z|^\rho\},$$

where (see Property 4.4.21, (5.0.2))

$$\{h_k\}_1^\infty \subset P_\rho(l), \quad h_k < h_{k+1},\ k = 1,2,\ldots,\ \lim_{k\to\infty} h_k = h. \tag{5.2.1}$$

In $[\rho, h(\theta)]$ we introduce the topology of projective limit of the sequence of $B$-spaces with the norms

$$\|f_k\| = \sup_{z\in\mathbb{C}} |f(z)|\exp\{-|z|^\rho[h(\arg z) + 1/k]\},\ k = 1,2,\ldots.$$

Let $[\rho,\infty) = [\rho, h_0(\theta))$, where $h_0(\theta) \equiv \infty$, and let $H_0(\infty) = \{g\}$ be the space of functions holomorphic in the vicinity of the point at infinity in $\overline{\mathbb{C}}$ and vanishing at infinity. In Ch. 5 (see (5.0.1)) we considered a certain operator $B_\alpha : [\rho,\infty) \to H_0(\infty)$. The convergence of the series (5.0.2) follows from the relationship between the order and type of an entire function (5.0.1) and its Taylor coefficients (see Theorem 1.10.3).

We establish important integral formulas relating $f$ and $\hat{f}$.

**Proposition 5.2.1.** *Let $l \in M_\rho$, and let $K \in P_\rho(l)$* (see Ch. 4). *We have the following integral formulas:*

$$f(\zeta) = \rho\int_0^{\infty(\operatorname{Arg}z=\theta)} \exp\{-\alpha(\zeta)z^\rho\}f(z)\,dz, \quad z \in \mathcal{D}_a(\theta;k)\ , \tag{5.2.2}$$

*where $\mathcal{D}_a(\theta;k) = \{\zeta \subset \mathcal{D}_\alpha(\theta) :\ \operatorname{Re}\alpha(\zeta)\exp\{i\rho\theta\} > k(\theta)\}$, $\theta \in (-\pi,\pi]$, $k$ is a finite function in $P_\rho(l)$, and $f \in [\rho, k(\theta)]$;*

$$f(\lambda) - \frac{1}{2\pi i}\int_\gamma K_\alpha(\lambda,z)\hat{f}(z)\,dz, \quad \lambda \in \mathbb{C}\ , \tag{5.2.3}$$

*where $K_\alpha(\lambda,z)$ is the function examined in* 5.1.1, *and $\gamma$ is a closed rectifiable Jordan contour encompassing the singularities of the function $\hat{f} \in H_0(\infty)$ and oriented counterclockwise.*

◀ 1. The formula (5.2.2) is proved by a classical procedure (see Proposition 1.10.5).

2. Taking $\lambda \neq 0$, we denote $\arg \lambda = \theta$. Let $\mathcal{D}$ be an unbounded domain in $\mathbb{C}$ such that $\partial \mathcal{D} = \gamma$, and let $C_R$ be a circle of radius $R$ as in formula (5.1.4) and such that $C_R \subset \mathcal{D}$. By elementary arguments, we justify the interchanging of the order of integration:

$$I := \frac{1}{2\pi i} \int\limits_{\gamma} K_\alpha(\lambda, z) \hat{f}(z)\, dz = \frac{1}{(2\pi i)^2} \int\limits_{C_R} E_\rho[\lambda u(\zeta)] u(\zeta)\, d\zeta \int\limits_{\gamma} \frac{\hat{f}(z)\, dz}{\zeta - z}\,. \quad (5.2.4)$$

Since $|z\hat{f}(z)| < const$ for $|z| > R_0\, (f(\infty) = 0)$, by the Cauchy integral formula we find

$$\frac{1}{2\pi i} \int\limits_{\gamma} \frac{\hat{f}(z)\, dz}{\zeta - z} = \hat{f}(\zeta), \quad \zeta \in Y_\lambda.$$

Applying (5.2.4) and changing the variables $q = u(\zeta)$, we obtain

$$I = \frac{1}{2\pi i} \int\limits_{\gamma_1} E_\rho(\lambda q) \hat{f}[u^{-1}(q)]\, dq,$$

where $\gamma_1 = \{q = u(\zeta) : \zeta \in \gamma\}$. Finally, using the series expansions (see (5.0.2), (5.1.5), and (5.0.1)) for $\hat{f}, E_\rho$ (which converge uniformly on the contour $\gamma_1$), we conclude that $I = f(\lambda)$. So, (5.2.2) is proved for $\lambda \neq 0$. If $\lambda = 0$, (5.2.2) follows from the continuity of the functions $f$ and $K_\alpha(\lambda, z)$ at the point $\lambda = 0$ (see Theorem 5.1.1) ▶

**5.2.2. The spaces $[\rho, h(\theta))$ and $[\rho, h(\theta)]$, and analytic functionals on $(\rho, \alpha)$-convex sets.** There are several important consequences of (5.2.2) and (5.2.3). For $h \in P_\rho(l)$, let $\Omega_h$ ($X_h$) be the $(\rho, \alpha)$-convex domain (a compact set) whose $(\rho, \alpha)$-support function is $h$ (see (4.4.14), Def. 4.4.15). The symbol $H_0(\overline{\mathbb{C}}) \setminus \Omega_h)$ $(H_0(\overline{\mathbb{C}}) \setminus X_h))$ denotes the space of functions holomorphic in the vicinity of the compact set $\overline{\mathbb{C}} \setminus \Omega_h$ on the Riemann sphere $\overline{\mathbb{C}}$ (respectively, holomorphic in the domain $\overline{\mathbb{C}} \setminus \Omega_h$) and equal to 0 at the point $\infty$. In $H(\overline{\mathbb{C}} \setminus \Omega_h)$ we introduce the topology of inductive limit generated by the sequence $\{\mathcal{F}_k\}_1^\infty$ of $B$-spaces of functions holomorphic in $H(\overline{\mathbb{C}} \setminus \overline{\Omega(h_k)})$, continuous in $F_k := \overline{\mathbb{C}} \setminus \Omega(h_k)$, $k \in \mathbb{N}$, and equal to 0 at the point $\infty$. The norms in these $B$-spaces are given by

$$\|g\|_k = \max\{|g(\zeta)| : \zeta \in F_k\}, \quad k \in \mathbb{N},$$

where $\{h_k\}_1^\infty$ is the sequence occurring in (5.2.1). In $H_0(\mathbb{C} \setminus X_h)$ we introduce the topology of projective limit of the sequence of $B$-spaces with the norms

$$\|g\|_k = \max\{|g(\zeta)| : \zeta = \overline{\mathbb{C}} \setminus \Omega(h + 1/k)\}, \quad k \in \mathbb{N}.$$

**Theorem 5.2.2.** *The linear operator $B_\alpha$ is a topological isomorphism of the following spaces:*

$$B_\alpha : [\rho, h(\theta)) \to H_0(\overline{\mathbb{C}} \setminus \Omega_h); \ B_\alpha : [\rho, h(\theta)] \to H_0(\overline{\mathbb{C}} \setminus X_h) \ . \tag{5.2.5}$$

◀ Let $g \in P_\rho(l)$ (see (4.3.18)). From Proposition 4.4.23 and formula (5.2.2) we deduce that $B_\alpha([\rho, g(\theta)]) \subset H_0(\overline{\mathbb{C}} \setminus X_g)$. Each function $\hat{f} \in H_0(\infty)$ can be expanded in a series (5.0.2), because $u' \neq 0$ in the exterior of some disk in $\mathbb{C}$. Let $g$ be a finite function in $P_\rho(l)$, and let $l < g$. We fix an arbitrary small number $\varepsilon > 0$. If $\hat{f} \in H_0(\overline{\mathbb{C}} \setminus X_g)$, then, taking $\gamma = \partial\Omega(g + \varepsilon)$ in (5.2.3), from Theorem 5.1.5 we conclude that $f := B_\alpha^{-1}\hat{f} \in [\rho, g(\theta)]$. We have taken into account the following estimate, which is uniform with respect to $\theta \in \mathbb{R}$:

$$\sup_{z\in\gamma} |K_\alpha(te^{i\theta}; z)| < C_\varepsilon \exp\left\{t^\rho \left[\varepsilon + \sup_{z\in\gamma} r_\alpha(\theta; z)\right]\right\} = C_\varepsilon \exp\{t^\rho[g(\theta) + 2\varepsilon]\}; \tag{5.2.6}$$

this estimate is a consequence of Theorem 5.1.5. Here $C_\varepsilon > 0$ is a constant. Consequently, $B_\alpha([\rho, g(\theta)]) = H_0(\overline{\mathbb{C}}\setminus X_g)$. Since $[\rho, h(\theta)] = \bigcup\limits_{i=1}^{\infty} [\rho, h_{k+i}(\theta)] \ \forall\, k \in \mathbb{N}$, we have $B_\alpha([\rho, h(\theta))) = H_0(\overline{\mathbb{C}} \setminus \Omega_g)$. So, $B_\alpha$ is an algebraic isomorphism of the spaces indicated in (5.2.5).

The fact that $B_\alpha$ is also a topological isomorphism is verified as in the proof of Theorem 1.10.8. For instance, the mapping $B_\alpha : \mathcal{A}_k \to \mathcal{F}_{k+1}$ (see 5.2.1)) is continuous because

$$\begin{aligned} &||\hat{f}||_{k+1} \leqslant \Gamma(1/\rho) \cdot \rho_k^{-1/\rho} ||f||_k, \ k \in \mathbb{N}, \\ &\rho_k = \inf\{[h_{k+1}(\theta) - h_k(\theta)] : \theta \in \mathbb{R}\} \ . \end{aligned} \tag{5.2.7}$$

The latter estimate follows from (5.2.2) and the inequalities

$$\begin{aligned} ||\hat{f}||_{k+1} &\leqslant \rho||f||_k \cdot \max\{H(\zeta, \theta) : \ \zeta \in \overline{\mathcal{D}_\alpha(\theta; h_{k+1})}; \ \theta \in \mathbb{R}\} \\ &\leqslant \rho||f||_k \int_0^\infty \exp\{-\rho_k r^\rho\}\, dr, \end{aligned}$$

where

$$H(\zeta, \theta) - \int_0^\infty |f(re^{i\theta})| \exp\{r^\rho[h_k(\theta) - \operatorname{Re} \alpha(\zeta)e^{i\rho\theta}]\}\, dr \ ▶$$

**Theorem 5.2.3.** *The generalized Laplace transformation*

$$\tilde{\mu}(\lambda) = \mu_z[K_\alpha(\lambda, z)], \ \mu \in H^*(\Omega_h) \ or \ H^*(X_h) \tag{5.2.8}$$

*is a topological isomorphism of $H^*(\Omega_h)$ onto $[\rho, h(\theta))$, and also of $H^*(X_h)$ onto $[\rho, h(\theta)]$, where $H^*(D)$ is the space conjugate to $H(D)$ and equipped with the strong topology* (see (4.4.14)).

◀ This follows from Theorem 5.2.2 and Theorem 1.10.7, because $\tilde{\mu} = B_\alpha^{-1}\hat{f}$ , where $\mu$ is generated by a function $\hat{f} \in H_0(\overline{\mathbb{C}} \setminus X_h)$ (see (5.2.3)) ▶

So, the generalized Borel transformation considered in (5.0.2) is generated by the function $K_\alpha(\lambda, z)$.

**Proposition 5.2.4.** *The space $H^*(X_h)$ is isomorphic to the subclass $D = \{\mu\}$ of the class $H^*(\mathbb{C})$ of analytic functionals whose determining set is $X_h$* (see Def. 3.2.2).

◀ Consider the mapping

$$A : H^*(X_h) \to H^*(\mathbb{C}); \quad A(\mu) = \mu|_{H(\mathbb{C})}.$$

We have $A[H^*(H_h)] \subset D$. Estimate (3.2.2) is a consequence of the representation of the functional $\mu \in H^*(X_h)$ in the following form (see Theorem 1.10.7):

$$\mu(\varphi) = \frac{1}{2\pi i} \int_\Gamma \varphi(z)\hat{f}(z)\, dz, \quad \varphi \in H(X_h), \tag{5.2.9}$$

where $\hat{f} \in H_0(\overline{\mathbb{C}} \setminus X_h)$, $\Gamma = \partial\omega$, $\omega$ is the domain containing $X_h$ with rectifiable Jordan boundary, and $\varphi \in H(\bar{\omega})$.

Conversely, suppose $\mu \in D$ and $\omega = \Omega(h + \varepsilon)$, $\varepsilon > 0$. The maximum principle for the function $|K_\alpha(\lambda, z)|$ (subharmonic in the variable $z$) and inequality (2.6) imply:

$$|K_\alpha(\lambda, z)| < E_\varepsilon \exp\{|\lambda|^\rho (h(\theta) + 2\varepsilon)\} \tag{5.2.10}$$

for all $\lambda \in \mathbb{C}$ and for all $z \in \omega$ . Substituting the function $\varphi_\lambda(z) = K_\alpha(\lambda, z)$ in formula (3.2.2) and using (5.2.10), we see that $\tilde{\mu} \in [\rho, h(\theta)]$ (see (5.2.8)). Then $\hat{f} = B_\alpha \tilde{\mu} \in H_0(\overline{\mathbb{C}} \setminus X_h)$ represents $\mu$ in the form (5.2.9). Since $\varepsilon > 0$ is arbitrarily small, $\mu \in H^*(X_h)$ ▶

From the foregoing and the arguments similar to the proof of Theorem 3.2.3 we obtain the following plane analog of the Polya theorem on the relationship between the indicator and the conjugate diagram of an entire function of exponential type.

**Theorem 5.2.5.** *Suppose $\rho > 0$, $l \in M_\rho$, and $\alpha \in \mathfrak{P}_\rho(l)$. Let $h$ be a finite function in $P_\rho(l)$, and let $X_h$ be a $(\rho, \alpha)$-convex compact set whose $(\rho, \alpha)$-support function is $h$. The following statements are equivalent:*

1) *the entire function $f$ is in $[\rho, \infty)$ and its $(\rho, \alpha)$-indicator diagram* (see Def. 4.4.2) *is equal to $X_h$;*

2) *the function $\hat{f} = B_\alpha f$ is in $H_0(\infty)$ and the smallest $(\rho, \alpha)$-convex compact set $K$ such that $\hat{f}$ (the conjugate diagram of $f$) extends analytically to the complement of $K$ coincides with $X_h$;*

3) *there exists an analytic functional $\mu$ such that $\mu[K_\alpha(\cdot, w)] = f$, and $X_h$ is a unique $(\rho, \alpha)$-convex support of $\mu$, i.e., $X_h$ belongs to any determining set of $\mu$ that is $(\rho, \alpha)$-convex and compact* (see 3.2.7).

We state another consequence of Theorem 5.2.2 and of Theorem 1.10.7.

**Theorem 5.2.6.** *Under the conditions and in the notation of Theorem 5.2.2, let* $[\rho, h(\theta))^*$ *and* $[\rho, h(\theta)]^*$ *be the spaces conjugate to* $[\rho, h(\theta))$ *and* $[\rho, h(\theta)]$, *respectively, and equipped with the strong topology. The adjoint* $B_\alpha^*$ *of* $B_\alpha$ *is a topological isomorphism of the following spaces:*

$$B_a^* : H(\Omega_h) \to [\rho, h(\theta))^*; \quad B_a^* : H(X_h) \to [\rho, h(\theta)]^*.$$

*Moreover,*

$$\langle \varphi, f \rangle = \langle F, \hat{f} \rangle = \frac{1}{2\pi i} \int\limits_T F(z) \hat{f}(z)\, dz, \tag{5.2.11}$$

*where* $F = (B_\alpha^*)^{-1} \varphi$; $\hat{f} = B_\alpha f$; $f$ *is in* $[\rho, h(\theta))$ *or in* $[\rho, h(\theta)]$; $\varphi$ *is in* $[\rho, h(\theta))^*$ *or in* $[\rho, h(\theta)]^*$; *and* $T$ *is a rectifiable Jordan contour oriented counterclockwise and encompassing the set of singularities of the function* $\hat{f}$. *The function* $K_\alpha$ (see (5.1.4)) *is related to the operators* $B_\alpha$ *and* $B_\alpha^*$ *as follows:*

$$K_\alpha(\lambda, z) = \left[ (B_\alpha^*)^{-1} \delta_\lambda \right](z) = \left[ B_\alpha^{-1}(1/(\zeta - z)) \right]_\zeta (\lambda), \tag{5.2.12}$$

*where* $\delta_\lambda$ *is the delta-function on* $\mathbb{C}$ *supported at* $\lambda$.

## 5.3. Applications of the analog of the Polya theorem

**5.3.1. Analytic realization of the generalized differentiation operator.** Suppose $h \in P_\rho$, and $\mathfrak{M}_\rho(h) \neq \varnothing$. Let $\Lambda : [\rho, h(\theta)) \to [\rho, h(\theta))$ be the operator of multiplication by the independent variable and $\Lambda^* : [\rho, h(\theta))^* \to [\rho, h(\theta))^*$ the operator adjoint to $\Lambda$. This adjoint is called the generalized differentiation operator because for $\rho = 1$ it has an analytic realization in the form of the differentiation operator, and for $h(\theta) \geqslant 0 \;\forall\, \theta \in \mathbb{R}$ it is representable as the Gelfond-Leont'ev operator. By Theorem 5.2.6, in general the role of such a realization may be played by the continuous operator $D^* = (B_\alpha^*)^{-1} \Lambda^* B_\alpha^* : H(\Omega_h) \to H(\Omega_h)$ adjoint to the operator $D = B_\alpha \Lambda B_\alpha^{-1} : H_0(C\Omega_h) \to H_0(C\Omega_h)$, where $H_0(C\Omega_h)$ is the space of functions holomorphic in the vicinity of the complement $C\Omega_h = \overline{\mathbb{C}} \setminus \Omega_h$ and equal to 0 at the point $\infty$.

**Theorem 5.3.1.** *Let* $\{h_n\}$ *be a sequence of functions in* $P_\rho(l)$ *satisfying* (5.2.1), *and let* $g_n = \dfrac{1}{2}(h_n + h_{n+1})$. *Take* $F \in H(\Omega_h)$, *and put* $\Gamma_n = \partial\Omega(g_n)$. *Then, in the previous notation, for all* $z \in \Omega(h_n)$ *we have*

$$(D^* F)(z) = \frac{1}{2\pi i} \int\limits_{\Gamma_n} F(\zeta) [B_\alpha(\lambda K_\alpha(\lambda, z))]_\lambda(\zeta)\, d\zeta, \tag{5.3.1}$$

*where* $B_\alpha$ (see (5.0.2)) *operates in the variable* $\lambda$, *and* $K_\alpha(\lambda, z)$ *is the function discussed in Section* 5.1 ($\Omega(k) := \Omega_k$).

◀ 1. We fix an integer $n > 0$ arbitrarily. Let $f$ be any element of $[r, h_n(\theta))$. By Proposition 5.2.1, $B_\alpha f$ is a function holomorphic outside a $(\rho, \alpha)$-convex compact set the $(\rho, \alpha)$-support function of which is less than $h$. We choose $\varepsilon > 0$ satisfying $6\varepsilon < \min_{\theta \in \mathbb{R}}[h_{n+1}(\theta) - h_n(\theta)]$, and put $T_n = \partial\Omega(h_n)$ . By Theorem 5.2.5, the curve $T_n$ is in the domain of holomophy of the functions $B_\alpha f$ and $B_\alpha \Lambda f$. Furthermore, $\Gamma_n \subset \mathbb{C} \setminus X(h_n + 3\varepsilon)$. By Property 4.4.24, , since $G_n$ is compact, in $\mathbb{R}$ there exist two sequences $\{\varphi_s\}_0^N$ ($\varphi_N = \varphi_0 + 2\pi$), and $\{\zeta_s\}_0^N$ ($\zeta_n = \zeta_0$), ordered in accordance with the orientation of the curve $\Gamma_n$ and such that for the closed segments $\Delta_s$ on $\Gamma_n$ with ends $\zeta_{s-1}, \zeta_s$ we have

$$\Delta_s \subset \mathcal{D}_\alpha(\varphi_s, h_n + 3\varepsilon), \quad s = 1, \ldots, N. \tag{5.3.2}$$

If $\zeta \in \Delta_s$ (see (5.2.2)), then

$$[B_\alpha \Lambda f](\zeta) = \frac{\rho}{2\pi i} \int_0^{\infty(\arg\lambda = \varphi_s)} \lambda \exp\{-\alpha(\zeta)\lambda^\rho\}\, d\lambda \int_{T_n} K_\alpha(\lambda, z) \hat{f}(z)\, dz. \tag{5.3.3}$$

But (see (5.2.6), (5.3.2))

$$|\lambda K_\alpha(\lambda, z) \exp\{-\alpha(\zeta)\lambda^\rho\}| < \exp\{-\varepsilon|\lambda|^\rho\} \quad \forall\, z \in T_n.$$

Therefore, the integral

$$\int_0^{\infty(\arg\lambda = \varphi_s)} \hat{f}(z) \exp\{-\alpha(\zeta)\lambda^\rho\} \lambda K_\alpha(\lambda, z)\, d\lambda$$

converges uniformly in $z$ on $T_n$ and, consequently, in (5.3.3) we can change the order of integration for $\zeta \in \Delta_s$:

$$[B_\alpha \Lambda f](\zeta) = \frac{1}{2\pi i} \int_{T_n} [B_\alpha(\lambda K_\alpha(\lambda, z))]_\lambda(\zeta) \cdot [B_\alpha f](z)\, dz.$$

2. Let $\langle F, \cdot \rangle$ be the element of $H_0^*(C\Omega_h)$ generated by $F$ (see (5.2.1)). By the above, changing the order of integration, we find:

$$\langle F, B_\alpha \Lambda f \rangle = \frac{1}{2\pi i} \int_{T_n} V_n(z) \cdot [B_\alpha f](z)\, dz\ , \tag{5.3.4}$$

where

$$V_n(z) = \frac{1}{2\pi i} \int_{\Gamma_n} F(\zeta) \cdot B_\alpha[\lambda \cdot K_\alpha(\lambda, z)]_\lambda(\zeta)\, d\zeta\ .$$

The integral

$$\int_0^{\infty(\arg\lambda=\varphi_s)} K_\alpha(\lambda,z)\exp\{-\alpha(\zeta)\lambda^\rho\}\,d\lambda \quad \text{at} \quad \zeta\in\Delta_s,\ s=1,\ldots,N,$$

converges uniformly not only on $T_n$, but on every compact subset of $\Omega(h_n)$. This can be seen from (5.2.6) by application of the maximum principle to the subharmonic function $|K_\alpha(\lambda,z)|$ in the variable $z$ for every fixed $\lambda$. Therefore, for $\zeta\in\Gamma_n$ the function $\Phi(\zeta,z):=B_\alpha[\lambda K_\alpha(\lambda,z)_\lambda](\zeta)$ (and, with it, the function $V_n$) is holomorphic in $z$ in $\Omega(h_n)$ $\forall\ n\in\mathbb{N}$. From the Cauchy principal theorem we deduce that the function $V(z):=V_n(z)$, $z\in\Omega(h_n)$, is holomorphic in $\Omega_h$. On the other hand,

$$\langle F, B_\alpha\Lambda f\rangle=\langle F, D(B_\alpha f)\rangle=\langle D^*F, B_\alpha f\rangle\ ,$$

i.e., the functions $V$ and $D^*F$ holomorphic in $\Omega_h$ determine one and the same element of $H_0^*(C\Omega_h)$. Together with Theorem 1.10.7, this implies Theorem 5.3.1 ▶

**Remark.** For $\rho=1$ and $\alpha(\zeta)=\zeta$ we have

$$K_\alpha(\lambda,z)=e^{\lambda z};\ [B_\alpha(\lambda e^{\lambda z})]_\lambda(\zeta)=(z-\zeta)^{-2};$$

$D^*F=F'$ is the classical differentiation operator. For $\rho>0$, $h>0$, and $\alpha(\zeta)=\zeta^\rho$ we have: $K_\alpha(\lambda,z)\equiv E_\rho(\lambda z)$, $(\lambda,z)\in\mathbb{C}^2$, is the Mittag - Leffler function;

$$[B_\alpha(\lambda K_\alpha(\lambda,z))]_\lambda(\zeta)=\zeta^{-2}\sum_{k=1}^{\infty}\frac{\Gamma((k+1)/\rho)}{\Gamma(k/\rho)}\left(\frac{z}{\zeta}\right)^{k-1};$$

$$(D^*F)(z)=\sum_{k=1}^{\infty}c_k\frac{\Gamma((k+1)/\rho)}{\Gamma(k/\rho)}z^{k-1}$$

is the generalized derivative (generated by $E_\rho(u)$) of the function $F(\zeta)=\sum_{k=1}^{\infty}c_k\zeta^k\in H(\Omega_h)$.

**5.3.2. Analytic realization of the generalized convolution operator, and of operator commuting with the generalized differentiation operator.** We fix an entire function $\varphi$ of normal type and of order $\rho$ with indicator $h_\varphi$, and introduce the operator $\Phi:[\rho,h(\theta))\to[\rho,h(\theta)+h_\varphi(\theta))$ of multiplication by $\varphi$. The adjoint operator $\Phi^*$ maps $[\rho,h(\theta)+h_\varphi(\theta))^*$ continuously onto $[\rho,h(\theta))^*$. It is called a generalized convolution operator, because for $\rho=1$ it coincides with the convolution operator up to the isomorphism $\Phi^*$.

In accordance with Subsection 4.4.3, we choose $l_i \in M_\rho$, $\alpha_i \in \mathfrak{P}_\rho(l_i)$ $(i = 1,2,3)$ such that $l_3 \leqslant l_1 + l_2$ and $h \in Q_\rho(l_1)$, $h_\varphi \in P_\rho(l_2)$. Let also $\alpha_i \in \mathfrak{P}_\rho(L_i)$, $i = 1,2,3$. For an analytic model of the operator $\Phi^*$, we may take the continuous operator

$$T^* = \left(B^*_{\alpha_1}\right)^{-1} \cdot \Phi^* \cdot B^*_{\alpha_3} : H(\Omega(h + h_\varphi)) \to H(\Omega_h),$$

adjoint to the operator

$$T := B_{\alpha_3} \cdot \Phi \cdot (B_{\alpha_1})^{-1} : H_0(C\Omega_h) \to H_0(C\Omega(h + h_\varphi)),$$

see (5.0.2).

**Theorem 5.3.2.** *Let $\{h_n\}$ be a sequence of functions satisfying* (5.2.1). *We put $g_n = \frac{1}{2}(h_n + h_{n+1})$ and $\Gamma_n = \partial\Omega(h_\varphi + g_n; \alpha_3)$. Then for every integer $n > 0$ and $z \in \Omega(h_n; \alpha_1)$ we have*

$$(T^* F)(z) = \frac{1}{2\pi i} \int\limits_{\gamma_n} [U_z(F)](v) \cdot [B_{\alpha_2}\varphi](v)\, dv, \tag{5.3.5}$$

*where $F \in H[\Omega(h + h_\varphi; \alpha_3)]$;*

$$[U_z(F)](v) = \frac{1}{2\pi i} \int\limits_{\Gamma_n} F(\zeta) \cdot B_{\alpha_3} \left[K_{\alpha_1}(\lambda, z) \cdot K_{\alpha_2}(\lambda, v)\right]_\lambda (\zeta)\, d\zeta; \tag{5.3.6}$$

*$\gamma_n$ is the oriented boundary of the $(\rho, \alpha_2)$-convex domain $G_n$ containing the $(r, \alpha_2)$-convex compact set $X = X(h_\varphi)$* (see (1.4.14)) *and so close to $X$ that the function $U_z(F)$ is holomorphic in the vicinity of $\overline{G_n}$ for all $z \in \Omega(h_n; \alpha_1)$.*

The proof is similar to that of Theorem 5.3.1.

**Corollary 5.3.5.** *For every $\lambda$ in $\mathbb{C}$ we have*

$$T^* \left[K_{\alpha_3}(\lambda, \zeta)\right]_\zeta (z) = \varphi(\lambda) K_{\alpha_1}(\lambda, z), \quad z \in \mathbb{C}. \tag{5.3.7}$$

◀ Formulas (5.3.6) and (5.2.3) imply the identity

$$U_z \left[K_{\alpha_3}(\lambda, \zeta)\right]_\zeta (v) = K_{\alpha_1}(\lambda, z) \cdot K_{\alpha_2}(\lambda, v), \quad v \in \gamma_n, \quad z \in \Omega(h_n; \alpha_1).$$

Then (5.3.7) is a consequence of (5.3.5) and (5.2.3) ▶

Theorem 5.3.2 acquires the simplest form if $h_\varphi \geqslant 0$, i.e., $l_2 = 0$, $l_1 = l_3 = l$, $\alpha_1 = \alpha_3 = \alpha \in \mathfrak{P}_\rho(l)$, $\alpha_2(\zeta) = \zeta^\rho$, $K_{\alpha_2}(\lambda, v) = E_\rho(\lambda v)$ is the Mittag-Leffler function.

By the well-known result of V.A. Tkachenko, in the above notation the operator $\Phi^* : [\rho, h(\theta) + h_\varphi(\theta))^* \to [\rho, h(\theta))^*$ is an epimorphism if $\varphi$ is an entire function of completely regular growth. From Theorem 5.3.2 we deduce the following statement.

**Theorem 5.3.4.** *Under the conditions of Theorem 5.3.2, if $\varphi$ is of completely regular growth, then the generalized convolution operator $T^*$ : $H[\Omega(h+h_\varphi)] \to H(\Omega_h)$ is an epimorphism.*

**5.3.3. Representing system of functions in a $(\rho, \alpha)$-convex domain.** Suppose $D$ is a bounded $(\rho, \alpha)$-convex domain and $h$ is its $(\rho, \alpha)$-support function, i.e., $D = \Omega_h$. We denote by $L$ an entire function of completely regular growth of order $\rho$, with indicator $h$, and having simple zeros $\lambda_1, \lambda_2, \dots, \lambda_n, \dots$, $|\lambda_1| \leqslant |\lambda_2| \leqslant \cdots$; $|\lambda_k| \uparrow \infty$. Suppose also that for every $\varepsilon > 0$ and $k > k_0(\varepsilon)$ we have

$$|L'(\lambda_k)| > \exp\{[h(\varphi_k) - \varepsilon]|\lambda_k|^\rho\}, \ \varphi_k = \arg\lambda_k \ . \tag{5.3.8}$$

We define

$$L_\nu(\lambda) = \frac{L(\lambda)}{(\lambda-\lambda_\nu)L'(\lambda_\nu)}; \quad \psi_\nu(z) = [B_\alpha L_\nu](z), \quad \nu = 1, 2, \dots . \tag{5.3.9}$$

Here we agree that $L_\nu(\lambda_\nu) = 1$ (by continuity). Let $F$ be a function analytic in some neighborhood $U$ of the closure $\overline{D}$ of the domain $D$, and let $\omega$ be a bounded domain with rectifiable Jordan boundary such that $\overline{D} \subset \omega \subset \overline{\omega} \subset U$. If $\varepsilon > 0$ is sufficiently small, these requirements are fulfilled, e.g., for $\omega = \Omega(h+\varepsilon)$ (see Property 4.4.22). For every $\nu \in \mathbb{N}$, the indicator of the function $L_\nu$ coincides with $h$. Consequently, for every $\nu \in \mathbb{N}$ the functions $\{\psi_\nu\}_1^\infty$ are holomorphic outside $\overline{D}$ (see Theorem 5.2.2). By (5.2.3), they form a biorthogonal system for the system $\{K_\alpha(\lambda_\nu, z)\}_1^\infty$, i.e.,

$$\frac{1}{2\pi i}\int_{\partial\omega} K_\alpha(\lambda_\nu, z)\psi_\mu(z)\,dz = \delta_{\nu\mu} \ ,$$

where $\delta_{\nu\mu} = 0$ for $\nu \neq \mu$ and $\delta_{\nu\mu} = 1$ otherwise. Any $F$ holomorphic in a neighborhood of $\overline{D}$ gives rise to the series

$$F \sim \sum_{\nu=1}^{\infty} \beta_\nu K_\alpha(\lambda_\nu, z), \tag{5.3.10}$$

where

$$\beta_\nu = \frac{1}{2\pi i}\int_\nu \psi_\nu(w)F(w)\,dw, \quad \gamma = \partial\omega. \tag{5.3.11}$$

**Theorem 5.3.5.** *For every $F \in H(\overline{D})$, the series* (5.3.10) *converges in the topology of the space $H(D)$ to $F$, i.e., $\{K_\alpha(\lambda_n, \cdot)\}_1^\infty$ is a representing system for the space $H(\overline{D})$.*

As a preliminary, we prove the following supplement to a theorem in Ch. 1.

**Lemma 5.3.6.** *Under the conditions of Theorem* 1.10.15 *the series* (1.10.18) *converges to $f$ in the topology of the space* $[\rho, H(\theta)]$.

◀ We estimate the remainder (see (5.3.9)) $R_n(\lambda) = \sum_{k=n+1}^{\infty} L_k(\lambda) f(\lambda_k)$ of the converging series (1.10.18).

1. Put $d = \inf\{[H(\theta) - h_f(\theta)] : \theta = \mathbb{R}\}$, and take $\delta \in (0, d)$. Let $\tau = d - \delta$ ($d > 0$ under the conditions of Theorem 1.10.15). From condition 3) of Theorem 1.10.13 (see (5.3.8)) we deduce that there exists a constant $B = B(\delta)$ such that

$$\left|\frac{f(\lambda_k)}{L'(\lambda_k)}\right| < B \exp\{|\lambda_k|^\rho [h_f(\varphi_k) - H(\varphi_k) + \delta]\} < Be^{-\tau|\lambda_k|^\rho}, \ \varphi_k = \arg \lambda_k$$

for $k > k_0$. By Theorem 1.10.14 and the Cauchy test, we see that the series $\sum_{k=1}^{\infty} \exp\{-\tau|\lambda_k|^\rho\}$ converges. Therefore, for $n > k_0$ we have

$$S_n := \sum_{k=n+1}^{\infty} \left|\frac{f(\lambda_k)}{L'(\lambda_k)}\right| < B \cdot \sum_{k=n+1}^{\infty} \exp\{-\tau|\lambda_k|^\rho\}; \quad \lim_{n\to\infty} S_n = 0. \tag{5.3.12}$$

2. Let $A_k(\lambda) = \dfrac{L(\lambda)}{\lambda - \lambda_k}$, $k \in \mathbb{N}$. We show that for every $\varepsilon > 0$ we can find a constant $C_\varepsilon > 0$ such that

$$|A_k(\lambda)| < C_\varepsilon \exp\{|\lambda|^\rho [H(\arg \lambda) + \varepsilon]\}, \ \lambda \in \mathbb{C} \quad \forall\, k \in \mathbb{N}. \tag{5.3.13}$$

First, $|A_k(\lambda)| \leqslant |L(\lambda)|$ for $|\lambda - \lambda_k| \geqslant 1$ and $k \in \mathbb{N}$. Since there exists a constant $D_\varepsilon > 0$ such that

$$|L(\lambda)| < D_\varepsilon \exp\{|\lambda|^\rho [H(\arg \lambda) + \varepsilon]\}, \quad \lambda \in \mathbb{C}\,, \tag{5.3.14}$$

estimate (5.3.13) is valid for $|\lambda - \lambda_k| \geqslant 1, k \in \mathbb{N}$.

Let $|\lambda - \lambda_k| < 1$. From the maximum principle we get

$$|A_k(\lambda)| < \max\{|L(t)| : |t - \lambda_k| \leqslant 1\} = |L(t_0)|, \tag{5.3.15}$$

where $|t_0 - \lambda_k| = 1$, $t_0 = t_0(\lambda; k)$. Let $\lambda = re^{i\varphi}$, $t_0 = r_0 e^{i\varphi_0}$. Now, (5.3.15) and (5.3.14) imply the inequality

$$|A_k\left(re^{i\varphi}\right)| < D(\varepsilon/2) \exp\{r_0^\rho [H(\varphi_0) + \varepsilon/2]\}. \tag{5.3.16}$$

We put $m = \max\{H(\theta); \theta \in \mathbb{R}\}$ and choose a sufficiently large number $R > 1$. Since $|t_0 - \lambda| < 2$ for every $k \in \mathbb{N}$, from Property 1.9.28 we deduce the estimate

$$r_0^\rho \left[H(\varphi_0) + \frac{\varepsilon}{2}\right] - r^\rho \left[H(\varphi) + \frac{\varepsilon}{2}\right] < \left(m + \frac{\varepsilon}{2}\right) \rho r^{\rho-1}(1 + \varepsilon(R)), \quad r > R.$$

Here $\varepsilon(R)$ tends to 0 (as $R \to \infty$) uniformly with respect to $t_0 - \lambda$. By (5.3.16), we have:

$$|A_k(re^{i\varphi})| < D(\varepsilon/2)\exp\{r^\rho[H(\varphi)+\varepsilon]\},\ r > R_\varepsilon;\ k > k_1. \tag{5.3.17}$$

We put $M_\varepsilon = D(\varepsilon/2)\exp\{R_\varepsilon^\rho(m+\varepsilon)\}$ and $G_\varepsilon = \inf\{r^\rho[H(\varphi)+\varepsilon] : |re^{i\varphi}| \leqslant R_\varepsilon\}$. Formula (5.3.17) and the maximum principle for analytic functions imply that inequality (5.3.13) is true for $k > k_1$ and $C_\varepsilon = \max\{D(\varepsilon/2), M_\varepsilon \exp\{-G_\varepsilon\}\}$. For every $k \in \mathbb{N}$, the function $A_k$ has order $\rho$ and indicator $H$. Consequently, for a suitable choice of $C_\varepsilon$, condition (5.3.13) is also valid for $k \leqslant k_1$.

3. Let $||R_n||_\varepsilon = \sup\{|R_n(\lambda)|\exp\{-|\lambda|^\rho[H(\arg\lambda)+\varepsilon]\} : \lambda \in \mathbb{C}\}$. By (5.3.12) and (5.3.13) we have: $||R_n||_\varepsilon < S_n \cdot C_\varepsilon$; $||R_n|| \to 0$ as $n \to \infty$ ▶

**Proof of Theorem 5.3.5.** It will be proved in two stages.

1. For $\varepsilon > 0$, let

$$||L||_\varepsilon = \sup\{|L(z)|\exp\{-|z|^\rho[h(\arg z)+\varepsilon]\} : z \in \mathbb{C}\}.$$

For sufficiently small $\varepsilon > 0$, the contour $\gamma$ in (5.3.11) can be chosen as follows: $\gamma = \partial\Omega(h+2\varepsilon)$ (see (4.4.14)). By the maximum principle for analytic functions and (5.2.7), we find (see (5.3.9)):

$$\sup_{\zeta\in\gamma}|\psi_\nu(\zeta)| = \sup\{|\psi_\nu(\zeta)| : \zeta \in \overline{\mathbb{C}}\setminus\Omega(h+2\varepsilon)\} \leqslant ||L_\nu||_\varepsilon \cdot \varepsilon^{-1/\rho}\Gamma(1/\rho).$$

From formula (5.3.13) (in which $H$ should be replaced by $h$) we get: $\sup_{\zeta\in\gamma}|\psi_\nu(\zeta)| \leqslant A_\varepsilon/|L_\nu'(\lambda_\nu)|$, $\nu \in \mathbb{N}$, where $A_\varepsilon > 0$ is a constant independent of $\nu$. Therefore (see (5.3.11)),

$$|\beta_\nu| < N_1|L'(\lambda_\nu)|^{-1}, \quad \nu \in \mathbb{N},$$

for some $N_1 > 0$. Consider the $(\rho,\alpha)$-convex compact set $X_\tau$ whose $(\rho,\alpha)$-support function $h_\tau$ is equal to $\tau h + (1-\tau)l$, where $\tau \in (0,1)$. By analogy with inequality (5.2.6), we prove that

$$|K_\alpha(\lambda_\nu, z) < N_2\exp\{[h_\tau(\varphi_\nu)+\varepsilon]\cdot|\lambda_\nu|^\rho\} \quad \forall\ z \in X_\tau,$$

for all $\nu \in \mathbb{N}$ and some $N_2 > 0$. The latter two inequalities and (5.3.8) imply that

$$|\beta_\nu K_\alpha(\lambda_\nu, z)| < N_1N_2\exp\{[h_\tau(\varphi_\nu)+2\varepsilon-h(\varphi_\nu)]\cdot|\lambda_\nu|^\rho\},\ \nu \geqslant \nu_1\ ,$$

for all $z \in X_\tau$. Denoting $\delta_0 = \inf\{h(\theta)-l(\theta) : \theta \in \mathbb{R}\}$, we observe that

$$|\beta_\nu K_\alpha(\lambda_\nu, z) < N_1N_2\exp\{[2\varepsilon-(1-\tau)\delta_0]\cdot|\lambda_\nu|^\rho\},$$

for $\nu > \nu_1$ and $z \in X_\tau$. Choosing $\varepsilon < (1-\tau)\delta_0/2$ and taking into account that $\overline{\lim\limits_{\nu\to\infty}}\ \nu|\lambda_\nu|^{-\rho} < \infty$ (see Theorem 1.10.14) we prove that the series (5.3.10) converges uniformly on every compact subset of $D$, i.e., in the topology of the space $H(D)$.

2. Now it suffices to show that for every fixed $z \in D$ the series (5.3.10) converges uniformly to $F(z)$. We put

$$\begin{aligned} R_n(z) &= F(z) - \sum_{\nu=1}^{n} \beta_\nu K_\alpha(\lambda_\nu, z); \qquad (5.3.18) \\ d_n(\lambda, z) &= K_\alpha(\lambda, z) - \sum_{\nu=1}^{n} L_\nu(\lambda) K_\alpha(\lambda_\nu, z). \end{aligned}$$

Using (5.2.12), (5.3.11), (5.3.9), and the Cauchy integral formula, we obtain

$$R_n(z) = \frac{1}{2\pi i} \int_\gamma F(t)\{B_\alpha[d_n(\lambda, z)]_\lambda\}(t)\, dt. \qquad (5.3.19)$$

If $z \in D$, then $r_\alpha(\theta; z) < h(\theta)$ $\forall\ \theta \in \mathbb{R}$ (see (4.4.13), Definition 4.4.15, Ch. 4). Therefore, the series $\sum_{\nu=1}^{\infty} L_\nu(\lambda) K_\alpha(\lambda_\nu, z)$ satisfies all conditions of Theorem 1.10.15 (see Theorem 5.1.4). By Lemma 5.3.6, this series converges to $K_\alpha(\lambda, z)$ for $\lambda \in \mathbb{C}$ in the topology of $[\rho, h(\theta)]$ if $z \in D$. Hence, $\lim_{n\to\infty} d_n(\lambda, z) = 0$ in the topology of $[\rho, h(\theta)]$ (see (5.3.18)). Since $B_\alpha$ is a continuous operator (see Theorem 5.2.2), we obtain $\{B_\alpha[d_n(\lambda, z)]_\lambda\}(t) \to 0$ as $n \to \infty$ uniformly in $t$ on every compact set $\gamma \subset \overline{\mathbb{C}} \setminus X_h$. Now, (5.3.19) implies that $R_n(z) \to 0$ as $n \to \infty$ ▶

**5.3.4. Continuation of the representing system $\{K_\alpha(\lambda_\nu, z)\}_1^\infty$ in the interior.** Suppose $\rho > 0$, $l \in M_\rho$, $\alpha \in \mathfrak{P}_\rho(l)$. The representing system of functions of the space $H(G)$, where $G$ is a domain in $\mathbb{C}$, is called *continuable in the interior* of a domain $D$ relatively compact in $G$ if this system is representing in $H(D)$. A $(\rho, \alpha)$-convex subdomain $D = \Omega_h$ of the $(\rho, \alpha)$-convex domain $G = \Omega_g$ is said to be *$(\rho, \alpha)$-convex complementary* if $g - h \in P_\rho$, where $h, g$ are the $(\rho, \alpha)$-support functions of the domains $D$ and $G$, respectively.

We have the following criterion of continuability in the interior.

**Theorem 5.3.7.** *Suppose $G$ is a bounded $(\rho, \alpha)$-convex domain; $\mathfrak{M} = \{K_\alpha(\lambda_\nu, z)\}_1^\infty$ is a representing system in $H(G)$. Then $\mathfrak{M}$ is continuable in the interior of every $(\rho, \alpha)$-convex complementary subdomain $D$ of $G$.*

◀ Put $G = \Omega_g$ and $D = \Omega_h$. Then $g \geqslant h$ and $h - g \in P_\rho$. Consider an entire function $\psi$ of completely regular growth with indicator $h_\psi = g - h$. We take the generalized convolution operator $T^*: H(G) \to H(D)$ defined by formula (5.3.5), where $\alpha_1 = \alpha_3 = \alpha$, $\alpha_2(\zeta) = \zeta^\rho$, and $K_{\alpha_2}(\lambda, v) = E_\rho(\lambda v)$ is the Mittag-Leffler function. By Theorem 5.3.4, for every function $f \in H(D)$

there exists a function $F \in H(G)$ such that $T^* F = f$. But, by assumption,

$$F(z) = \sum_{\nu=1}^{\infty} c_\nu K_\alpha(\lambda_\nu, z), \quad z \in G, \tag{5.3.20}$$

where the series converges in the topology of $H(G)$. Since $T^*$ is continuous (see Theorem 5.2.6), with the aid of (5.3.7) we deduce from (5.3.20) that

$$f(w) = \sum_{\nu=1}^{\infty} c_\nu \{T_z^*[K_\alpha(\lambda_\nu, z)]\}(w) = \sum_{\nu=1}^{\infty} c_\nu \psi(\lambda_\nu) K_\alpha(\lambda_\nu, w)$$

for all $w \in D$, where the series converges in the topology of $H(D)$ ▶

# Chapter 6

# Geometrical Analysis of Asymptotics of Functions Plurisubharmonic in $\mathbb{C}^n$

This chapter considers various growth calibers (orders, types) and associated growth scales of the functions of class $\mathfrak{U}$ of the functions plurisubharmonic in $\mathbb{C}^n$ and dependent only on the moduli of variables $z_1, \ldots, z_n$. Belonging to this class are used in numerous studies functions associated with prescribed entire function $f$ as the maximum of modulus $M_f$, the maximum term $\mu_f$, Nevanlinna characteristics $m(r; f)$, etc. Presentations are peculiar in applying the geometrical approach to examine the growth indicators of the entire function (see Introduction). The most common method – convex analysis – is employed because the elements of the class $\mathfrak{U}$ are convex functions of logarithms of variables in $\mathbb{C}^n$ (see theorem 1.8.14). Generally, the function $\Phi$ from the class $\mathfrak{U}$ is considered prescribed in $\mathbb{R}^n_+$, sometimes in $\mathbb{C}^n$, using notations $\Phi(r)$, $\Phi(z)$, respectively.

## 6.1. Simplest properties of functions belonging to the classes $\mathfrak{P}$ and $\mathfrak{U}$.

Consider the class $\mathfrak{P}$ of functions plurisubharmonic in $\mathbb{C}^n$; in particular, it contains the modulus, and the logarithm of the modulus of every entire function. The growth of every function $u \in \mathfrak{P}$ is controlled by its majorant:

$$M_u(r) = \max\{u(z),\ |z_j| \leqslant r_j,\ j = 1, \ldots, n\},\ r \in \mathbb{R}^n_+.$$

**Property 6.1.1.** $M_u \in \mathfrak{U}\ \forall\ u \in \mathfrak{P}$.

◄ On $\mathbb{C}^n$, consider the function

$$m_u(z) = \max\{u(z_1 e^{i\theta_1}, \ldots, z_n e^{i\theta_n});\ |\theta_i| \leqslant \pi,\ i = 1, \ldots, n\}.$$

Obviously, $m_u(z) \equiv m_u(|z_1|, \ldots, |z_n|)$, $z \in \mathbb{C}^n$. Therefore, if $u$ is continuous, then $M_u = m_u$ is a continuous function of class $\mathfrak{U}$ (see Exercises 1.2.7 and

1.8.4). To complete the proof, it only remains to apply assertion a) of Theorem 1.8.1 and Exercise 1.8.4 ▶

Theorem 1.8.14, Corollary 1.8.15, and Property 6.1.1 imply the following statement.

**Property 6.1.2.** *For every entire function $f$, the functions $M_f$, $\ln^+ M_f$, $\mu_f$, $\ln^+ \mu_f$, and $m(r; f)$ belong to* $\mathfrak{U}$ (see the introduction to Ch. 6). *If $\Phi \in \mathfrak{U}$, then*

$$\Phi(t_1^\lambda s_1^\mu, \ldots, t_n^\lambda s_n^\mu) \leqslant \lambda\Phi(t) + \mu\Phi(s), \ \mu = 1 - \lambda \tag{6.1.1}$$

*for every $t, s \in \mathbb{R}_0^n$* and $\lambda \in [0, 1]$.

**Proposition 6.1.3.** *Every function $\Phi(r)$ in $\mathfrak{U}$ is continuous and convex with respect to $\ln r_1, \ldots, \ln r_n$ everywhere on $\mathbb{R}_+^n$* *).

◀ The function $\Phi$ is upper semicontinuous on $\mathbb{R}_+^n$ and is monotone nondecreasing in each variable (see Theorem 1.8.14). Hence,

$$\Phi(a) \geqslant \varlimsup_{r \to a;\, r \neq a} \Phi(r) = \varlimsup_{r \to a;\, r > a} \Phi(r) \geqslant \Phi(a) \quad \forall\, a \in \mathbb{R}_+^n,$$

where the symbol $r > a$ means that $r \neq a$, $r_i \geqslant a_i$, $i = 1, \ldots, n$. Therefore, (6.1.1) is valid for all $t, s \in \mathbb{R}_+^n$. By Theorem 1.2.2, to prove Proposition 6.1.3, it suffices to prove that the set $B_c = \{r \in \mathbb{R}_+^n : \ \Phi(r) > c\}$ is open in the relative topology of the octant $\mathbb{R}_+^n$ for every $c \in \mathbb{R}$. Let $a \in B_c$ . If $a = 0$, then $\mathbb{R}_+^n \subset B_c$. Suppose $a \neq 0$ and, for instance, $(a_1, \ldots, a_k, 0, \ldots, 0) = a$, where $1 \leqslant k \leqslant n$, $a_i > 0$, $i = 1, \ldots, k$. The function $\varphi(r_1, \ldots, r_k) := \Phi(r_1, \ldots, r_k, 0, \ldots, 0)$ is a convex function of $\ln r_1, \ldots, \ln r_k$ and, consequently, it is continuous on $\mathbb{R}_0^k$. There exists a number $\varepsilon \in (0, \delta)$, where $\delta = \min\{a_i \ : \ i = 1, \ldots, k\}$, such that $(a_1 - \varepsilon, \ldots, a_k - \varepsilon, 0, \ldots, 0) \in B_c$. The set $\{r \in \mathbb{R}_+^n : r_i > a_i - \varepsilon, i = 1, \ldots, k; r_i \geqslant 0; i = k+1, \ldots, n\}$ is included in $B_c$, because $\Phi(r)$ is monotone nondecreasing in each variable. So, $B_c$ is open for every $c \in \mathbb{R}$ ▶

**Property 6.1.4.** *If $\Phi \in \mathfrak{U}$ or $\Phi \in \mathfrak{P}$ and $\Phi$ is upper bounded, then $\Phi(z) \equiv$* const, $z \in \mathbb{C}^n$.

Property 6.1.4 is a consequence of Property 6.1.1 and Exercise 1.4.7.

**Remark.** Below, we examine only the nonconstant functions in $\mathfrak{U}, \mathfrak{P}$; consequently (see Property 1.4), they are not upper bounded.

Along with $\mathfrak{U}$, we also consider its subclass $\mathfrak{C}$ consisting of all functions in $\mathfrak{U}$ that are nonnegative on $\mathbb{R}_+^n$ and logarithmically convex on $\mathbb{R}_0^n$ as functions of $\ln r_1, \ldots, \ln r_n$. We observe that if $\Phi(r) \in \mathfrak{C}$ and $\Phi(r) \not\equiv 0$, then $\Phi(r) > 0$ $\forall\, r \in \mathbb{R}_0^n$ (see Property 1.4.19).

**Proposition 6.1.5.** *Suppose $\Phi(r) \in \mathfrak{C}$ and $\Phi \not\equiv 0$. Let $v^*$ be the Young transform of the function $v(u) = \ln \Phi(e^u)$. Then* dom $v^* \subset \mathbb{R}_+^n$ *and, if* $\overline{\text{dom } v^*}$ *is compact, then there exists a unique function $\psi : \mathbb{R}^n \to (-\infty, \infty]$, $\psi \not\equiv \infty$, which is convex and lower semicontinuous, and satisfies*

$$\Phi(r) = \sup_{y \in \mathbb{R}^n} r_1^{y_1} \cdots r_n^{y_n} \cdot e^{-\psi(y)} \quad \forall\, r \in \mathbb{R}_+^n\,. \tag{6.1.2}$$

---

*) This means that inequality (6.1.1) is valid in $\mathbb{R}_+^n$.

*Moreover,* $\psi = v^*$.

◀ For $c > \inf\{v(u) : u \in \mathbb{R}^n\}$, we put $A_c = \{u \in \mathbb{R}^n : v(u) \leqslant c\}$. If $z \in A_c$, then

$$\{u \in \mathbb{R}^n : u_i \leqslant z_i,\ i = 1, \ldots, n\} \subset$$
$$A_c \subset \{u \in \mathbb{R}^n :< u, y > -v^*(y) \leqslant c\}$$

for every $y \in \operatorname{dom} v^*$, because $v(u)$ is monotone nondecreasing in each variable. Therefore, if $\alpha_j = u_j - z_j$, $j = 1, \ldots, n$, then $\alpha_j y_j + < z, y > -v^*(y) \leqslant c$ $\forall\ \alpha_j \leqslant 0$; $j = 1, \ldots, n$. So, $\operatorname{dom} v^* \subset \mathbb{R}^n_+$. By Corollary 1.5.11 there exists a unique function $\psi$ with the properties described above for which (6.1.2) is valid for all $r \in \mathbb{R}^n_0$. Since $K = \overline{\operatorname{dom} v^*} \subset \mathbb{R}^n_+$, we see that $F(r, y) = r_1^{y_1} \cdots r_n^{y_n} e^{-\psi(y)}$ is a function continuous in $r$ on $\mathbb{R}^n_+$ for every $y \in K$ and upper semicontinuous in the totality of variables. If $K$ is compact, then from Exercise 1.2.7 we deduce that the functinon on the right in (6.1.2) is continuous on $\mathbb{R}^n_+$. But $\Phi(r)$ also has this property (see Proposition 6.1.3). Therefore, (6.1.2) is valid ▶

**Exercise 6.1.6.** Let $h(u)$ be a convex function on $\mathbb{R}^n$ monotone nondecreasing in each variable; let

$$\tilde{h}(r) = \varlimsup_{a \to r} \exp\{h(\ln a_1, \ldots, \ln a_n)\}, \quad \text{where} \quad a \in \mathbb{R}^n_0;\ r \in \mathbb{R}^n_+ .$$

Then $\tilde{h} \in \mathfrak{C}$.

Various growth characteristics (order, type) of a function $u \in \mathfrak{P}$ are defined to be the same growth calibers of its majorant $\Phi = M_u$. Since $\Phi \in M_u$ (see Property 6.1.1), in the remaining part of this chapter we consider only the functions of class $\mathfrak{U}$. The indicators of a function $u \in \mathfrak{P}$ are examined in Section 8.1.

## 6.2. Various definitions of orders for functions of class $\mathfrak{U}$

### 6.2.1. The order function and the $x$-order.

**Definition 6.2.1.** A subset $G$ of $\mathbb{R}^n$ (of $\mathbb{R}^n_+$) is said to be *complete* in $\mathbb{R}^n$ (in $\mathbb{R}^n_+$) if, together with each its point $x$, it contains the set $\Pi_x = \{x \in \mathbb{R}^n : u_i \leqslant x_i,\ i = 1, \ldots, n\}$ (respectively, $\Pi_x^+ = \{u \in \mathbb{R}^n_+ : 0 \leqslant u_i \leqslant x_i,\ i = 1, \ldots, n\}$). A set $G \subset \mathbb{R}^n$ is said to be *octant-like* if $\{u \in \mathbb{R}^n : u_i \geqslant x_i,\ i = 1, \ldots, n\} \subset G$ for every $x \in G$.

**Exercise 6.2.2.** If $G$ is a closed domain complete in $\mathbb{R}^n$ (in $\mathbb{R}^n_+$) and such that $G \neq \mathbb{R}^n$ ($G \neq \mathbb{R}^n_+$), then $G = \bigcup \Pi_x | x \in \partial G$ $\quad (G = \bigcup \Pi_x^+ | x \in \partial G)$.

**Definition 6.2.3.** Let $\mathfrak{B} = \{G\}$ denote the class of closed bounded domains complete in $\mathbb{R}^n_+$. For $x \in \mathbb{R}^n_0$, $\Phi \in \mathfrak{U}$, we put $M_\Phi(t; G; x) = \max\{\Phi(ct^x) : c \in G\}$. The order $\rho = \rho(G; x)$ of the function $M_\Phi(t; G; x)$ is called the *$x$-order* of the function $\Phi$ (see Def. 1.1.1).

**Definition 6.2.4.** The *order function* of a function $\Phi \in \mathfrak{U}$ is the function on $\mathbb{R}^n$ defined by

$$\rho_\Phi(u) = \overline{\lim_{t\to\infty}} (\ln t)^{-1} \cdot \ln^+ \Phi(t^{u_1}, \dots, t^{u_n}).$$

If $n = 1$, then $\rho_\Phi(u) = \max\{0, \gamma u\}$, where $\gamma$ is the order of $\Phi(t)$, i.e., the order function $\rho_\Phi(u)$ and the order $\gamma$ determine each other.

**Proposition 6.2.5.** *Let $\Phi \in \mathfrak{U}$. For all $G \in \mathfrak{B}$ and $x \in \mathbb{R}_0^n$, we have3 $\rho(G; x) = \rho_\Phi(x)$.*

◀ Since $G$ is bounded and complete in $\mathbb{R}_+^n$, we can find numbers $0 < s < S < \infty$ such that $G_s \subset G \subset G_S$, where $G_\alpha = \{r \in \mathbb{R}_+^n : r_i \leqslant \alpha^{x_i},\ i = 1, \dots, n\}$, $\alpha > 0$. Therefore,

$$M_\Phi(t; G_s; x) \leqslant M_\Phi(t; G; x) \leqslant M_\Phi(t; G_S; x). \tag{6.2.1}$$

The function $\Phi$ is in $\mathfrak{U}$ and, consequently, it is monotone increasing in each variable. Therefore, $M_\Phi(t; G_\alpha; x) = \Phi((t\alpha)^x)$. From (6.2.1) we deduce that

$$\rho_\Phi(x) = \rho(G_s; x) \leqslant \rho(G; x) \leqslant \rho(G_S; x) = \rho_\Phi(x) \ ▶$$

Unlike the system of $x$-orders $\{\rho(G; x) : x \in \mathbb{R}_0^n\}$, the order function takes into account all "parabolic" directions of growth of the function $\Phi \in \mathfrak{U}$, in particular, those in which some of the coordinates $r_1, \dots, r_n$ decrease.

**Proposition 6.2.6.** *Let $\Phi \in \mathfrak{U}$. The order function $\rho_\Phi(u)$ is a nonnegative sublinear function on $\mathbb{R}^n$ monotone nondecreasing in each variable* (it may take infinite values).

◀ We show that $\rho_\Phi$ is convex. The other properties of it immediately follow from the definitions of $\rho_\Phi$ and $\mathfrak{U}$. We fix arbitrarily $x, y \in \mathbb{R}^n$, $x \neq y$, and put $\tau = \rho_\Phi(x) + \rho_\Phi(y)$. The function $\rho_\Phi$ is quasiconvex in $\mathbb{R}^n$, because $\rho_\Phi(u) = \overline{\lim_{t\to\infty}}\, t^{-1} \cdot \ln^+ \Phi(e^{tu_1}, \dots, e^{tu_n})$ and $\ln^+ \Phi(e^u)$ is a qusiconvex function (see the example in Subsection 1.4.2). If $\tau < \infty$ and $\rho_\Phi(x)$, $\rho_\Phi(y) > 0$, then, putting $\lambda = \tau^{-1} \cdot \rho_\Phi(x)$ and $\mu = \tau^{-1} \cdot \rho_\Phi(y)$, and taking into account that

$$\rho_\Phi(\alpha u) = \alpha \rho_\Phi(u) \quad \forall\, \alpha > 0, \quad u \in \mathbb{R}^n, \tag{6.2.2}$$

we obtain

$$\rho_\Phi(x + y) \leqslant \max\left\{\rho_\Phi\left(\frac{x}{\lambda}\right), \rho_\Phi\left(\frac{y}{\mu}\right)\right\} = \rho_\Phi(x) + \rho_\Phi(y). \tag{6.2.3}$$

Inequality (6.2.3) is also fulfiled under the condition $\rho_\Phi(x) \cdot \rho_\Phi(y) = 0$. We show this similarly, requiring only that $\lambda + \mu = 1$, $\lambda, \mu > 0$. Finally, if $\tau = \infty$, then (6.2.3) is obvious. By (6.2.2), (6.2.3), we conclude that $\rho_\Phi$ is sublinear ▶

**6.2.2. The order in each variable and the order in the totality of variables** Usually, the study of the asymptotics of a function of several variables is confined to examination of the growth in each variable (with the other variables fixed) or along the "bisector" of the octant $\mathbb{R}^n_+$.

Consider the order $\rho_i(\Phi; r_1, \ldots, [r_i], \ldots, r_n)$ of a function $\Phi \in \mathfrak{U}$ in the variable $r_i$ with the values of the other variables fixed.

**Borel theorem 6.2.7.** *Let $\Phi \in \mathfrak{U}$, and let $\rho_\Phi(u)$ be the order function of $\Phi$. For every $i \in \{1, \ldots, n\}$ and every $r_j > 0$, $j = 1, \ldots, [i], \ldots, n$, we have $\rho_\Phi(e_i) \equiv \rho_i(\Phi; r_1, \ldots, [r_i], \ldots, r_n)$.*

The following modification of Theorem 6.2.7 is useful for the subsequent discussion.

**Lemma 6.2.8.** *For $\Phi \in \mathfrak{U}$, put $\varphi_x(r) = \overline{\lim_{t\to\infty}} (\ln t)^{-1} \ln^+ \Phi(rt^x)$. Then for every $x \in \mathbb{R}^n$ we have*

$$\rho_\Phi(x) = \varphi_x(r) \quad \forall\, r \in \mathbb{R}^n_0. \tag{6.2.4}$$

◀ Formula 6.2.4 is obvious if $x \in \mathbb{R}^n_- := \{u \in \mathbb{R}^n : u_1, \ldots, u_n \leqslant 0\}$. Let $x \in \mathbb{R}^n \setminus \mathbb{R}^n_-$. If $\varphi_x(r) \not\equiv \infty$ for $r \in \mathbb{R}^n_0$, then there exists a point $a \in \mathbb{R}^n_0$ such that $\varphi_x(a) < \infty$. Consider the octant $\Pi_b = \{r \in \mathbb{R}^n : r \geqslant b\}$, where $b \in \mathbb{R}^n_0$ and $b \leqslant a$ (here and below $a \leqslant b$ means that $a_i \leqslant b_i$, $i = 1, \ldots, n$). Since $\varphi_x(r)$ is monotone nondecreasing in each variable, $\varphi_x(b) < \infty$.

Using the notation of formula (6.1.1), we deduce from it the following inequality:

$$\Phi(t_1^\lambda s_1^\mu, \ldots, t_n^\lambda s_n^\mu) \leqslant \max\{\Phi(t), \Phi(s)\}. \tag{6.2.5}$$

Suppose $\lambda \in (0,1)$, $\mu = 1 - \lambda$, and $r \in \Pi_b$ . Since $\Phi \in \mathfrak{U}$, putting $t_i = b_i \cdot t^{x_i/\lambda}$ and $s_i = b_i^{-\lambda/\mu} \cdot r_i^{1/\mu}$, $i = 1, \ldots, n$, by (6.1.1) we find:

$$\Phi(rt^x) \leqslant \max\{\Phi(bt^{x/\lambda}), \Phi(r_1^{1/\mu} \cdot b_1^{-\lambda/\mu}, \ldots, r_n^{1/\mu} \cdot b_n^{-\lambda/\mu})\}. \tag{6.2.6}$$

From this and the definition of $\varphi_x$ we see that $\varphi_x(r) \leqslant \varphi(b)/\lambda$ or $\varphi_x(r) \leqslant \varphi_x(b)$ for every $\lambda < 1$ and $r \in \Pi_b$. But $\varphi(r)$ is monotone nondecreasing in each variable. Therefore, $\varphi_x(r) \equiv \varphi(b) = \varphi(\mathbb{I}) = \rho_\Phi(x)$, $r \in \Pi_b$. Since $\mathbb{R}^n_0 = \bigcup \Pi_b |\, b \leqslant a;\ b \leqslant \mathbb{I}$, the lemma is proved ▶

**Definition 6.2.9.** Let $\Phi \in \mathfrak{U}$. The value $\rho_i(\Phi) = \rho_\Phi(e_i)$ is called the *order* of $\Phi$ *in the variable* $r_i$, $i = 1, \ldots, n$, and the value $\rho(\Phi) = \rho_\Phi(\mathbb{I}) := \rho_\Phi(1, ..., 1)$ is called the *order of $\Phi$ in the totality of variables.* A function $\Phi$ is said to have *finite order* if $\rho(\Phi) < \infty$.

**Proposition 6.2.10.** *For every function $\Phi \in \mathfrak{U}$ we have*

$$\max_{1\leqslant i\leqslant n} \{\rho_i(\Phi)\} \leqslant \rho(\Phi) \leqslant \sum_{i=1}^n \rho_i(\Phi).$$

◀ By Proposition 6.2.6, $\rho_\Phi$ is sublinear and monotone nondecreasing in each variable. Therefore, since $x = x_1e_1 + \cdots + x_ne_n$, from formulas (6.2.2),

(6.2.3) we conclude that

$$\max_{1\leqslant i\leqslant n}\{x_i\rho_i(\Phi)\}\leqslant\rho_\Phi(x)\leqslant\sum_{i=1}^{n}x_i\rho_i(\Phi),\ x\in\mathbb{R}^n_+ \blacktriangleright \tag{6.2.7}$$

It should be noted that $\rho_\Phi(\mathbb{I})\geqslant\rho_\Phi(e_i)$, $i=1,\ldots,n$. Since $\rho_\Phi(u)\geqslant 0$ for $u\in\mathbb{R}^n$, from (6.2.7) and Proposition 6.2.10 we deduce the following statement.

**Corollary 6.2.11.** *Let $\Phi\in\mathfrak{U}$. The following conditions are equivalent:*

i) $\rho_1(\Phi),\ldots,\rho_n(\Phi)<\infty$ (*respectively,* $\rho_1(\Phi)=\cdots=\rho_n(\Phi)=0$);

ii) $\rho(\Phi)<\infty$ (*respectively,* $\rho(\Phi)=0$);

iii) $\rho_\Phi$ *is a finite function in* $\mathbb{R}^n$ (*respectively,* $\rho_\Phi(u)\equiv 0$ *for* $u\in\mathbb{R}^n$).

**6.2.3. The hypersurface of conjugate orders and the order hypersurface** Let $\Phi\in\mathfrak{U}$. Consider the set $B(\Phi)$ of all points $a\in\mathbb{R}^n_+$ with the property

$$\Phi(r)<r_1^{a_1}+\cdots+r_n^{a_n},\ |r|>r_0=r_0(a;\Phi). \tag{6.2.8}$$

If $B(\Phi)$ is not empty, it is octant-like.

**Theorem 6.2.12.** *Let $\Phi\in\mathfrak{U}$, and let $B(\Phi)\neq\varnothing$. We put $B^-=\{(a_1^{-1},\ldots,a_n^{-1}):\ a\in B(\Phi)\cap\mathbb{R}^n_0\}$, denote by $\rho_\Phi$ the order function of $\Phi$, and define $\Omega^+=\{b\in\mathbb{R}^n_+:\ \rho_\Phi(b)\leqslant 1\}$. Then $\overline{B^-}$ is a closed convex domain in $\mathbb{R}^n_+$ and*

$$\overline{B^-}=\Omega^+. \tag{6.2.9}$$

◀ Let $a\in\mathbb{R}^n_0\cap B(\Phi)$, and let $b=(a_1^{-1},\ldots,a_n^{-1})$. Then, since $\Phi\in\mathfrak{U}$, we have

$$1\geqslant\varlimsup_{|r|\to\infty}\frac{\ln^+\Phi(r)}{\ln(r_1^{a_1}+\cdots+r_n^{a_n})}=\varlimsup_{|r|\to\infty}\frac{\ln^+\Phi(r)}{\max\limits_{1\leqslant i\leqslant n}\{a_i\ln r_i\}}=$$
$$=\varlimsup_{t\to\infty}(\ln t)^{-1}\cdot\sup\{\ln^+\Phi(r):\ r\in A_t\}=\rho_\Phi(b), \tag{6.2.10}$$

where $A_t=\{r\in\mathbb{R}^n_0:\ a_i\ln r_i\leqslant\ln t,\ 1\leqslant i\leqslant n\}$. Hence, for $t>\max\{a_i:\ 1\leqslant i\leqslant n\}$ we find

$$t^{-1}\cdot\rho(\Phi)=\rho_\Phi(t^{-1},\ldots,t^{-1})\leqslant\rho_\Phi(b)\leqslant 1,$$

i.e., $\rho(\Phi)<\infty$ ($\rho_\Phi$ is a function monotone nondecreasing in each variable). Then $\rho_\Phi$ is finite and convex; therefore, it is continuous in $\mathbb{R}^n$ (see Corollary 6.2.11 and Proposition 6.2.6). So, $\Omega^+$ is a closed convex domain complete in $\mathbb{R}^n_+$, and $\overline{B^-}\subset\Omega^+$. Reversing the arguments, we see that (see (6.2.10)) if $b\in\mathbb{R}^n_0$ and $\rho_\Phi(b)<1$, then $b\in B^-$, i.e., $\Omega^+\subset\overline{B^-}$. So, formula (6.2.9) is proved and, with it, Theorem 6.2.12.

**Exercise 6.2.13.** Let $\Phi\in\mathfrak{U}$. The following assertions are equivalent:

**Exercise 6.2.13.** Let $\Phi \in \mathfrak{U}$. The following assertions are equivalent:
i) $B(\Phi) \neq \varnothing$ (respectively, $\overline{B(\Phi)} = \mathbb{R}^n_+$);
ii) $\rho(\Phi) < \infty$ (respectively, $\rho(\Phi) = 0$).

**Definition 6.2.14.** If the set (see (6.2.8)) $B(\Phi)$ is not empty, then its boundary $S = S(\Phi)$ is called the *hypersurface of conjugate orders* of $\Phi$, and the collection of coordinates $(\gamma_1, \ldots, \gamma_n)$ of an arbitrary point $\gamma \in S$ is called a system of conjugate orders of $\Phi$ .

**Exercise 6.2.15.** A system $(\gamma_1, \ldots, \gamma_n)$ of positive numbers is a system of conjugate orders of a given function $\Phi \in \mathfrak{U}$ if and only if

$$\varlimsup_{|r|\to\infty} \frac{\ln^+ \Phi(r)}{\ln(r_1^{\gamma_1} + \cdots + r_n^{\gamma_n})} = 1. \tag{6.2.11}$$

**Definition 6.2.16.** Let $\rho_\Phi$ be the order function for $\Phi \in \mathfrak{U}$. The set $T_\Phi = \{u \in \mathbb{R}^n : \rho_\Phi(u) = 1\}$ is called the *order hypersurface* of $\Phi$.

The following statement shows the relationship between the hypersurfaces $S(\Phi)$ and $T_\Phi$.

**Proposition 6.2.17.** *If* $\Phi \in \mathfrak{U}$, $S(\Phi) \cap \mathbb{R}^n_0 \neq \varnothing$, *and* $S^- = \{(\gamma_1^{-1}, \ldots, \gamma_n^{-1}) : \gamma \in S(\Phi) \cap \mathbb{R}^n_0\}$, *then* $T_\Phi \cap \mathbb{R}^n_0 = S^-$ (see Fig. 12).

◀ By assumption, there exists an element $\alpha$ in $S^-$. Since $S^- \subset \partial B^-$ (see (6.2.9)), $B^-$ is a star-like set complete in $\mathbb{R}^n_0$, and $\rho_\Phi$ is a positively homogeneous function (see Proposition 6.2.6), we conclude that $p(\alpha) = \rho_\Phi(\alpha) = 1$, where $p$ is the Minkowski functional of the set $\Omega^+$. Therefore,

$$T_\Phi \neq \varnothing, \quad \text{and} \quad \Omega := \{u \in \mathbb{R}^n : \rho_\Phi(u) \leqslant 1\} \neq \mathbb{R}^n.$$

The Minkowski functional $p_\Omega$ of $\Omega$ is equal to $\rho_\Phi$; so, by Proposition 6.2.6, $\Omega$ satisfies the conditions of Theorem 1.3.17 and, consequently, $T_\Phi = \partial\Omega$. But $\overline{B^-} = \Omega \cap \mathbb{R}^n_+$ (see (6.2.9)). This implies the required statement ▶

**Exercise 6.2.18.** The set $S(\Phi) \cap \mathbb{R}^n_0$ is not empty if and only if $0 < \rho_\Phi < \infty$.

Let $Y$ be the class of nonnegative sublinear functions on $\mathbb{R}^n$ monotone nondecreasing in each variable; we agree that $\varphi_0(u) \equiv 0 \notin Y$. Next, let

$$\mathfrak{M}_n = \{\Phi \in \mathfrak{U} : 0 < \rho(\Phi) < \infty\}. \tag{6.2.12}$$

**Theorem 6.2.19.**

i) *Let* $\rho_\Phi$ *be the order function of a function* $\Phi \in \mathfrak{U}$. *Then* $\{\rho_\Phi : \Phi \in \mathfrak{M}_n\} = Y$.

ii) *The hypersurface* $S$ *in* $\mathbb{R}^n_+$ ( *in* $\mathbb{R}^n$) *bounding a closed domain* $B$ *in* $\mathbb{R}^n_+$ ($\Omega$ *in* $\mathbb{R}^n$) *and satisfying* $S \cap \mathbb{R}^n_0 \neq \varnothing$ *is the hypersurface of conjugate orders (order hypersurface) of some function* $\Phi \in \mathfrak{M}_n$ *if and only if the set* $\overline{B^-}$, *where* $B^- = \{(a_1^{-1}, \ldots, a_n^{-1}) : a \in B \cap \mathbb{R}^n_0\}$, (*respectively, the set* $\Omega$) *is a convex domain* (see Def. 6.2.1) *complete in* $\mathbb{R}^n_+$ (*complete in* $\mathbb{R}^n$ *and containing* 0).

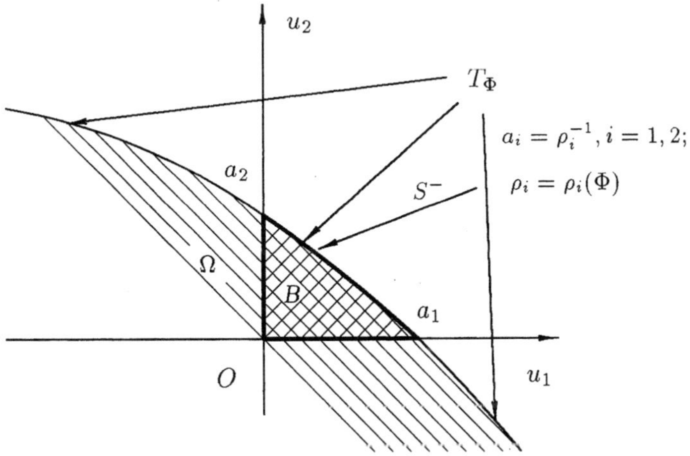


Fig. 1

◀ 1. Proposition 6.2.6 implies that $\{\rho_\Phi : \Phi \in \mathfrak{M}_n\} \subset Y$ (see Corollary 6.2.11). Let $\varphi \in Y$. Consider the function

$$F(r) = \overline{\lim_{a \to r;\, a \in \mathbb{R}_0^n}} \exp\{\varphi(\ln a_1, \dots, \ln a_n)\},$$

upper semicontinuous in $\mathbb{R}_+^n$. By Exercise 6.1.6 and Corollary 6.2.11, $F \in \mathfrak{U}$ and $\rho_F = \varphi$, i.e., $F \in \mathfrak{M}_n$.

2. The "only if" part in statement ii) follows from Propositions 6.2.17 and 6.2.6. We prove the "if" part. Consider the complete (in $\mathbb{R}^n$) hull $\tilde{B}$ of the convex set $\overline{B^-}$ complete in $\mathbb{R}_+^n$, and denote by $\varphi$ the Minkowski functional of $\tilde{B}$. Then $\partial\tilde{B} \cap \mathbb{R}_0^n = S^-$, where $S^- = \{b \in \mathbb{R}^n : (b_1^{-1}, \dots, b_n^{-1}) \in S \cap \mathbb{R}_0^n\}$. By Theorem 1.3.16, $\varphi$ is a nonnegative sublinear function on $\mathbb{R}^n$. But $0 \in \operatorname{int} \tilde{B}$, therefore (see 1.3.14, ii)), $\varphi$ is a finite function on $\mathbb{R}^n$.

Let $\{u, v\} \subset \mathbb{R}^n$, and let $u \leqslant v$. For every $\varepsilon > 0$ we have $v[p(v) + \varepsilon]^{-1} \in \tilde{B}$. Since $\tilde{B}$ is complete, we see that $u[p(v) + \varepsilon]^{-1} \in \tilde{B}$. Consequently, $\varphi(u) \leqslant \varphi(v) + \varepsilon$, i.e., $\varphi(u) \leqslant \varphi(v)$. So, $\varphi \in Y$. It only remains to use assertion i) of the theorem and Proposition 6.2.17. The other part of assertion ii) is proved similarly, by invoking the Minkowski functional of $\Omega$.

**Example**. Let $\rho_1, \dots, \rho_n > 0$, and let $\Phi(r) = r_1^{\rho_1} \dots r_n^{\rho_n}$. The definition of the function $\Phi$ directly gives $\rho_\Phi(u) = \max\{\rho_1 u_1 + \dots + \rho_n u_n, 0\}$. Therefore, $T_\Phi = \{u \subset \mathbb{R}^n : <\rho, u>= 1\}$ and, by Proposition 6.2.17,

$$S(\Phi) = \left\{\gamma \in \mathbb{R}_0^n : \sum_{i=1}^n \rho_i/\gamma_i = 1\right\}; \quad \rho_i(\Phi) = \rho_i;\ \rho(\Phi) = \sum_{i=1}^n \rho_i.$$

**6.2.4. Geometric meaning of the order function** If $\Phi \in \mathfrak{U}$ and $V(u) = \ln^+ \Phi(e^u)$ is a convex function, then the order function $\rho_\Phi$ is its asymptotic function and the epigraph epi $\rho_\Phi$ is the asymptotic cone of the convex set epi $V$ (see Def. 1.6.2). In the general case when $V(u)$ is a quasiconvex function (see Section 1.4; Introduction), the role of the order function $\rho_\Phi$ is similar.

**Definition 6.2.20.** Let $\Omega_V$ be the collection of all cones $K$ with vertex at $0 \in \mathbb{R}^{n+1}$ such that $K$ can be shifted into the epigraph epi $V$ of a quasiconvex function $V$ on $\mathbb{R}^n$. *The asymptotic cone* $\Pi(V)$ *of the epigraph* epi $V$ is defined to be the set $\overline{\bigcup K | K \in \Omega_V}$.

**Theorem 6.2.21.** *Let $\Phi$ be an arbitrary function in $\mathfrak{U}$ of finite order. Then the asymptotic cone $\Pi(V)$ of the epigraph* epi $V$ *of the function $V(u) = \ln^+ \Phi(e^u)$ is convex, and $\Pi(V) =$* epi $\rho_\Phi$.

◀ Let $K$ be an arbitrary cone with vertex at $0 \in \mathbb{R}^{n+1}$ and such that $(v, b) + K \subset$ epi $V$, where $(v, b) \in \mathbb{R}^n \times \mathbb{R}$; $(x, \alpha) = (x_1, \dots, x_n, \alpha) \in K$. Then $(xt + v, \alpha t + b) \in$ epi $V$ $\forall\, t > 0$ and

$$t^{-1} \cdot V(xt + v) \leqslant b \cdot t^{-1} + \alpha \quad \forall\, t > 0. \tag{6.2.13}$$

Passing to the upper limit as $t \to \infty$ and applying Lemma 6.2.8, we obtain: $\alpha \geqslant \rho_\Phi(x)$ $\forall\, (x, \alpha) \in K$. Since $\rho(\Phi) < \infty$, by Corollary 6.2.11 and Proposition 6.2.6 epi $\rho_\Phi$ is a closed convex set in $\mathbb{R}^{n+1}$ (see Theorem 1.4.4). Therefore, $\Pi(V) \subset$ epi $\rho_\Phi$. Let $(x, \alpha) \in$ epi $\rho_\Phi$, and let $\beta = \alpha + \varepsilon$, where $\varepsilon > 0$ is an arbitrarily fixed number. Then $V(xt) < \beta t$, $t > t_0$. Hence, the ray (see Def. 6.2.20) $\{(x\tau, \beta\tau) \subset \mathbb{R}^n \times \mathbb{R} : \tau \geqslant 0\}$ is included in $\Omega_V$. Thus, $(x, \alpha) \in \Pi(V)$; epi $\rho_\Phi \subset \Pi(V)$, i.e., $\Pi(V) =$ epi $\rho_\Phi$ ▶

**6.2.5. Order function and the growth scales for the functions of class $\mathfrak{M}_n$.** If $\Phi \in \mathfrak{C}$, then the function $V(u) = \ln^+ \Phi(e^u)$ is majorized by the order function $\rho_\Phi$ (see (1.6.7)): $V(u) \leqslant V(0) + \rho_\Phi(u)$ $\forall\, u \in \mathbb{R}^n$. A similar statement is valid in the general case.

**Lemma 6.2.22.** *Let $\{\varphi(t) : t > 0\}$ be the collection of quasiconvex functions defined in a convex domain $\Omega \subset \mathbb{R}^n$ and such that $\varphi(u) = \overline{\lim\limits_{t \to \infty}}\, \varphi_t(u)$ is a finite continuous function in $\Omega$; we put $\tilde{\varphi}(u) = \overline{\lim\limits_{t \to \infty;\, v \to u}}\, \varphi_t(v)$. Then*

i)

$$\tilde{\varphi}(u) = \varphi(u), \; u \in \Omega; \tag{6.2.14}$$

ii) *for every $\varepsilon > 0$ and every compact set $K \subset \Omega$ there exists a number $t_0 = t_0(\varepsilon; K)$ such that*

$$\varphi_t(u) \leqslant \varphi(u) + \varepsilon \quad \forall\, u \in K;\; t > t_0. \tag{6.2.15}$$

◀ 1. Suppose $x \in \Omega$, $\Delta(x) \subset \Omega$ is an arbitrary $n$-dimensional simplex having $x$ in the interior, and $x_0, \dots, x_n$ are the vertices of this simplex. The function $\varphi_t(u)$ is quasiconvex in $\Omega$; therefore, $\varphi_t(u) \leqslant \max\limits_{0\leqslant i\leqslant n} \varphi_t(x_i)$ $\forall\ u \in \Delta(x)$, $t > 0$. Passing to the upper limit, we find: $\varphi(x) \leqslant \tilde{\varphi}(x) \leqslant \max\limits_{o\leqslant i\leqslant n} \varphi(x_i)$. Now, (6.2.14) follows from the continuity of $\varphi$: the vertices $\{x\}_0^n$ can be arbitrarily close to $x$.

2. Since $\varphi$ is continuous, from (6.2.14) it follows that for every point $x \in K$ and every $\varepsilon > 0$ there exists an open neighborhood $U(x)$ of $x$ such that $\varphi_t(u) < \varphi(u) + \varepsilon$ $\forall\ t > t_0(x)$, $u \in U(x)$. The system $\{U(x), x \in K\}$ forms an open covering of the compact set $K$. By the Heine-Borel lemma, $K$ is covered by a finite number of $U(x)$'s and, consequently, (10.15) is valid for some $t_0 > 0$ ▶

**Corollary 6.2.23.** *Let $\Phi \in \mathfrak{U}$, and let $\varphi_t(u; a) = (\ln t)^{-1} \ln^+ \Phi(at^u)$. If $\Phi$ is a function of finite order* (see Def. 6.2.9), *then*

$$\rho_\Phi(u) = \varlimsup_{t\to\infty;\, v\to u} \varphi_t(v; \mathbb{I}) \quad \forall\ u \in \mathbb{R}^n. \tag{6.2.16}$$

◀ By Corollary 6.2.11 and Proposition 6.2.6, $\rho_\Phi$ is finite and convex and, consequently, continuous on $\mathbb{R}^n$. Applying Lemma 6.2.22 to the sequence $\varphi_t(u; \mathbb{I})$, we verify (6.2.16) ▶

**Theorem 6.2.24.** *Let $\Phi$ be an arbitrary function in $\mathfrak{U}$ of finite order. For every $\varepsilon > 0$ there exists a constant $C_\varepsilon > 0$ such that*

$$\Phi(e^u) < C_\varepsilon \exp\{\rho_\Phi(u) + \varepsilon|u|\} \quad \forall\ u \in \mathbb{R}^n. \tag{6.2.17}$$

◀ We apply assertion ii) of Lemma 6.2.22 to the sequence $\varphi_t(u) = t^{-1}V(ut)$, where $V(u) = \ln^+ \Phi(e^u)$, and to the compact set $K = \{u \in \mathbb{R}^n : \ |u| = 1\}$:

$$\varphi_t(y) < \rho_\Phi(y) + \varepsilon \quad \forall\ y \in K;\ t > t_0 = t_0(\varepsilon; K).$$

Since the continuous function $V$ is bounded in the ball $\{u \in \mathbb{R}^n : \ |u| \leqslant t_0\}$ (see Proposition 6.1.3), we arrive at (6.2.17) ▶

So, the order function $\rho_\Phi$ of a function $\Phi \in \mathfrak{U}$ has the same properties as the other growth characteristics of $\Phi$ and, therefore, also can be classified as such a characteristic. We construct a growth scale for $\mathfrak{M}_n$ (see (6.2.12)) employing the fact that the order function $\rho_\Phi$ serves as a comparison function for $\Phi$ (see (6.2.17)).

Consider the following growth scale for $\mathfrak{M}_n$:

$$M_n = \{H(r; \varphi) = \tilde{\varphi}(r),\ r \subset \mathbb{R}_+^n : \ \ \varphi \in Y\},$$

(see Exercise 6.1.6), where $Y$ is the class of functions occurring in Theorem 6.2.19. For $n = 1$ we obtain $M_1 = \{\max\{r^\gamma, 1\} : \gamma > 0\}$; this scale is equivalent, as $r \to \infty$, to the scale $\{r^\gamma : \gamma > 0\}$. For $n > 1$ the parameter of the scale $M_n$ is a function rather than a number or a system of numbers (see (6.2.8)). For $\varphi \in Y$, we put $D_\varphi = \{u \in \mathbb{R}^n : \ \varphi(u) > 0\}$ and $S^n = \{u \in \mathbb{R} : \ |u| = 1\}$.

A function $H(r;\varphi)$ in the scale $M_n$ is said to be *asymptotically equivalent* to $\Phi \in \mathfrak{U}$ if $T_\varphi(y) \equiv 1$, $y \in S^n \cap D_\varphi$, where

$$T_\varphi(y) = \overline{\lim_{|u|\to\infty;\ \frac{u}{|u|}\to y}} [\varphi(u)]^{-1} \cdot \ln^+ \Phi(e^u).$$

By Corollary 6.2.23, from the continuity and positive homogeneity of the functions $\varphi$ and $\rho_\Phi \in Y$ (see Theorem 6.2.19; Theorem 1.4.4) we deduce the following fact.

**Theorem 6.2.25.** *Let $\Phi \in \mathfrak{M}_n$, and let $\rho_\Phi$ be the order function of $\Phi$. Then $H(r;\rho_\Phi)$ is a unique element of the scale $M_n$ asymptotically equivalent to $\Phi$, and $\{H(r;\rho_\Phi) : \Phi \in \mathfrak{M}_n = M_n\}$, i.e., $M_n$ is a complete growth scale for $\mathfrak{M}_n$ .*

## 6.3. Local $\mathcal{A}_\Phi$-type structure for function $\Phi$ in the class $\mathfrak{U}$

There exist quite a few various definitions of types for functions of class $\mathfrak{U}$. In Section 6.2 we saw that numerous definitions of orders of such functions are focused around a single notion, namely, the order functions; however, the situation with the types is much more complicated, because they depend on the method of exhausting $\mathbb{R}^n_+$. Taking this into account, for the sake of simplicity we confine ourselves to examination of only one notion of "type" for $\Phi \in \mathfrak{U}$, namely, to the system of type functions. This notion has a clear geometric meaning.

The order function $\rho_\Phi$ for a function $\Phi \in \mathfrak{U}$ determines the concave cone $D_\rho = \{u \in \mathbb{R}^n : \rho_\phi(u) > 0\}$ of growth directions for $\Phi$. For each growth direction $x \in D_\rho$, we introduce a finer growth characteristic, namely, the type function $\sigma_f(r;x)$, $r \in \mathbb{R}^n_+$. Its geometric nature is clarified by the following property: the epigraph of the function $\delta_\Phi(u;x) := \ln \sigma_f(e^u;x)$ is the envelope of the upper asymptotes of the function $\ln^+ \Phi(e^u)$ in the direction $x$. In the one-dimensional case, such a geometric meaning is pertinent to the order and type of a function (see the Introduction). We study general properties of type functions, and the dependence of the structure of a type function on the corresponding growth directions.

### 6.3.1. Definition of type functions and their simplest properties.

**Definition 6.3.1.** Let $\Phi \in \mathfrak{U}$, let $\rho_\Phi$ be the order function of the function $\Phi$, and let $D_\rho = \{u \in \mathbb{R}^n : \infty > \rho_\Phi(u) > 0\}$ be the collection of "directions"of the growth of normal order for $\Phi$. Suppose $D_\rho \neq \varnothing$, $x \in D_\rho$. The function

$$\sigma_\Phi(r;x) = \overline{\lim_{b\to r}}\ \overline{\lim_{t\to\infty}}\ t^{-\rho_\Phi(x)}\Phi(bt^x), \quad r \in \mathbb{R}^n_+, \quad b \in \mathbb{R}^n_+,$$

defined on $\mathbb{R}^n_+$ is called *the $x$-type function for* $\Phi$, and the value $\sigma_\Phi(x) := \sigma_\Phi(\mathbb{I};x)$ is called *the $x$-type for* $\Phi$. Next, the system $\mathcal{A}_\Phi = \{\sigma_\Phi(r;x) : x \in D_\rho\}$ is called the *type of* $\Phi$.

**Example.** Let $\Phi(r_1, r_2) = \varphi(r_1) + \varphi(r_2)$, where $\ln \varphi(e^\alpha) = \sqrt{\alpha^2+1} + \alpha$, $\alpha \in \mathbb{R}$; $\varphi(0) = 1$. We have: $\Phi \in \mathfrak{C}$ ; $\rho_\Phi(u) = 2 \cdot \max\{u_1, u_2, 0\}$;

$$\begin{array}{lll} \sigma_\Phi(r;x) = r_1^2,\ r = (r_1,r_2) \in \mathbb{R}_+^2 & \text{if} & x_1 > x_2;\ x_1 > 0; \\ \sigma_\Phi(r;x) = r_1^2 + r_2^2,\ r \in \mathbb{R}_+^2 & \text{if} & x_1 = x_2 > 0; \\ \sigma_\Phi(r;x) = r_2^2,\ r \in \mathbb{R}_+^2 & \text{if} & x_2 > x_1;\ x_2 > 0. \end{array}$$

The type $\mathcal{A}_\Phi = \{\sigma_\Phi(r;x) : x \in D_\rho\}$ of a function $\Phi$ in $\mathfrak{M}_n$ (see (6.2.12)) is a multidimensional analog of the type of a function of one variable. Indeed, if $n = 1$ and $\Phi \in \mathfrak{M}_1$, then $\rho_\Phi(x) = \max\{\gamma x, 0\}$, where $\gamma > 0$; $D_\rho = \{x \in \mathbb{R} : x > 0\}$; and $\sigma_\Phi(r;x) = \sigma r^\gamma$, where $\gamma$, $\sigma$ are the order and the type of $\Phi$, respectively. So, for $n = 1$ the $x$-type function $\sigma_\Phi(r;x)$, $x > 0$, determines the order $\gamma$ and the type $\sigma$ of $\Phi$, and *vice versa*.

Obviously, $\sigma_\Phi(r;\lambda x) = \sigma_\Phi(r;x)\ \forall\ \lambda > 0$. Therefore, $\mathcal{A}_\Phi = \{\sigma_\Phi(r;x) : x \in T_\Phi\}$, where $T_\Phi$ is the order hypersurface of $\Phi$ (see Def. 6.2.16).

We mention some properties of the $x$-type function $\sigma_\Phi(r;x)$, the first of which immediately follows from the definition.

**Proposition 6.3.2.** *In the notation of Definition* 6.3.1, *for* $x \in D_\rho$ *we put*

$$\tau_\Phi(r;x) = \underline{\lim}_{t\to\infty} t^{-\rho_\Phi(x)} \Phi(rt^x), \quad r \in \mathbb{R}_+^n; \tag{6.3.1}$$
$$A_x = \{r \in \mathbb{R}_+^n : \sigma_\Phi(r;x) < \infty\};\ B_x = \{r \in \mathbb{R}_+^n : \tau_\Phi(r;x) < \infty\}.$$

*If* $A_x \neq \varnothing$, *then:*

i) *the functions* $\sigma_\Phi(r;x)$ *and* $\tau_\Phi(r;x)$ *share all properties of the traces on* $A_x$ *of the functions belonging to* $\mathfrak{C}$; *moreover, they have the property*

$$\varphi(r\lambda^x) = \lambda^{\rho_\Phi(x)} \cdot \varphi(r) \quad \forall\ \lambda > 0; \quad r \in \mathbb{R}_+^n; \tag{6.3.2}$$

ii) $A_x = \overset{\circ}{B}_x$ *and*

$$\sigma_\Phi(r;x) \equiv \tau_\Phi(r;x), \quad r \in \mathbb{R}_0^n \cap A_x. \tag{6.3.3}$$

◄ By Definition 6.3.1 and formula (6.3.1), we have

$$0 \leqslant \tau_\Phi(r;x) \leqslant \sigma_\Phi(r;x) \quad \forall\ r \in \mathbb{R}_+^n$$

Formula (6.3.2), say, for $\sigma_\Phi(r;x)$, is verified immediately, furthermore, $V(u) = \sigma_\Phi(e^u;x)$ is a nonnegative convex function on $\mathbb{R}^n$. Then

$$V(u) \cdot e^{\langle u,a\rangle} = V\left(\frac{x\langle u,a\rangle}{\rho_\Phi(x)} + u\right)$$

is a convex function on $\mathbb{R}^n$ for every $a \in \mathbb{R}^n$. By Theorem 1.4.20, $V(u)$ is a logarithmically convex function on the convex set $\ln A_x$ (see (6.3.1)). Since

$\sigma_\Phi(r;x)$ is an upper semicontinuous function, $A_x$ is an open set. But $A_x \subset B_x$, therefore $A_x \subset \overset{\circ}{B}_x$, $\overset{\circ}{B}_x \neq \varnothing$. The function $V_1(u) = \tau_\Phi(e^u;x)$ is convex and, consequently, continuous on the set $\Omega = \{u \in \mathbb{R}^n : (e^{u_1},\dots,e^{u_n}) \subset \ln \overset{\circ}{B}_x\}$. We see that $\tau_\Phi(e^u;x) \equiv \sigma_\Phi(e^u;x)$, $u \in \Omega$. So, $\overset{\circ}{B}_x \subset A_x$, and $\overset{\circ}{B}_x = A_x$ ▶

**Proposition 6.3.3.** *Let $x \in \mathbb{R}^n_0$, and let $\Phi \in \mathfrak{M}_n$. The following assertions are equivalent:*

a) *the $x$-type $\sigma_\Phi(x)$ is finite (respectively, $\sigma_\Phi(x) = 0$; respectively, $\sigma_\Phi(x) = \infty$);*

b) $\sigma_\Phi(r;x) < \infty\ \forall\ r \in \mathbb{R}^n_+$ *(respectively, $\sigma_\Phi(r;x) \equiv 0$, $r \in \mathbb{R}^n_+$; respectively, $\sigma_\Phi(r;x) \equiv \infty$, $r \in \mathbb{R}^n_+$), where $\sigma_\Phi(r;x)$ is the $x$-type function of $\Phi$.*

◀ Since $\Phi \in \mathfrak{M}_n$, there exists an index $j$ such that $\rho_j(\Phi) > 0$ (see Corollary 6.2.11)). Inequality (6.2.7) implies that $\mathbb{R}^n_0 \subset D_\rho$. We prove that a)→b). By Proposition 6.3.2, $\tau_\Phi(\alpha^x;x) = \alpha^{\rho_\Phi(x)}\sigma_\Phi(x) < \infty\ \forall\ \alpha > 0$. Since $x \in \mathbb{R}^n_0$ and the function $\tau_\Phi(r;x)$ is monotone nondecreasing in each variable, we have $\tau_\Phi(r;x) < \infty\ \ \forall\ r \in \mathbb{R}^n_+$. It only remains to use formula (6.3.3) ▶

**6.3.2. Geometric characteristic of the $x$-type function.** We fix an arbitrary point $x$ on the order hypersurface $T_\Phi$ of a function $\Phi \in \mathfrak{M}_n$. Our goal is to find out how the properties of the type function $\sigma_\Phi(r;x)$ depend on the structure of the order hypersurface $T_\Phi$ in the vicinity of $x$, i.e., how the type $\mathcal{A}_\Phi$ behaves locally. Below we use the following notation: $Y$ is the class of nonnegative sublinear functions on $\mathbb{R}^n$ monotone nondecreasing in each variable; we agree that $\varphi_0(u) \equiv 0 \notin Y$; for $\rho \in Y$, we put

$$\mathfrak{M}_n(\rho) = \{\Phi \in \mathfrak{M}_n : \rho_\Phi(u) \equiv \rho(u),\ u \in \mathbb{R}^n\}. \tag{6.3.4}$$

Here $\rho_\Phi$ is the order function of $\Phi$; $\mathfrak{M}_n(\rho)$ is the class of functions with the given order hypersurface $T^\rho = \{u \in \mathbb{R}^n : \rho(u) = 1\}$;

$$\mathfrak{N}^x_n(\rho) = \{\Phi \in \mathfrak{M}_n(\rho) :\ 0 < \sigma_\Phi(x) < \infty\}, \tag{6.3.5}$$

where $x \in T^\rho$; and $\sigma_\Phi(x)$ is the $x$-type of $\Phi$ (see Def. 6.3.1).

**Proposition 6.3.4.** *In the notation of Definition 6.3.1 and Proposition 6.3.2, if the $x$-type $\sigma_\Phi(x)$ is less than $\infty$, then $A_x$ is a logarithmically convex domain complete in $\mathbb{R}^n_+$ that coincides with $\mathbb{R}^n_+$ if $x \in \mathbb{R}^n_0$. If $\sigma_\Phi(x) = 0$, then $\sigma_\Phi(r;x) \equiv 0\ \forall\ r \in A_x$.*

◀ Formula (6.3.2) shows that $\sigma_\Phi(t^x;x) = t^{\rho_\Phi(x)} \cdot \sigma_\Phi(x)\ \forall\ t > 0$. Therefore, the parabolic ray $\{(t^{x_1},\dots,t^{x_n}) : t > 0\}$ is included in $\mathcal{A}_x$. From Proposition 6.3.2 we deduce the first assertion of Proposition 6.3.4. The second assertion is a consequence of Propositions 6.3.2 and 1.4.19 ▶

We examine a geometric property of the $x$-type function $\sigma_\Phi(r;x)$.

**Definition 6.3.5.** The nonvertical straight line

$$E = \{(xt + y, \gamma t + \delta) \in \mathbb{R}^n \times \mathbb{R} :\ t \in \mathbb{R}\}$$

is called an (upper one-dimensional) *$x$-asymptote of the function* $V(u) = \ln^+ \Phi(e^u)$, where $\Phi \in \mathfrak{M}_n(\rho), \rho \in Y$ (see Proposition 6.3.4), if

$$\overline{\lim_{t\to\infty}}\,[V(xt+y) - \gamma t - \delta] = 0.$$

Obviously, $\gamma = \rho(x)$, i.e., the $x$-asymptote is parallel to the ray $L_x = \{(xt, \rho(x)t) \in \mathbb{R}^n \times \mathbb{R} : t \geqslant 0\}$ of the asymptotic cone $\Pi(V)$ of the set epi $V$ (see Theorem 6.2.21). In what follows we are interested in the asymptotics of the function $V$ on the directions $D_\rho$ of nonzero order of growth for $\Phi$. If $\Phi \in \mathfrak{N}_n^x(\rho)$, then $V$ has at least one $x$-asymptote (for $y = 0, x \in T^\rho$).

**Definition 6.3.6.** Suppose $\rho \in Y$, $x \in T^\rho$, and $\Phi \in \mathfrak{N}_n^x(\rho)$. *The asymptotic $x$-cylinder $I_x(V)$ of the epigraph* epi $V$ *of the function* $V(u) = \ln^+\Phi(e^u)$ is defined to be the set epi $\delta_\Phi^x$, where

$$\delta_\Phi^x(u) = \overline{\lim_{t\to\infty}}\,[V(xt+y) - t\rho(x)]. \tag{6.3.6}$$

Hence, $I_x(V)$ is the epigraph of the envelope of the $x$-asymptotes for $V$. Obviously, $\delta_\Phi^x = \ln \tau_\Phi(e^u; x) = \ln \sigma_\Phi(e^u; x)\ \forall\ u \in \Omega$, where

$$\Omega = \{u \in \mathbb{R}^n : \sigma_\Phi(e^u; x) < \infty\}. \tag{6.3.7}$$

Later we shall see that $\Omega = \mathbb{R}^n$. Therefore, from (6.3.2) and Proposition 6.3.4 it follows that $I_x(V)$ is a convex cylinder.

### 6.3.3. Formulation and the geometric meaning of the main theorem on the structure of the $x$-type function.

With a function $\rho \in Y$, we associate the compact convex set

$$K_\rho = \{y \in \mathbb{R}^n :\ \langle y, u\rangle \leqslant \rho(u)\ \forall\ u \in \mathbb{R}^n\}, \tag{6.3.8}$$

the support function of which is $\rho$. Let

$$\partial\rho(x) = \{y \in K_\rho :\ \langle y, x\rangle = \rho(x)\} \tag{6.3.9}$$

be a face of $K_\rho$ orthogonal to the vector $x \in \mathbb{R}^n \setminus \{0\}$, (see Fig. 13) , or

$$\partial\rho(x) = \{y \in \mathbb{R}^n :\ \rho(u) \geqslant \rho(x) + \langle u - x, y\rangle\ \forall\ u \in \mathbb{R}^n\} \tag{6.3.10}$$

is the subdifferential of $\rho$ at $x$ (see Proposition 1.5.13; Example 1 in Section 1.5).

**Proposition 6.3.7.** *A sublinear function $\rho$ belongs to $Y$ if and only if $\{0\} \in K_\rho \subset \mathbb{R}^n_+$ and $K_\rho \setminus \{0\} \neq \varnothing$.*

◀ Let $S = \{u \in \mathbb{R}^n :\ \rho(u) = 0\}$. Since the function $\rho$ is monotone nondecreasing in each variable, we have $\mathbb{R}^n_- = \{u \in \mathbb{R}^n :\ u \leqslant 0\} \subset S$. Hence, $\{0\} \in K_\rho \subset \{y \in \mathbb{R}^n :\ \langle y, u\rangle \leqslant 0\ \forall\ u \in S\} \subset \{y \in \mathbb{R}^n :\ \langle y, u\rangle \leqslant 0\ \ \forall\ u \in \mathbb{R}^n_-\} = \mathbb{R}^n_+$. The "if" part is obvious ▶

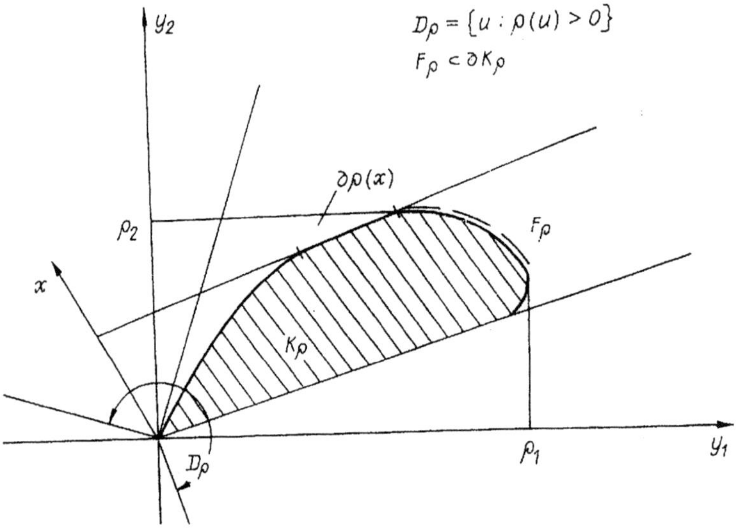


Fig. 13.

For $\rho \in Y$, let $x \in T^\rho = \{u \in \mathbb{R}^n : \rho(u) = 1\}$. Consider the subclass $\mathfrak{N}_n^x(\rho)$ of $\mathfrak{C}$ that consists of the functions $\varphi$ satisfying

$$\varnothing \neq \operatorname{dom} v^* \subset \partial\rho(x), \tag{6.3.11}$$

where $v^*$ is the Young transform of the function $v(u) = \ln\varphi(e^u)$. We also agree that $\varphi_0(u) \equiv 0 \notin N_n^x(\rho)$.

**Proposition 6.3.8.** *The functions of class* $N_n^x(\rho) = \{\varphi(r)\}$ *have the following properties:*

i) $\varphi(rt^x) \equiv t\varphi(r) \quad \forall\, t \geqslant 0$ *for* $r \in \mathbb{R}_+^n$*;*

ii) *for every* $\varphi \in N_n^x(\rho)$*, there exists a unique function* $\psi : \partial\rho(x) \to (-\infty, \infty]$*,* $\psi \not\equiv \infty$*, such that* $\psi$ *is convex, lower semicontinuous, and*

$$\varphi(r) \equiv \sup\{r_1^{y_1} \dots r_n^{y_n} e^{-\psi(y)} : y \in \partial\rho(x)\}, \; r \in \mathbb{R}_+^n; \tag{6.3.12}$$

*moreover,* $\psi = v^*|_{\partial\rho(x)}$*;* $0^0 = 1$*. In particular, if* $\rho$ *is differentiable at* $x$*, there exists a constant* $C_\varphi > 0$ *such that*

$$\varphi(r) \equiv C_\varphi \cdot \prod_{i=1}^n r_i^{\nu_i}, \quad r \in \mathbb{R}_+^n, \quad \textit{where } \nu_i = \frac{\partial\rho}{\partial x_i}. \tag{6.3.13}$$

*Conversely, for* $\varphi \in \mathfrak{C}$ *identity* (6.3.12) *implies that* $\varphi \in N_n^x(\rho)$*.*

◀ Formula (6.3.12) follows from formula (6.3.11) and Proposition 1.5. In its turn, (6.3.12) implies property i) (see (6.3.9)). The representation (3.13) is a consequence of Theorem 1.4.26. The final assertion follows from Proposition 6.1.5 ▶

**Theorem 6.3.9.** *The following statements are equivalent:*

i) *In the above notation,*

$$N_n^x(\rho) = \{\sigma_\Phi(r; x) : \ \Phi \in \mathfrak{N}_n^x(\rho)\} = \{\sigma_\Phi(r; x) : \ \Phi \in \mathfrak{N}_n^x(\rho) \cap \mathfrak{C}\},$$

*i.e., for any $\rho \in Y$ and $\varphi \in N_n^x(\rho)$ there exists $\Phi \in \mathfrak{N}_n^x(\rho)$ such that $\rho_\Phi = \rho$, $x \in T^\rho$, and the $x$-type function $\sigma_\Phi(r; x)$ is identically equal to $\varphi(r)$, $r \in \mathbb{R}_+^n$.*

ii) *If $M_n^x(\rho) = \{\Phi \in \mathfrak{M}_n(\rho) : \ \sigma_\Phi(x) = 0\}$, where $\sigma_\Phi(x)$ is the $x$-type of the function $\Phi$ and $x \in T^\rho$, then $\{\sigma_\Phi(r; x) : \ \Phi \in M_n^x(\rho)\} = \{\sigma_\Phi(r; x) : \ \Phi \in M_n^x(\rho) \cap \mathfrak{C}\} = \{\varphi_0(r) \equiv 0\}$.*

We explain the geometric meaning of the theorem (see Proposition 1.5.3 and Corollary 1.5.4).

1. It can be shown that:

a) every support hyperplane to the asymptotic $x$-cylinder $I_x(V)$ (see Definition 6.3.6) is parallel to some hyperplane that supports the asymptotic cone $\Pi(V) = \operatorname{epi} \rho$ and passes through its ray $\{(xt, t) \in \mathbb{R}^n \times \mathbb{R} : \ t > 0\}$ (see Theorem 6.2.21);

b) every support hyperplane to the convex set $G_x = \{u \in \mathbb{R}^n : \ \sigma_\Phi(e^u; x) \leqslant 1\}$ (i.e., to the directrix of the cylinder $I_x(V)$) is parallel to some hyperplane that passes through $x$ and supports the order hypersurface $T^\rho$. In particular, if the latter hyperplane is unique, then it is parallel to the boundary of $G_x$ (in this case, $G_x$ is a half-space).

2. All of the boundary of the asymptotic $x$-cylinder $I_x(V)$ is formed by the $x$-asymptotes of $V$, i.e., $\operatorname{dom} \delta_\Phi^x = \mathbb{R}^n$ (see (6.3.6) and Definition 6.3.5).

The geometric property 1 b) is a sufficient condition for (6.3.11).

**Lemma 6.3.10.** *Let $x \in \mathbb{R}^n \setminus \{0\}$, and let $v$ be a convex function in $\mathbb{R}^n$ monotone nondecreasing in each variable and such that*

$$v(xt + u) \equiv v(u) + t, \ (u, t) \in \mathbb{R}^n \times \mathbb{R}. \tag{6.3.14}$$

*Then $\operatorname{dom} v^* \subset \{y \in \mathbb{R}_+^n : \langle y, x\rangle = 1\}$, where $v^*$ is the Young transform of $v$.*

◀ Let $y \in \operatorname{dom} v^*$. By (6.3.14), for every $t \in \mathbb{R}$ we have

$$v^*(y) = \sup_{u \in \mathbb{R}^n} (\langle xt + u, y\rangle - v(xt + u)) = v^*(y) + t(\langle y, x\rangle - 1).$$

This is possible only if $\langle y, x\rangle = 1$. On the other hand, $\operatorname{dom} v^* \subset \mathbb{R}_+^n$ (see Proposition 1.5) ▶

**Lemma 6.3.11.** *For $\rho \in Y$, take $x \in T^\rho \cap \mathbb{R}_0^n$. Let $\varphi \in \mathfrak{C}$, and let $D = \{r \in \mathbb{R}_+^n : \varphi(r) \leqslant 1\}$. Then $\varphi \in N_n^x(\rho)$ if and only if $\varphi(r) \equiv \inf\{\alpha > 0 : r\alpha^{-x} \subset D\}$, $r \in \mathbb{R}_+^n$, is the Minkowski $x$-functional of the set $D$ complete in $\mathbb{R}_+^n$, and every support hyperplane to the convex set $\ln D = \{(\ln r_1, \ldots, \ln r_n) : \ r \in D \cap \mathbb{R}_0^n\}$ complete in $\mathbb{R}^n$ is parallel to some support hyperplane to $\Omega = \{u \in \mathbb{R}^n : \ \rho(u) \leqslant 1\}$ passing through the point $x$.*

◀ *The "only if" part.* Assertion i) of Proposition 6.3.8 implies that $\varphi$ is the Minkowski $x$-functional of $D$. Other geometric properties of the function $\varphi$ have already been mentioned (see (6.3.9); Corollary 1.5.4).

*The "if" part.* The function $v(u) = \ln \varphi(e^u)$ is convex in $\mathbb{R}^n$, and $\ln D = \{u \in \mathbb{R}^n : \ v(u) \leqslant 0\}$. Since $\varphi(rt^x) \equiv t\varphi(r) \ \forall \ t \geqslant 0, \ r \in \mathbb{R}^n_+$, from Lemma 6.3.10 we deduce the formula

$$\operatorname{dom} v^* \subset \{y \in \mathbb{R}^n : \ \langle y, x \rangle = 1\}. \tag{6.3.15}$$

Every hyperplane $\Pi$ support to $\ln D$ admits the following representation: $\Pi = \{u \in \mathbb{R}^n : \ \langle u, y \rangle = d\}$, where $y \in \operatorname{dom} v^*$ and $d \geqslant v^*(y)$ (see Corollary 1.5.4). By assumption, it is parallel to some hyperplane $\Gamma_y = \{u \in \mathbb{R}^n : \ \langle u, y \rangle = c\}$ passing through $x$ and support to $\Omega$. From formula (6.3.15) we find: $c = 1$. Since $\{0\} \in \Omega \subset \{u \in \mathbb{R}^n : \ \langle y, u \rangle \leqslant 1\}$, we have $\langle y, u \rangle \leqslant \rho(u) \ \forall \ u \in \mathbb{R}^n$, i.e., $y \in K_\rho$ (see (6.3.8)). Now, formulas (6.3.15) and (6.3.9) imply $\operatorname{dom} v^* \subset \partial\rho(x)$. Thus, $\varphi \in N^x_n(\rho)$ ▶

### 6.3.4. Proof of Theorem 6.3.9.

**Lemma 6.3.12.** *Suppose $W$ is a convex function in $\mathbb{R}^n$, $x \in \mathbb{R}^n \setminus \{0\}$, and $\rho$ is an asymptotic function for $W$* (see Def. 1.6.4) *finite at $x$. Then the sequence of functions $g_t(u) = W(u + xt) - t\rho(x)$ decreases as $t \to \infty$.*

◀ Consider the function $\omega(\alpha) = \alpha W(u + x/\alpha)$, $\alpha > 0$; $\omega(0) = \rho(x)$. For every fixed $u \in \mathbb{R}^n$ the function $\varphi(t) = W(u+xt)$ is convex for $t > 0$. Therefore (see Exercise 1.4.8), the function $\alpha\varphi(1/\alpha) = \omega(\alpha)$ is convex for $\alpha > 0$. The existence of the finite limit

$$\lim_{t \to \infty} t^{-1} \cdot W(xt + u) = \rho(x)$$

implies that $\omega(\alpha)$ is a continuous convex function on $[0, \infty]$. Consequently, the ratio $[\alpha - 0]^{-1}[\omega(\alpha) - \omega(0)] = g_t(u)$ decreases as $\alpha \to +0$ or as $t \to \infty$ ▶

**Proof of Theorem 6.3.9.**

1. For $\Phi \in \mathfrak{N}^x_n(\rho)$, we put $\beta = \ln \sigma_\Phi(x)$. For every $\varepsilon > 0$ there exists a number $\tau_\varepsilon > 0$ such that the function $V_\Phi(u) = \ln^+ \Phi(e^u)$ satisfies

$$V_\Phi(x\tau) - \tau < \beta + \varepsilon \quad \forall \, \tau \geqslant \tau_\varepsilon \, . \tag{6.3.16}$$

We denote $t_\varepsilon = e^{\tau_\varepsilon}$ and $G(r) = \Phi(rt^x_\varepsilon)$. The epigraphs of the quasiconvex functions $V_\Phi$ and $V_G$ have one and the same asymptotic cone $\Pi$ (see Def. 6.2.20) because $V_\Phi(u) \equiv V_G(u - a)$, $u \in \mathbb{R}^n$, where $a = x\tau_\varepsilon$; moreover, $\Pi = \operatorname{epi} \rho$ (see Theorem 6.2.21). From Theorem 6.2.24 we deduce that $V_G(v) < \rho(v) + \varepsilon|v| + \alpha_\varepsilon$ $\forall \ v \in \mathbb{R}^n$ for some $a_\varepsilon \geqslant \beta + \varepsilon + \tau_\varepsilon$. Let $\varphi_\varepsilon(u) = \rho(u-a) + \varepsilon|u-a| + \alpha_\varepsilon$, $u \in \mathbb{R}^n$. Then

$$V_\Phi(u) \equiv V_G(u - a) < \varphi_\varepsilon(u) \qquad \forall \ u \in \mathbb{R}^n_+ \tag{6.3.17}$$

So, the convex cone $K$ = epi $\varphi_\varepsilon$ is a subset of epi $V_\Phi$; on the other hand, formulas (6.3.16), (6.3.17) imply that the ray $L = \{(xt+a, t+\alpha_\varepsilon) \in \mathbb{R}^n \times \mathbb{R} : t \geqslant 0\}$ has the vertex $m_0 = (a, \alpha_\varepsilon)$ common with the cone $K$ and is outside $K$, and $L \subset$ epi $V_\Phi$ $(\rho(x) = 1)$.

2. Consider the set $H$ = conv $(K \cup L)$, where conv $M$ is the convex hull of $M$. We show that

$$H = \bigcup_{\alpha \geqslant \alpha_\varepsilon} \text{conv}\ [(K \cap \Gamma_\alpha) \cup (L \cap \Gamma_\alpha)], \tag{6.3.18}$$

where $\Gamma_\alpha = \{(u, b) \in \mathbb{R}^n \times \mathbb{R} : \ b = a\}$ is a horizontal hyperplane in $\mathbb{R}^{n+1}$. Since $L$ and $K$ are convex sets, $H = \bigcup\limits_{0 \leqslant \lambda \leqslant 1} (\lambda K + (1-\lambda)L)$. Therefore, if $m = (u, \alpha) \in H$ and $m \notin L, K$, then we can find points $m_i = (u_i, \alpha_i)$, $i = 1, 2$, and a number $\lambda_0 \in (0,1)$ such that $m_1 \in K$, $m_2 \in L$, and $m = \lambda_0 m_1 + (1-\lambda_0) m_2$. The points $m_1, m_0, m_2$ determine a two-dimensional convex cone $K_1$ containing $m$ in the interior. The cone $K$ and the ray $L$ do not contain horizontal rays and are located in the half-space $\{(u, \alpha) \in \mathbb{R}^n \times \mathbb{R} : \ \alpha \geqslant \alpha_\varepsilon\}$. Therefore, the cone $K_1$ also possesses these properties, and in the hyperplane $\Gamma_\alpha$ passing through $m$ there exist points $m_1' \in K$, $m_2' \in L$ such that for some $\lambda_1 \in (0,1)$ we have $m = \lambda_1 m_1' + (1-\lambda_1) m_2'$, i.e., $H \subset T$, where $T$ is the right-hand side of (6.3.18). The reverse inclusion is obvious, i.e., $H = T$. Since $V_\Phi$ is a quasiconvex function, $\Gamma_\alpha \cap$ epi $V_\Phi$ is a convex set (see Exercise 1.4.15). Now, (6.3.18) implies $H \subset$ epi $V_\Phi$.

3. We consider the function $\Psi_\varepsilon$ such that epi $\Psi_\varepsilon = H$. This is a convex function, and $\Psi_\varepsilon \geqslant V_\Phi$,

$$\lim_{t\to\infty} t^{-1}\Psi_\varepsilon(xt+a) = 1, \quad (\Psi_\varepsilon(xt+a) \equiv t + \alpha_\varepsilon). \tag{6.3.19}$$

Therefore (see (1.6.3)),

$$\lim_{t\to\infty} t^{-1}\Psi_\varepsilon(xt+u) = 1 \quad \forall\ u \in \mathbb{R}^n. \tag{6.3.20}$$

Since $x \subset T^\rho$, by (6.3.6) we find

$$\delta_\Phi^x(u) = \inf_{\gamma>0} \sup_{t>\gamma} g_t(u); \qquad g_t(u) = V_\Phi(xt+u) - t.$$

From (6.3.19) and Lemma 6.3.12 we conclude that

$$\delta_\Phi^x(u) \leqslant \sup_{t>\gamma} g_t(u) \leqslant \sup_{t>\gamma} P(t,u) = P(\gamma, u), \tag{6.3.21}$$

where $P(t,u) = \Psi_\varepsilon(xt+u) - t$ . But $\varphi_\varepsilon \geqslant \Psi_\varepsilon$ (see (6.3.17)). Hence $\delta_\Phi^x \leqslant F_\varepsilon$ , where

$$F_\varepsilon(u) = \rho(u + x\gamma - a) + \varepsilon|u + x\gamma - a| + \alpha_\varepsilon. \tag{6.3.22}$$

So, dom $\delta_\Phi^x = \mathbb{R}^n$, $A_x = B_x = \mathbb{R}_+^n$ (see Proposition 6.3.2) and $\delta_\Phi^x(u) > -\infty$ $\forall\ u \in \mathbb{R}^n$ because $\delta_\Phi^x(0) = \ln \sigma_\Phi(x) > -\infty$ (see Proposition 1.4.5). From inequality (6.3.22) we obtain: dom $(\delta_\Phi^x)^* \subset$ dom $F_\varepsilon^* = O_\varepsilon(K_\rho)$ $\forall\ \varepsilon > 0$, where $O_\varepsilon(K_\rho)$ is the $\varepsilon$-neighborhood of $K_\rho$ (see (6.3.8), Example 1 in Section 1.5). Therefore,

$$\text{dom } (\delta_\Phi^x)^* \subset K_\rho. \tag{6.3.23}$$

On the other hand, by Lemma 6.3.10, dom $(\delta_\Phi^x)^* \subset \{y \in \mathbb{R}^n : \ \langle y, x\rangle = 1\}$. Combining this with (6.3.23) and (6.3.9), we see that

$$\text{dom } (\delta_\Phi^x)^* \subset \partial\rho(x). \tag{6.3.24}$$

So, $\{\sigma_\Phi(r;x);\ \Phi \in \mathfrak{N}_n^x(\rho)\} \subset N_n^x(\rho)$ (see Proposition 6.3.2).

4. Let $\varphi \in N_n^x(\rho)$. We put $\alpha(s) = \exp\{s - 2\sqrt{s}\}$ and $A = \{r \in \mathbb{R}_+^n : \rho(\ln r_1, \dots, \ln r_n) > 1\}$. Consider the function $\Phi(r) = \max\{\varphi(r), g(r)\}$, where

$$g(r) = \begin{cases} \alpha(\rho(\ln r_1, \dots, \ln r_n)), & r \in A; \\ e^{-1}, \quad r \in \mathbb{R}_+^n \setminus A. \end{cases} \tag{6.3.25}$$

Then $\Phi \in \mathfrak{C}$ (see Exercise 1.4.9). By Theorem 1.6.5 and Condition 6.3.11, for the function $v(u) = \ln \varphi(e^u)$ the asymptotic function $vO^+$ satisfies the inequality $vO^+(u) \leqslant \sup\{\langle y, u\rangle : y \in \partial\rho(x)\} \leqslant \rho(u) \quad \forall\, u \in \mathbb{R}^n \ (\partial\rho(x) \subset K_\rho)$. Therefore, the order function $\rho_\Phi$ coincides with $\rho$, i.e., $\Phi \in \mathfrak{M}_n(\rho)$. Now, the Gateaux differential $\partial\rho(x;v)$ (see Def. 1.4.24) of $\rho$ is a finite function on $\mathbb{R}^n \times \mathbb{R}^n$; consequently, the $x$-type function $\sigma_g(r;x)$ is identically equal to 0. Therefore, $\sigma_\Phi(r;x) \equiv \varphi(r)$. Thus, $N_n^x(\rho) \subset \{\sigma_\Phi(r;x) : \Phi \in \mathfrak{N}_n^x(\rho) \cap \mathfrak{C}\}$. Assertion i) of the theorem is proved (see Proposition 3).

5. Assertion ii) is proved similarly. Indeed, $\beta$ may be arbitrary in (6.3.16). Since $\ln \sigma_\Phi(x) = \delta_\Phi^x(0) = -\infty$, from inequality (6.3.2), Proposition 6.3.4, and Proposition 1.4.5 we deduce that $\delta_\Phi^x(u) \equiv -\infty$ or $\sigma_\Phi(r;x) \equiv 0$, $r \in \mathbb{R}_+^n$. To prove that $M_n^x(\rho)$ is nonempty, it suffices to consider the function $g(r)$ (see (6.3.25)).

**Corollary 6.3.13.** *If $\rho \in Y$, $\Phi \in \mathfrak{M}_n(\rho)$, $x \in T^\rho$, and the $x$-type $\sigma_\Phi(x)$ of the function $\Phi$ is finite, then* (see (6.3.1), (6.3.3))

$$\sigma_\Phi(r;x) \equiv \tau_\Phi(r;x) < \infty \qquad \forall\ r \in \mathbb{R}_+^n \tag{6.3.26}$$

*and assertions* a), b) *of Proposition* 6.3.3 *are equivalent for every $x \in D_\rho$.*

### 6.3.5. Type in one variable. An estimate for the $x$-type function.

**Definition 6.3.14.** Suppose that a function $\Phi \in \mathfrak{U}$ is of order $\rho_n = \rho_n(\Phi) < \infty$ in the variable $r_n$. Then the function

$$\sigma_n(r') = \varlimsup_{r_n \to \infty} r_n^{-\rho_n} \Phi(r', r_n), \qquad r' := (r_1, \dots, r_{n-1}) \in \mathbb{R}^{n-1}$$

is called the *type of $\Phi$ in the variable $r_n$*.

The type of $\Phi$ in other variables is defined similarly. Simple examples show that, for instance, the type $\sigma_n(r')$ depends essentially on the variables $r_1, \dots, r_{n-1}$, unlike the order of $\Phi$ in one of the variables (see Theorem 6.2.7).

**Theorem 6.3.15.** *Suppose that a function $\Phi \in \mathfrak{U}$ satisfies the following conditions: $0 < \rho_n$, $\sigma_n(1, \dots, 1)$, $\rho < \infty$, where $\rho_n = \rho_n(\Phi)$ and $\rho = \rho(\Phi)$ is the order of $\Phi$ in the totality of variables. Then the type $\sigma_n(r')$ of $\Phi$ satisfies the inequality*

$$\sigma_n(t, \dots, t) \leqslant At^{\rho - \rho_n} \quad \forall\, t > 0; \quad A = \text{const} > 0, \tag{6.3.27}$$

*and*

$$\sigma_n(r') \equiv \sigma_\Phi((r', 1); e_n) \tag{6.3.28}$$

*in $\mathbb{R}_0^{n-1}$, where $\sigma_\Phi(r; e_n)$ is the $e_n$-type function of $\Phi$; $e_n = (0, \dots, 0, 1)$.*

◀ By Corollary 6.2.11, the function $\Phi$ has finite order function $\rho(u) = \rho_\Phi(u)$, $\rho(u) \not\equiv 0$, $u \in \mathbb{R}^n$; moreover (see item 6.3.2), $\sigma_\Phi(e_n) = \sigma_n(1, \dots, 1)$. Therefore, $\Phi \in \mathfrak{N}_n^{e_n}(\rho)$. By Corollary 6.3.13, we have $\sigma_n(r') \equiv \sigma_\Phi((r', 1); e_n)$. Next,

$$\sup_{y \in \partial\rho(e_n)} (y_1 + \dots + y_{n-1}) \leqslant \sup_{y \in K_\rho} ||y|| - \rho_n = \rho - \rho_n, \tag{6.3.29}$$

because $\partial\rho(e_n) = \{y \in K_\rho : y_n = \rho_n\}$ (see (6.3.8) and (6.3.9)). Now, (6.3.27) follows from fromulas (6.3.28) and (6.3.29) if we observe that $\sigma_\Phi(r; e_n) = \sigma_\Phi(r; \lambda e_n)$ for all $\lambda > 0$, in particular, for $\lambda = \rho_n^{-1}$ ▶

Some other properties of the function $\sigma_n(r')$ can also be deduced from (6.3.28) (see, e.g., Theorem 6.3.9). Similarly, it is possible to estimate the $x$-type function $\sigma_\Phi(r; x)$ for $\Phi \in \mathfrak{N}_n^x(\rho)$:

$$\sigma_\Phi(e^u; x) \leqslant A_x \exp\{\partial\rho(x; u)\}, \tag{6.3.30}$$

where $\partial\rho(x; u)$ is the Gateaux differential of $\rho$ at the point $x$ (see Theorem 1.4.25), $\rho \in Y$, and $x \in D_\rho$

### 6.3.6. Asymptotic properties of type functions and the related growth scales.

**Proposition 6.3.16.** *Suppose $\rho \in Y$, $x \in T^\rho$, and $\Phi \in \mathfrak{N}_n^x(\rho)$. Let $\sigma_\Phi(r; x)$ be the $x$-type function for $\Phi$. Then for every $\varepsilon > 0$ and every compact set $K \subset \{v \in \mathbb{R}^n : \langle v, x\rangle = 0\}$ there is a semicylinder*

$$\Pi_x = \{xt + v \in \mathbb{R}^n : v \in K;\ t > t_0(\varepsilon)\} \tag{6.3.31}$$

*with base $K$ and direction vector $x$ such that*

$$\Phi(e^u) \leqslant \sigma_\Phi(e^u; x)(1 + \varepsilon) \qquad \forall\, u \in \Pi_x. \tag{6.3.32}$$

*If, moreover, $\Phi \in \mathfrak{C}$, then*

$$\sigma_\Phi(r;x) \leqslant \Phi(r) \qquad \forall\, r \in \mathbb{R}^n_+. \tag{6.3.33}$$

◀ The system of quasiconvex functions $\{\varphi_t(v) = V_\Phi(xt+v) - t,\ t > 0\}$ (where $V_\Phi(u) = \ln^+ \Phi(e^u))$ satisfies the conditions of Lemma 6.2.22: the function $\delta^x_\Phi(v) = \varlimsup\limits_{t\to\infty} \varphi_t(v)$ is convex and, consequently, continuous on $\mathbb{R}^n$ (see item 6.3.2, Corollary 6.3.13). Therefore, there exists a number $t_0 = t_0(\varepsilon; K)$ such that $V_\Phi(v+xt) < t + \delta^x_\Phi(v) + \varepsilon = \delta^x_\Phi(v+xt) + \varepsilon\ \forall\, t > t_0$. This implies (6.3.32). Inequality (6.3.33) follows from Lemma 6.3.12 ▶

So, the $x$-type function $\sigma_\Phi(r;x)$ may serve as a comparison function for $\Phi \in \mathfrak{N}^x_n(\rho)$. Taking this and other properties of $\sigma_\Phi(r;x)$ into account, we construct a growth scale for $\mathfrak{N}^x_n(\rho)$. For $\rho \in Y$ and $x \in T^\rho$, we denote by $N^x_n(\rho) = \{\varphi\}$ the class of functions considered in Proposition 6.3.8. If $\rho$ is differentiable at $x$, then

$$N^x_n(\rho) = \left\{\tau \cdot \prod_{i=1}^n r_i^{\nu_i},\ \ \tau > 0\right\}, \quad \text{where} \quad \nu_i = \frac{\partial \rho}{\partial x_i}.$$

For $n = 1$, we have $\rho(u) = \max\{0, \gamma u\}$, $\gamma > 0$; therefore, $N^x_1(\rho) = \{\tau r^\gamma, \tau > 0\}$, where $x = \gamma^{-1}$. This is the well-known growth scale for the class $\mathfrak{N}^x_1(\rho)$ (see Section 1.1). A function $\varphi$ in the scale $N^x_n(\rho)$ is said to be *asymptotically equivalent* to a given function $\Phi \in \mathfrak{N}^x_n(\rho)$ if

$$\Delta_x(r) := \varlimsup_{a\to r;t\to\infty} \Phi(at^x)/\varphi(at^x) = 1 \quad \forall\, r \in \mathbb{R}^n_0. \tag{6.3.34}$$

Since $\varphi \in \mathfrak{C}$, we see that $\varphi$ is continuous on $\mathbb{R}^n_0$, and, moreover, $\varphi(r) > 0$ $\forall\, r \in \mathbb{R}^n_0$ (see Proposition 1.4.19). Since $\Phi \in \mathfrak{N}^x_n(\rho)$, Corollary 6.3.13 and Lemma 6.2.22 imply that

$$\tau_\Phi(r;x) = \sigma_\Phi(r;x) = \varlimsup_{a\to r;t\to\infty} t^{-1} \cdot \Phi(at^x) \qquad \forall\, r \in \mathbb{R}^n_0.$$

We use (6.3.34) and Proposition 6.3.8 to obtain

$$\Delta_x(r) = [\varphi(r)]^{-1} \cdot \sigma_\Phi(r;x) = 1 \qquad \forall\, r \in \mathbb{R}^n_0.$$

Finally, from Theorem 6.3.9 we deduce the following statement.

**Theorem 6.3.17.** *Suppose $\rho \in Y$, $x \in T^\rho$, and $\Phi \in \mathfrak{N}^x_n(\rho)$. Then in the scale $N^x_n(\rho)$ there is a unique function $\varphi(r) \equiv \sigma_\Phi(r;x)$, $r \in \mathbb{R}^n_+$, asymptotically equivalent to $\Phi$, and the scale $N^x_n(\rho)$ is complete, i.e., $\{\sigma_\Phi(\cdot;x) : \Phi \in \mathfrak{N}^x_n(\rho)\} = N^x_n(\rho)$* (cf. Theorem 6.2.25).

**Remark.** Formula (6.3.34) implies the following asymptotic property of the function $\varphi = \sigma_\Phi(\cdot;x)$ under the conditions of Theorem 6.3.17:

$$\varlimsup_{\langle u,x\rangle\to\infty;u\in\Pi_x} [\varphi(e^u)]^{-1}\Phi(e^u) = 1$$

for every closed semicylinder $\Pi_x$ (see (6.3.31)) with directional vector $x$ and bounded base (cf. Proposition 6.3.16). A similar property is fulfilled for the function $\tilde{\varphi} = H(\cdot; \rho_\Phi)$ under the conditions and in the notation of Theorem 6.2.25:

$$\overline{\lim_{|u|\to\infty; u\in K}} [\varphi(e^u)]^{-1} \ln^+ \Phi(e^u) = 1$$

for every closed cone $K$ with vertex at 0 and such that $K \setminus \{0\} \subset D_\varphi$.

## 6.4. Global structure of the type $\mathcal{A}_\Phi$ for function $\Phi \in \mathfrak{U}$

**6.4.1. The type $\mathcal{A}_\Phi$ junction condition.** Let $\mathcal{A}_\Phi = \{\sigma_\Phi(\nabla; \S) : \S \in \mathcal{T}_\Phi\}$ be the type of a function $\Phi \in \mathfrak{M}_n$ (see item 6.3.1). We want to know how the type $\mathcal{A}_\Phi$ is structured globally, i.e., how the type functions corresponding to various positions of the point $x$ on the order hyperplane $T_\Phi$ of $\Phi$ are related to one another ("joined"). From the geometric standpoint, this is equivalent to finding the relationship between asymptotic $x$-cylinders of the set epi $V$, where $V_\Phi(u) = \ln^+ \Phi(e^u)$, $x \in T_\Phi$ (see (6.3.6)), or between their directrices

$$G_x = \{u \in \mathbb{R}^n : \sigma_\Phi(e^u; x) \leqslant 1\}, \qquad x \in T_\Phi. \tag{6.4.1}$$

For $\rho \in Y$, let $K_\rho$ be the compact convex set the support function of which is $\rho$ (see (6.3.8)). We let $C = \{\lambda y \in \mathbb{R}^n : \lambda \geqslant 0;\ y \in K_\rho\}$ denote the smallest cone with vertex at zero and containing $K_\rho$. Finally $\Gamma_\rho = K_\rho \cap \overline{C \setminus K_\rho}$ is the boundary of $K_\rho$ with respect to the cone $C$, $T^\rho = \{u \in \mathbb{R}^n : \rho(u) = 1\}$, $B_\rho = \bigcup \partial\rho(x) | x \in T^\rho$, and $E_\rho = \overline{B_\rho}$ .

**Lemma 6.4.1.** *We have $\Gamma_\rho = E_\rho\ \forall\ \rho \in Y$; moreover, $\Gamma_\rho \cap$ri $C \subset B_\rho \subset E_\rho$ .*

◀ Let $k = \dim K_\rho > 1$ (if $k = 1$, the result is obvious). First, $E_\rho \subset \Gamma_\rho$. Indeed, if $y \in \partial\rho(x)$, where $x \in T^\rho$, then $y \neq 0$ and $\lambda y \in K_\rho$ for $0 \leqslant \lambda \leqslant 1$ (see Proposition 6.3.7). On the other hand, if $\lambda > 1$, then $\langle \lambda y, x\rangle = \lambda > 1$, i.e., $\lambda y \notin K_\rho$ (see (6.3.8)).

We prove the reverse inclusion. Let $k = n$. Since $C$ is a convex cone, $\overline{C} = \overline{\overset{\circ}{C}}$ (see Theorem 1.3.12). Consequently, $\Gamma_\rho = \overset{\circ}{C} \cap \Gamma_\rho$. We fix an arbitrary $y \in \Gamma_\rho$. Then $y = \lim\limits_{m\to\infty} y_m$ , where $y_m \in \overset{\circ}{C} \cap \Gamma_\rho$ . We draw a support hyperplane $\Pi_m$ to $K_\rho$ through $y_m$ . Since $y_m \neq 0$, there exists a point $u_m \neq 0$ such that $\Pi_m = \{y \subset \mathbb{R}^n : \langle u_m, y\rangle - 1\}$ . Therefore, $y_m \in \partial\rho(u_m)$, $m = 1, 2, \ldots$, and $y \in E_\rho$ . Thus, $\Gamma_\rho \subset E_\rho$ . Let $k < n$, and let $M = \operatorname{aff} K_\rho$ (see Def. 1.3.4). In accordance with Definition 1.3.6, $M$ is a $k$-dimensional subspace. In $M$, we consider an orthonormal basis $f_1, \ldots, f_k \in \mathbb{R}^n$ such that

$$H : \mathbb{R}^k \to M; \qquad H(\alpha) = \sum_{i=1}^{k} \alpha_i f_i; \qquad \langle f_i, f_j\rangle = \delta_{ij} \tag{6.4.2}$$

($\delta_{ij}$ is the Kronecker symbol). Denote $\tilde{K} = H^{-1}(K_\rho)$, $\tilde{\Gamma}_\rho = H^{-1}(\Gamma_\rho)$, $L = \{\gamma \in \mathbb{R}^k : \varphi(\gamma) = 1\}$, and

$$\varphi(\beta) = \sup\left\{\sum_{i=1}^{k} \alpha_i\beta_i : (\alpha_1, \dots, \alpha_k) \in \tilde{K}\right\}.$$

By the above, $\tilde{\Gamma}_\rho \subset \overline{\bigcup \partial\varphi(\beta)}|\beta \subset L$. But for $x \in M$ we have: $\partial\rho(x) = H[\partial\varphi(\beta)]$ where $\beta = H^{-1}(x)$ (see (6.4.2)). Therefore, $\Gamma_\rho \subset \overline{\bigcup \partial\rho(x)}|x \in T^\rho \cap M \subset E_\rho$. The last formula in the lemma follows from Theorem 1.3.12 ▶

**Definition 6.4.2.** For $\rho \in Y$, let $\Gamma_\rho$ be the part of the boundary of $K_\rho$ mentioned in Lemma 6.4.1. Let $H(\Gamma_\rho)$ denote the class of lower semicontinuous functions on $\Gamma_\rho$ that are convex on all convex portions of $\Gamma_\rho$. Next, let $E_\rho(\psi) = \{\varphi_x(r) : x \in T^\rho\}$ be the system of functions on $\mathbb{R}^n_+$ corresponding to $\psi \in H(\Gamma_\rho)$, specifically,

$$\varphi_x(r) \equiv \sup_{y \in \partial\rho(x)} r_1^{y_1} \dots r_n^{y_n} e^{-\psi(y)}, \qquad r \in \mathbb{R}^N_+, \tag{6.4.3}$$

for every $x \in T^\rho$. We say that the $\mathcal{A}_\Phi$-type of $\Phi \in \mathfrak{M}_n(\rho)$ satisfies *the junction condition* if in $H(\Gamma_\rho)$ there is a function $\psi$ with the property $\mathcal{A}_\Phi = \mathcal{E}_\rho(\psi)$.

**Remark.** Corollary 1.5.11 implies that the system $E_\rho(\psi) = \{\varphi_x(r) : x \in T^\rho\}$ uniquely determines the values of $\psi \in H(\Gamma_\rho)$ on the set $M = \bigcup \partial\rho(x)|x \in T^\rho$, i.e., if $E_\rho(\psi_1) = E_\rho(\psi_2)$, then $\psi_1|_M = \psi_2|_M$, where $\psi_i \in H(\Gamma_\rho)$, $i = 1, 2$, and $M \subset \Gamma_\rho$ (see Lemma 6.4.1). Moreover, from Corollary 1.5.11 we deduce that, if $\varphi_x$ is an arbitrary function in $E_\rho(\psi)$ other than 0, then

1) the set $\{(u, u_{n+1}) \in \mathbb{R}^n \times \mathbb{R} : u_{n+1} = \langle u, y\rangle - \psi(y)\}$, $y \in \partial\rho(x) \cap \operatorname{dom}\psi$, is the collection of all support hyperplanes to the epigraph epi $\beta_x$ of the convex function $\beta_x(u) = \ln\varphi_x(e^u)$;

2) if the sets epi $\beta_{x_i}$, $i = 1, 2$, have support hyperplanes $\Pi_i$, $i = 1, 2$, parallel to each other, then $\Pi_1 = \Pi_2$.

The junction condition is illustrated by the following example.

**Example.** Let $\Phi$ be the function occurring in the example to Definition 6.3.1, and let

$$\rho(u) = \max\{2u_1, 2u_2, 0\}; \; \psi(u) = \ln\frac{1}{2}\sqrt{y_1^{y_1} y_2^{y_2}}, \qquad y \in \Gamma_\rho,$$

where $\Gamma_\rho = \{y \in \mathbb{R}^2_+ : y_1 + y_2 = 2\}$, $0^0 = 1$. It can be shown that $\mathcal{A}_\Phi = \mathcal{E}_\rho(\psi)$ (by using Corollary 1.5.11).

### 6.4.2. Theorem on the junction of type functions, and the Dirichlet problem for convex functions.

**Theorem 6.4.3.** *Let $\rho \in Y$, and let $M = \bigcup \partial\rho(x)|x \in T^\rho$. If $\Phi \in \mathfrak{M}_n(\rho) \cap \mathfrak{C}$, then the $\mathcal{A}_\Phi$-type satisfies the junction condition and $\mathcal{A}_\Phi = E_\rho(\psi)$, where $\psi$ is a function in $H(\Gamma_\rho)$* (see Def. 6.4.2) *such that $\psi|_M = V^*|_M = W^*|_M$,* *and* $V^*$, $W^*$ *are the Young transforms of the functions* $V(u) = \ln^+ \Phi(e^u)$ *and* $W(u) = \ln \Phi(e^u)$, *respectively. Moreover,* $\{\mathcal{A}_\Phi : \Phi \in \mathfrak{M}_n(\rho) \cap \mathfrak{C}\} = \{E_\rho(\psi) : \psi \in H(\Gamma_\rho)\}$.

In addition to the remark to Definition 6.4.2, we clarify the geometric meaning of Theorem 6.4.3.

**Definition 6.4.4.** Suppose $\rho \in Y$ and $\Phi \in \mathfrak{M}_n(\rho) \cap \mathfrak{C}$. Put $V(u) = \ln^+ \Phi(e^u)$ and $D = \operatorname{dom} V^*$. Let $\Gamma = \partial D$ be the boundary of $D$ with respect to its affine hull aff $D$, and let $E = \Gamma \cap D$. Suppose $E \neq \varnothing$. A *hyperasymptote of the epigraph* epi $V$ is defined to be any hyperplane in the set

$$B = \{(u, u_{n+1}) \in \mathbb{R}^n \times \mathbb{R} : u_{n+1} = \langle u, y\rangle - V^*(y),\ y \in E\}. \qquad (6.4.4)$$

This name for the elements of $B$ is motivated by the geometric meaning of the conjugate function $V^*$ (see Proposition 1.5.3).

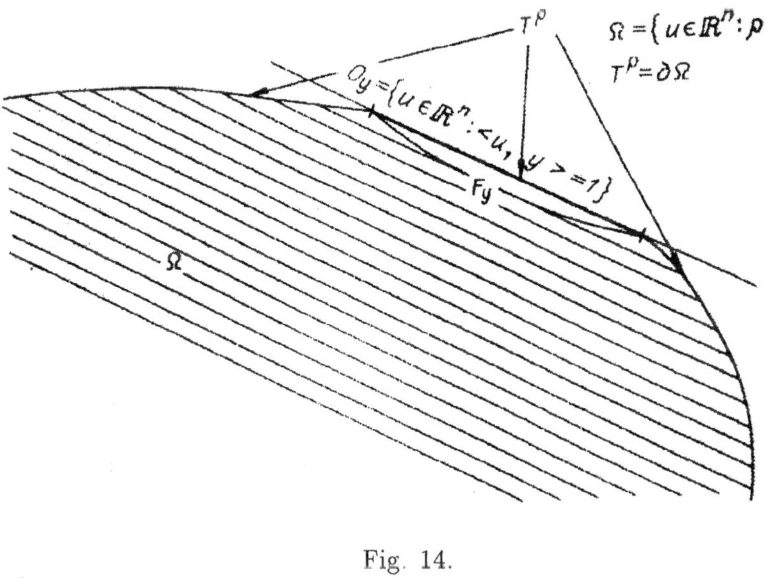


Fig. 14.

If $x \in T^\rho$ and $\Phi \in N_n^x(\rho) \cap C$ (see (6.3.5)), then in the notation of Definition 6.4.4 (see Lemma 6.4.1, (1.6.5), Theorem 1.6.5, and Theorem 6.4.3) $\partial\rho(x) \subset \Gamma$, $\partial\rho(x) \cap D \neq \varnothing$. Then (see Def. 6.3.6) the surface of the asymptotic $x$-cylinder $I_x(V)$ of the set epi $V$ is the envelope of all hyperasymptotes of epi $V$ parallel to some hyperplanes support to the asymptotic cone $\Pi(V)$ and passing through the ray $L_x = \{(x\tau, \tau) \in \mathbb{R}^n \times \mathbb{R} : \tau \geqslant 0\}$ of this cone. If the cone $\Pi(V) = \operatorname{epi} \rho$ is not strictly convex, or the order hypersurface $T^\rho$ is not strictly convex, then one and the same hyperasymptote $\Pi_y \in B$ (see (6.4.4)) may happen to be a support hyperplane to more than one asymptotic cylinder for the set epi $V$. Namely, if the set $O_y = \{u \in T^\rho : y \in \partial\rho(u)\}$ is not a singleton (Fig. 14), then the hyperasymptote $\Pi_y$ is a support hyperplane to the cylinder $I_x(V)$ of the set epi $V$ for every $x \in O_y$. Besides the hyperasymptotes of the set epi $V$, the cylinder $I_x(V)$ has no other support hyperplanes with $x \in T^\rho$ (see Corollary 1.5.11 and the remark to Definition 6.4.2).

To prove Theorem 6.4.3, we need the solution of the Dirichlet problem for convex functions.

**Lemma 6.4.5.** *Let $\varphi$ be a lower semicontinuous function defined on a compact set $K \subset \mathbb{R}^n$, let* dom $\varphi \neq \varnothing$, *and let $S =$* epi $\varphi$. *Then* conv $S$ *is a closed set in $\mathbb{R}^{n+1}$.*

◀ Let $b = \inf\{\varphi(y) : y \in K\}$, and let $a < b$. The mapping $A : \mathbb{R}^{n+1} \to \mathbb{R}^{n+1}$; $u_j = y_j(y_{n+1} - a)^{-1}$, $j = 1, \dots, n$; $u_{n+1} = (y_{n+1} - a)^{-1}$ is a projective transformation of $\mathbb{R}^{n+1}$. It maps homeomorphically the set $L = \{(y, y_{n+1}) \in \mathbb{R}^n \times \mathbb{R} : y_{n+1} > a\}$ onto the set $M = \{(u, u_{n+1}) \in \mathbb{R}^n \times \mathbb{R} : u_{n+1} > 0\}$. But $S$ is closed (see Theorem 1.2.2), therefore, $A(S)$ is also closed in the topology of $M$ induced from $\mathbb{R}^{n+1}$ $(S \subset L)$. The set $T = A(S) \cup \{0\}$ is compact in $\mathbb{R}^{n+1}$, because $K$ is compact. Let $u \in (\text{conv } T) \setminus \{0\}$. Then for some natural $r$ we have $u = \sum_{i=1}^{r} \lambda_i u^{(i)}$, where $u^{(i)} \in T \setminus \{0\}$ and $0 < \sigma := \sum_{i=1}^{r} \lambda_i \leqslant 1$, $\lambda_i \geqslant 0$ (see Exercise 1.3.5). If $\sigma < 1$, then $\sum_{i=1}^{r} \frac{\lambda_i}{\sigma} \cdot \sigma u^{(i)} = u$. The set $T$ is star-like relative to the point 0: the restriction of the mapping $A$ to $L$ takes segments to segments, and $S =$ epi $\varphi$. Consequently, $\sigma u^{(i)} \in T$. So, $u \in \text{conv } (T \setminus \{0\})$ and $(\text{conv } T) \setminus \{0\} = \text{conv } (T \setminus \{0\}) = A(\text{conv } S)$. The set conv $T$ is compact (see Theorem 1.3.9) and the set $(\text{conv } T) \setminus \{0\}$ is closed in the relative topology of $M$. Therefore, the set conv $S = A^{-1}[(\text{conv } T) \setminus \{0\}]$ is closed in the relative topology of $L$, and in $\mathbb{R}^{n+1}$ $(\text{conv } S \subset \{(y, y_{n+1}) \in \mathbb{R}^n \times \mathbb{R} : y_{n+1} \geqslant b\})$ ▶

The following lemma solves the Dirichlet problem for convex functions.

**Lemma 6.4.6.** *Let $K$ be a compact convex set in $\mathbb{R}^n$, and let* $1 \leqslant \dim K \leqslant n$. *Denote by $\Gamma = \partial K$ the boundary of $K$ with respect to* aff $K$; *let $q : \Gamma \to (-\infty, \infty]$ be a lower semicontinuous function on $\Gamma$ convex on all convex portions of $\Gamma$. Then on the compact set $K$ there exists a convex lower semicontinuous function $\Theta$ such that* ri $K \subset$ dom $\Theta$ and $\Theta|_\Gamma = q$.

◀ The assertion is obvious if $\dim K = 1$. Suppose $z \in$ ri $K$, denote by $p(x)$ the Minkowski functional of the set $K - z$, and put $\alpha(y) = (1 - p(y - z))^{-1}$ for $y \in$ ri $K$, $\alpha(y) = q(y)$, $y \in \Gamma$. The function conv $\alpha$ is the solution of the Dirichlet problem. Indeed, $p(y - z)$ is convex and, consequently, continuous on aff $K$, and $p(y - z) < 1$ for $y \in$ ri $K$, $p(y - z) = 1$ for $y \in \Gamma$ (see Theorems 1.3.16, 1.3.17). Therefore, $\alpha(y)$ is a lower semicontinuous function on $K$ and, by Lemma 6.4.5, $I :=$ conv epi $\alpha$ is closed in $\mathbb{R}^{n+1}$. But $I =$ epi conv $\alpha$ and, consequently, conv $\alpha$ is a lower semicontinuous function on $K$ (see Exercise 1.4.12). Finally, $(\text{conv } \alpha)|_\Gamma = q$, because $q$ is a function convex on every convex subset of $\Gamma$ (see Exercise 1.4.13) ▶

**Proof of Theorem 6.4.3.**

1. *The "if" part.* Let $\delta_\Phi^x(u) = \ln \sigma_\Phi(e^u; x)$, $x \in T^\rho$. By Lemma 6.3.12, $\delta_\Phi^x =$

$\inf_{t>0} g_t$, where $g_t(u) = V(xt+u) - t$ (see (6.3.26)), $V(u) = \ln^+ \Phi(e^u)$. Then (see Theorem 1.5.12) $\sup_{t>0} g_t^* = (\delta_\Phi^x)^*$. But $g_t^*(y) = V^*(y) + t(1 - \langle y, x\rangle)$, therefore, $\operatorname{dom} g_t^* = \operatorname{dom} V^* \ \forall\, t > 0$. Since $\overline{\operatorname{dom} V^*} = K_\rho$ (see (6.3.8); (6.5) Ch. 1) and $\rho(x) = 1$, for $\varphi_x := (\delta_\Phi^x)^*$ we have: $\operatorname{dom} \varphi_x \subset \partial\rho(x)$, and $\varphi_x = V^*|_{\partial\rho(x)}$. On the other hand, for $x \in T^\rho$ and $W(u) = \ln \Phi(e^u)$, we have

$$\delta_\Phi^x(u) = \lim_{t\to\infty} [W(xt+u) - t\rho(x)], \tag{6.4.5}$$

and for $W$ the asymptotic function $\rho_1(x)$ is such that $\max\{\rho_1(x), 0\} \equiv \rho(x)$, $x \in \mathbb{R}^n$. Therefore, if $x \in T^\rho$, then $\partial\rho_1(X) = \partial\rho(x)$. Taking this into account, in the same way as before, from formula (6.4.5) we deduce that $\varphi_x = W^*|_{\partial\rho(x)} = V^*|_{\partial\rho(x)} \ \forall\, x \in T^\rho$. It only remains to use formula (6.3.12) for the functions of class $N_n^x(\rho)$. The function $\psi := V^*|_{\Gamma_\rho}$ is required (see Lemma 6.4.1).

2. *The "only if" part.* Let $\psi \in H(\Gamma_\rho)$. The function $q(y) := \psi(y)$, $y \in \Gamma_\rho$; $q(y) := \infty$, $y \in \partial K_\rho \backslash \Gamma_\rho$, satisfies the conditions of Lemma 6.4.6 if $K = K_\rho$. Let $\Theta(y)$ be a solution of the Dirichlet problem for the compact set $K_\rho$ such that $\Theta|_\Gamma = q$ and $\operatorname{ri} K_\rho \subset \operatorname{dom} \Theta \subset K_\rho \subset \mathbb{R}_+^n$. By Theorem 1.3.12 , $\overline{\operatorname{dom} \Theta} = K_\rho$. Now, formula (1.6.6) implies that the function $h := \Theta^*$ has $hO^+ = \rho$ in the role of the asymptotic function. But $\rho \geqslant 0$, consequently, $hO^+ = h^+O^+ = \rho$, where $h^+(u) = \max\{0, h(u)\}$. The function $\Phi(r) = \exp\{h(\ln r_1, \dots \ln r_n)\}$ extends continuously to $\mathbb{R}_+^n$ (see Exercise 6.1.6: $\partial K_\rho \subset \mathbb{R}_+^n$). Thus, $\Phi \in \mathfrak{M}_n(\rho) \cap \mathfrak{C}$. Then, by Theorem 6.3.9, for every $x \in T^\rho$ the $x$-type function $\sigma_\Phi(r; x)$ must be equal identically to $\varphi_x(r)$, $r \in \mathbb{R}_+^n$ (see (6.4.3)), because $(h^+)^*(y) = h^*(y) = \Theta(y) = \psi(y) \ \forall\, y \in \partial\rho(x)$ (see (6.3.9), Theorem 1.5.6). So, the type of $\Phi$ is $\mathcal{A}_\Phi = \mathcal{E}_\rho(\psi)$ ▶

## Chapter 7

# Growth Characteristics (Orders, Types) of Entire Functions of Several Variables, and Their Applications

In this chapter we study the growth of certain functions associated with an entire function $f(z) = f(z_1, \ldots, z_n)$ of complex variables $z_1, \ldots, z_n$: the maximum $M_f$ of its modulus, the maximal term $\mu_f$, the Nevanlinna characteristic $m(r, f)$, etc. These functions and the logarithms of the first two belong to the class $\mathfrak{U}$ (see Property 6.1.2); so, the results of Chapter 6 are applicable. For instance, the growth scales for the classes $\mathfrak{M}_n$ and $\mathfrak{N}_n^x$ considered in that chapter are also growth scales for the corresponding classes of entire functions. It is generally agreed that an entire function $f$ is in $\mathfrak{M}_n$, $\mathfrak{M}_n(\rho)$, or $\mathfrak{N}_n^x(\rho)$ if $\ln M_f$ is in the same class.

The growth characteristics of $\ln M_f$ are at the same time considered to be growth characteristics of the function $f$ itself. For instance, the order of an entire function $f$ in the totality of variables is the order $\rho(\ln^+ M_f)$ of $\ln^+ M_f$ which we denote by $\rho(f)$. Other characteristics treated in Chapter 6 (conjugate orders, orders and types in one variable, the order function, the system of type functions, etc.) are carried over to entire functions in a similar way. For an entire function $f$, these growth characteristics do satisfy all conditions necessary for $f$ to belong to a particular class, as described in Chapter 6. The additional question arising here is about the existence of an entire function with a prescribed set of growth characteristics (the order function, the system of type functions). This question is analyzed in Section 7.2. The relationship between the growth characteristics of an entire function and its Taylor coefficients is studied in Section 7.2. The formulas considered there are of fundamental importance for constructing entire functions with prescribed growth characteristics. Sections 7.3 and 7.4 are devoted to comparison of the growth of $M_f$, $\mu_f$, and $m(r, f)$.

## 7.1. Relationship between growth characteristics of an entire function and its Taylor coefficients

**7.1.1. Simplest properties of functions $M_f$.**

1) If $f(z) \neq$ *const*, then $M_f(r) > M_f(0) \geqslant 0 \ \forall\, r \in \mathbb{R}_0^n$.

2) The function $M_f(r)$ is monotone nondecreasing in each variable $r_1, \ldots, r_n$ and is continuous in $\mathbb{R}_+^n$ (see Exercise 1.2.7).

The following *Hadamard three circles theorem* is well known.

*Let $F(z)$ be a function holomorphic in the annulus $\{z \in \mathbb{C} : 0 \leqslant A < |r| < B\}$. Then $\ln M_f(r)$ is a convex function of $\ln r$ on the interval $(\ln A, \ln B)$, and if $F(z)$ differs from the monomial $az^m$, where $m$ is an integer, $a \neq 0$, then $\ln M_F(r)$ is a strictly convex function of $\ln r$.*

We state a multidimensional analog of this theorem (see Property 6.1.1).

**Theorem 7.1.1 (J. Valiron).** *For an entire function $f(z)$ of $n$ complex variables, denote by $M_f(r)$ be the maximum of its modulus on the polydisc $\{z \in \mathbb{C}^n : |z_i| \leqslant r_i, i = 1, \ldots, n\}$. Then $V_f(u) := \ln M_f(e^u)$ is a convex function on $\mathbb{R}^n$.*

**Remark.** The entire function $\exp\{z^m w^k\}$, where $m, k \geqslant 0$ are integers, shows that, unlike the case of one variable, $V_f$ may fail to be strictly convex even if $f$ is different from a monomial.

**7.1.2. Taylor coefficients and the growth of the maximum of modulus.**

**Lemma 7.1.2.** *The series*

$$f(z) = \sum_{\|k\| \geqslant 0}^{\infty} a_k z^k, \ z^k = z_1^{k_1} \ldots z_n^{k_n},$$

*represents an entire function in $\mathbb{C}^n$ if and only if*

$$\lim_{\|k\| \to \infty} |a_k|^{1/\|k\|} = 0. \tag{7.1.1}$$

◄ Put $|a_m| = \max_{\|k\|=s} |a_k|$, $m = m(s)$. Condition (7.1.1) is equivalent to the condition

$$\lim_{s \to \infty} |a_m|^{1/s} = 0. \tag{7.1.2}$$

Let $b_s = \sum_{\|k\|=s} |a_k|$. For every natural $s$ we have $|a_m| \leq b_s \leqslant (s+1)^n \cdot |a_m|$. Since $\lim_{s \to \infty} (s+1)^{n/s} = 1$, formula (7.1.2) is equivalent to the relation $\lim_{s \to \infty} b_s^{1/s} = 0$. In its turn, by the Cauchy-Hadamard theorem, the later is equivalent to the absolute convergence of the series $\sum\limits_{s=0}^{\infty} b_s R^s$ in $\mathbb{R}_+ (b_s \geqslant 0)$, or to the absolute convergence of the series $\sum\limits_{\|k\|=0}^{\infty} |a_k| r^k$ in $\mathbb{R}_+^n$. The latter assertion is valid if and only if $f$ is an entire function (if a holomorphic function is represented by a series, then this series converges absolutely inside every convergence polydisc ) ▶

**Lemma 7.1.3.** *Let $x \in \mathbb{R}^n$, and let $\max\{x_1, \dots, x_n\} > 0$. Suppose that for some $b \in \mathbb{R}_0^n$ there exist constants $\Delta, A > 0$ such that*

$$M_f(bt^x) < \exp\{At^\Delta\} \quad \forall\, t > t_0, \tag{7.1.3}$$

*where $f(z) = \sum\limits_{\|k\|\geqslant 0} a_k z^k$ is an entire function. Then*

$$|a_k| b^k < (Ae\Delta/\lambda)^{\lambda/\Delta} \quad \forall\, \lambda > \lambda_0, \tag{7.1.4}$$

*where $\lambda = \langle k, x\rangle$. Conversely, if for the Taylor coefficients of an entire function $f$ inequality* (7.1.4) *is valid, then for every $\alpha_1, \dots, \alpha_n \in (0,1)$ there is a constant $B$ such that*

$$M_f(b_1\alpha_1 t^{x_1}, \dots, b_n\alpha_n t^{x_n}) < Be^{At^\Delta} \quad \forall\, t > t_1. \tag{7.1.5}$$

◀ From the Cauchy estimate $|a_k| r^k \leqslant M_f(r)\ \forall\, r \in \mathbb{R}_0^n$ and inequality (7.1.3) with sufficiently large $\lambda = \langle k, x\rangle$, we obtain for $\tau = \ln t$:

$$-\ln(|a_k| b^k) > \sup_{\tau > \tau_0} [\lambda\tau - Ae^{\Delta\tau}] = \frac{\lambda}{\Delta} \ln \frac{\lambda}{Ae\Delta}.$$

This is equivalent to (7.1.4) (see Example 2 in item 1.5). Consider the inequality $M_f \leqslant S$, where $S_f(r) = \sum\limits_{\|k\|\geqslant 0} |a_k| r^k$. Let

$$\sum = S_f(b_1\alpha_1 t^{x_1}, \dots, b_n\alpha_n t^{x_n}).$$

Then

$$M_f(b_1\alpha_1 t^{x_1}, \dots, b_n\alpha_n t^{x_n}) < \Sigma = \Sigma_1 + \Sigma_2. \tag{7.1.6}$$

Here $\Sigma_1$ is the part of the sum $\Sigma$ involving the summands the multiindices of which satisfy the condition $\langle k, x\rangle > \lambda_0$, and $\Sigma_2 = \Sigma - \Sigma_1$. Obviously,

$$\Sigma_2 < Ct^{\lambda_0} \quad \forall\, t > 0, \tag{7.1.7}$$

where $C = S_f(b)$. Next, by (7.1.4), (7.1.6) with sufficiently large $t > 0$ we have, putting $T = \prod\limits_{i=0}^{n} (1 - \alpha_i)^{-1}$:

$$\Sigma_1 \leqslant \sum_{\|k\|\geqslant 0} \alpha^k \cdot \sup_{\langle k,x\rangle > \lambda_0} |a_k| b^k t^{\langle k,x\rangle} \leqslant T \cdot \sup_{\lambda > \lambda_0} t^\lambda \left(\frac{Ae\Delta}{\lambda}\right)^{\lambda/\Delta} = Te^{At^\Delta}.$$

Combining this with (7.1.7), we arrive at (7.1.5) ►

**Theorem 7.1.4.** *Let* $D = \{u \in \mathbb{R}^n : \max\{u_1, \dots, u_n\} > 0\}$. *For an entire function* $f(z) = \sum_{\|k\| \geqslant 0} a_k z^k$, $z^k = z_1^{k_1} \dots z_n^{k_n}$, *we put*

$$\beta_f(u) = \overline{\lim_{\langle k,u\rangle \to \infty}} \ \langle k,u\rangle \ln\langle k,u\rangle - \ln|a_k|, \ u \in D,$$

*where* $(-\ln|a_k|)^{-1} = 0$ *if* $a_k = 0$. *Then*

*1) the order function* $\rho_f(u)$ *coincides with* $\beta_f(u)$, $u \in D$;

*2) if* $\rho \in Y$, $f \in \mathfrak{M}_n(\rho)$, *then for every* $x \in \{u \in \mathbb{R}^n : \rho(u) > 0\}$ *the* $x$*-type function* $\sigma_f(r;x)$ *of* $f$ *is determined by the formula*

$$(\sigma_f(r;x)e\rho(x))^{1/\rho(x)} = \overline{\lim_{\langle k,x\rangle \to \infty}} \ \langle k,x\rangle^{1/\rho(x)} \cdot (|a_k|r^k)^{1/\langle k,x\rangle} \tag{7.1.8}$$

*is valid in* $\mathbb{R}_0^n$.

◄ If $\rho_f(x) < \infty$ for $x \in D$, then (7.1.3) is valid for $A = 1$, $b = \mathbb{I}$, $\Delta = \rho_f(x) + \varepsilon$, where $\varepsilon > 0$. By Lemmas 7.1.2, 7.1.3 and inequality (7.1.4), we have

$$\beta_f(x) \leqslant \rho_f(x) + \varepsilon \quad \forall\, \varepsilon > 0, \text{ or } \beta_f(x) \leqslant \rho_f(x).$$

On the other hand, the definition of $\beta_f$ implies that if $\beta_f(x) < \infty$, then (7.1.4) is true for $\Delta = \beta_f(x) + \varepsilon$, $b = \mathbb{I}$, $A = 1$. Therefore, taking Property 6.1.2 and Lemma 6.2.8 into account, from (7.1.5) we deduce that $\rho_f(x) \leqslant \beta_f(x)$. So, $\rho_f(x) = \beta_f(x)\ \forall\, x \in D$. Assertion 2) is proved in exactly the same manner (when using Lemma 7.1.3, we take $b = r$ and refer to the continuity of the function $\sigma_\Phi(r;x)$ in $\mathbb{R}_+^n$, see Corollary 6.3.13, Propositions 6.3.2 and 6.1.3) ►

**Example.** Let $\Omega$ be a convex domain complete in $\mathbb{R}^n$ and containing 0. Denote by $p_\Omega, h_\Omega$ its Minkowski functional and support function, respectively, and put $C = \text{dom } h_\Omega$ (then dim $C = n$). For instance, if $\Omega = \{u \in \mathbb{R}^n;\ u_1 \leqslant 1, \dots, u_n \leqslant 1\}$, then $p_\Omega(u) = \max\{0, u_1, \dots, u_n\}$, $h_\Omega(y) = \|y\|$ for $y \in \mathbb{R}_+^n$, and $C = \mathbb{R}_+^n$. Putting $f(z) = \sum_{k \in C} \|k\|^{-h_\Omega(k)} z^k$, we see that this is an entire function the order function of which coincides with $p_\Omega$ (see Theorem 7.1.4). Proposition 6.2.17, Theorem 7.1.4, and the formula

$$\lim_{\|k\| \to \infty} [\langle k,x\rangle \ln\langle k,x\rangle] / \sum_{i=1}^n x_i k_i \ln k_i = 1, \tag{7.1.9}$$

where $x \in \mathbb{R}_0^n$, imply the following statement.

**Corollary 7.1.5.** *A system* $(\gamma_1, \dots, \gamma_n)$ *of positive numbers is the system of conjugate orders of an entire function* $f(z) = \sum_{\|k\| \geqslant 0} a_k z^k$ *if and only if*

$$\overline{\lim_{\|k\| \to \infty}} \ (-\ln|a_k|)^{-1} \left( \sum_{i=1}^n \frac{k_i}{\gamma_i} \ln k_i \right) = 1.$$

Omitting the proof, we mention the following result.

**Theorem 7.1.6.** *Suppose $\rho \in Y$ and $x \in D_\rho$ . Let $f \in \mathfrak{M}_n(\rho)$, let $\{a_k\}$ be the Taylor coefficients of $f$, and $\sigma_f(r;x)$ its the $x$-type function. Then for all $r \in \mathbb{R}_0^n$ we have*

$$\sigma_f(r;x)e\rho(x) \equiv \sup_{y\in\partial\rho(x)} P_x(y)r_1^{y_1}\dots r_n^{y_n},$$

*where*

$$P_x(y) = \overline{\lim_{\|k\|\to\infty;k\to y}} \exp\left\{\frac{\rho(x)\ln|a_k|}{\langle k,x\rangle}\right\};$$

*$k \to y$ means that $k_j\cdot|k|^{-1} \to y_j\cdot|y|^{-1}$, $j = 1,\dots,n$. In particular, if $\rho$ is differentiable at $x$, then*

$$\sigma_f(r,x)e\rho(x) \equiv P_x(\text{grad }\rho(x))r_1^{\nu_1}\dots r_n^{\nu_n}\ r\in\mathbb{R}_+^n\ ,$$

where $\nu_j = \frac{\partial\rho}{\partial x_j}$, $j = 1,...,n$.

## 7.2. Existence of entire functions with prescribed growth characteristics

Let $H(\mathbb{C}^n)$ be the class of entire functions in $\mathbb{C}^n$. We examine growth characteristics for functions in the following subclass of $\mathfrak{U}$ (see Theorem 7.1.1):

$$\mathfrak{U}_0 = \{\ln^+ M_f :\ f\in H(\mathbb{C}^n)\}. \tag{7.2.1}$$

### 7.2.1. Existence of an entire function with prescribed order function.

**Lemma 7.2.1.** *Denote by $O_\varepsilon(K)$ the $\varepsilon$-neighborhood of a compact convex set $K$ in $\mathbb{R}^n$ containing 0. Let $p_\varepsilon$ and $p$ be the Minkowski functionals of the sets $O_\varepsilon(K)$ and $K$, respectively. We put $U_\varepsilon(y) = \{x\in\mathbb{R}^n :\ |x-y| < \varepsilon p(y)\}$. Then for all $y \neq 0$ we have*

$$p_\varepsilon(x)\to p(y)\quad \text{as } \varepsilon\to 0\quad \text{and } x\to y,\ x\in U_\varepsilon(y). \tag{7.2.2}$$

◂ Since $O_\varepsilon(K) = K + S_\varepsilon$, where $S_\varepsilon = \{y\in\mathbb{R}^n :\ |y| < \varepsilon\}$, the set $O_\varepsilon(K)$ contains 0 and is open and convex in $\mathbb{R}^n$. Therefore, for all $\varepsilon > 0$, $p_\varepsilon$ is a finite convex function; consequently, it is continuous on $\mathbb{R}^n$ (see Theorems 1.3.15 and 1.3.16). Since $\bigcap_{\varepsilon>0} O_\varepsilon(K) = K$ and the function $p_\varepsilon(y)$ is monotone nondecreasing as $\varepsilon\to 0$ for every fixed $y$, Proposition 1.3.14 implies

$$p(y) = \sup_{\varepsilon>0} p_\varepsilon(y) = \lim_{\varepsilon\to 0} p_\varepsilon(y)\ \ \forall\, y\in\mathbb{R}^n.$$

Hence (see Exercise 1.2.6), $p$ is lower semicontinuous on $\mathbb{R}^n$ and

$$\sup_{\varepsilon>0} \inf\{p_\varepsilon(x) : |x-y| < \varepsilon, \ x \in U_\varepsilon(y)\} = p(y). \tag{7.2.3}$$

So, (7.2.2) is valid for $y \in \{x \in \mathbb{R}^n : p(x) = \infty\}$. Let $C$ be the smallest convex cone with vertex at 0 and containing $K$. If $y \in C \setminus \{0\}$, then $0 < p(y) < \infty$ ($K$ is compact, see Theorem 1.3.15). Then for all $x \in U_\varepsilon(y)$ we have $p_\varepsilon(x) = p(y) \cdot p_\varepsilon(\tau) \leqslant p(y)$, where $\tau = x/p(y)$, because $\tau_0 = y/p(y) \in K$ (see Theorem 1.3.17) and $S_\varepsilon(\tau_0) = \{\tau \in \mathbb{R}^n : |\tau - \tau_0| < \varepsilon\} \subset O_\varepsilon(K) \subset \{\tau \in \mathbb{R}^n : p_\varepsilon(\tau) < 1\}$. Combining this with (7.2.3), we arrive at (7.2.2) ▶

**Theorem 7.2.2.** *In the notation of Theorem* 6.2.19, *let* $\rho \in Y$. *Then there exists an entire function* $f$ *the order function* $\rho_f(u)$ *of which coincides with* $\rho(u)$, $u \in \mathbb{R}^n$.

◀ Let $\rho_s(y)$ be the Minkowski functional of the set $O_{1/\sqrt{s}}(K_\rho)$ (see Lemma 7.2.1), where $K_\rho$ is the compact convex set associated with $\rho$ (see (6.3.8)). The required function can be defined by the formula

$$f(z) = \sum_{s=o}^{\infty} \sum_{\|k\|=s} a_k z^k,$$

where

$$a_k = \exp\{-p_s(k) \cdot \ln s\} = s^{-p_s(k)}. \tag{7.2.4}$$

A. The set $\overline{O_\varepsilon(K_\rho)}$ is convex and compact for every $\varepsilon > 0$. Let $\alpha_s = \inf\{p_s(y) : \|y\| = 1; \ y \in \mathbb{R}^n_+\}$. Then $\alpha_s > 0$, and $\alpha_s$ is monotone nondecreasing as $s \to \infty$ (see Proposition 1.3.14). Hence, $-s^{-1} \cdot \ln|a_k| \geqslant \alpha_s \cdot \ln s$, $s = \|k\|$ and, by Lemma 7.1.2, $f$ is an entire function.

B. Let $u \in \{v \in \mathbb{R}^n : \max\{v_1, \ldots, v_n\} > 0\}$. We put $y_k = k[p_s(k)]^{-1}$. Since $p_s(y_k) = 1$, we have $y_k \in O_{1/\sqrt{s}}$ and $\langle y_k, u\rangle \leqslant \rho(u) + |u|/\sqrt{s}$ (see Exercise 1.3.20). Hence, by (7.2.4),

$$\frac{\langle k, u\rangle \ln\langle k, u\rangle}{-\ln|a_k|} \leqslant \left(\rho(u) + \frac{|u|}{\sqrt{s}}\right) \frac{\ln s + \max\{u_1, \ldots, u_n\}}{\ln s}.$$

Therefore, from Theorem 7.1.4 we deduce that $\rho_f(u) \leqslant \rho(u)$. So, if $\rho(u) = 0$, then $\rho_f(u) = 0$. Suppose $\rho(u) > 0$, and take $y_0 \in \partial\rho(u)$ (see (6.3.9)). Then $p(y_0) = 1$, where $p(y)$ is the Minkowski functional of $K_\rho$, and $\|y_0\| > 0$ (see Propositions 6.4.1 and 6.3.7). If $s$ is a sufficiently large natural number, the inequality

$$\sum_{i=1}^{n} (y_i - s\tau_i^{(0)})^2 < [p(\tau_0)]^2 s, \quad \tau_0 = y_0/\|y_0\|$$

has at least one integral solution $(k_1^{(s)}, \ldots, k_n^{(s)}) = k^{(s)} \in \mathbb{R}^n_+$ satisfying the condition $\|k^{(s)}\| = s$. Consequently, $|k^{(s)} \cdot \|k^{(s)}\|^{-1} - \tau_0| < p(\tau_0) \cdot (\sqrt{s})^{-1}$. By

Lemma 7.2.1,

$$\rho_f(u) \geqslant \varlimsup_{\langle k^{(s)},u\rangle\to\infty} \frac{\langle k^{(s)},u\rangle \cdot \ln\langle k^{(s)},u\rangle}{p_s(k^{(s)})\ln s} = \rho(u).$$

Thus, $\rho_f = \rho$ ▶

**7.2.2. Existence of an entire function with a given junction condition for type functions.**

**Theorem 7.2.3.** *In the notation of Definition* 6.4.2, *let* $\rho \in Y$, $\Phi \in \mathfrak{M}_n(\rho) \cap \mathfrak{C}$. *Then there exists an entire function* $f \in \mathfrak{M}_n(\rho)$ *the type* $\mathcal{A}_f$ *of which coincides with the type* $\mathcal{A}_\Phi$ *of* $\Phi$ (see Definition 6.3.1).

By Theorem 6.4.3, this means that there exists an entire function with a prescribed junction condition for the type function. To prove Theorem 7.2.3, we need several auxiliary assertions.

**Lemma 7.2.4.** *For* $\rho \in Y$, *take* $\Psi \in \mathfrak{M}_n(\rho)$ *and put* $V(u) = \ln\Psi(e^u)$. *Let* $G = V^*$ *be the Young transform of* $V$. *Then*

*a)*

$$\alpha := \rho(x) = \varlimsup_{t\to\infty} t^{-1}\cdot\ln^+ V(xt) = \varlimsup_{\langle y,x\rangle\to\infty} \frac{\langle y,x\rangle\ln\langle y,x\rangle}{G(y)}; \tag{7.2.5}$$

*b) if* $0 < \alpha < \infty$, $u \in \mathbb{R}^n$, *then*

$$\varlimsup_{t\to\infty} e^{t\alpha}\cdot V(xt+u) = (\alpha e)^{-1}\times$$

$$\times\left[\varlimsup_{\langle y,x\rangle\to\infty} \langle y,x\rangle \exp\{[\langle u,y\rangle - G(y)]\langle y,x\rangle^{-1}\}\right]^\alpha.$$

◀ By assumption, $\alpha < \infty$. We can find $t_0 = t_0(x)$ with the property

$$G(y) \geqslant \langle y,x\rangle t - V(xt) > \langle y,x\rangle t - e^{(\alpha+\varepsilon)t}\ \forall\, t > t_0, \tag{7.2.6}$$

where $\varepsilon > 0$ is any fixed number. Now, using Example 2 in Section 1.5, with sufficiently large $R$, we obtain:

$$G(y) \geqslant \frac{s}{\alpha+\varepsilon}\ln\frac{s}{(\alpha+\varepsilon)e},\quad s > R:\ s = \langle y,x\rangle. \tag{7.2.7}$$

By (7.2.7), the number $R$ can be chosen so that $G(y) > 0$ for all $s > R$. Denoting by $\tau$ the second limit in (7.2.5), from (7.2.7) we deduce that $\tau \leqslant \alpha$. So, if $\alpha = 0$, then (7.2.5) is valid.

If $\alpha > 0$, then $\varlimsup_{t\to\infty} t^{-1}\cdot\Phi(t) = \infty$, where $\Phi(t) = V(xt)$. By Theorem 1.6.5, for some $s_0 > 0$ we have: $(s_0,\infty) \subset \operatorname{dom}\Phi^*$. But for every $s \in (s_0,\infty)$ there exists $y \in \mathbb{R}^n$ such that $\langle y,x\rangle = s$, $\Phi^*(\langle y,x\rangle) = G(y)$ (see Proposition 1.5.3:

through a support straight line to the epigraph epi $\Phi \subset \mathbb{R}^{n+1}$, we can always draw a support hyperplane to epi $V$). Therefore,

$$\delta := \overline{\lim_{s\to\infty}} \frac{s\ln s}{\Phi^*(s)} \leqslant \tau \leqslant \alpha < \infty.$$

Hence, taking $\Delta = \delta + \varepsilon$, we find:

$$\Phi^*(s) \geqslant \frac{s}{\Delta} \ln \frac{s}{\Delta e} \quad \forall s > s_0,\ s_0 = s(\varepsilon). \tag{7.2.8}$$

By Propositions 1.5.13 and 1.4.22,

$$\Phi(t) = st - \Phi^*(s),\ s \in \partial\Phi(t) = [\Phi'_-(t), \Phi'_+(t)],$$

and $\partial\Phi(t) \subset (s_0, \infty)$ for sufficiently large $t$ . Using (7.2.8), estimating $\Phi$ from above, and passing to the limit as $t \to \infty$, we see that $\alpha \leqslant \delta$, $\alpha = \tau = \delta$, and formula (7.2.5) is true. Assertion b) is proved similarly. Here $\Phi(t) = V(xt + u)$, $\Phi^*(\langle y, x\rangle) = G(y) - \langle u, y >$, and we use the formula

$$\overline{\lim_{t\to\infty}} \Phi(t) = (\alpha e)^{-1} \left[\overline{\lim_{s\to\infty}} \exp\left\{-\frac{\Phi^*(s)}{s}\right\}\right]^\alpha,$$

which is verified as above ►

**Lemma 7.2.5.** *Let $\Phi, \rho$ be the functions occurring in Theorem 7.2.3. We put $V(u) = \Phi(e^u)$, $G = V^*, C =$ dom $G$, and denote by $p$ the Minkowski functional of the compact set $K_\rho$ the support function of which is $\rho$. Next, let $T =$ dom $p$. Then* ri $T =$ ri $C$,

$$\lim_{\lambda\to\infty} \frac{G(\lambda y)}{\lambda \ln \lambda} = p(y) \quad \forall\ y \subset \text{ri } C, \tag{7.2.9}$$

*(in particular, this limit exists), and $C$ is a cone in $\mathbb{R}^n$ with vertex at 0. If* dim $C - 1$, *then $T = C$ and formula (7.2.9) is valid for $y \in C$.*

◄ 1. The following formula is similar to assertion a) of Lemma 7.2.4:

$$A(y) := \overline{\lim_{\lambda\to\infty}} \frac{G(\lambda y)}{\lambda\ln\lambda} = \overline{\lim_{\langle u,y\rangle\to\infty}} \frac{\langle u, y\rangle}{W(u)},\quad y \in C, \tag{7.2.10}$$

where $W = \ln^+ V$ (see the proof of (7.2.5); in this case the upper limit is replaced by the lower limit). Since $V$ is a logarithmically convex function, $\Psi(u,t) := V(xt + u)e^{-t\rho(x)}$ is a convex function in the totality of the variables $u, t$ in $\mathbb{R}^n \times \mathbb{R}$ for every fixed $x$ (see Theorem 1.4.20). The Young transform of $\Psi$ in the variable $u$

$$\Psi^*(y,t) = G(ye^{\rho t})e^{-\rho t} - t\langle y, x\rangle,\ \rho = \rho(x)$$

is a concave function in the variable $t$. Therefore, the ratio $(\ln\lambda)^{-1}\times[\varphi(\lambda)-\varphi(1)]$, where $\varphi(\lambda):=\lambda^{-1}\cdot G(\lambda y)$, is monotone nonincreasing (see P. 1.4.22) and the limits on the left-hand sides of (7.2.9) and (7.2.10) exist for all $y\in C$.

2. We examine geometric properties of the set $C=\operatorname{dom} G$. From the property of $\varphi$ mentioned above it follows that $\lambda y\in C$ for $\lambda>1$ if $y\in C$. But $V\geqslant 0$ and, consequently, $0\in C$. Since $C$ convex, it is a cone in $\mathbb{R}^n$ with vertex at 0. We consider the asymptotic function $VO^+$ of $V$. Obviously, $VO^+(x)=\infty$ for all $x\in D_\rho$. Let $M=\mathbb{R}^n\setminus D_\rho=\{u\in\mathbb{R}^n:\ \rho(u)=0\}$, and let $x\in M$. By Lemma 6.3.12, the limit $\lim\limits_{t\to\infty}V(xt)$ exists and is finite. So, $VO^+(x)=0$. The function $VO^+$ is support for the cone $K$ polar to $M$, i.e.,

$$K=\{y\in\mathbb{R}^n:\ \langle y,x\rangle\leqslant 0\quad\forall\, x\in M\}.$$

By Theorem 1.6.5, we have $K=\overline{C}$, where $K_\rho=\overline{\operatorname{dom} W^*}$, $W=\ln^+V$. Therefore, if $T$ is the smallest convex cone with the vertex at zero and containing the compact set $K_\rho$, then $\operatorname{ri} T=\operatorname{ri} C$ and this cone is generated by the set $H=\{y\in\operatorname{ri} C:\ p(y)=1\}$, where $p$ is the Minkowski functional of $K_\rho$

.

3. It suffices to prove (7.2.9) for $y\in H$, because the functions $p$ and $A$ are homogeneous (see (7.2.10)). Let $y_0\in H$. Since $0\in\operatorname{dom} W^*$, we see that $\lambda y_0\in\operatorname{dom} W^*\quad\forall\,\lambda\in(0,1)$. By (7.2.10) and the inequality $W(u)+W^*(\lambda y_0)\geqslant\langle u,\lambda y_0\rangle\ \forall\, u\in\mathbb{R}^n$, we have $A(y_0)\leqslant 1$. On the other hand, $W(u)\leqslant W(0)+\rho(u)\ \forall\, u\in\mathbb{R}^n$, because epi $\rho$ is an asymptotic cone of the convex cone epi $W$ (see Theorem 1.6.5 and Proposition 1.6.1) and $(0,W(0))+\operatorname{epi}\rho\subset\operatorname{epi} W$. By (7.2.10), this inequality implies the relation

$$A(y_0)\geqslant\varlimsup_{\langle u,y_0\rangle\to\infty}\langle u,y\rangle[\rho(u)]^{-1}.$$

The limit on the right is equal to 1, because $y_0\in\partial K_\rho\cap\operatorname{ri}C$ and, by Lemma 6.4.1, $y_0\in\partial\rho(x_0)$ for some $x_0\in D_\rho$ (see (6.3.6) and (6.3.10)); moreover,

$$\langle u,y_0\rangle\leqslant\rho(u)\quad\forall\, u\in\mathbb{R}^n,\qquad\langle x_0,y_0\rangle=\rho(x_0)>0.$$

Thus, $A(y_0)\geqslant 1$, and, consequently, $A(y_0)=1$.

The last assertion of the lemma is proved similarly and is based on the identity $\rho(u)=\max\{0,\langle u,\alpha\rangle\}$, $u\in\mathbb{R}^n$ valid for the element $\alpha\in\mathbb{R}^n\setminus\{0\}$ such that $K_\rho=[0,\alpha]$ is the segment with the ends 0 and $\alpha$ ▶

**Lemma 7.2.6.** *In the notation of Theorem 7.2.3, let $V(u)=\Phi(e^u)$, and let $A_\Phi=\{\sigma_\Phi(\cdot;x),\ x\in D_\rho\}$ be the type of $\Phi$. We put $G=V^*,C=\operatorname{dom} V^*$, and denote by $\varphi_x$ the Young transform of the function $\sigma_x(u)=\sigma_\Phi(e^u;x)$. Then* $\operatorname{dom}\varphi_x\subset C$ *and the following limit exists:*

$$\lim_{\lambda\to\infty}F_\lambda(y;v)=\varphi_x(y)\quad\forall\, y\in C;\ v\subset B(y),$$

*where*

$$F_\lambda(y;v)=\frac{G(\lambda y+v)}{\lambda}-\frac{\langle y,x\rangle}{\rho(x)}\ln\lambda,$$

$$B(y) = \{\beta \in \mathbb{R}^n : yt + \beta \in C \;\; \forall\, t \geqslant t_0(\beta)\}.$$

◀ 1. Let $g_t(u,v) = e^{-t\rho(x)} \cdot [V(xt+u) - \langle u, v\rangle]$ . Then

$$\sigma_x(u) = \inf_{\delta>0} \sup_{t>\delta} g_t(u,v) \quad \forall\, (u,v) \in \mathbb{R}^n \times \mathbb{R}^n$$

By Theorem 1.5.12, we have

$$\varphi_x(y) = \sup_{\delta>0}[\sup_{t>\delta} g_t(u,v)]^* \leqslant \sup_{\delta>0} \inf_{t>\delta} \; g_t^*(y,v).$$

This immediately implies the estimate

$$\varphi_x(y) \leqslant \varliminf_{\lambda\to\infty} F_\lambda(y;v), \quad (u,v) \in \mathbb{R}^n \times \mathbb{R}^n. \tag{7.2.11}$$

2. Let $v = 0$. By Lemma 6.3.12, the limit

$$\sigma_x(u) = \lim_{t\to\infty} e^{-t\rho(x)} \cdot V(xt+u),$$

exists, and $\sigma_x(u) = \inf_{t>0} g_t(u,0)$. Hence, by Theorem 1.5.12,

$$\varphi_x(y) = \sup_{t>0} g_t^*(y,0) = \lim_{\lambda\to\infty} F_\lambda(y;0); \text{ dom } \varphi_x \subset C.$$

3. Let $\Phi(y,v) = \varlimsup_{\lambda\to\infty} F_\lambda(y;v), \quad y \in C,\; v \in B(y)$ (see the conditions of the lemma). We can find $\tau > 0$ such that $p = y\tau + v \in C$. It is easy to check that $\Phi(y,v) = \Phi(y,p)$. Since $G$ is a convex function, and $C$ is a cone (see Lemma 7.2.5), we have

$$\lambda^{-1} \cdot G(\lambda y + p) \leqslant \frac{\alpha}{\lambda} \cdot G\left(\frac{\lambda}{\alpha} y\right) + \frac{\beta}{\lambda} \cdot G\left(\frac{p}{\beta}\right), \alpha + \beta = 1;\ \alpha, \beta > 0.$$

Letting $\lambda \to \infty$ and assuming that $y_0 \in C$, by the arguments of item 2 we have:

$$\Phi(y,p) \leqslant \Phi(y,0) - \frac{\langle y, x\rangle}{\rho(x)} \ln\alpha \quad \forall \alpha \in (0,1),$$

or $\Phi(u,v) \leqslant \varphi_x(y),\; y \in C$. Together with (7.2.11), this yields Lemma 7.2.6 ▶

**Lemma 7.2.7.** *In the notation of Lemma 7.2.6, let $x \in D_\rho$ . We put $\delta_x(u) = \ln\sigma_x(u)$ and $\Gamma = \{y \in \mathbb{R}^n : \langle y, x\rangle = \rho(x)\}$. Let $\delta_x^*$ and $\sigma_x^*$ be the Young transforms of the functions $\delta_x$ and $\sigma_x$, respectively. Then $\sigma_x^*(y) = \delta_x^*(y) - 1 \;\; \forall\, y \in \Gamma$.*

◀ The function $\sigma_x$ is logarithmically convex; consequently, either $\sigma_x(u) \equiv 0,\ u \in \mathbb{R}^n$, or $\sigma_x(u) > 0,\ u \in \mathbb{R}^n$ (see Propositions 6.3.2, 6.1.5). In the first case the lemma is obvious. In the second case we have $\sigma_x^*(y) \geqslant \delta_x^*(y) - 1 \;\; \forall\, y \in \mathbb{R}^n$, because $e^\alpha \geqslant 1 + \alpha \;\; \forall\, \alpha \in \mathbb{R}$. The relation

$$\delta_x(u + xt) \equiv t\rho(x) + \delta_x(u),\ (u,t) \in \mathbb{R}^n \times \mathbb{R}, \tag{7.2.12}$$

implies that for every $v \in \mathbb{R}^n$ there exist $t \in \mathbb{R}$, $u \in \mathbb{R}^n$ such that $v = u + xt$ and $\delta_x(u) = 0$. Let $S = \{u \in \mathbb{R}^n : \delta_x(u) = 0\}$. Then, taking (7.2.12) into account, we have:

$$\delta_x^*(y) = \sup_{u \in S}\langle u, y\rangle \quad \forall\, y \in \Gamma; \sigma^*(y) \geqslant \sup_{u \in S}\{\langle u, y\rangle - e^0\} = \delta_x^*(y) - 1 \blacktriangleright$$

**Proof of Theorem 7.2.3.** 1. Let $X_1$ be the affine hull of the set $C = \operatorname{dom} G$, where $G = V^*$. We put $V(u) = \Phi(e^u)$ and denote by $X_2 = X_1^{\perp}$ the orthogonal complement of the subspace $X_1$ ($0 \in C$ by Lemma 7.2.5) in $\mathbb{R}^n$. Next, we put $S = X_2 \cap \{y \in \mathbb{R}^n : |y| \leqslant \delta\}$, where $\delta > 0$ is such that $C + S$ contains an infinite number of integral points (if $\dim X_1 = n$, then $S = \{0\}$). On the set $C + S$ we define the function $N(y) = N(p \oplus q) := G(p)$, where $p \in C$ and $q \in S$. Consider the function

$$f(z) = \sum_{||k|| \geqslant 0} c_k z^k = \sum_{k \in T} e^{-N(k)} z^k; \; z^k = z_1^{k_1} \dots z_n^{k_n}, \tag{7.2.13}$$

where $T = \mathbb{R}_+^n \cap (C + S)$. Then $T \neq \varnothing$, because $C \subset \mathbb{R}_+^n$ (see Proposition 1.5.13 and (1.4.7): the partial derivatives (left and right) of $V$ are nonnegative because $V$ is monotone increasing in each variable). By Theorem 1.5.6 we have $G^* = V$. Since $\operatorname{dom} V = \mathbb{R}^n$, the asymptotic function $GO^+(x)$ for G is identically equal to $\infty$, $x \in \mathbb{R}^n \setminus 0$ (see Theorem 1.6.5). Using formula (1.6.6) for the function $GO^+$, we find

$$\lim_{||k|| \to \infty} ||k||^{-1} \cdot N(k) = \lim_{||p^{(k)}|| \to \infty} ||p^{(k)}||^{-1} \cdot G(p^{(k)}) = \infty; \; k = p^{(k)} \oplus q^{(k)}.$$

Therefore, (7.1.1) is fulfilled for $f$ (see (7.2.13)); consequently, $f$ is an entire function.

2. We define the order function $\rho_f$ for $f$ with the aid of Theorem 7.1.4. By Lemma 7.2.4 with $u \in D = \mathbb{R}^n \setminus \mathbb{R}_-^n$ we have:

$$\rho_f(u) = \overline{\lim_{\langle k, u\rangle \to \infty}} \frac{\langle k, u\rangle \ln\langle k, u\rangle}{G(p^{(k)})}$$

$$= \overline{\lim_{\langle p^{(k)}, u\rangle \to \infty}} \frac{\langle p^{(k)}, u\rangle \ln\langle p^{(k)}, u\rangle}{G(p^{(k)})} \leqslant \rho(u), \tag{7.2.14}$$

where $\mathbb{R}_-^n = \{r \in \mathbb{R}^n : r_1 \leqslant 0; r_n \leqslant 0\}$. Suppose $u \in D_\rho$. For the compact convex set $K_\rho$, we denote by $F_\rho$ the part of its boundary in ri $C$, and put $\Pi = \{y \in \mathbb{R}^n : \langle y, u\rangle > 0\} \cap \operatorname{ri} C$ ($\Pi \neq \varnothing$ by the proof of Lemma 7.2.5, item 2). For $y \in F_\rho \cap \Pi$, consider the set

$$E = \{x \in \mathbb{R}^n : |\langle y, u\rangle x - \langle x, u\rangle y| < \varepsilon\}, \tag{7.2.15}$$

where $\varepsilon > 0$ is such that $\{x \in \mathbb{R}^n : \langle y, x\rangle > M\} \cap E \subset C + S$ for some $M > 0$. The set $E$ is an $n$-dimensional convex cylinder having directrix $y$ and symmetric relative to 0; consequently, $E$ is of infinite volume. By the Minkowski theorem (see Section 1.7), $E$ contains infinitely many integral points $\{k^{(s)}, s = 1, 2, \dots\}$, because $y \in K_\rho \subset \mathbb{R}^n_+$ (see Proposition 6.3.8). We order these points in such a manner that $\langle k^{(s)}, u\rangle \to \infty$ as $s \to \infty$. Let $k^{(s)} = p^{(s)} \oplus q^{(s)}, p^{(s)} \in C, q^{(s)} \in S$. By (7.2.15), we have: $p^{(s)} \cdot \langle k^{(s)}, u\rangle^{-1} \to y \cdot \langle y, u\rangle^{-1}$ as $s \to \infty$. In Lemma 7.2.5 the sequence of convex functions $\varphi_\lambda(y) := (\lambda \ln \lambda)^{-1} G(\lambda y)$ converges uniformly in $y$ on the compact subsets of ri $C$ as $\lambda \to \infty$ (see Theorem 1.4.27). From (7.2.9) and the continuity of the sublinear function $p$ in ri $C$ we deduce that $\lim\limits_{\lambda\to\infty; v\to y} \varphi_\lambda(v) = p(y)$, where $v, y \in$ ri $C$ (see Theorem 1.3.16, Theorem 1.4.4, and Lemma 6.2.22). Recalling that $p(y) = 1$ for $y \in (\text{ri } C) \cap \partial K_\rho$ (see Theorem 1.3.17), from the above and Theorem 7.1.4 (see (7.2.13)) for $y \in F_\rho \cap \Pi$ we obtain:

$$\rho_f(u) = \overline{\lim_{\langle k,u\rangle\to\infty}} \frac{\langle k, u\rangle \ln\langle k, u\rangle}{N(k)} \geqslant \lim_{s\to\infty} \frac{\langle k^{(s)}, u\rangle \ln\langle k^{(s)}, u\rangle}{G(p^{(s)})} = \langle y, u\rangle. \quad (7.2.16)$$

Then $\rho_f(u) \geqslant \langle u, y\rangle \;\; \forall\, y \in \Gamma_\rho \cap \Pi$, where $\Gamma_\rho = \overline{F_\rho}$ . Let $y_0 \in \partial\rho(u)$. By Lemma 6.4.1, $y_0 \in \Gamma_\rho$ . Therefore (see (6.3.6)), $\rho_f(u) \geqslant \langle u, y_0\rangle = \rho(u)$. If dim $C = 1$, this identity is verified similarly and is based on the last assertion of Lemma 7.2.5. Combining this with (7.2.14), we conclude that $\rho_f = \rho$, i.e., $f \in \mathfrak{M}_n(\rho)$.

3. The type $\mathcal{A}_\Phi$ of a function $\Phi \in \mathfrak{M}_n(\rho)$ is determined by the system $\mathcal{A}_\Phi = \{\sigma_\Phi(\cdot; x) : x \in T_\rho\}$, where $T_\rho = \{u \in \mathbb{R}^n : \rho(u) = 1\}$ (see Section 6.3). Let $\varphi_f(u, x) = \ln \sigma_f(e^u; x)$, where $\sigma_f(\cdot; x)$ is the $x$-type function of a function $f$ (see (7.2.13)). We use (7.1.8), taking into account that $f \in \mathfrak{M}_n(\rho)$:

$$\begin{aligned} 1 + \varphi_f(u, x) &= \overline{\lim_{\langle k,x\rangle\to\infty}} \left[\tfrac{\langle k,u\rangle - N(k)}{\langle k,x\rangle} + \ln\langle k, x\rangle\right] = \\ &= \overline{\lim_{\langle p^{(k)},x\rangle\to\infty}} \left[\tfrac{\langle p^{(k)},u\rangle - G(p^{(k)})}{\langle p^{(k)},x\rangle} + \ln\langle p^{(k)}, x\rangle\right], \end{aligned}$$

where $k = p^{(k)} \oplus q^{(k)}$. As in the proof of inequality (7.2.14), by Lemma 7.2.4, in the notation of Lemmas 7.2.6 and 7.2.7 we obtain:

$$\varphi_f(u, x) \leqslant \delta_x(u) \;\;\forall\, u \in \mathbb{R}^n. \quad (7.2.17)$$

As was mentioned in the proof of Lemma 7.2.7, we have $|\delta_x(u)| < \infty \;\forall\, u \in \mathbb{R}^n$. Fix $u \in \mathbb{R}^n$. In the proof of the "only if" part of Theorem 6.4.3, (see also Theorem 6.3.10), it was established that dom $\delta_x^* \subset \partial\rho(x)$. Therefore, there exists a point $y_0 \in$ dom $\delta_x^* \subset \partial\rho(x)$ such that $\delta_x^*(u) = \langle u, y_0\rangle - \delta_x^*(y_0)$, $y_0 = y_0(u)$.

Let $m$ be an integral point in $T$ (see item 1) such that $p \in$ ri $C$, where $p \oplus q = m$, $q \in S$, and $v_0 = m - \langle m, x\rangle y_0$. We consider the convex cylinder $Q = \{y \in \mathbb{R}^n : |y - \langle y, x\rangle y_0 - v_0| \leqslant \varepsilon\}$ symmetric with respect to $m$, where

$\varepsilon > 0$ is such that $Q \cap \{y \in \mathbb{R}^n : \langle x, y\rangle > M\} \subset \text{ri } C + S$ for some $M > 0$ (see item 1). As in item 2, we apply the Minkowski theorem to find a sequence $k^{(s)}, s = 1, 2, \ldots$, of integral points in $O \cap T$ such that $\langle k^{(s)}, x\rangle \to \infty$ as $s \to \infty$, and $k^{(s)} = p^{(s)} \oplus q^{(s)}$ $\forall\ s \in \mathbb{N}$. We write $v_0 = a \oplus b$, where $a \in X_1, b \in X_2, H = \{v \in X_1 : |v - a| \leqslant \varepsilon\}$, and $v^{(s)} = p^{(s)} - y_0\langle k^{(s)}, x\rangle$, $s \in \mathbb{N}$. By the Pythagorean theorem, $||u_1 + u_2||^2 = ||u_1||^2 + ||u_2||^2, u_i \in X_i, i = 1, 2$, where $||\cdot||$ is the norm in $\mathbb{R}^n$. Taking the structure of the cylinder $Q$ into account, we see that $v^{(s)} \in H$. Lemmas 7.2.6 and 7.2.7 imply $y_0 \in \text{dom } \sigma_x^* \subset C$. By Theorem 1.4.27, the sequence $F_\lambda(y_0; v)$ considered in Lemma 7.2.6 converges uniformly in the variable $v$ on every compact subset of ri $B(y_0)$ and, consequently, by Lemmas 7.2.6 and 7.2.7,

$$\lim_{s\to\infty} F_\lambda(y_0; v^{(s)}) = \sigma_x^*(y_0) = \delta_x^*(y_0) - 1,$$

where $\lambda_s = \langle k^{(s)}, x\rangle$, $s = 1, 2, \ldots$ . On the other hand, by formula (7.1.8), in the notation adopted we have

$$\varphi_f(u; x) + 1 = \overline{\lim_{\langle k,x\rangle\to\infty}} \left[\tfrac{\langle k,u\rangle - G(p^{(k)})}{\langle k,x\rangle} + \ln\langle k, x\rangle\right] \geqslant$$
$$-\lim_{s\to\infty} F_{\lambda_s}(y_0; v^{(s)}) + \langle u, y_0\rangle = \langle u, y_0\rangle - \delta_x^*(y_0) + 1 = \delta_x(u)$$

i.e., $\sigma_f(e^u; x) \geqslant \sigma_x(u)$ for all $u \in \mathbb{R}^n$. Together with (7.2.17), this finishes the proof ($x$ is an arbitrary element of $T_\rho$).

We observe that if $\dim\ C = 1$, then (see item 2) $\max\{\langle u, \alpha\rangle, 0\} = \rho(u)$, where $\alpha \in \mathbb{R}^n_+$, and in the above notation we have:

$$y_0 = \alpha, \quad \sigma_x(u) = A_x \exp\{\langle u, \alpha\rangle\}, \quad A_x = \exp\{-\delta_x^*(\alpha)\} \quad \forall\ x \in D_\rho \ \blacktriangleright$$

## 7.3. The maximum of the modulus and the maximal term of an entire function: comparative growth

The following one variable fact is famous for its numerous applications.

**Proposition 7.3.1.** *If* $f(z) = \sum_{k=0}^{\infty} a_k z^k$ *is an entire function of finite order,* $M_f(r) = \max_{|z|=r} |f(z)|$ *is the maximum of its modulus, and* $\mu_f(r) = \max_{k\geqslant 0} |a_k| r^k$ *is its maximal term, then*

$$\lim_{r\to\infty} (\ln \mu_f(r))^{-1} \cdot \ln M_f(r) = 1. \tag{7.3.1}$$

In the case of several variables no immediate analog of (7.3.1) is available, as is shown by the example below.

**Example 1.** Let $f(z_1, z_2) = z_2 e^{z_1}$, and let

$$\Gamma = \{r \in \mathbb{R}^2_+ \,:\, r_1 = e^t;\, r_2 = \exp\{t - e^t\},\, t \geqslant 0\}. \tag{7.3.2}$$

Then $M_f(r) = r_2 e^{r_1}$, $\mu_f(r) = r_2 r_1^{[r_1]}([r_1]!)^{-1}$, where $[r_1]$ is the integral part of $r_1$. By using the Stirling formula, it is easily seen that

$$\lim_{r\in\Gamma;\|r\|\to\infty} (\ln \mu_f(r))^{-1} \cdot \ln M_f(r) = 2.$$

However another analog of (7.3.1) is valid, as we shall see shortly. Clearly, if $P(z)$ is a polynomial of $n$ complex variables $z = (z_1, \dots, z_n)$, then

$$\lim_{\mu_P(r)\to\infty} (\ln \mu_P(r))^{-1} \cdot \ln M_P(r) = 1.$$

The following statement is central in this section.

**Theorem 7.3.2.** *Let $f$ be a transcendental entire function (i.e., not a polynomial) of finite order $\rho(f)$ in the totality of variables. Put*

$$\gamma_f(u) = \lim_{t\to\infty} (\ln t)^{-1} \cdot \ln^+ M_f(t^u), \quad u \in \mathbb{R}^n, \tag{7.3.3}$$

*$D_\gamma = \{u \in \mathbb{R}^n \,:\, \gamma_f(u) > 0\}$. Let $K$ be an arbitrary cone in $\mathbb{R}^n$ with vertex at $0$ and such that $K \setminus \{0\} \subset D_\gamma$. Then*

$$\lim_{|u|\to\infty;u\in K} [\ln \mu_f(e^u)]^{-1} \cdot \ln M_f(e^u) = 1. \tag{7.3.4}$$

*This is true, in particular, if $f \in \mathfrak{M}_n$ and $K \setminus \{0\} \subset D_\rho \subset D_\gamma$, where $D_\rho = \{u \in \mathbb{R}^n \,:\, \rho_f(u) > 0\}$ and $\rho_f(u)$ is the order function for $f$.*

Before proving Theorem 7.3.2, we make several remarks. In the general case the cone $D_\gamma$ does not coincide with the cone $G_\gamma = \{x \in \mathbb{R}^n \,:\, \gamma_f(x) = \infty\}$. For instance, in Example 1 we have $D_\gamma = \{u \in \mathbb{R}^2 \,:\, \max\{u_1, u_2\} > 0\}$, and $G_\gamma = \{u \in \mathbb{R}^2 \,:\, u_1 > 0\}$. The set $D_\gamma$ is the growth cone for the convex (see the Valiron theorem) function $V_f(u) = \ln^+ M_f(e^u)$ (and for the function $\ln \mu_f(e^u)$, see Property 6.1.2), i.e., $D_\gamma$ is the union of all rays in $\mathbb{R}^n$ outgoing from $0$ and such that $V_f$ is strictly monotone increasing on each particular ray as the variable tends to $\infty$.

One of the reasons why the "immediate" multidimensional analog of formula (7.3.1) may fail for $n > 1$ (see (7.3.2)) is that in general the cone $\{u \in \mathbb{R}^2 \,:\, \max\{u_1, u_2\} > 0\}$ may contain paths going to $\infty$ on which the functions $V_f$ and $\ln^+ \mu_f(e^u)$ are not monotone increasing. This is shown by

**Example 2.** Let $f = 1 + z_1 z_2$. The relation

$$\lim_{\|r\|\to\infty} (\ln \mu_f(r))^{-1} \cdot \ln^+ M_f(r) = 1$$

is invalid: the functions $M_f, \mu_f$ are constant, e.g., on the set $\{r \in \mathbb{R}^2_+ \,:\, r_1 r_2 = 10\}$.

**Lemma 7.3.3.** *In the notation and under the conditions of Theorem* 7.3.2, *we have*

$$\begin{aligned}\lim_{t\to\infty}(\ln t)^{-1}\cdot\ln\mu_f(t^x) &= \gamma_f(x),\ x\in\mathbb{R}^n\\ \lim_{t\to\infty}(\ln t)^{-1}\cdot\ln^+\ln^+\mu_f(t^x) &= \rho_f(x),\ x\in\mathbb{R}^n.\end{aligned}\tag{7.3.5}$$

◀ Let

$$S_f(r)=\sum_{||k||\geqslant 0}|a_k|r_1^{k_1}\dots r_n^{k_n};\ B=\prod_{i=1}^{n}(1-\alpha_i^{-1})^{-1}.$$

For all $\alpha_1,\dots,\alpha_n>1$ we have the estimates

$$\mu_f(r)\leqslant M_f(r)\leqslant S_f(r)\leqslant B\mu_f(\alpha_1 r_1,\dots,\alpha_n r_n)\ \forall\, r\in\mathbb{R}^n_+.\tag{7.3.6}$$

The lemma now follows the fact that the epigraphs of the convex function $v(u)=\ln^+\mu_f(e^u)$ and of its translate $v(u_1+\ln\alpha_1,\dots,u_n+\ \ln\ \alpha_n)$ have one and the same asymptotic cone and, consequently, one and the same asymptotic functions.

A similar statement is also true for the pair of quasiconvex functions $\ln^+\ln^+\mu_f(e^u)$ and $\ln^+\ln^+M_f(e^u)$ (see Definitions 1.6.2, 1.6.4, and Theorem 6.2.21) ▶

**Lemma 7.3.4.** *Let $X$ be a closed convex subset of $\mathbb{R}^n$, and $A(X)$ the asymptotic cone of $X$. Put $||x||=\sum_{i=1}^{n}|x_i|$, $X_c=X\bigcap\{x\in\ \mathbb{R}^n:\ ||x||\geqslant c\}$, and $P=\bigcap\overline{P(X_c)}|c>0$. Then $A(X)=P$.*

◀ Clearly, $0\in A(X)\cap P$. Thus, it suffices to show that $A(X)\setminus\{0\}=P\setminus\{0\}$. Let $x\in A(X)\setminus\{0\}$. There exist two sequences $\{\lambda_n>0\}_1^\infty$, and $\{x_n\in X\}_1^n$ such that $\lim_{n\to\infty}\lambda_n x_n=x$, $\lim_{n\to\infty}\lambda_n=0$. Since

$$\lim_{n\to\infty}||\lambda_n x_n||=\lim_{n\to\infty}\lambda_n||x_n||=||x||\neq 0,$$

we see that $\lim_{n\to\infty}||x_n||=\infty$. Therefore, for every $\varepsilon>0$ we have $\lambda_n x_n\in P(X_c)$, $\forall\, n>n_0(c)$, i.e. $x\in\overline{P(x_c)}$. Thus, $x\in P\setminus\{0\}$ and $A(X)\subset P$. Now, let $x\in P\setminus\{0\}$. Take two sequences $\{\varepsilon_n>0\}_1^\infty$ and $\{c_n>0\}_1^\infty$ such that $\lim_{n\to\infty}\varepsilon_n=0$, $\lim_{n\to\infty}c_n=\infty$, and the $c_n$ are monotone increasing. Since $P(X_c)\supset P(X_d)$ for $0<c<d$, by the definition of $P$ we obtain: $P=\bigcap\overline{P[X(c_n)]}|n\in\mathbb{N}$, where $X(d):=X_d$. Consequently, $x\in\overline{P[X(c_n)]}\ \forall\, n\in\mathbb{N}$. Therefore, there is a sequence $\{y_n\in P[X(c_n)]:\ n\in\mathbb{N}\}$ such that $||x-y_n||<\varepsilon_n$. Then $y_n=\lambda_n x_n$, $\lim_{n\to\infty}y_n=x$, where $\lambda_n>0$, $x_n\in X(c_n)\ \forall\, n\in\mathbb{N}$. We see that

$$0<\lambda_n c_n\leqslant||\lambda_n x_n||=||y_n||<||x||+\varepsilon_n.$$

Thus, $\lambda_n\to 0$ as $n\to\infty$; by Proposition 1.6.3, we have $x\in A(X)\setminus\{0\}$. So, $P\subset A(X)$, $P=A(X)$ ▶

**Lemma 7.3.5.** *Suppose $V$ is a lower semicontinuous convex function on $\mathbb{R}^n$ not identicaly equal to infinity. Let $X = \overline{\text{dom } V^*}$, let $VO^+$ be the asymptotic function for $V$, and let $T = \overline{\text{dom } VO^+}$. Then*

$$T = \{u \in \mathbb{R}^n : \quad \langle u, y \rangle \leqslant 0 \quad \forall\, y \in A(X)\}. \tag{7.3.7}$$

◀ Consider the indicator function of $X$: $\delta_X(u) = 0$ for $u \in X$, $\delta_X(u) = \infty$ for $u \in \mathbb{R}^n \setminus X$. The conjugate function $\delta_X^*$ is the support function for $X$ (see Example 1 in Section 1.5). But $X = \overline{\text{dom } V^*}$; therefore,

$$\delta_X^* = VO^+; \quad \overline{\text{dom } \delta_X^*} = T \tag{7.3.8}$$

(see (1.6.5)). For the same reason, the support function $\nu$ of the set dom $\delta_X^*$ coincides with the asymptotic function $\delta_X O^+ = \delta_{A(X)}$ for $\delta_X$. Taking the remark to Theorem 1.3.19 into account, we arrive at the formula

$$\begin{aligned} T &= \{u \in \mathbb{R}^n : \quad \langle u, y \rangle \leqslant \nu(y) \quad \forall\, y \in \mathbb{R}^n\} \\ &= \{u \in \mathbb{R}^n : \quad \langle u, y \rangle \leqslant 0 \quad \forall\, y \in A(X)\}, \end{aligned}$$

because $\nu = \delta_{A(X)}$ (see (7.3.8)) ▶

**Proof of Theorem 7.3.2.** Since $\gamma_f : \mathbb{R}^n \to (-\infty, \infty]$ is a lower semicontinuous function (see Theorem 1.6.5; (7.3.3)), the cone $D_\gamma$ is open. Therefore, (7.3.4) is equivalent to the condition

$$\lim_{u \to x; t \to \infty} \left(\ln \mu_f\left(e^{ut}\right)\right)^{-1} \cdot \ln M_f\left(e^{ut}\right) = 1 \quad \forall\, x \in D_\gamma. \tag{7.3.9}$$

Suppose that $\overset{\circ}{\Gamma} \neq \varnothing$, where

$$\Gamma = \{x \in \mathbb{R}^n : 0 < \gamma_f(x) < \infty\}. \tag{7.3.10}$$

Since $V_f(u) = \ln M_f(e^u)$ is a convex function, the sequence $\Phi_t(u) = t^{-1}[V_f(ut) - V_f(0)]$ increases as $t \to \infty$, and its limit $\gamma_f(u)$ (see (7.3.3)) is a convex (therefore, continuous) function on $\overset{\circ}{\Gamma}$. Consequently, $\Phi_t(u) \to \gamma_f(u)$ uniformly on every compact set $F \subset \overset{\circ}{\Gamma}$. The same is true for the function $v(u) = \ln\, \mu_f(e^u)$. Now, (7.3.9) for $x \in \overset{\circ}{\Gamma}$ follows from Lemma 7.3.3.

2. Since $f$ is of finite order $\rho(f)$ in the totality of variables, for every $\gamma > \rho(f)$ there exists $s > 0$ such that the Taylor coefficients $\{a_k\}$ of $f$ satisfy

$$|a_k| < \|k\|^{-\|k\|/\gamma}, \qquad \|k\| > s. \tag{7.3.11}$$

(see Theorem 7.1.4). We denote $N_f = \{k : a_k \neq 0\}$. Let $X$ be the closed convex hull of $N_f$, and $A(X)$ the asymptotic cone of $X$. We put $\Pi = \{y \in \mathbb{R}^n : \|y\| = \gamma\}$ and consider the $\varepsilon$-neighborhood $E_\varepsilon$ of the set $\Pi \cap A(X)$. Since

$X \subset \mathbb{R}^n_+$, by Lemma 7.3.4 there is a positive number $c = c(\varepsilon)$ (see (7.3.11)) such that

$$\overline{P(X_c)} \cap \Pi \subset E_\varepsilon. \tag{7.3.12}$$

Let

$$S(\varepsilon, u) = \max_{\alpha \in E_\varepsilon} e^{\langle \alpha, u \rangle}; \ T_\varepsilon(u) = \max\{c, 2^\gamma S(\varepsilon; u)\}. \tag{7.3.13}$$

Putting $B = \{k \in N_f : \|k\| > T_\varepsilon(u)\}$, from (7.3.11) we deduce that

$$M_f(e^u) \leqslant \sum_{k \in N_f} |a_k| e^{\langle k, u \rangle} \leqslant$$

$$\sum_{\|k\| \leqslant T_\varepsilon(u)} |a_k| e^{\langle k, u \rangle} + \sum_{k \in B} e^{\langle k, u \rangle} \cdot \ \|k\|^{-\|k\|/\gamma}. \tag{7.3.14}$$

Since $\gamma k/\|k\| \in E_\varepsilon \ \forall\ k \in B$, formulas (7.3.12), (7.3.13) imply:

$$T_\varepsilon(u) \geqslant \sup_{\alpha \in E_\varepsilon} 2^\gamma e^{\langle \alpha, u \rangle} > 2^\gamma e^{\gamma \langle k, u \rangle / \|k\|} \quad \forall\ k \in N_f.$$

Therefore,

$$e^{\langle k, u \rangle} \|k\|^{-\|k\|/\gamma} \leqslant 2^{-\|k\|}, \ \|k\| > T_\varepsilon(u),$$

and by (7.3.14) we have

$$\mu_f(e^u) \leqslant M_f(e^u) \leqslant (T_\varepsilon(u) + 1)^n \cdot \mu_f(e^u) + 1. \tag{7.3.15}$$

3. We put $v(u) = \ln \mu_f(e^u)$, $Q(u) = \ln[T_\varepsilon(u) + 1]$. Consider the limit

$$B_\varepsilon(x) := \overline{\lim_{u \to x; t \to \infty}} \frac{Q(u)}{v(u)}, \ x \in D_\gamma.$$

If $vO^+(x) = \gamma_f(x) = \infty$ (see Lemma 7.3.3), then by formula (1.6.3) we have $B_\varepsilon(x) = 0$. But if $x \in \Gamma$, then (see Exercise 1.3.20)

$$B_\varepsilon(x) = (\gamma_f(x))^{-1} \cdot \max\{0, \sup_{y \in S} \langle x, y \rangle + \varepsilon |x|\}, \quad x \in \Gamma, \tag{7.3.16}$$

where $S = \Pi \cap A(X)$. We observe that $v = g^*$, where $g(y) = -\ln |a_y|$ for $y \in N_f$, and $g(y) = \infty$ for $y \in \mathbb{R}^n \setminus N_f$. Therefore (see Proposition 1.5.10), in the notation of item 2 of the proof of Theorem 7.3.2, we have $\overline{\mathrm{dom}\ v^*} = X$.

Let $T = \overline{\mathrm{dom}\ v0^+}$. Since $A(X) = \{\lambda y : \lambda \geqslant 0;\ y \in A(X) \cap \Pi\} \subset \mathbb{R}^n_+$, Lemma 7.3.5 shows that

$$T = \{u \in \mathbb{R}^n : \langle u, y \rangle \leqslant 0 \quad \forall\ y \in A(X) \cap \Pi\}. \tag{7.3.17}$$

Consequently, if $x \in T$, then $\sup\{\langle x,y\rangle : y \in \Pi \cap A(X)\} \leqslant 0$. From (7.3.16) we deduce that

$$B_\varepsilon(x) = 0 \quad \forall\, x \in D_\gamma \setminus \Gamma;\ B_\varepsilon(x) = \frac{\varepsilon|x|}{\gamma_f(x)} \quad \forall\, x \in \Gamma. \tag{7.3.18}$$

Since $\varepsilon > 0$ is arbitrarily small, by (7.3.18) it is easily seen that (7.3.15) implies (7.3.9).

4. The last statement of the theorem is justified by the fact that $D_\rho \subset \{x \in \mathbb{R}^n : \gamma_f(x) = \infty\}$, and that the order $\rho_f$ of an entire function $f$ is a finite and convex (therefore, continuous) function on $\mathbb{R}^n$ (Proposition 6.2.6, Corollary 6.2.11).

**Corollary 7.3.6.** *Under the conditions of Theorem 7.3.2, we have*

$$\lim_{r_1,\dots,r_n\to\infty} [\ln \mu_f(r)]^{-1} \cdot \ln M_f(r) = 1$$

*if $f(z)$ is nonconstant in each of the variables $z_1, ..., z_n$.*

◀ It suffices to show that $\mathbb{R}^n_+ \setminus \{0\} \subset D_\gamma$. Since the entire function $f(z)$ depends genuinely on $n$ variables $(z_1, ..., z_n) = z$, its Taylor expansion involves nonzero coefficients $a_{k^{(j)}} \neq 0$, $k^{(j)} = (k_{j_1}, \dots, k_{j_n})$, $j = 1, \dots, n$, such that $k_{j_j} > 0$ for $j = 1, \dots, n$. We put $\varphi(r) = \max\limits_{1\leqslant j\leqslant n} |a_{k^{(j)}}| r^{k^{(j)}}$. Then $\varphi \leqslant \mu_f$ and $\nu \leqslant \gamma_f$, where $\nu(u) = \max\limits_{1\leqslant j\leqslant n} \langle k^{(j)}, u\rangle$ is the asymptotic function for $\ln \varphi(e^u)$. Therefore,

$$\mathbb{R}^n_+ \setminus \{0\} \subset \{u \in \mathbb{R}^n : \nu(u) > 0\} \subset D_\gamma \quad ▶$$

**Remark.** The complement $\mathbb{R}^n \setminus D_\gamma$ of the set $D_\gamma$ is a convex cone containing the negative octant $\{u \in \mathbb{R}^n : u \leqslant 0\}$. In Theorem 7.3.2 the condition $\overline{K} \setminus \{0\} \subset D_\gamma$ cannot be relaxed, as is shown by Example 1. Here $D_\gamma = \{u \in \mathbb{R}^2 : \max\{u_1, u_2\} > 0\}$. The path $\Gamma$ (see (7.3.2)) cannot be placed in a cone $K$ with vertex at 0 and such that $\overline{K} \setminus \{0\} \subset D_\gamma$. This example can easily be generalized to the case of an arbitrary number of variables: it suffices to take the function $f(z) = z_n exp\{z_1 + \cdots + z_{n-1}\}$.

## 7.4. On the growth of the Nevanlinna characteristic for an entire function of several variables.

### 7.4.1. The order function of the Nevanlinna characteristic.

**Definition 7.4.1.** *The Nevanlinna characteristic of an entire function $f$ in $\mathbb{C}^n$* is introduced by the formula

$$m(r; f) = \frac{1}{(2\pi)^n} \int_0^{2\pi} \cdots \int_0^{2\pi} \ln^+ |f(r_1 e^{i\theta_1}, \dots, r_n e^{i\theta_n})| d\theta_1 \dots d\theta_n.$$

**Theorem 7.4.2.** *For every entire function $f$ on $\mathbb{C}^n$ its Nevanlinna characteristic $m(r; f)$ belongs to the class $\mathfrak{U}$* (see Property 6.1.2).

**Theorem 7.4.3.** *The order functions, order hypersurfaces, hypersurfaces of conjugate orders of the functions $\ln^+ M_f(r)$ and $m(r; f)$ coincide for every entire function $f(z)$ on $\mathbb{C}^n$.*

◀ For every $r \in \mathbb{R}^n_+$ there is $w \in \mathbb{C}^n$ such that $|w_i| = r_i$ and $M_f(r) = |f(w)|$. The function $u(z) = \ln^+ |f(z)|$ is plurisubharmonic in $\mathbb{C}^n$; since $u(z) \geqslant 0$ for $z \in \mathbb{C}^n$, and $|\zeta - z| \geqslant |\zeta| - |z|$, inequality (1.8.4) yields

$$u(w) \leqslant \frac{1}{(2\pi)^n} \prod_{j=1}^{n} \frac{R_j + r_j}{R_j - r_j} \int_0^{2\pi} \cdots \int_0^{2\pi} u(R_1 e^{i\theta_1}, \dots, R_n e^{i\theta_n}) d\theta_1 \dots d\theta_n,$$

where $R_j > r_j$. Now, putting $\alpha_j = R_j / r_j$, we see that

$$m(r; f) \leqslant \ln^+ M_f(r) \leqslant \prod_{j=1}^{n} \frac{\alpha_j + 1}{\alpha_j - 1} m(\alpha_1 r_1, \dots, \alpha_n r_n; f) \tag{7.4.1}$$

for every $\alpha_1, \alpha_2, \dots, \alpha_n > 1, r \in \mathbb{R}^n_+$. Applying Lemma 6.2.8, and the fact that $m(r; f) \in \mathfrak{U}$ (see Theorem 7.4.2), we deduce Theorem 7.4.3 ▶

**7.4.2. On the growth category for the Nevanlinna characteristic.**

**Definition 7.4.4.** Two nonnegative functions $\Phi, \chi \in \mathfrak{U}$ have *the same growth category* if

1) up to a constant factor, $\Phi(e^u)$ and $\chi(e^u)$ have the same asymptotic function (see Def. 1.6.4);

2) the functions $\Phi$ and $\chi$ have one and the same order function $\rho := \rho_\Phi = \rho_\chi$;

3) if $\Phi, \chi \in \mathfrak{M}_n$, these functions are simultaneously of minimal, normal, or maximal $x$-type (see Def. 6.3.1) for every $x \in T^\rho$, where $T^\rho = \{u \in \mathbb{R}^n : \rho(u) = 1\}$ is their common order hypersurface; also, they are in the same convergence class, i.e., the functions $\varphi_x(t) = \Phi(t^x)$ and $\phi_x(t) = \chi(t^x)$, belong to one and the same class of convergence or divergence for every $x \in T^\rho$ (see Def. 1.1.4).

**Theorem 7.4.5.** *For every entire function $f$, the functions*

$$\ln^+ M_f(r), \ \ln^+ \mu_f(r), \ m(r, f), \ \ln^+ S_f(r)$$

(see (7.3.6)) *have the same growth category.*

◀ Since inequalities (7.3.6), (7.4.1) are of similar nature, it suffices to prove the theorem for the functions $\Phi = \ln^+ M_f(r)$ and $\chi = m(r, f)$. For them, condition 1) of Definition 7.4.4 is verified as in the proof of Lemma 7.3.3. Condition 2) follows from Theorem 7.4.3. If $f \in \mathfrak{M}_n$, then the functions $\Phi, \chi$ have common order functions $\rho(u) < \infty, \forall\, u \in \mathbb{R}^n$ (see Corollary 6.2.11), and

$\rho \in Y$ (see Theorem 6.2.19). By (7.4.1), for all $\alpha_1, \dots, \alpha_n > 1$, $x \in T^\rho$, and $r \in \mathbb{R}^n_+$ we have

$$\sigma_\chi(r,x) \leqslant \sigma_\Phi(r,x) \leqslant \prod_{j=1}^{n} \frac{\alpha_j+1}{\alpha_j-1}\sigma_\chi(\alpha_1 r_1, \dots, \alpha_n r_n), \tag{7.4.2}$$

where $T^\rho$ is the common order hypersurface for $\Phi, \chi$, and $\sigma_\Phi(r,x)$, $\sigma_\chi(r,x)$ are the $x$-type functions for $\Phi$ and $\chi$, respectively. By Corollary 6.3.13, we see that the first part of condition 3) in Definition 7.4.2 is valid for $\Phi$ and $\chi$. To complete the proof, it only remains to apply the following lemma.

**Lemma 7.4.6.** *For $\Phi \in \mathfrak{M}_n$, let $D_\rho = \{u \in \mathbb{R}^n : \rho_\Phi(u) > 0\}$, where $\rho_\Phi$ is the order function of $\Phi$, and let $x \in D_\rho$. If the integral $I(r) = \int\limits_1^\infty \Phi(rt^x) \cdot t^{-\rho_\Phi(x)-1}dt$ converges (diverges) at some point $b \in \mathbb{R}^n_0$, then it converges (diverges) for all $r \in \mathbb{R}^n_0$.*

◀ For all $b, r \subset \mathbb{R}^n_0$, $t \geqslant 0$, $\lambda + \mu - 1$, $\lambda$, $\mu > 0$, we have

$$\Phi(rt^x) \leqslant \lambda\Phi(bt^{x/\lambda}) + \mu\Phi(a_1^{1/\mu}, \dots, a_n^{1/\mu}), \tag{7.4.3}$$

where $a_i = r_i b_i^{-\lambda}$, $i = 1, \dots, n$ (see (6.2.6)). By Corollary 6.2.11, the order function $\rho_\Phi$ of $\Phi$ is finite in $\mathbb{R}^n$. Let $\alpha = \rho_\Phi$. Fix $\varepsilon > 0$ and $\delta \in (0, \alpha)$. In inequality (7.4.3) we take $t \geqslant t_0 > e^{2/\gamma}, 1/\mu = \gamma \ln t$, where the number $\gamma > 0$ is such that

$$\gamma \cdot [\rho_\Phi(\ln a) + \varepsilon |\ln a|] < \alpha - \delta$$

(here $\ln a = (\ln a_1, \dots, \ln a_n)$). Since $\Phi \in \mathfrak{M}_n$, inequality (6.2.16) implies:

$$\Phi(a_1^{1/\mu}, \dots, a_n^{1/\mu}) < C_\varepsilon t^{\alpha-\delta} \ \forall\, t \geqslant t_0. \tag{7.4.4}$$

Consider the integral

$$B = \int\limits_0^\infty \Phi(bt^{x/\lambda}) t^{-\alpha-1} dt, \text{ where } \lambda = \lambda(t) = 1 - (\gamma \ln t)^{-1} = 1 - \mu.$$

Changing the variables $\tau = t^{1/\lambda}$ and taking into account that $t^{1/\lambda} \cdot [1 - (\gamma \ln t - 1)^{-2}]dt = d\tau$, $t_0 > e^{2/\gamma}$, we arrive at the formula

$$B = \int\limits_0^\infty \Phi(b\tau^x)\tau^{-\alpha-1}\psi(\tau)d\tau,$$

where $\tau_0^{\lambda(t_0)} = t_0$, $\psi(\tau) = \tau^{\alpha/\gamma \ln t}[1 - (\gamma \ln t - 1)^{-2}]^{-1}$, and $t = \beta(\tau)$ is an implicit function defined by the equation $\tau^{\lambda(t)} = t$. Since $\ln \tau (\ln t)^{-1} \to 1$ as $\tau \to \infty$, it follows that $f$ is bounded on the interval $(\tau_0, \infty)$. If the integral $I(b)$ converges, then the integral $B$ under consideration also converges ($\Phi(b\tau^x) \geqslant 0$ for $\tau \geqslant \tau_1 \geqslant \tau_0$). Combined with (7.4.3), (7.4.4), this yields the convergence of the integral $I(r)$ for all $r \in \mathbb{R}_0^n$.

If the integral $I(b)$ diverges, then the divergence of the integral $I(r)$ is proved similarly: the roles of $r$ and $b$ in the above arguments must be interchanged ▶

**Remark.** If $f$ is an entire function of finite order in the totality of variables, for the functions $\ln^+ M_f(r)$ and $\ln^+ \mu_f(r)$ Theorem 7.4.5 is weaker than Theorem 7.3.2. For the functions $\ln^+ M_f(r)$ and $m(r; f)$ no analog of Theorem 7.3.2 is available even for $n = 1$, as is shown by the function $e^z$.

From (7.4.2), Proposition 6.3.8, and Theorem 6.3.9 we obtain the following statement.

**Corollary 7.4.7.** *Suppose $\rho \in Y, x \in T^\rho$, and $f$ is an entire function in $\mathfrak{N}_n^x(\rho)$. Then for all $r \in \mathbb{R}_+^n$ we have*

$$\sigma_m(r; x) \leqslant \sigma_f(r; x) \leqslant A_x \sigma_m(r; x), \tag{7.4.5}$$

*where*

$$A_x = \inf_{\alpha_1, \ldots, \alpha_n > 1} \sup_{y \in \partial\rho(x)} \prod_{i=1}^n \frac{\alpha_i + 1}{\alpha_i - 1} \alpha_i^{y_i};$$

*$\sigma_m(r; x)$, $\sigma_f(r; x)$ are the $x$-type functions for $m(r; f)$ and $f$, respectively. In particular, if $\rho$ is differentiable at $x$, then $\partial\rho(x) = \left\{\left(\frac{\partial\rho}{\partial x_1}, \ldots, \frac{\partial\rho}{\partial x_n}\right)\right\}$ and $A_x = \prod_{i=1}^n \Phi\left(\frac{\partial\rho}{\partial x_i}\right)$, where*

$$\Phi(y) = \frac{1 + y + \sqrt{1 + y^2}}{1 - y + \sqrt{1 + y^2}} \cdot \left(\frac{1 + \sqrt{1 + y^2}}{y}\right)^y.$$

Chapter 8

# Indicator Diagram of an Entire Function of Several Variables with Nonnegative Indicator

Among other things, in this chapter we study the geometric image in $\mathbb{C}^n$, $n > 1$, of the nonnegative radial indicator of an entire function $f(z)$ of $n$ complex variables. This image is the *indicator diagram.* The radial indicator describes the growth of $f$ in one direction only, while actually all directions of nonzero growth order for $f$ are described by the concave cone $D_\rho = \{y \in \mathbb{R}^n : \rho_f(y) > 0\}$, where

$$\rho_f(y) = \overline{\lim_{t\to\infty}} (\ln t)^{-1} \cdot \ln^+ \ln^+ M_f(t^y), \quad y \in \mathbb{R}^n, \tag{8.0.1}$$

is the order function of $f$ (it is assumed that $0 < \rho(f) < \infty$, where $\rho(f)$ is the order of $f$ in the totality of variables; see Definitions 6.2.4, 6.2.9, and the beginning of Ch. 7).

Let us restrict ourselves to the growth directions of $f$ in which all variables $z_1, \dots, z_n$ grow simultaneously. In this setting, the growth of $f$ is determined by the hypersurface

$$S_f = \{\rho \in \mathbb{R}_0^n : \rho_f(\rho_1^{-1}, \dots, \rho_n^{-1}) = 1\} \tag{8.0.2}$$

of its positive conjugate orders (see Def. 6.2.14, Proposition 6.2.17). The object under study in this chapter is the corresponding system of regularized indicators $\{h_f(\cdot;\rho), \rho \in S_f\}$, where

$$h_f(z;\rho) = \overline{\lim_{w\to z}} \overline{\lim_{t\to\infty}} t^{-1} \cdot \ln |f(t^x w)|, \quad z \in \mathbb{C}^n; \tag{8.0.3}$$

$x_i = \rho_i^{-1}$, $i = 1, \dots, n$. Precisely for the elements of this system we shall resolve the problems mentioned in the Introduction. For simplicity, the ideas of the proofs are explained by the example of the radial indicator for an entire function of exponential type, except for the cases where this is impossible in view of essential difference (see Theorem 8.2.12). Investigation of this system of indicators employs the notion of the system of regularized indicators for a function plurisubharmonic in $\mathbb{C}^n$, and the properties of this system.

## 8.1. System of indicators and indicator diagrams of an entire function of several variables

**System of regularized indicators of a function of class $\mathfrak{P}$.** For $u \in \mathfrak{P}$, let $M_u(r)$ be the maximum of $u$ in the polydisc $\Pi_r = \{z \in \mathbb{C}^n : |z_j| \leqslant r_j,\ j = 1, \dots, n\}$ (see Section 6.1), and let $\rho_u(y)$ be the order function of $u$, i.e., the order function of $M_u$ (see the above reference). If $u \in \mathfrak{M}_n$ (see (6.2.12)), then, by Theorem 6.2.19, $\rho_u \in Y$ and the set $D_u = \{y \in \mathbb{R}^n : \rho_u(y) > 0\}$ is a concave cone in $\mathbb{R}^n$ of full dimension and with vertex at 0. This cone characterizes the totality of all directions of nonzero growth order for $u$. We consider the system of regularized indicators for $u$ corresponding to the set $D_u$:

$$L_u(z; y) = \overline{\lim_{w \to z}}\ \overline{\lim_{t \to \infty}}\ t^{-\rho_u(y)} u(t^y w), \quad z \in \mathbb{C}^n;\ y \in D_u, \tag{8.1.1}$$

where $t^y w = (t^{y_1} w_1, \dots, t^{y_n} w_n)$. In particular, $L_f(z, \mathbb{I}) := L_{\ln|f|}(z, \mathbb{I})$ is the well-known radial indicator of an entire function $f$ in $\mathbb{C}^n$. Since $L_u(z; y) \equiv L_u(z, x)$, $z \in \mathbb{C}^n$, where $x = y \cdot [\rho_u(y)]^{-1}$, we see that the system (8.1.1) can be written compactly as follows:

$$\{L_u(\cdot; x),\ x \in T_u\}, \tag{8.1.2}$$

where $T_u$ is the order hypersurface of $u$. We mention several properties of the system (8.1.2). The majorant in the angular coordinates for $L_u(\cdot; x)$, where $x \in T_u$, coincides with the $x$-type function

$$\sigma_u(r; x) = \overline{\lim_{a \to r}}\ \overline{\lim_{t \to \infty}}\ t^{-1} \cdot M_u(t^x a), \quad r \in \mathbb{R}^n_+.$$

**Property 8.1.1.** *Suppose the $x$-type $\sigma_u(x) := \sigma_u(\mathbb{I}; x)$ is finite, where $x \in T_u$* (see (8.1.2)). *Then* (see (8.1.1))

$$\gamma_u(r; x) := \sup_{|\varphi_j| \leqslant \pi} L_u\left[\left(r_1 e^{i\varphi_1}, \dots, r_n e^{i\varphi_n}\right); x\right] \equiv \sigma_u(r; x).$$

Property 8.1.1 is a multidimensional analog of Proposition 1.10.1.

◀ 1. Let $|w_j| = a_j,\ j = 1, \dots, n$. By (8.1.1) we have

$$\gamma_u(r; x) \leqslant \sigma_u(r; x) \quad \forall\, r \in \mathbb{R}^n_+.$$

To prove the reverse inequality, we need the formula

$$k_u(z) := \overline{\lim_{t \to \infty, w \to z}}\ u_t(w) \equiv L_u(z; x), \quad z \in \mathbb{C}^n;\ u_t(w) = t^{-1} \cdot u(t^x w). \tag{8.1.3}$$

Obviously, $L_u(\cdot; x) \leqslant k_u$. On the other hand, for every $\varepsilon > 0$ we can find a neighborhood $V_\varepsilon$ of $z$ such that

$$\overline{\lim_{t \to \infty}}\ u_t(w) \leqslant L_u(z; x) + \varepsilon \quad \forall\, w \in V_\varepsilon.$$

Since $\sigma_u(x) < \infty$, it follows that $\sigma_u(r;x) < \infty \; \forall \; r \in \mathbb{R}_+^n$ (see Corollary 6.3.13). Therefore, the sequence $\{u_t, \; t \geqslant t_0 > 0\}$ of plurisubharmonic functions is uniformly bounded from above on every compact subset of $\mathbb{C}^n$. Using Theorem 1.8.8, we conclude that $k_u \leqslant L_u(\cdot;x)$, and formula (8.1.3) follows.

2. Formula (8.1.3) applied to $M_u$ implies that for any vector $r \in \mathbb{R}_+^n$ we can find sequences $\{t_n\}$ and $\{w^{(n)}\}$ in $\mathbb{R}$ and $\mathbb{C}^n$, respectively, with the properties

$$\lim_{n\to\infty} t_n = \infty; \quad \lim_{n\to\infty} w^{(n)} = z; \quad \sigma_u(r;x) = \lim_{n\to\infty} u_{t_n}(w^{(n)}),$$

where $r = (|z_1|, \ldots, |z_n|)$. Therefore, by (8.1.3), $\sigma_u(r;x) \leqslant L_u(r;x) \leqslant \gamma_u(r;x)$ ▶

**Property 8.1.2.** *Under the conditions of Property* 8.1.1, *if the indicator* $L_u(z;x)$ *of* $u \in \mathfrak{P}$ *depends only on the absolute values of the variables* $z_1, \ldots, z_n$, *then* $L_u(z;x) \equiv \sigma_u(|z|;x)$, $z \in \mathbb{C}^n$, $x \in T_u$.

From the Hartogs lemma (see Theorem 1.8.8) we deduce the following fact.

**Property 8.1.3.** *Under the conditions of Property* 8.1.1, *the indicator* $L_u(z;x)$ *is a plurisubharmonic function in* $\mathbb{C}^n$, *and* $L_u(t^x, z;x) \equiv tL_u(z;x)$, $z \in \mathbb{C}^n$, $t > 0$; $x \in T_u$.

The results of this subsection will be used below in the case of $u = \ln|f|$, where $f$ is an entire function. Then, in the notation for the growth indicators, the symbol $u$ will be replaced by $f$.

**8.1.2. Indicator diagram of an entire function with prescribed system of positive conjugate orders.** Let $f$ be an entire function in $\mathbb{C}^n$, and let $0 < \rho(f) < \infty$. Then (see (8.1.2) and Proposition 6.2.17) $\{(x_1^{-1}, \ldots, x_n^{-1}), \; x \in T_f\} \cap \mathbb{R}_0^n = S_f$ is the hypersurface of positive conjugate orders for $f$. Therefore, a necessary and sufficient condition for a system $\rho = (\rho_1, \ldots, \rho_n)$ to belong to $S_f$ and for $f$ to satisfy the inequality

$$|f(z)| < A \exp\{a(|z_1|^{\rho_1} + \cdots + |z_n|^{\rho_n})\}, \quad z \in \mathbb{C}^n,$$

with some constants $A, a > 0$ is $\rho_f(x) = 1$, where $x = (\rho_1^{-1}, \ldots, \rho_n^{-1})$, and the $x$-type $\sigma_f(x)$ is finite (see Property 8.1.2). Taking this into account, we introduce the following definition.

**Definition 8.1.4.** In the previous notation, the function (see (8.1.4))

$$h_f(z;\rho) := L_f(z;x), \; z \in \mathbb{C}^n, \text{ where } \rho \in S_f, \; x = (\rho_1^{-1}, \ldots, \rho_n^{-1}),$$

is called the *$\rho$-indicator* of an entire function $f$.

Consider the geometric image of the $\rho$-indicator in $\mathbb{C}^n$.

**Definition 8.1.5.** In the notation of Definition 8.1.4, put $h = \max\{0, h_f(\cdot;\rho)\}$. The set

$$\Omega_h = \bigcup \Pi_\tau | \operatorname{Re} \tau > 0, \quad \Pi_\tau = \{z \in \mathbb{C}^n : \; h(\tau^x z) < \operatorname{Re} \tau\}, \qquad (8.1.4)$$

where $\tau^{x_k} = |\tau|^{x_k} \exp\{ix_k \arg\tau\}$, $k = 1, \ldots, n$, is called the *$\rho$-indicator diagram* of $f$.

If $n = 1$, then $x = \rho^{-1}$ and $h(z) = h_f(z;\rho) = |z|^\rho h_f(\theta)$, where $\rho$ and $h_f(\theta)$ are the order and the indicator of $f$, respectively. Next (see (8.1.4)), $\Omega_h = \{w^{-1} \in \mathbb{C} : w \in \overline{\mathbb{C}} \setminus K\}$ is the conjugate set for the $\rho$-convex compact set

$$K = \{w \in \mathbb{C} : \Phi_\rho(zw) \leqslant h(z) \ \forall\, z \in \mathbb{C}\} \tag{8.1.5}$$

(the function $\Phi_\rho$ is defined by formula (3.1.9)) associated with $h$ and playing the role of the indicator diagram of $f$. If $\rho = 1$, then $K$ is a compact convex set. Therefore (see (8.1.4)), $\Omega_h$ is a union of disks the boundaries of which contain 0.

Now we pass to the geometric structure of the $\rho$-indicator diagram of an entire function of several variables.

## 8.2. Circular sets and their properties

**8.2.1. Minkowski duality for circular sets and functions of class $B_n(x)$.** Let $x \in \mathbb{R}^n_0$. Consider the closed disk $K = \{\tau \in \mathbb{C}\setminus\{0\} : \operatorname{Re}\tau^{-1} \geqslant 1\} \cup \{0\}$ with diameter $[0,1]$. For $w \in \mathbb{C}^n$, we put

$$O_x(w) = \{\tau^x w = (\tau^{x_1} w_1, \ldots, \tau^{x_n} w_n) \in \mathbb{C}^n : \tau \in K\}. \tag{8.2.1}$$

If $x_1 = \cdots = x_n = a > 0$, the set $O_x(w)$ lies on the complex line $\Pi = \{\lambda w : \lambda \in \mathbb{C}\}$, and 0 lies in the interior of $O_x(w)$ in the topology of $\Pi$ if $w \neq 0$, $a > 2$. If $a$ is 1 or 2, the set $O_x(w)$ is a disk (respectively, a cardiod) with diameter $[0, w]$.

**Definition 8.2.1.** A set $\Omega$ in $\mathbb{C}^n$ is said to be *$x$-circular* if, together with every point $z$, $\Omega$ contains the set $O_x(w)$ (see (8.2.1)) with some $w = w(z) \in \mathbb{C}^n$. The set $\omega = \{w \in \mathbb{C}^n : O_x(w) \subset \Omega\}$ is called the *kernel* of $\Omega$.

An important special case of $x$-circular domains is the class of complete Reinhardt domains. Each of these domains coincides with its kernel. Every $x$-circular set $\Omega$ and its kernel are $x$-parabolically star-like, i.e., if $z \in \Omega$, then $t^x \cdot z \in \Omega \ \forall\, t \in [0,1]$.

A circular set can be described analytically in terms of the Minkowski functional of its kernel.

**Definition 8.2.2.** Let $x \in \mathbb{R}^n_0$, and let $0 \in \Omega \subset \mathbb{C}^n$. The function

$$p_\Omega(z) = \inf\{s > 0 : zs^{-x} \in \Omega\}, \quad z \in \mathbb{C}^n,$$

is called the *$x$-parabolic Minkowski functional* of $\Omega$.

Let $A_n(x)$ be the class of $x$-circular domains in $\mathbb{C}^n$, and let $B_n(x)$ be the class of upper semicontinuous functions on $\mathbb{C}^n$ such that

1)

$$0 \leqslant h(z) < \infty; \ h(t^x z) = th(z) \quad \forall\, z \in \mathbb{C}^n; \ t \geqslant 0; \tag{8.2.2}$$

2) $\psi(\lambda) := h(\lambda^x a)$ is a convex function in the half-plane

$$\Pi(\lambda) = \{|\lambda| e^{i \mathrm{Arg}\ \lambda} : \ |\lambda| \geqslant 0; \ |\mathrm{Arg}\ \lambda + \varphi| \leqslant \pi/2\}$$

for all $a \in \mathbb{C}^n \setminus \{0\}$, $\varphi \in \mathbb{R}$. We partially order these sets in a natural way (by inclusion and by pointwise domination, respectively).

**Theorem 8.2.3.** *There exists a natural antiisomorphism $\nu : A_n(x) \to B_n(x)$ between the sets $A_n(x)$ and $B_n(x)$. It is given by the formula $\nu(\Omega) = p_\omega$, where $p_\omega$ is the $x$-parabolic Minkowski functional of the kernel $\omega$ of $\Omega \in A_n(x)$. The inverse to $\nu$ is given by the formula*

$$\nu^{-1}(p) = \Omega_p := \bigcup \Pi_\tau(p) | \operatorname{Re} \tau > 0, \tag{8.2.3}$$

*where $p \in B_n(x)$, $\Pi_\tau(p) = \{z \in \mathbb{C}^n : \ p(\tau^x z) < \operatorname{Re} \tau\}$* (cf. (8.1.4)) *and, moreover, $p$ is the $x$-parabolic Minkowski functional of the kernel $\omega$ of $\Omega$.*

We prove the theorem for $x = \mathbb{I} = (1, \dots, 1)$.

**Lemma 8.2.4.** *If $A_n = A_n(\mathbb{I})$, $\Omega \in A_n$, $\omega$ is the kernel of $\Omega$, and $L_a = \{\lambda a : \ \lambda \in \mathbb{C}\}$, where $a \in \mathbb{C}^n \setminus \{0\}$, then the intersection $\omega \cap L_a$ is a convex set.*

◀ Let $\lambda_i a \in \omega$, $i = 1, 2$, and let $\lambda = (1-\alpha)\lambda_1 + \alpha\lambda_2 \neq 0$, where $\alpha \in (0,1)$. We put $O(w) = O_{\mathbb{I}}(w)$, $w \in \mathbb{C}^n$. Assume that $\lambda a \in L_a \setminus \{\omega\}$. Then there exists a point $z_0 \in O(\lambda a) \setminus \Omega$. Hence, $z_0 \neq 0$, and $z_0 = \tau\lambda a$, where $\operatorname{Re} 1/\tau_0 = \operatorname{Re} \lambda/u$, $u := \lambda\tau_0 \neq 0$. On the other hand, $z_0 \notin O(\lambda_1 a) \cup O(\lambda_2 a)$ (see Def. 8.2.1), i.e., $\operatorname{Re} \lambda_j/u < 1$, $j = 1, 2$. So, $\operatorname{Re} \lambda/u < 1$, a contradiction. Therefore, $\lambda a \in \omega$ ▶

**Proof of Theorem 8.2.3.** 1. The kernel $\omega$ of every set $\Omega \in A_n$ is an open star-like set. Therefore, the formula

$$\omega = \{w \in \mathbb{C}^n : \ p_\omega(w) < 1\}, \tag{8.2.4}$$

where $p_\omega$ is the Minkowski functional of $\omega$, implies that $p_\omega$ is an upper semicontinuous function. Hence in the notation of Lemma 8.2.4 we have: $B := \omega \cap L_a = \{\lambda \in \mathbb{C} : \ \psi(\lambda) < 1\}$, where $\psi(\lambda) = p_\omega(\lambda a)$ is the Minkowski functional of $B$. By Lemma 8.2.4, $B$ is a convex set; consequently, $\psi$ is a convex function on $\mathbb{C}$ (see Theorem 1.3.15). So, $\nu[A_n] \subset B_n$, where $B_n := B_n(\mathbb{I})$.

2. We show that the mapping $\nu$ is a monomorphism. Let $z \in \Omega \setminus \{0\}$. There exist elements $\tau \in K \setminus \{0\}$ (see (8.2.1)) and $w \in \omega$ such that

$w = z\tau^{-1}$ and $p_\omega(z\tau^{-1}) < 1$ (see (8.2.4)). Taking into account that $|\tau|^2 \leqslant \mathrm{Re}\ \tau$ (see (8.2.3)), from (8.2.2) we deduce that $z \in \Pi_{\bar\tau}(p)$ for $p = p_\omega$, i.e., $\Omega \subset \Omega_p$. But $\Omega_p = \bigcup \Pi_\tau(p) | \mathrm{Re}\ (1/\tau) = 1$ (see (8.2.3)). Now, let $z \in \Omega_p \setminus \{0\}$. Then $z \in \Pi_{\tau_0}(p)$ , where $\tau_0 \neq 0$ and $\mathrm{Re}\ (1/\tau_0) = 1$. Since $\mathrm{Re}\ \tau_0 = |\tau_0|^2$, we see that $p(w) = p_\omega(w) < 1$, where $w = z(\bar\tau_0)^{-1}$. Therefore, $w \in \omega$ and $z \in O_{\mathbb{I}}(w) \subset \Omega$. Thus, $\Omega = \Omega_p$.

3. Let $h \in B_n$. Then $E = \{w \in \mathbb{C}^n : h(w) < 1\}$ is an open star-like set. Therefore, $\Omega = \bigcup O_{\mathbb{I}}(w) | w \in E$ is an open circular set. Arguing as in item 2, we can show that $\Omega = \Omega_h = \Omega_p$, where $p = p_\omega$ is the Minkowski functional of the kernel $\omega$ of the domain $\Omega$. It only remains to show that $h = p$. For this it suffices to prove that $E = \omega$. Suppose that there exists a point $w_0 \in \omega \setminus E$. Then $h(w_0) \geqslant 1$ and $w_0 \neq 0$. By assumption, the function $\Phi(\lambda) = h(\lambda w_0)$, $\lambda \in \mathbb{C}$, is convex. Consequently, in $\mathbb{C}$ we can find a number $b$ such that

$$\mathrm{Re}\ \bar b\lambda \leqslant \Phi(\lambda) \quad \forall\ \lambda \in \mathbb{C}, \quad \mathrm{Re}\ b = \Phi(1) = h(w_0) \geqslant 1 \tag{8.2.5}$$

($b$ is an element of the subdifferential of $\Phi$ at 1, see Def. 1.4.21). Therefore, $\mathrm{Re}\ \tau \leqslant h[w_0 \cdot \tau \bar b^{-1}]$, $\mathrm{Re}\ \tau > 0$, i.e., $w_0 \bar b^{-1} \notin \Omega_h = \Omega$ (see (8.2.3)). On the other hand, $w_0 \bar b^{-1} \in O(w_0) \subset \Omega = \bigcup O(w) | w \in \omega$ (see (8.2.1) and (8.2.5)). We have arrived at a contradiction. So, $\Omega = \nu^{-1}(h)$. The antiisomorphism of the sets $A_n(x)$ and $B_n(x)$ follows from the standard properties of the Minkowski functional ▶

Formula (8.2.3) allows us to give an analytic description of the set dual to a compact convex subset of $C$.

**Proposition 8.2.5.** *Let $K$ be a compact convex subset of $C$ with $0 \in K$. Put $h_K(w) = \sup\{\mathrm{Re}\ \langle z, w\rangle : z \in K\}$ for $w \in \mathbb{C}^n$. Then the set $\tilde K = \{w \in \mathbb{C}^n : \langle z, w\rangle \neq 1\ \forall\ z \in \mathbb{C}^n\}$ dual to $K$ is a $\mathbb{I}$-circular domain, $h := h_K$ is the Minkowski functional of the kernel $\omega$ of $K$, and $\tilde K = \Omega_h$ (see (8.2.3)).*

◀ Clearly, $0 \in \Omega_h \cap \tilde K$. Let $w \in \Omega_h \setminus \{0\}$. Then (see (8.2.3)) there is a number $\tau \in \mathbb{C} \setminus \{0\}$ such that $h_K(\tau w) < \mathrm{Re}\ \tau$. Therefore, the analytic hyperplane $\Gamma_w = \{z \in \mathbb{C}^n : \langle z, w\rangle = 1\}$ lies in the half-space $\{z \in \mathbb{C}^n : \mathrm{Re}\ \langle z, \tau w\rangle > h_K(\tau w)\}$. The complex version of Theorem 1.3.19 for $\mathbb{C}^n = \mathbb{R}^{2n}$ shows that $K = \{z \in \mathbb{C}^n : \mathrm{Re}\ \langle z, u\rangle \leqslant h_K(u)\ \forall\ u \in \mathbb{C}^n\}$. Consequently, $\Gamma_w \subset \mathbb{C}^n \setminus \{0\}$, $w \in \tilde K$. Thus, $\Omega_h \subset \tilde K$.

Let $w \in \tilde K \setminus \{0\}$. In the above notation, $\Gamma_w \subset \mathbb{C}^n \setminus K$. Since $0 \in K$, Theorem 1.3.22 implies the existence of a real hyperplane $\{z \in \mathbb{C}^n : \mathrm{Re}\ \langle z, u\rangle = 1$, where $u \in \mathbb{C}^n \setminus \{0\}$, such that

$$K \subset \{z \in \mathbb{C}^n : \mathrm{Re}\ \langle z, u\rangle < 1\}; \quad \Gamma_w \subset \{z \in \mathbb{C}^n : \mathrm{Re}\ \langle z, w\rangle > 1\}.$$

Thus, $h_K(u) < 1$ and $u = \tau w$ with some $\tau \in \mathbb{C} \setminus \{0\}$ such that $\mathrm{Re}\ \tau > 1$. Consequently, $h_K(\tau w) < \mathrm{Re}\ \tau$, i.e. (see (8.2.3)), $w \in \Omega_h$. So, $\tilde K \subset \Omega_h$, and $\tilde K = \Omega_h$ ▶

A criterion for the boundedness of a circular set is provided by the following statement.

**Corollary 8.2.6.** *An $x$-circular domain $\Omega$ is bounded if and only if the function $h = \nu(\Omega)$* (see Theorem 8.2.3) *satisfies the condition* $\inf\{h(z) : z \in S\} > 0$, *where*

$$S = \{z \in \mathbb{C}^n : |z_1|^{\rho_1} + \cdots + |z_n|^{\rho_n} = 1\}; \quad \rho_i = x_i^{-1},\ i = 1, \ldots, n.$$

The assumption of Corollary 8.2.6 is fulfilled, e.g., for the continuous functions of class $B_n(x)$ positive in $\mathbb{C}^n \setminus \{0\}$.

For further examination, it is convenient to use an equivalent functional characteristic of circular sets.

**Definition 8.2.7.** Let $\Omega$ be an $x$-circular set in $\mathbb{C}^n$, where $x \in \mathbb{R}_0^n$. The function

$$q_\Omega(z) = \inf\{\alpha > 0 : zs^{-x} \in \Omega,\ \operatorname{Re} s \geqslant \alpha\}, \quad z \in \mathbb{C}^n, \tag{8.2.6}$$

is called the *$x$-circular Minkowski functional* of $\Omega$.

From the geometric structure of an $x$-circular set it follows that the numerical sets whose lower bounds are considered in (8.2.6) are not empty. The following proposition is elementary.

**Proposition 8.2.8.** *Let $x \in \mathbb{R}_0^n$. For every $x$-circular set $\Omega$ in $\mathbb{C}^n$, the functionals $q_\Omega$ and $p_\omega$ coincide, where $\omega$ is the kernel of $\Omega$* (see Definitions 8.2.2, 8.2.7).

**8.2.2. Circular holomorphy domains.** In the class $B_n(x)$, consider the subclass $P_n(x)$ of functions plurisubharmonic in $\mathbb{C}^n$ (we have indeed $P_n(x) \subset B_n(x)$, see Theorem 1.8.5 and Propositions 1.9.4, 1.9.11). The class $P_n(x)$ contains the $\rho$-indicators of entire functions, where $\rho = (x_1^{-1}, \ldots, x_n^{-1})$ (see Def. 8.1.4 and Property 8.1.3). The following statement is an analog of Theorem 8.2.3.

**Theorem 8.2.9.** *Let $E_n$ be the class of complete logarithmically convex Reinhardt domains in $\mathbb{C}^n$ with center at 0, and let $x \in \mathbb{R}^n$. There exists an antiisomorphism $\nu : E_n \to P_n(x) \cap \mathfrak{C}$* (see Chapter 6) *such that $\nu(D) = p_D$, where $p_D$ is the $x$-parabolic Minkowski functional of $D \in E_n$. The inverse for $\nu$ is given by the formula*

$$\nu^{-1}(p) = D := \{z \in \mathbb{C}^n : p(z) < 1\}, \quad p \in P_n(x) \cap \mathfrak{C}, \tag{8.2.7}$$

*where $p = p_D$; $p(z) = p(|z_1|, \ldots, |z_n|)$, $z \in \mathbb{C}^n$.*

The proof leans upon the following converse to Theorem 1.8.14.

**Lemma 8.2.10.** *Let $\Phi$ be a finite upper semicontinuous real-valued function on $R$ monotone nondecreasing in each variable and convex relative to the logarithms of the variables. Put $u(z) = \Phi(|z_1|, \ldots, |z_n|)$, $z \in \mathbb{C}^n$. Then $u \in \mathfrak{A}$* (see Chapter 6).

◀ 1. Consider the manifold $\Gamma = \{z \in \mathbb{C}^n : z_1 \dots z_n = 0\}$. Let $w \in \mathbb{C}^n \setminus \Gamma$. Then there exists a complete $n$-circular domain $U$ centered at $z$ and included in $\mathbb{C}^n \setminus \Gamma$. In $U$, an analytic mapping $\omega = (\ln z_1, ..., \ln z_n)$ is well defined (certain branches of all logarithms are fixed; for instance, we may take the principal values). By Theorem 1.8.13,

$$V(\omega) := u(e^{\omega_1}, ..., e^{\omega_n}) = \Phi(e^{\mathrm{Re}\ \omega_1}, ..., e^{\mathrm{Re}\ \omega_n}),\ \omega \in \mathbb{C}^n$$

is a plurisubharmonic function in $\mathbb{C}^n$. Consequently (see Theorem 1.8.5), $u(z) = V(\ln z_1, ..., \ln z_n) \in PSH_n(U)$. So, $u \in PSH_n(\mathbb{C}^n \setminus \Gamma)$.

2. Suppose $w \in \Gamma$ and, say, $w = (w_1, ..., w_k, 0, ..., 0)$, where $1 \leqslant k < n$. Consider the function

$$u_k(z) := u(z_1, ..., z_k, 0, ..., 0), \quad z \in \mathbb{C}^n.$$

Applying Proposition 6.1.3 and the first part of the proof, we see that $u_k$ is a plurisubharmonic function in a sufficiently small neighborhood of an arbitrary point $w$ in $\mathbb{C}^k$. Thus, for every fixed vector $a \in \mathbb{C}^n \setminus \{0\}$ there exists a number $\varepsilon > 0$ such that

$$u_k(w) \leqslant \frac{1}{2\pi} \int_0^{2\pi} \Phi(|w_1 + re^{i\theta} a_1|, ..., |w_k + re^{i\theta} a_k|, 0, ..., 0) d\theta \quad \forall\, r \in (0, \varepsilon).$$

Since $\Phi$ is monotone nondecreasing in each variable, we arrive at the inequality

$$u(w) \leqslant \frac{1}{2\pi} \int_0^{2\pi} u(w + are^{i\theta}).$$

For the same reason, the latter is true with $w = 0$ $(k = 0)$. But $w$ is an arbitrary point of $\Gamma$. Thus, $u \in \mathfrak{A}$ ▶

**Proof of Theorem 8.2.9.** From the definition of the functional $p_D$ for a Reinhardt domain $D$, we deduce that $D$ is representable as in (8.2.7), and formula (8.2.2) for $p_D$ is valid. Therefore, $E_c = \{z \in D : p_D(z) < c\}$ is open for every $c \in \mathbb{R}$ $(E_c = \varnothing\ \ \forall\, c \leqslant 0)$. Consequently, by Theorem 1.2.2, $p_D$ is upper semicontinuous in $\mathbb{C}^n$.

Let $B = \{r = (|z_1|, \dots, |z_n|) : z \in D\}$. Then

$$p_D(r) = p_B(r) \quad \forall\, r \in \mathbb{R}^n_+, \tag{8.2.8}$$

where $p_B$ is the $x$-parabolic Minkowski functional of the complete logarithmically convex domain $B$ in $\mathbb{R}^n_+$. Consequently, $p_B$ is monotone nondecreasing in each variable (see the proof of Theorem 6.2.19).

For $\{b,c\} \subset \mathbb{R}_0^n$ and $\varepsilon > 0$, we put $u = b[p_B(b)+\varepsilon]^{-x}$, $v = c[p_B(c)+\varepsilon]^{-x}$. Then $u, v \in B$, and the logarithmic convexity of $B$ implies that

$$(u_1^\lambda v_1^\mu, \ldots, u_n^\lambda v_n^\mu) \in B \quad \forall\ \lambda, \mu > 0,\ \lambda + \mu = 1.$$

Consequently, since $y = \exp\{\ln\ y\}$ $(y > 0)$ is a convex function of $\ln\ y$, we find:

$$p_B(b_1^\lambda c_1^\mu, \ldots, b_n^\lambda c_n^\mu) \leqslant [p_B(b)+\varepsilon]^\lambda \cdot [p_B(c)+\varepsilon]^\mu \leqslant \lambda p_B(b) + \mu p_B(c) + \varepsilon$$

for all $\varepsilon > 0$. Therefore, $p_B(r)$ is a convex function of $\ln\ r_1, \ldots, \ln\ r_n$. So, $p_D \in \mathfrak{A}$ (see (8.2.7), (8.2.8)). Now, Lemma 8.2.10 implies that $p_D \in P_n(x) \cap \mathfrak{A}$.

We have $P_n(x) \cap \mathfrak{A} = P_n(x) \cap \mathfrak{C}$, i.e., if $p \in P_n(x) \cap \mathfrak{A}$, then $p(e^u)$ $(u \in \mathbb{R}^n)$ is a logarithmically convex function (see the proof of Property 6.3.2).

The remaining part of the theorem is a direct consequence of formulas (8.2.7), (8.2.8).

Theorem 8.2.9 implies that $P_n(x) \neq B_n(x)$. We find a condition ensuring that $\nu^{-1}(h) = \Omega_h$ (see (8.2.3)) is a domain of holomorphy in $C$. For this, we need a lemma.

**Lemma 8.2.11.** *Let $\rho_1 > 0, \ldots, \rho_n > 0$, and let $0 < \gamma < 3^{-1} \cdot \min\{\rho_1, \ldots, \rho_n\}$. We put $\Gamma = \{z \in \mathbb{C}^n : z_1 \cdots z_n = 0\}$, and*

$$\Phi(r) = \left(\sum_{j=1}^n r_j^{\rho_j/\gamma}\right)^\gamma, \quad r \in \mathbb{R}_+^n.$$

*The function $\|z\|_\rho := \Phi(|z_1|, \ldots, |z_n|)$ is three times continuously differentiable in $\mathbb{C}^n \setminus \{0\}$ and belongs to $P_n(x) \cap \mathfrak{C}$. Moreover,*

*1) the Levi form of this function is positive definite on $\mathbb{C}^n \setminus \Gamma$, i.e.,*

$$L(\|z\|_\rho) := \sum_{j,k=1}^n \frac{\partial^2 \|z\|_\rho}{\partial z_j \partial \bar{z}_k} s_j \bar{s}_k > 0 \quad \forall\ z \in \mathbb{C}^n \setminus \Gamma,\ s \in \mathbb{C}^n \setminus \{0\};$$

*2) the trace of $\|z\|_\rho$ on every coordinate hyperplane $z_{j_1} = 0, \ldots, z_{j_r} = 0$, where $p < n$, $j_1, \ldots, j_p \subset \{1, 2, \ldots, N\}$, has the same property.*

◄ We put $\lambda_i = \rho_i/\gamma$, $i = 1, \ldots, n$. The function

$$\varphi_a(u) = \ln\left(\sum_{i=1}^n e^{\lambda_i u_i}\right) + \langle u, a\rangle, \quad u \in \mathbb{R}^n,$$

is quasiconvex for every $a \in \mathbb{R}^n$ (see Exercise 1.4.17). By Theorem 1.4.16, $\varphi_0(u)$ is a convex function on $\mathbb{R}^n$, and the function

$$\Phi(r) = \exp\{\gamma\varphi_0(\ln r_1, ..., \ln r_n)\}, \quad r \in \mathbb{R}_0^n,$$

is convex relative to $\ln r_1, ..., \ln r_n$ (see Exercise 1.4.9). Since $\Phi(r)$ is monotone nondecreasing in each varaible on $\mathbb{R}_+^n$, Lemma 8.2.10 shows that $\|z\|_\rho \in P_n(x) \cap \mathfrak{C} \cap C^3 \setminus \{0\}$.

We reshape the Levi form $L(\|z\|_\rho)$, introducing the polar coordinates $z_j = r_j e^{i\theta_j}$, $j = 1, ..., n$. For every $j, k = 1, ..., n$ we have

$$h_{jj} = \frac{\partial^2 \|z\|_\rho}{\partial z_j \partial \bar{z}_j} = \frac{1}{4}\frac{\partial^2 \Phi}{\partial^2 r_j} + \frac{1}{4r_j}\frac{\partial \Phi}{\partial r_j} = \frac{1}{4}\gamma\lambda_j^2 r_j^{\lambda_j - 2} M^{\gamma-2}[M + (\gamma - 1) r_j^{\lambda_j}], \tag{8.2.9}$$

$$h_{jk} = \frac{\partial^2 \|z\|_\rho}{\partial z_j \partial \bar{z}_k} = \frac{1}{4}\frac{\partial^2 \Phi}{\partial r_j \partial r_k} e^{i(\theta_k - \theta_j)} = \frac{1}{4}\gamma(\gamma - 1)\lambda_j\lambda_k r_j^{\lambda_j - 1} r_k^{\lambda_k - 1} M^{\gamma - 2},$$

where $j \neq k$ and $M = \sum_{j=1}^{n} r_j^{\lambda_j}$. This implies that if $z \in \mathbb{C}^n \setminus \{0\}$ has zero coordinates, for example, $z = (z_1, ..., z_m, 0, ..., 0)$, and $z_j \neq 0$, $j = 1, \ldots, m$, where $m < n$, then $h_{jk} = 0$, $j, k = m + 1, ..., n$. At such a point, the form $L(\|z\|_\rho)$ looks like this:

$$L(\|z\|_\rho) = \sum_{j=1,k}^{m} h_{jk} s_j \bar{s}_k, \quad s \in \mathbb{C}^m, \tag{8.2.10}$$

and coincides with the Levi form for the trace of $\|z\|_\rho$ on the plane $z_{m+1} = \cdots = z_n = 0$.

As it is well known, a Hermitian form $L = \sum_{j=1,k}^{m} h_{jk} s_j \bar{s}_k$ is positive definite only if all principal minors

$$D_p = \begin{vmatrix} h_{11} & h_{12} & \ldots & h_{1p} \\ h_{21} & h_{22} & \ldots & h_{2p} \\ \vdots & \vdots & \ddots & \vdots \\ h_{p1} & h_{p2} & \ldots & h_{pp} \end{vmatrix}, \qquad p = 1, 2, ..., m,$$

of the matrix $\|h_{jk}\|_1^m$ are positive. Therefore, for the proof of the lemma, it suffices to check property 1). By (8.2.9), (8.2.10), for $z \in \mathbb{C}^n \setminus \Gamma$ we obtain

$$D_p = (\gamma/4)^p M^{p(\gamma-2)} \cdot \prod_{j=1}^{p} \lambda_j^2 r_j^{\lambda_j - 2} \cdot E_p, \quad p = 1, 2, ..., n, \tag{8.2.11}$$

where

$$E_p = \begin{vmatrix} M+a_1 & a_2 & \dots & a_{p-1} & a_p \\ a_1 & M+a_2 & \dots & a_{p-1} & a_p \\ \vdots & \vdots & \ddots & \vdots & \vdots \\ a_1 & a_2 & \dots & M+a_{p-1} & a_p \\ a_1 & a_2 & \dots & a_{p-1} & M+a_p \\ , & & & & \end{vmatrix},$$

$a_j = (\gamma - 1)r_j^{\lambda_j}$, $j = 1, 2, ..., p$. Elementary transformations of the determinant $E_p$ yield the recursive formula $E_p = ME_{p-1} + M^{p-1}a_p$, $p = 2, ..., n$. By induction, we deduce that (see (8.2.9))

$$E_p = M^{p-1}\left(M + \sum_{i=1}^{p} a_i\right) = M^{p-1}\left(\gamma \sum_{i=1}^{p} r_i^{\lambda_i} + \sum_{i=p+1}^{n} r_i^{\lambda_i}\right).$$

Now, (8.2.11) implies that $D_p > 0$, $\forall\, p = 1, \dots, n$ ▶

**Remark.** If $\rho_1 = \rho_2 = \dots = \rho_n = \rho > 0$, $\gamma = \rho/2$, then the function

$$\|z\|_\rho = |z|^\rho = \left(\sum_{j=1}^{n} z_j \bar{z}_j\right)^{\rho/2}, \qquad z \in \mathbb{C}^n,$$

belongs to the class $P_n(x) \cap C^\infty \setminus \{0\}$, where $x = (1/\rho, ..., 1/\rho)$, ; moreover, $L(|z|^\rho) > 0$ $\forall\, z, s \in \mathbb{C}^n \setminus \{0\}$.

**Theorem 8.2.12.** *Suppose* $x \in \mathbb{R}_0^n$, $h \in P_n(x)$, *and there exists a sequence* $h_k \in C^3(\mathbb{C}^n \setminus \{0\}) \cap P_n(x)$ *such that* $h_k \downarrow h$. *Then* $\Omega_h$ *is a holomorphy domain in* $\mathbb{C}^n$.

◀ We put $\rho_i = x_i^{-1}$, $i = 1, 2, \dots, n$, and denote by $\|z\|_\rho$ the function considered in Lemma 8.2.11. The sequence $\{h_k(z) + \|z\|_\rho / k\}$ also satisfies the conditions of the theorem. Furthermore,

$$\Omega(h_k) \subset \Omega(h_{k+1}),\ k = 1, 2, \dots,\ \Omega_h = \bigcup_{k=1}^{\infty} \Omega(h_k) \quad (\Omega(g) = \Omega_g).$$

Therefore, it suffices to prove the theorem for the function $H$ such that $H \in C^3(\mathbb{C}^n \setminus \{0\})$ and for some small $\varepsilon > 0$ we have

$$H(z) - \varepsilon\|z\|_\rho \in P_n(x), \quad H(z) > \varepsilon\|z\|_\rho \quad \forall\, z \in \mathbb{C}^n \setminus \{0\}$$

1. By Theorem 8.2.3, $\Omega_H$ is an $x$-circular domain in $\mathbb{C}^n$ with $0 \in \Omega_H$. We put $L = \{\tau \in \mathbb{C} : \operatorname{Re} \tau \geqslant 0\}$, $(G(\tau, z) = H(\tau^x z) - \operatorname{Re} \tau$ $(\tau \in L,\ z \in \mathbb{C}^n)$, and $M = \{\tau \in L : |\tau| = 1\}$. Then

$$\Omega_H = \{z \in \mathbb{C}^n : g(z) < 0\} \text{ for } g(z) = \inf\{G(\tau^x z) : \tau \in M\}, \quad (8.2.12)$$

moreover, $g \in C^2(\omega_\varepsilon)$, where $\omega_\varepsilon = \{z \in \mathbb{C}^n : G(z) < \varepsilon\|z\|_\rho, z \neq 0\}$. We show that the minimum in (8.2.12) is attained at a unique point $t = \alpha(z)$, $z \in \omega_\varepsilon$, $\alpha(\omega_\varepsilon) \subset \operatorname{int} M$. Suppose that for fixed $z \in \Omega_H$ there are two points $\tau_1, \tau_2 \in M$ such that $g(z) = G(\tau_i, z)$, $i = 1, 2$. Since $H \in P_n(x)$, it follows that $G(\tau, z)$ is a convex function in $\tau$ on $L$. Then $G(\tau, z) \leqslant g(z)\ \forall\ \tau \in [\tau_1, \tau_2]$. If $\tau_2 \neq -\tau_1$, then for every $\tau \in (\tau_1, \tau_2)$ we can find a point $m \in M$ such that $m = t\tau$, where $t = t(\tau) > 1$. Since $G$ is positive homogeneous in $\tau$ and $g(z) < 0$, we have $G(m, z) = tG(\tau, z) \leqslant tg(z) < g(z)$. On the other hand, $G(m, z) \geqslant g(z)$, a contradiction. Let $\tau_2 = -\tau_1$. Since $G$ is sublinear in $\tau$, we conclude that $G(\tau_1, z) + G(\tau_2, z) \geqslant 0$ and, consequently, $g(z) = G(\tau_1, z) \geqslant 0$, which is impossible.

We show that $g \in C^2(\omega_\varepsilon)$, where $g(\zeta) = G(\alpha(\zeta), \zeta)$, $\zeta \in \omega_\varepsilon$ (see (8.2.12)). It suffices to prove that $\alpha \in C^2(\omega_\varepsilon)$. The function $G$ is positively homogeneous in $t$ on $L$ for every fixed $\zeta \in \mathbb{C}^n$, and three times continuously differentiable on the set $\{(\tau, \zeta) \in L \times \mathbb{C}^n : \tau\zeta \neq 0\}$. By the Euler identity, we obtain

$$\tau \frac{\partial G}{\partial \tau} + \overline{\tau} \frac{\partial G}{\partial \overline{\tau}} = G, \qquad \tau \frac{\partial^2 G}{\partial \tau^2} + \overline{\tau} \frac{\partial^2 G}{\partial \tau \partial \overline{\tau}} = 0. \tag{8.2.13}$$

Observing that $H(\zeta) - \varepsilon\|\zeta\|_\rho \in P_n(x)$, we see that the function

$$F(\tau, \zeta) = G(\tau, \zeta) - \varepsilon\|\tau^x \zeta\|_\rho = G(\tau, \zeta) - \varepsilon|\tau|\|\zeta\|_\rho$$

is convex in $\tau$. Writing the Laplace operator in the complex variable form and applying Theorem 1.8.18, for $|\tau| = 1$ we obtain

$$\frac{1}{4}\Delta F = \frac{\partial^2 F}{\partial \tau \partial \overline{\tau}} = \frac{\partial^2 G}{\partial \tau \partial \overline{\tau}} - \frac{1}{4}\varepsilon\|\zeta\|_\rho \geqslant 0. \tag{8.2.14}$$

As a function of $\tau$, $G$ attains its minimum on $M$ at the point $\tau = \alpha(\zeta) \in M$. Therefore (see (8.2.13)),

$$\tau \frac{\partial G}{\partial \tau} - \overline{\tau} \frac{\partial G}{\partial \overline{\tau}} = 0, \qquad \tau \frac{\partial G}{\partial \tau} = \frac{G}{2}, \quad \text{where} \quad \tau = \alpha(\zeta). \tag{8.2.15}$$

Consider the functions

$$\Phi(\tau, \zeta) = \operatorname{Im} \tau \frac{\partial G(\tau, \zeta)}{\partial \tau}; \quad \varphi(s) = \Phi(e^{is}, \zeta), \quad |s| < \pi/2$$

(see (8.2.10)). Clearly, some solution $s = s_0$ of the equation $\varphi(s, \zeta) = 0$ satisfies $e^{is_0} = \alpha(\zeta)$ (see (8.2.15)). By (8.2.14) and (8.2.15) for $\tau = e^{is_0} = \alpha(\zeta)$, $\zeta \in \omega_\varepsilon$, we have

$$\frac{\partial h}{\partial s} = \operatorname{Im} \left( ie^{is_0} \frac{\partial G}{\partial \tau} - 2i \frac{\partial^2 G}{\partial \tau \partial \overline{\tau}} \right) = \frac{G}{2} - 2\frac{\partial^2 G}{\partial \tau \partial \overline{\tau}} < \frac{\varepsilon\|\zeta\|_\rho}{2} - 2\frac{\varepsilon\|\zeta\|_\rho}{4} = 0.$$

Since $\frac{\partial h}{\partial s}$ is continuous near $(s_0, \zeta)$, the implicit function theorem shows that $\alpha \in C^2(\omega_\varepsilon)$.

By (8.2.12), $\partial\Omega_H = \{z \in \mathbb{C}^n : g(z) = 0\}$. We show that $\operatorname{grad} g(\zeta) \neq 0 \ \forall\, \zeta \in \partial\Omega_H$. Using (8.2.15) and (8.2.13), we arrive at

$$\frac{\partial G}{\partial \tau} = \sum_{j=1}^{n} x_j \zeta_j \alpha^{x_j - 1} \frac{\partial H}{\partial u_j}[\alpha^x(\zeta)\zeta] - \frac{1}{2} = \frac{\partial G}{\partial \bar{\tau}} = 0 \tag{8.2.16}$$

for $\tau = \alpha(\zeta)$, $\zeta \in \partial\Omega_H$. Therefore,

$$\frac{\partial g}{\partial \zeta_j} = \frac{\partial G}{\partial \tau} \cdot \frac{\partial \alpha}{\partial \zeta_j} + \frac{\partial G}{\partial \bar{\tau}} \cdot \frac{\partial \bar{\alpha}_j}{\partial \zeta_j} + \frac{\partial G}{\partial \zeta_j} = \frac{\partial G}{\partial \zeta_j} = \alpha^{x_j} \frac{\partial H}{\partial u_j}[\alpha^x(\zeta)\zeta]. \tag{8.2.17}$$

Combining this with (8.2.16), we conclude that

$$\sum_{j=1}^{n} x_j \zeta_j \frac{\partial g}{\partial \zeta_j} = \frac{\alpha}{2} \qquad \forall\, \zeta \in \partial\Omega_H. \tag{8.2.18}$$

Consequently, $\operatorname{grad} g(\zeta) \neq 0 \quad \forall\, \zeta \in \partial\Omega_H$.

2. We put $N(h; s; xw) = \sum\limits_{k,j=1}^{n} \frac{\partial^2 H(w)}{\partial u_k \partial u_j} s_j x_k u_k;$

$$L(h; s; xw) = \sum_{k,j=1}^{n} \frac{\partial^2 H(w)}{\partial u_j \partial \bar{u}_k} s_j x_k \bar{u}_k; \quad xw = (x_1 w_1, \ldots, x_n w_n).$$

Then for all $(s, w) \in \mathbb{C}^n \times \mathbb{C}^n$ we have

$$\sum_{j=1}^{n} s_j (1 - x_j) \frac{\partial H}{\partial u_j}(w) \equiv N(H; s; xw) + L(H; s; xw). \tag{8.2.19}$$

To see this, it suffices to differentiate the Euler identity

$$H(w) = \sum_{j=1}^{n} x_j \left( \frac{\partial H}{\partial u_j}(w) w_j + \frac{\partial H}{\partial \bar{u}_j}(w) \bar{w}_j \right), \quad w \in \mathbb{C}^n,$$

in each variable. Taking $w = \alpha^\tau(\zeta)\zeta$, $s = \alpha^x(\zeta)v$ in (8.2.19) and using (8.2.17), we obtain for $\zeta \in \partial\Omega_H$:

$$\sum_{j=1}^{n} v_j (1 - x_j) \frac{\partial g}{\partial \zeta_j}(w) = N(H; s; xw) + L(H; s; xv). \tag{8.2.20}$$

Next, we transform the "scalar product" $L(g;v;s) = \sum_{j,k=1}^{n} \frac{\partial^2 g}{\partial\zeta_j \partial\bar{\zeta}_k} v_j \bar{s}_k$, $\zeta \in \partial\Omega_H$, using (8.2.17):

$$L(g;v;s) = \bar{\alpha}A \cdot \left[\sum_{j=1}^{n} x_j v_j \frac{\partial g}{\partial\zeta_j} + N(H;\alpha^x v;\alpha^x x\zeta)\right] +$$

$$+L(H;\alpha^x v;\alpha^x s) + 0,5BL(H;\alpha^x v;\alpha^x x\zeta)$$

where $A = \sum_{k=1}^{n} \frac{\partial\alpha}{\partial\bar{\zeta}_k} \cdot \bar{s}_k$, $\alpha = \alpha(\zeta)$, and $B = 2\alpha \cdot \sum_{k=1}^{n} \frac{\partial\bar{\alpha}}{\partial\bar{\zeta}_k} \cdot \bar{s}_k$. From (8.2.20) and the relations $\alpha\bar{\alpha} \equiv 1$ and $\bar{\alpha}A = -0,5B$, we deduce that

$$L(g;v;s) = \sum_{j=1}^{n} \bar{\alpha}Av_j \frac{\partial g}{\partial\zeta_j} + BL(H;\alpha^x v;\alpha^x x\zeta) + L(H;\alpha^x v;\alpha^x s). \tag{8.2.21}$$

In particular, for $v = x\zeta$ by (8.2.17) we have

$$L(g;x\zeta;s) = A/2 + B \cdot L(H;x\zeta;\alpha^x) + L(H;x\zeta\alpha^x;\alpha^x s),$$

where $L(H;v) := L(H;v;v)$. On the other hand,

$$L(g;x\zeta;s) = \sum_{k=1}^{n} \frac{\partial}{\partial\bar{\zeta}_k} \left(\sum_{j=1}^{n} \frac{\partial g}{\partial\zeta_j} x_j\zeta_j\right) \bar{s}_k = A/2.$$

Since $H(z) - \varepsilon||z||_\rho \in P_n(x)$ and $0 \notin \partial\Omega_H$, we can use Theorem 1.8.18 in and Lemma 8.2.11 to conclude that $L(H;\alpha^x x\zeta) > 0 \quad \forall\ \zeta \in \partial\Omega_H$. Therefore, the last two inequalities yield

$$B = -L(H;\alpha^x\zeta;\alpha^x s) \cdot [L(H;\alpha^x\zeta)]^{-1}. \tag{8.2.22}$$

Now, we reshape the Levi form $L(g) = L(g;s;s)$ under the additional condition $\sum_{k=1}^{n} \frac{\partial g}{\partial\zeta_k} s_k = 0$ (see (8.2.20), (8.2.21)):

$$L(g) = B \cdot L(H;\alpha^x s;\alpha^x x\zeta) + L(H;\alpha^x s) = [L(H;\alpha^x x\zeta)]^{-1} \times$$

$$[L(H;\alpha^x s) \cdot L(H;\alpha^x x\zeta) - L(H;\alpha^x s;\alpha^x x\zeta) \times L(H;\alpha^x x\zeta;\alpha^x s)]$$

This implies that $L(g) \geqslant 0$, because the scalar product $L(H;v;t)$ satisfies the Cauchy-Bunyakovski inequality

$$|L(H;v;t)|^2 = L(H;v;t)L(H;t;v) \leqslant L(H;v)L(H;t), \quad (v,t) \in \mathbb{C}^n \times \mathbb{C}^n.$$

To show that $\Omega_H$ is a holomorphy domain, it remains to refer to the following result of the multivariable complex analysis.

**Theorem A.** *Let* $\Omega \subset \mathbb{C}^n$ *be an open set with* $C^2$*-boundary, and let* $\Omega = \{z \in \mathbb{C}^n : \rho(z) < 0\}$, *where* $\rho$ *is of class* $C^2$ *in some open set containing* $\overline{\Omega}$, *and* $\operatorname{grad} \rho \neq 0$ *on* $\Omega$. *Then* $\Omega$ *is a holomorphy domain in* $\mathbb{C}^n$ *if and only if*

$$\sum_{j,k=1}^{n} \frac{\partial^2 \rho}{\partial z_j \partial \overline{z}_k} s_j \overline{s}_k \geqslant 0$$

*for all* $z \in \partial\Omega$ *and all* $s$, *satisfying*

$$\sum_{j=1}^{n} \frac{\partial \rho}{\partial z_j} s_j = 0.$$

**8.2.3. Topological properties of $x$-circular domains.** We consider the class $A_n(x)$ of $x$-circular domains in $\mathbb{C}^n$.

**Property 8.2.13.** *Every $x$-circular domain is simply connected.*

◀ Let $D \in A_n(x)$. Consider a closed path $\gamma = \{z = z(t) \in D : t \in [0,1]\}$ in $D$ such that $z(0) = z(1) = 0$. Since $D$ is an $x$-parabolic star-like domain (see Def. 8.2.1), $D$ is arcwise connected and the mapping

$$\Phi : \; [0,1] \times [0,1] \to D; \quad \Phi(t,\tau) = \tau^x z(t), \quad t, \tau \in [0,1],$$

is continuous. Consequently, $\gamma$ is homotopically equivalent to the path $\gamma_0 = \{z(t) \equiv 0 : \; t \in [0,1]\}$. Therefore, the fundamental group $\pi_1(D,0)$ with basic point 0 is trivial, and so is the group $\pi_1(D)$ (see Subsection A.1.5) ▶

## 8.3. An analog of the Polya-Martineau-Ehrenprice theorem

**8.3.1. Algebraic isomorphism of the spaces $[\rho, h]$ and $H(\Omega_h)$.** Now, we formulate the following multidimensional analog of the Polya theorem (see the Introduction, Theorem 3.3.1).

**Theorem 3.1.** *For* $x \in \mathbb{R}^n_0$ *and* $h \in P_n(x)$, *put* $\rho = (x_1^{-1}, \ldots, x_n^{-1})$ *and denote by* $[\rho, h]$ *the space of entire functions* $f$ *in* $\mathbb{C}^n$ *such that*

$$\overline{\lim_{w \to z}}\; \overline{\lim_{t \to \infty}}\; t^{-1} \cdot \ln |(t^x w)| \leqslant h(z) \quad \forall\, z \in \mathbb{C}^n.$$

*Let* $H(\Omega_h)$ *be the space of functions holomorphic in* $\Omega_h$ (see (8.2.3)). *The operator* $\mathcal{A}: [\rho, h] \to H(\Omega_h)$, $\mathcal{A}f = g$, *where*

$$f(z) = \sum_{||k|| \geqslant 0} \frac{a_k z^k}{\Gamma(\langle k, x\rangle + 1)}, \quad g(z) = \sum_{||k|| \geqslant 0} a_k z^k \tag{8.3.1}$$

*and* $z^k = z_1^{k_1} \dots z_n^{k_n}$, $||k|| = k_1 + \dots + k_n$, *is an algebraic isomorphism between* $[\rho, h]$ *and* $H(\Omega_h)$. *Moreover, the following statements are equivalent:*

1) *for a function* $f \in [\rho, h]$, *its* $\rho$*-indicator diagram is equal to* $\Omega_h$, *or the nonnegative truncation of its* $\rho$*-indicator* $h_f^+(z; \rho) := \max\{h_f(z;\rho), 0\}$ *is identically equal to* $h(z)$, $z \in \mathbb{C}^n$ (see Definitions 8.1.4, 8.1.5);

2) *the largest* $x$*-circular domain* $\Omega$ *to which the function* $g = \mathcal{A}f$ *extends analytically (the "conjugate diagram") coincides with* $\Omega_h$.

**Remarks.** 1. For every $f \in [\rho, h]$, the order function satisfies $\rho_f(x) \leqslant 1$, and if $\rho_f(x) = 1$, then $f$ is of finite $x$-type. Therefore (see 8.1.2), Theorem 8.3.1 is a substantial generalization of Theorem $A_2$.

2. The domain $D$ described in condition 2) of Theorem 8.3.1 actually exists. Consider the class $G = \{\Omega_i : i \in I\}$ of all $x$-circular domains such that the power series of $g$ admits analytic continuation into each of them (see (8.3.1)). The set $\Omega := \bigcup \Omega_i | i \in I$ is an $x$-circular (consequently, simply connected) domain in $\mathbb{C}^n$. By the monodromy theorem, in the domain $\Omega$ there exists a univalent function that is an analytic extension of $g$.

Theorem 8.3.1 is a direct consequence of Theorem 8.2.3 and the following two lemmas.

**Lemma 8.3.2.** *If* $f \in [\rho, h]$, *then the function* $g = \mathcal{A}f$ *extends analytically to the* $x$*-circular domain* $\Omega_h$ (see (8.2.3)).

◀ From 8.3.1 and Lemma 7.1.3 we deduce that the Taylor coefficients $\{a_k\}$ of $g$ satisfy the inequality

$$|a_k| < (ae/\lambda)^\lambda \cdot \Gamma(\lambda + 1) \quad \forall\, \lambda > \lambda_0, \text{where } a > 0,\ \lambda = \langle k, x\rangle.$$

Hence $\varlimsup_{||k|| \to \infty} |a_k|^{1/||k||} < \infty$, i.e., $g$ is holomorphic in the vicinity of 0 in $\mathbb{C}^n$. We fix an arbitrary $\theta \in (-\pi/2, \pi/2)$. As in the classical proof of Theorem $A_1$ (see the Introduction), we easily see that the integral

$$G_\theta(z) = \int_0^{\infty (\arg \tau = \theta)} e^{-\tau} f(\tau^x z)\, d\tau \tag{8.3.2}$$

is a holomoprhic function in $\Pi(\theta) = \{z \in \mathbb{C}^n : h[z(e^{i\theta})^x] < \cos\theta\}$ that extends $g$. But the domain $\bigcup \Pi(\theta) | \, |\theta| < \pi/2 = \Omega_h$ (see (8.2.3), Property 8.2.13) is $x$-circular and, therefore, simply connected. By the monodromy theorem, we prove Lemma 8.3.2 ▶

**Lemma 8.3.3.** *Let $x \in R_0^n$, and let $h$ be a function in $P_n(x)$* (see 8.2.2). *If $g = H(\Omega_h)$, then* (see Theorem 8.3.1) $f := \mathcal{A}^{-1}g \in [\rho, h]$, *where $\rho = (\rho_1, \dots, \rho_n)$, $\rho_j = x_j^{-1}$, $j = 1, \dots, n$..*

◀ Reversing the arguments at the beginning of the proof of Lemma 8.3.2, we see that $f$ is an entire function such that either $\rho_f(x) < 1$ or $\rho_f(x) = 1$, and $\sigma_f(x) < \infty$. Fixing $z \neq 0$, we consider the function $\varphi_z(\tau) = f(\tau^x z)$ holomorphic inside the angle $L_\alpha = \{\tau \in \mathbb{C} : |\arg\tau| \leqslant \alpha\pi/2\}$, $\alpha \in [0, 1]$, and continuous in its closure. From the foregoing it follows that this function is of exponential type on $L_\alpha$. Let $\mathbb{Z}_+^n \setminus \{0\}$ be the set of all points in $\mathbb{R}_+^n$ with integral coordinates; we put $\lambda = \inf\{\langle k, x\rangle : k \in \mathbb{Z}_+^n \setminus \{0\}\}$. Then $\lambda > 0$ and

$$|\varphi_z(\tau) - \varphi_z(0)| = O(|\tau|^\lambda), \quad \tau \to 0.$$

Therefore, if the order of the function $\varphi_z$ is equal to 1 and $h_\varphi$ is the indicator of $\varphi_z$, then, by Theorem 1.10.17, the Laplace transform

$$\widehat{\varphi}_z(s) = \int_0^\infty e^{-st} \cdot \varphi_z(t)\, dt \tag{8.3.3}$$

is holomorphic in the half-plane $\{s \in \mathbb{C} : \operatorname{Re} s > h_\varphi(0)\}$. Substituting $\tau = st$ with real $s > h_\varphi(0)$ in (8.3.3) and using formula (8.3.2) with $\theta = 0$, for some $N > 0$ we find:

$$\widehat{\varphi}_z(s) = s^{-1} g(zs^{-x}), \quad s > N. \tag{8.3.4}$$

Since $h \in P_n(x)$, it follows that $zs^{-x} \in \Omega_h$ for $\operatorname{Re} s > h(z)$ (see Property 8.2.8). But $g \subset H(\Omega_h)$, and, by the uniqueness theorem, we obtain (8.3.4) for $\operatorname{Re} s > \max\{h_\varphi(0), h(z)\}$. Since the straight line $\{s \in \mathbb{C} : \operatorname{Re} s = h_\varphi(0)\}$ passes through singular points of $\widehat{\varphi}_z$ (see Theorem 1.10.17), we see that $h_\varphi(0) \leqslant h(z)$. But if the order of $\varphi_z$ on $L_\alpha$ is strictly less than 1, then

$$\overline{\lim_{t\to\infty}}\, t^{-1} \cdot \ln|\varphi_z(te^{i\theta})| \leqslant 0 \leqslant h(z).$$

The upper semicontinuity of $h$ implies that $f \in [\rho, h]$ ▶

**Proof of Theorem 8.3.1.** Isomorphism of the spaces $[\rho, h]$ and $H(\Omega_h)$ follows from Lemmas 8.3.2, 8.3.3.

1) → 2). By Lemma 8.3.2, $g \in H(\Omega_h)$. Let $k$ be the $x$-parabolic Minkowski functional for the kernel of the domain $\Omega$. Since $\Omega_h \subset \Omega =$

$\Omega_k$, we have $h \geqslant k$ (see Theorem 8.2.3). On the other hand, $g \in H(\Omega_k)$ and, by Lemma 8.3.3, $f = \mathcal{A}^{-1}g \in [\rho, k]$. Hence, $h = h_f^+ \leqslant k$. So, $h = k$ and $\Omega_h = \Omega_k = \Omega$. The inverse implication 2) $\to$ 1) is proved similarly.

**8.3.2. Topological isomorphism of the spaces $[\rho, h]$ and $H(\Omega_h)$.** Suppose that for a function $h \in P_n(x)$ there exist a sequence of continuous functions in $P_n(x)$ monotonically decreasing to $h$. This is so, e.g., for every continuous function $h \in P_n(x)$. We endow $[\rho, h]$ with the topology of projective limit of the sequence $\{B_k\}$ of Banach spaces of entire functions with the norms

$$\|f\|_k = \sup\{|f(z)| \exp\{-h_k(z)\},\ z \in \mathbb{C}^n\}, \quad k = 1, 2, \dots;$$

in $H(\Omega_h)$, we introduce the topology of uniform convergence on all compact subsets of $\Omega_h$ (we may choose, for instance, $\|g\|_k = \sup\{|g(z)| : z \in \Omega(h_k)\}$, $k = 1, 2\dots$, as seminorms determining this topology; by Corollary 8.2.5, $\{\Omega(h_k)\}$ determines the collection of bounded sets in $\Omega_h$.

**Theorem 8.3.4.** *Under the conditions and in the notation of Theorem* 8.3.1, *and under the above restrictions on* $h$, *the isomorphism* $\mathcal{A} : [\rho, h] \to h(\Omega_h)$ *is topological.*

◀ We prove the theorem for $x = \mathbb{I}$ *). Observe that $\Omega(h_k) = \{z \in \mathbb{C}^n : \alpha_k(z) < 0\}$, where $\alpha_k(z) = \inf\{h_k(ze^{i\theta}) - \cos\theta : |\theta| \leqslant \pi/2\}$ (see (8.2.12)). Since $h_k$ is continuous, so is $\alpha_k$, and $M_k = \overline{\Omega(h_k)}$ is compact, $M_k \subset \overset{\circ}{M}_{k+1}$, $\bigcup\limits_{k=1}^{\infty} M_k = \Omega_h$ (see Exercise 1.2.7).

Let $k \in \mathbb{N}$, and let $f \in [\rho, h]$. We put $g = \mathcal{A}f$ and

$$\Delta_\theta = \{\zeta \in \partial M_k : h_k(\zeta e^{i\theta}) = \cos\theta)\}, \ \Delta = \{\theta \in (-\pi/2, \pi/2) : \Delta_\theta \neq \varnothing\}.$$

By the maximum principle for holomorphic functions, we have

$$\|\mathcal{A}f\|_k = \sup_{\zeta \in \partial M_k} |g(\zeta)| = \sup_{\theta \in \Delta} \sup_{\zeta \in \Delta_\theta} |g(\zeta)|. \tag{8.3.5}$$

But $g(z) = G_\theta(z)\ \forall\ z \in \Pi(\theta)$, where $\Pi(\theta) = \{z \in \mathbb{C}^n : h(ze^{i\theta}) < \cos\theta\}$ (see (8.3.2)). Also, $\Delta_\theta \subset \Pi(\theta)$ and $f \in B_{k+1}$; consequently, for all $\zeta \in \Delta_\theta$ we obtain

$$|g(\zeta)| \leqslant \|f\|_{k+1} \cdot \int_0^\infty \exp\{-r[\cos\theta - h_{k+1}(\zeta e^{i\theta})]\}\, dr =$$

*) If $x = \mathbb{I}$, then $\rho = \mathbb{I}$.

$$\|f\|_{k+1} \cdot \int_0^\infty \exp\{-r[h_k(\zeta e^{i\theta}) - h_{k+1}(\zeta e^{i\theta})]\}\, dr. \qquad (8.3.6)$$

Let $S = \{z \in \mathbb{C}^n : \|z\| = 1\}$, where $\|z\| = |z_1| + \cdots + |z_n|$, and let $m_k = \min\{h_k(w) - h_{k+1}(w),\ w \in S\}$. Then $m_k > 0$ , and for $\varepsilon_k \in (0, m_k)$ we have: $h_k(w) - h_{k+1}(w) > \varepsilon\|w\|\ \forall\ w \in \mathbb{C}^n \setminus \{0\}$. By (8.3.6), we see that

$$|g(\zeta)| \leqslant \|f\|_{k+1} \cdot \int_0^\infty \exp\{-r\varepsilon_k\|\zeta\|\}\, dr = \|f\|_{k+1}(\varepsilon_k\|\zeta\|)^{-1}. \qquad (8.3.7)$$

There exists a constant $\gamma_k > 0$ such that $\|\zeta\| > \gamma_k\ \forall\ \zeta \in \partial M_k$ because $0 \notin \partial M_k$. Now from (8.3.5) – (8.3.7) we get

$$\|\mathcal{A}f\|_k \leqslant A_k\|f\|_{k+1},\ k = 1, 2, \ldots, \qquad (8.3.8)$$

where the constant $A_k > 0$ does not depend on $f$.

Combining this with Proposition 1.10.10, we conclude that the linear operator $\mathcal{A}$ is continuous. But $[\rho, h]$ and $H(\Omega_h)$ are $F$-spaces. Therefore, by the Banach theorem, $\mathcal{A} : [\rho, h] \to H(\Omega_h)$ is an open mapping. Consequently, the inverse operator $\mathcal{A}^{-1}$ is also continuous ▶

## 8.4. Consequences of Theorem 8.3.1

**8.4.1. Existence of entire functions with a given positive $\rho$-indicator of class $P_n(x)$.** The $\rho$-indicator of an entire function $f$ (see Definition 8.1.4) is said to be *positive* if $h_f(z; \rho) > 0\ \forall\ z \in \mathbb{C}^n \setminus \{0\}$. From Corollary 8.2.6 it follows that the $\rho$-indicator of a function with bounded $\rho$-indicator diagram is such. Theorems 8.2.12 and 8.3.1 imply some conditions ensuring the existence of an entire function with a given system of $\rho$-positive conjugate orders and with a given positive $\rho$-indicator.

**Theorem 8.4.1.** *Let $x \in \mathbb{R}^n_0$, and let $\rho = (x_1^{-1}, \ldots, x_n^{-1})$. If $h$ is a positive function in $P_n(x)$ satisfying the conditions of Theorem* 8.2.12, *then there exists an entire function $f$ in $\mathbb{C}^n$ whose $r$-indicator $h_f$ is $h$.*

The required function $f$ is equal to $\mathcal{A}^{-1}g$ (see (8.3.1)), where $g$ is a function holomorphic in the pseudoconvex domain $\Omega_h$ (see (8.2.3)) and nonextendible beyond the boundary. Since a $\rho$-indicator is upper semicontinuous, Theorem 8.4.1 is valid for every function $h \in P_n(x)$ with the property $\operatorname{int}\{z \in \mathbb{C}^n : h(z) = 0\} = \varnothing$. If $x \in \mathbb{R}^n_0$ has equal coordinates, the approximation discussed in Theorem 8.2.12 exists for every function in $P_n(x)$ (see Theorem 1.8.11). Therefore, Theorem 8.4.1 implies the following weak version of the Martineau theorem about the existence of an entire function with a given indicator.

**Corollary 8.4.2.** *For every positive plurisubharmonic function* $h$ *on* $\mathbb{C}^n$ *positively homogeneous of order* $\rho > 0$, *there exists an entire function of order* $\rho$ *in the totality of variables the regularized radial indicator* $L_f(z,\mathbf{1})$ *of which is* $h$.

**8.4.2. System of integral summation methods for multiple power series.** We dwell on the multidimensional generalization of the results of Subsection 3.3.2. Suppose

$$g(z) = \sum_{\|k\|\geqslant 0} a_k z^k \tag{8.4.1}$$

is a power series convergent in the vicinity of 0 in $\mathbb{C}^n$; let $f = \mathcal{A}^{-1}g$ be the associated entire function (see (8.3.1)). Consider the system of integrals (see (8.3.2)) $\{G_\theta : |\theta| < \pi/2\}$. It has been shown that $G_\theta$ is an analytic extension of $g$ into the interior $B_\theta$ of the set of points of absolute convergence for the integral $G_\theta$, and $0 \in B_\theta$.

Elementary arguments yield the following statement.

**Proposition 8.4.3.** *In the above notation, we have* $B_\theta = \{z(e^{i\theta}\cos\theta)^x : z \in B\}$ $\forall\ \theta \in (-\pi/2, \pi/2)$, *and* $M := \bigcup B_\theta \,|\, |\theta| < \pi/2$ *is an* $x$*-circular domain in* $\mathbb{C}^n$. *If* $p$ *is the* $x$*-parabolic Minkowski functional of the* $x$*-parabolically star-like set* $B = B_0$ (see Definition 8.2.2, (8.2.3)), *then*

$$B_\theta = \{w \in \mathbb{C}^n : p[w(e^{i\theta})^x] < \cos\theta\}, \quad |\theta| < \pi/2;\ M = \Omega_p.$$

We state a multidimensional version of Definition 3.3.3.

**Definition 8.4.4.** *The generalized Borel* $x$*-polygon of the series* $g$ (see (8.4.1)) is the set $M$ (see Proposition 8.4.2) equal to the domain of summability of the series $g$ by the system of integrals $\{G_\theta, |\theta| < \pi/2\}$.

From Theorem 8.2.3, Proposition 8.4.3, Lemmas 8.3.2 and 8.3.3 we deduce the following supplement to Theorem 8.3.1.

**Theorem 8.4.5.** *In the above notation, conditions* 1) *and* 2) *of Theorem* 8.3.1 *are equivalent to condition* 3): *the generalized Borel polygon* $M$ *of* $g$ (see (8.4.1)) *coincides with* $\Omega_h$, *and* $B = \{z \in \mathbb{C}^n : h(z) < 1\}$, $p = h$.

◄ 2) →3). We have $f = \mathcal{A}^{-1}g \in [\rho, h]$ (see Lemma 8.3.3). But $\Pi(\theta) \subset B_\theta$, $\theta < \pi/2$ (see the proof of Lemma 8.3.2), i.e., $\Omega_h \subset M$. On the other hand, from condition 2) we have: $M \subset \Omega = \Omega_h$. Therefore, $M = \Omega_h$. Now, by Theorem 8.2.3 we conclude (see (8.2.3)) that $p = \nu^{-1}(M) = \nu^{-1}(\Omega_h) = h$. Hence, $B = \{z \in \mathbb{C}^n : h(z) < 1\}$.

3) $\to$ 2). By assumption, $g \in H(\Omega_h)$. By Lemma 8.3.3, $f = \mathcal{A}^{-1}g \in [\rho, h]$. Let $k = h_f^+$ (see Theorem 8.3.1). Then $k \leqslant h$ and, therefore, $f \in [\rho, k]$. In the course of the proof of Lemma 8.3.2 it was shown that $\Omega_k \subset M = \Omega_h$. By Theorem 8.2.3, we have $k \geqslant h$. Thus, $k = h$, and condition 1) follows. Now, Theorem 8.3.1 yields $\Omega = \Omega_h = M$ ▶

**8.4.3. Multidimensional analog of Borel's theorem.** The multidimensional analog of Borel's theorem given below (see Proposition 1.10.4) makes it possible to use the $x$-type function of an entire function (see Subsection 8.1.1) for the description of the convergence domain $D_g$ of the series $g$.

**Theorem 8.4.6.** *Let $f$ be an entire function in $\mathbb{C}^n$ with a system of positive conjugate orders $\rho = (\rho_1, \dots, \rho_n)$. We put $x = (\rho_1^{-1}, \dots, \rho_n^{-1})$ and consider the n-fold power series $g = \mathcal{A}f$ associated with $f$ (see (8.3.1)). Then the $x$-type function $\sigma_f(|z|; x)$ of $f$ is the $x$-parabolic Minkowski functional for the convergence domain $D_g$ of $g$,*

$$D_g = \{z \in \mathbb{C}^n : \sigma_f[(|z_1|, \dots, |z_n|); x] < 1\},$$

*and $S = \{r \in \mathbb{R}_0^n : \sigma_f(r; x) = 1\}$ is the hypersurface of conjugate convergence radii for $g$.*

◀ We apply Theorem 8.3.1 to the function $h(z) = \sigma_f(|z|; x)$, taking into account that $h \in P_n(x)$ (see Propositions 6.3.2, 6.3.3, Theorem 1.8.14). Then $D_g = \Omega_h = \{z \in \mathbb{C}^n : h(z) < 1\}$ is a complete logarithmically convex Reinhardt domain with center at 0 (see Theorem 8.2.9) ▶

**8.4.4. Property of the functional associated with a circular holomorphy domain.** The following supplement to Theorem 8.2.3 is a consequence of Theorem 8.3.1 (see the proof of Theorem 8.4.1).

**Theorem 8.4.6.** *Let $x \in \mathbb{R}_0^n$, let $\Omega$ be an $x$-circular holomorphy domain in $\mathbb{C}^n$, and let $h = \nu(\Omega)$ be the functional associated with $\Omega$. Then $h \in P_n(x)$.*

Geometrically this means that, under the conditions of the theorem, the kernel of $\Omega$ is a holomorphy domain.

**Remark.** Let $D_n(x)$ be the class of $x$-circular holomorphy domains in $\mathbb{C}^n$. From Theorems 8.2.12 and 8.4.6 it follows that the mapping $\nu : D_n(x) \to P_n(x)$ (see (8.2.3)) is a monomorphism. This is an epimorphism if the assumptions of Theorem 8.2.12 are fulfilled for every function $h \in P_n(x)$. For $x = \mathbb{I}$, this condition is valid indeed (see Theorem 1.8.11).

# Appendix A

# Riemann Surface of the Inverse Function for a Polynomial of Fractional Order

We examine the structure of the Riemann surface for the inverse of the mapping

$$\alpha(z) = z^\rho \left[1 + \frac{a_1}{z} + \cdots + \frac{a_n}{z^n}\right] \tag{A.0.1}$$

in the case of nonintegral $\rho > 0$. This surface was used in Chapter 4 for construction of the plane indicator diagram of an entire function of order $\rho$. For simplicity and by analogy with the case where $\rho$ is an integer the mapping (0.1) will always be called a polynomial; in particular, we use this term for $\rho < n$.

A Riemann surface is a central notion in the exposition below. We recall the definition.

A *two-dimensional manifold* is a Hausdorff space $X$ each point of which has a neighborhood homemorphic to an open subset of $\mathbb{R}^2$.

Let $X$ be a two-dimensional manifold. A *complex chart* on $X$ is a homeomorphism $\varphi : U \to V$ of some open subset $U \subset X$ onto some open subset $V \subset \mathbb{C}$. Two complex charts $\varphi_i : U_i \to V_i$, $i = 1, 2,$ are said to be *biholomorphically compatible* if the mapping

$$\varphi_2 \circ \varphi_1^{-1} : \quad \varphi_1(U_1 \cap U_2) \to \varphi_2(U_1 \cap U_2)$$

is biholomorphic. A *complex atlas* on $X$ is any system $\mathfrak{A} = \{\varphi_i : U_i \to V_i,\ i \in I\}$ of pairwise compatible charts covering $X$, i.e., $\bigcup U_i | i \in I = X$. Two complex atlases $\mathfrak{A}$ and $\mathfrak{A}'$ on $X$ are said to be *biholomorphically compatible* if every chart in $\mathfrak{A}$ is biholomorphically compatible with every chart in $\mathfrak{A}'$.

Biholomorphic compatibility of complex atlases is an equivalence relation. The following definition is based on this fact.

**Definition A.0.1.** A *complex structure* on a two-dimensional manifold $X$ is an equivalence class of biholomorphically compatible atlases on $X$.

So, a complex structure on $X$ can be defined by indicating one atlas.

**Definition A.0.2.** A *Riemann surface* is the pair $(X, \sum)$ consisting of a two-dimensional manifold $X$ and a complex structure $\sum$ on $X$.

Usualy, a Riemann surface is denoted by $X$ instead of $(X, \sum)$, when it is clear which complex structure is meant. Sometimes the Riemann surface is denoted by $(X, \mathfrak{A})$, when the atlas $\mathfrak{A}$ represents the structure $\sum$.

The Riemann surface of the inverse to the polynomial (A.0.1) belongs to a special class of coverings of $\mathbb{C}$. To describe the properties of this class, we need some general information about coverings, and other topological notions. Partly, this information was used also in the main part of the book.

## A.1. Some topological information

We begin with some important notions and facts of the theory of coverings. For our purposes it is convenient to use the terminology accepted in the monograph by R. Forster (see References).

**A.1.1. Holomorphic coverings.** A subset $A$ of a topological space is said to be *discrete* if every point $a \in A$ has a neighborhood $V$ such that $V \cap A = \{a\}$. A mapping $p : \mathcal{Y} \to X$ between two topological spaces $X, \mathcal{Y}$ is said to be *discrete* if the preimage $p^{-1}(x)$ of every point $x \in X$ is a discrete subset in $\mathcal{Y}$.

**Definition A.1.1.** Let $X, \mathcal{Y}$ be topological spaces. A mapping $p : \mathcal{Y} \to X$ is called a *covering (a covering mapping)* if it is continuous, open and discrete. The space $X$ is called the *base*, and $\mathcal{Y}$ is called the *covering space*.

If $p : \mathcal{Y} \to X$ and $q : Z \to X$ are two coverings, then a mapping $f : \mathcal{Y} \to Z$ is said to be *fiberwise* if $p = q \circ f$. If, moreover, $f$ is a homeomorphism, then the coverings $p$ and $q$ are called *equivalent*.

**Proposition A.1.2.** *Let $X, \mathcal{Y}$ be Riemann surfaces, and let $p : \mathcal{Y} \to X$ be a nonconstant holomorphic mapping. Then $p$ is a covering mapping.*

**Definition A.1.3.** Let $X, \mathcal{Y}$ be topological spaces, and let $p : \mathcal{Y} \to X$ be a covering. A point $y \in \mathcal{Y}$ is called a *branching point* for $p$ if $y$ has no neighborhood $V$ such that $p|_V$ is one-to-one. The covering $p$ is said to be *unbranched* if $p$ has no branching points. Otherwise, $p$ is said to be *branched*.

**Proposition A.1.4.** *A covering $p : \mathcal{Y} \to X$ is unbranched if and only if it is locally topological.*

**Definition A.1.5.** Under the conditions of Definition A.1.1, the mapping $p : \mathcal{Y} \to X$ is called an *unbounded unbranched covering* if every point $x \in X$ has an open neighborhood $U$ such that the preimage $p^{-1}(U)$ can be represented in the form $p^{-1}(U) = \bigcup V_j | \, j \in I$, where the $V_j, j \in I$, are mutually disjoint open subsets in $\mathcal{Y}$ and all mappings $p : V_j \to U$ are homeomorphisms. The nature of branching points of holomorphic mappings is clarified by the following fact.

**Proposition A.1.6.** *Suppose $X, \mathcal{Y}$ are Riemann surfaces, $p : \mathcal{Y} \to X$ is a nonconstant holomorphic mapping (holomorphic covering), and $a \in \mathcal{Y}$. We put $b := p(a)$. Then there exists a natural number $k \geqslant 1$ and mappings $\varphi : U \to V$ on $\mathcal{Y}$ and $\psi : U' \to V'$ on $X$ with the following properties:*

1) $a \in U$; $\varphi(a) = 0$; $b \in U'$; $\psi(b) = 0$; 2) $p(U) \subset U'$;
3) *the mapping* $q := \psi \circ p \circ \varphi^{-1}$ *has the form* $q(z) = z^k$ *for all* $z \in V$.
The number $k \geqslant 1$ is called the *multiplicity of* $p$ *at* $a$.

**A.1.2. On automorphism groups on a Riemann surface.** Let $X$ be a topological space, and $R$ an equivalence relation in $X$. We denote by $X/R$ the factor-set, and by $\pi : X \to X/R$ the canonical mapping (projection) taking each element $x$ in $X$ to the equivalence class of $x$. The set $X/R$ is equipped with the factor-topology: a set $V \subset X/R$ is open if and only if the set $\pi^{-1}(V)$ is open in $X$. The set $X/R$ equipped with the factor-topology is called the *factor-space.*

Let $G$ be a group of conformal automorphisms operating on a Riemann surface $X$. The role of the unit of the group is played by the identity automorphism. The group $G$ is said to be *free and totally disconnected* (or to operate freely and totally disconnectedly) if every point $x \in X$ has a neighborhood $U$ such that the sets $\{g(U),\ g \in G\}$ are mutually disjoint.

The *orbit of a point* $x \in X$ *under the action of the group* $G$ is the set $G_x = \{g(x);\ g \in G\}$. Two orbits $G_x, G_y$ either do not intersect or coincide. This makes it possible to introduce an equivalence relation $R$ in $X$: $x \sim y$ if $G_x = G_y$. The factor-space $X/R$ is called the *space of orbits of* $G$ and is denoted by $X/G$. A *fundamental set for the group* $G$ is a subset $F$ in $X$ containing exactly one point of each orbit. A *fundamental domain for* $G$ is a domain in $X$ that, together with a part of its boundary, forms a fundamental set for $G$.

**Proposition A.1.7.** *If* $G$ *operates on an open Riemann surface* $X$ *totally disconnectedly and freely, then* $\pi : X \to X/G$ *is an unbranched unbounded covering, and the space of orbits* $X/G$ *is a Riemann surface.*

The structure of $X/G$ is induced by the structure of $X$ as follows. Let $\{(V, \varphi_V) :\ V \in A\}$ be an atlas on $X$ with the following properties:

a) the set $A$ consists of open neighborhoods of arbitrary points in $X$;
b) if $V \in A$, then $g(V) \in A\ \forall\ g \in G$;
c) the mapping $\pi_V := \pi|_V :\ V \to \pi(V)$ is a homeomorphism.

Then $\{(\pi(V), \varphi_V \circ \pi_V^{-1})\}$ is an atlas on $X/G$.

**Definition A.1.8.** Let $X$ and $X'$ be open Riemann surfaces, let $G$ and $G'$ be free and totally disconnected groups of conformal automorphisms operating on $X$ and $X'$, respectively, and let $F : X \to X'$ be a conformal homeomorphism. The mapping $F$ is said to *agree with the action of* $G$ and $G'$ if $F$ induces an isomorphism $F_* : G \to G'$ of the groups in question in accordance with the rule

$$F \circ g \circ F^{-1} = g',\ g \in G,\ g' \in G'.$$

**Proposition A.1.9.** *Under the conditions of Definition* A.1.8, *there exists* *a unique conformal homeomorphism* $\widehat{F} : X/G \to X'/G'$ *such that the diagram*

$$\begin{array}{ccc} X & \xrightarrow{F} & X' \\ \downarrow \pi & & \downarrow \pi' \\ X/G & \xrightarrow{\widehat{F}} & X'/G' \end{array}$$

*is commutative, where* $\pi$ *and* $\pi'$ *are the canonical mappings.*

The mapping $\widehat{F}$ is called the *factor-mapping* for $F$. Proposition A.1.9 remains valid for homeomorphic mappings, i.e., if Definition A.1.8 is formulated for homeomorphisms without the requirement of conformity.

**A.1.3. Lifting of mappings.** Let $X, \mathcal{Y}, Z$ be topological spaces, $p : \mathcal{Y} \to X$ a covering, and $f : Z \to X$ a continuous mapping. A *lifting* of $f$ with respect to $p$ is a continuous mapping $g : Z \to \mathcal{Y}$ such that $f = p \circ g$, i.e., for every $x \in X$ the fiber $f^{-1}(x)$ is taken by $g$ to the fiber $p^{-1}(x)$. For this reason, $g$ is a continuous fiberwise mapping (see Definition A.1.1).

**Proposition A.1.10.** *Let* $X, \mathcal{Y}, Z$ *be Riemann surfaces,* $p : \mathcal{Y} \to X$ *a holomorphic unbranched covering, and* $f : Z \to X$ *a holomorphic mapping. Then every lifting* $g : Z \to \mathcal{Y}$ *of* $f$ *is holomorphic.*

**Proposition A.1.11.** *Let* $X, \mathcal{Y}$ *be Hausdorff spaces and* $p : \mathcal{Y} \to X$ *an unbranched unbounded covering. Next, let* $Z$ *be a simply connected, arcwise connected and locally arcwise connected topological space* *), *and let* $f : Z \to X$ *be a continuous mapping. Then, for every choice of points* $z_0 \in Z$, $y_0 \in \mathcal{Y}$ *with the property* $f(z_0) = p(y_0)$, *there exists exactly one lifting* $f : Z \to \mathcal{Y}$ *such that* $\tilde{f}(z_0) = y_0$.

**Definition A.1.12.** An unbranched unbounded covering $p : \mathcal{Y} \to X$ is said to be *universal* if $\mathcal{Y}$ and $X$ are connected manifolds, and $\mathcal{Y}$ is a simply connected manifold.

**Proposition A.1.13.** *Two universal coverings over the same base are equivalent* (see Def. A.1.1).

**A.1.4. The number of covering sheets.**

**Proposition A.1.14.** *Let* $X$ *and* $\mathcal{Y}$ *be Hausdorff spaces. Suppose* $X$ *is arcwise connected and* $p : \mathcal{Y} \to X$ *is an unbranched unbounded covering. Then for every two points* $x_0, x_1 \in X$ *the fibers* $p^{-1}(x_0)$ *and* $p^{-1}(x_1)$ *are of equal cardinality.*

The cardinality of the fiber $p^{-1}(x)$ is called the *number of covering sheets*. A continuous mapping $f : \mathcal{Y} \to X$ of locally compact spaces $X, \mathcal{Y}$ is said to be *proper* if the preimage of every compact set is compact.

Let $\mathcal{Y}$, $X$ be Riemann surfaces, and let $f : \mathcal{Y} \to X$ be a nonconstant holomorphic mapping. For $x \in X$ we denote by $v(f, x)$ the multiplicity of $f$

*) The latter term is used for the spaces the topology of which has a base of arcwise connected neighborhoods.

at $x$ (see Proposition A.1.6). A value $c \in \mathcal{Y}$ is said to be taken by $f$ on $X$ $m$ times with account of multiplicity if

$$m = \sum_{x \in p^{-1}(c)} v(f, x).$$

**Proposition A.1.15.** *Let $X, \mathcal{Y}$ be Riemann surfaces, and $f : \mathcal{Y} \to X$ a proper nonconstant holomorphic mapping. Then there exists a natural number $n$ such that $f$ takes each value $c \in \mathcal{Y}$ precisely $n$ times with account of multiplicity.*

**Proposition A.1.16.** *On a compact Riemann surface $X$, every nonconstant meromorphic function $f : X \to \overline{\mathbb{C}}$ has as many zeros as poles with account of multiplicity.*

**A.1.5. Some more topological information.**

1. *Symmetric product.* Let $\Sigma_n$ be the class of permutations of the numbers $(1, 2, \ldots, n)$. For a topological space $X$, let $X^n = X \times \cdots \times X$ denote the product of $n$ copies of $X$. In $X^n$, we introduce the equivalence relation $R$ : $(x_1, \ldots, x_n) \sim (x_{\sigma(1)}, \ldots, x_{\sigma(n)})$, $\sigma \in \Sigma_n$. The factor-space $SP^n(X) = X^n/R$ equipped with the factor-topology is called the *$n$-fold symmetric product of the space $X$*. By $\{x_1, \ldots, x_n\}$ we denote an equivalence class in $SP^n(X)$.

In the case of $X = \mathbb{C}$, this structure helps to describe the relationship between the roots of a polynomial and its coefficients.

**Theorem A.1.17.** *Let $SP^n(\mathbb{C})$ be the $n$-fold symmetric product of $\mathbb{C}$. The mapping*

$$\Phi : SP^n(\mathbb{C}) \to \mathbb{C}^n, \quad \Phi(\{q_1, \ldots, q_n\}) = (p_1, \ldots, p_n),$$

*defined by the formula*

$$\prod_{i=1}^{n} (x - q_i) = \sum_{i=0}^{n} p_i x^{n-i}, \quad p_0 = 1,$$

*is a homeomorphism.*

The space $SP^n(\mathbb{C})$ is metrizable. The distance between two poits can be defined as follows:

$$\rho(\{x\}, \{y\}) = \inf\{|x - y_\sigma| : \sigma \in \Sigma_n\}.$$

Here $y_\sigma = (y_{\sigma(1)}, \ldots, y_{\sigma(n)})$ and $|a - b| = \sqrt{\sum_{i=1}^{n} (a_i - b_i)^2}$.

The axioms of a metric space are verified immediately. We clarify the meaning of the convergence in this metric. Let $\{\alpha\}$ be an element in $SP^n(\mathbb{C})$ with different coordiantes $\alpha_1, \ldots, \alpha_m$ of multiplicities $r_1, \ldots, r_m$, where $r_1 + r_2 + \cdots + r_m = n$. If $\rho(\{z\}^{(k)}, \{\alpha\}) \to 0$ as $k \to \infty$, then the coordinates of the element $\{z\}^{(k)}$ break into $m$ nonintersecting groups $\left\{z_{ij}^{(k)},\ j = 1, \ldots, r_i\right\}$, $i = 1, \ldots, m$, with the following property:

$$z_{ij}^{(k)} \to \alpha_i \quad \text{as} \quad k \to \infty \quad \text{for} \quad j = 1, \ldots, r_i,\ i = 1, \ldots, m.$$

2. *Simplicial spaces and their fundamental groups.* Let $\Delta$ be a Euclidean simplex of dimension $k \leqslant n$, i.e., the convex hull of a set $A$ in $\mathbb{R}^n$ consisting of $k+1$ points and not included in to any $(k-1)$-dimensional plane (if $k = 0$, then $\Delta = A$, see Exercise 1.3.7). If $k \geqslant 1$, and $B$ is a proper subset of $A$, then the convex hull of $B$ is called a *face of the simplex* $\Delta$.

Let $X$ be a topological space. Its subset $T$ is called a *topological simplex of dimension* $k$ if there exists a homeomorphism $\varphi : T \to \Delta$, where $\Delta$ is a Euclidean simplex of dimension $k$. By a *face of the simplex* $T$ we mean the preimage of a face of $\Delta$ under $\varphi$. By a *triangulation of the space* $X$ we mean a system of topological simplexes $\Lambda = \{T_j : j \in J\}$ that form a covering of $X$ ( i.e., $X = \bigcup T_j | \, j \in J$) and satisfy the following conditions:

1) the faces of each simplex in $\Lambda$ belong to $\Lambda$;

2) if the intersection $T_i \cap T_j$ of two different simplexes $T_i$ and $T_j$ is nonempty, then either some of these simplexes is a face of the other, or $T_i \cap T_j$ is a common face of $T_i$ and $T_j$.

The space $X$ for which a triangulation $\Lambda$ exists is called a *simplicial space*. A subset $F$ of a simplicial space $X$ is closed if and only if so are the sets $F \cap T_j\ \forall\, j \in J$.

We mention a property of one-dimensional homotopy groups of simplicial spaces. Let $X$ be a topological space, and let $x \in X$. We denote by $\pi_1(X, x)$ the set of all classes of homotopically equivalent closed paths in $X$ with beginning and end at $x$. On $\pi_1(X, x)$, we introduce a multiplication by putting $[\alpha] * [\beta] = [\alpha\beta]$ (where $[\gamma]$ is the equivalence class with representative $\gamma$); here

$$\alpha\beta = \begin{cases} \alpha(2t), & 0 \leqslant t \leqslant 1/2; \\ \beta(2t-1), & 1/2 \leqslant t \leqslant 1. \end{cases}$$

With respect to this operation, $\pi_1(X, x)$ is a group the unit of which is the class of paths equivalent to the path $U_0(t) \equiv x\ \forall\, t \in [0, 1]$. The set $\psi_1(x, x)$ is called the *fundamental group of* $X$ *with basic point* $x$. For an arcwise connected topological space $X$, the fundamental groups with different basic points are isomorphic to one another. Therefore, in this case the notation $\pi_1(X) = \pi_1(X, x)$ is used and $\pi_1(X)$ is called the *fundamental group of* $X$. An arcwise connected topological space $X$ is said to be *simply connected* if $\pi_1(X) = 0$.

We state a particular case of the so-called addition theorem.

**Theorem A.1.18.** *Let $X$ be a simplicial space, and let $A, B$ be its simply connected subspaces such that $A \cup B = X$. If $A \cap B$ is connected, then $X$ is simply connected.*

## A.2. The construction of the Riemann surface and its properties

To construct the Riemann surface for the inverse of a polynomial (0.1), a well-known procedure, usually applied to algebraic functions, is employed. However, in the case of a noninteger $\rho$, a peculiarity arises: if an element of the inverse to the polynomial in question admits analytic continuation along a closed path encompassing the point 0 in $\mathbb{C}$, then we arrive at the initial element after a fractional number of turns, i.e., after "rotation" about 0 through an angle that is not a multiple $2\pi$.

**A.2.1. The space $L$ (operations, holomoprhic mappings).** Let $L = (0,\infty) \times \mathbb{R}$ be the Riemann surface of the logarithm. The complex structure on $L$ is determined by the local homeomorphism

$$\pi : L \to \mathbb{C}^*; \quad \pi[(r,\theta)] = re^{i\theta}, \tag{A.2.1}$$

where $\mathbb{C}^* = \mathbb{C} \setminus \{0\}$. In $L$, we introduce a multiplication:

$$Z_1 \cdot Z_2 = (|Z_1| \cdot |Z_2|, Arg\, Z_1 + Arg\, Z_2),\ Z_1, Z_2 \in L.$$

With respect to this operation, $L$ is an abelian group with the unit $\mathbb{I} = (1,0)$. This leads to the notion of the $\rho$th power of an element $Z$ for $\rho \in \mathbb{R}$:

$$Z^\rho = (|Z|^\rho, \rho\, Arg\, Z),\ Z \in L.$$

Also, we define the operation of multiplication by complex numbers on $L$:

$$a \cdot Z = (|a| \cdot |Z|, \arg a + Arg\, Z),\ a \in \mathbb{C}^*,\ Z \in L.$$

In the above notation,

$$\pi(Z_1 \cdot Z_2) = \pi(Z_1) \cdot \pi(Z_2),\ \pi(aZ) = a\pi(Z),$$

so, the mapping $\pi$ is a homomorphism between the multiplicative groups of $L$ and $\mathbb{C}^*$.

We consider automorphism groups on $L$. Most usual is the cyclic group $G_L$ with generator $A$, where

$$AZ = (|Z|, Arg\, Z + 2\pi),\ Z \in L. \tag{A.2.2}$$

Along with $G_L$, we also consider the group $G_\rho$ with generator $A_\rho$ , where

$$A_\rho Z = (|Z|, Arg\, Z + 2\pi\rho),\ Z \in L, \tag{A.2.3}$$

$\rho > 0$. If $\rho$ is a noninteger, then, in general, elements of $G_\rho$ do not act fiberwise with respect to $\pi$:

$$\pi(A_\rho Z) \neq \pi(Z) \quad \forall\, Z \in L.$$

Besides automorphisms, we shall use other holomorphic mappings defined in some domain $V \subset L$ and taking values, for instance, in $\mathbb{C}$. Note that every such mapping $w = F(Z)$, $Z = (r, \theta) \in V$, is characterized by the Cauchy-Riemann condition $\frac{\partial F}{\partial r} = -\frac{i}{r}\frac{\partial F}{\partial \theta}$, and its derivative is computed by the formula $F'(Z) = e^{-i\theta}\frac{\partial F}{\partial r}$, $Z \in V$.

**A.2.2. The Riemann surface of the power function.** To illustrate the idea of the construction of the Riemann surface for the inverse of the polynomial (0.1), we use the example of a power function. Let

$$F_\rho : L \to L;\ F_\rho(r, \theta) = (r^{1/\rho}, \theta/\rho) \quad \forall\ (r, \theta) \in L.$$

Usually, the function

$$z = \pi \circ F_\rho(r, \theta) = \exp\left\{\frac{1}{\rho}(\ln r + i\theta)\right\}, \quad (r, \theta) \in L, \tag{A.2.4}$$

is viewed as a solution of the equation $p = z^\rho$, where $p = \pi[r, \theta]$ (see (A.2.1)). Let $L_\rho = L/G_\rho$ be the space of orbits of the group $G_\rho$ operating on $L$ ($L_\rho$ is equipped with the factor-topology). In particular, $L_1 = L/G_L \approx \mathbb{C}^*$ (see item 1°: $G_L = G_1$). Since

$$F_\rho \circ A_\rho^k = A^k \circ F_\rho \quad \forall\ k \in \mathbb{Z}$$

and $F_\rho$ is a biholomoprhic mapping, the factor-mapping (see A.1) $f_\rho : L_\rho \to \mathbb{C}^*$ is also biholomorphic. Thus, analytic continuation of an element of the function (A.2.4) yields the same element after turning about $p = 0$ through the angle of $2\pi\rho$, i.e., after a "fractional" number of turns.

The mapping $f_\rho$ makes it possible to compactify $L_\rho$ . We consider the compactification $\widehat{L}_\rho$ with the help of two points, corresponding to the "north and south poles" of the Riemann sphere $\mathbb{C}$. In $\widehat{L}_\rho$ , we intorduce the topology induced from $\mathbb{C}$. For $\rho = n \in \mathbb{N}$, the compact set $\widehat{L}_n$ is called the Reimann surface of the function $f_n$ . By analogy, for a noninteger $\rho$ it is but natural to call the set $\widehat{L}_\rho$ the *Riemann surface of the function* $f_\rho$ .

Similar ideas will be used in the much more complicated situation of the Riemann surface for the inverse of a polynomial of fractional order:

$$\alpha(Z) = \pi(Z^\rho) + a_1\pi(Z^{\rho-1}) + \cdots + a_n\pi(Z^{\rho-n}) -$$

$$= \pi(Z^\rho) \cdot \left[1 + \frac{a_1}{z} + \cdots + \frac{a_n}{z^n}\right], \quad Z \in L,$$

where $a_n \neq 0$, $\rho$ is a noninteger, and for the integer $n > 0$ the case where $n > \rho > 0$, is not excluded. In the sequel, along with the symbol $\pi(Z^\rho)$, we also use the symbol $Z^\rho$ if this does not lead to confusion. In the literature, the latter symbol is often used for the power function when, e.g., it is understood in the sense of the principal value:

$$Z^\rho = \exp\{\rho \ln Z\}, \; \ln Z = \ln r + i\theta, \; (r,\theta) \in L.$$

This simplifies the formula for a polynomial $\alpha : L \to \mathbb{C}$ :

$$\alpha(Z) = Z^\rho + a_1 Z^{\rho-1} + \cdots + a_n Z^{\rho-n}, \; Z = (r,\theta) \in L. \tag{A.2.5}$$

**A.2.3. Construction of the Riemann surface for the inverse of a polynomial of fractional order.** For an open set $U \subset \mathbb{C}$, we consider the multiplicative abelian group $Ln^*(U) = \{f\}$ of holomorphic mappings $f : U \to L$. If $f \in Ln^*(U)$, then $\pi \circ f$ (see (A.2.1)) is a holomorphic function in $U$ the lifting of which to $L$ is $f$. Starting with the system of sets $\{Ln^*(U), \; U \subset \mathbb{C}\}$ and using the operations of restriction and passage to a direct limit, in a natural way we arrive at the definition of the bundle $Ln^*$ of germs of holomorphic mappings over $\mathbb{C}$ with values in $L$. The topology in $Ln^*$ is introduced as usual: the base of this topology is the system of sets $\{[U,f] : U \subset \mathbb{C}, \; f \in Ln^*(U)\}$, where $U$ is an open set in $\mathbb{C}$, $[U,f] := \{\rho_p(f) : p \in U\}$, and $\rho_p(f)$ is a germ of $f$ at the point $p$. Moreover, the projection

$$\varphi : Ln^* \to \mathbb{C}; \quad \varphi[\rho_p(f)] = p \quad \forall\, \rho_p(f) \in Ln^* \tag{A.2.6}$$

is a locally topological mapping.

Let $\alpha$ be a polynomial on $L$ defined by formula (A.2.5). As in the theory of algebraic functions, we uniformize locally the inverse of $\alpha$ (which is a multivalued function with values in $L$). Consider the set $N_\alpha = \{Z \in L : \alpha'(Z) = 0\}$, where (see Subsection A.2.1)

$$\alpha'(Z) = \rho[\pi(Z)]^{\rho-1} + \sum_{j=1}^{n} a_j(\rho - j)[\pi(Z)]^{\rho-j-1}$$

is the derivative of $\alpha$. We have

$$N_\alpha = \bigcup_{j=1}^{k} \pi^{-1}(z_j), \tag{A.2.7}$$

where $\{z_1, \ldots, z_k\}$ is the set of all zeros of the polynomial

$$\beta(z) = \rho z^n + (\rho-1)a_1 z^{n-1} + \cdots + (\rho-n)a_n \, .$$

Therefore, $N_\alpha$ is a closed discrete set. Note that if $\nu_j$ is the multiplicity of the zero $z_j$, $j = 1, \ldots, k$, then $\nu_1 + \cdots + \nu_k = n$.

For every point $Z \in L \setminus N_\alpha$, the biholomorphic mapping $f \in Ln^*(V_Z)$ defined in a circular neighborhood $V_Z(p)$ of $p = \alpha(Z)$ and satisfying

$$f_Z(p) = Z; \ \alpha \circ f_Z(q) \equiv q, \ q \in V_Z(p), \tag{A.2.8}$$

has a unique germ $f_Z \in Ln^*$. Moreover, if $V := V_Z \cap V_W \neq \varnothing$, $q_0 \in V$, and $f_Z(q_0) = F_W(q_0)$, then $f_Z(q) \equiv f_W(q)$, $q \in V$.

Let $Z \in N_\alpha$ . Then $\pi(Z) = z_j$ , $j = \{1, \dots, k\}$ (see (A.2.7)), and $\mu = \nu_j + 1$ is the multiplicity of the mapping $\alpha$ at the point $Z$. We can find $\lambda > 0$ and a biholomorphic mapping $f \in Ln^*(S_\lambda)$, where $S_\lambda = \{t \in \mathbb{C} : |t| < \lambda\}$, such that $f$ determines a local uniformization of the set

$$\{(w, p) \in L \times \mathbb{C} : \alpha(W) = p\}$$

near the point $(Z, \alpha(Z))$:

$$W = f(t), \ p = \alpha(Z) + t^\mu, \ t \in S_\lambda; \ f(0) = Z. \tag{A.2.9}$$

Equivalently, such a uniformization is given by the mapping $f(\varepsilon t)$, $t \in S_\lambda$, where $\varepsilon$ is an arbitrary element of the set $\{\varepsilon \in \mathbb{C} : \varepsilon \neq 1, \ \varepsilon^\mu = 1\}$.

With the mapping $f$, we associate its germ $f_Z$ at the point 0. Consider the sets

$$H'_\alpha = \{f_Z : Z \in L \setminus N_\alpha\}, \quad O_\alpha = \{f_Z : Z \in N_\alpha\}$$

of "regular" and "singular" elements of the inverse to the polynomial $\alpha$ and form their union $H_\alpha = H'_\alpha \cup O_\alpha$. So, the operator

$$\gamma_\alpha : L \to H_\alpha; \quad \gamma_\alpha(Z) = f_Z \ , \ Z \in L, \tag{A.2.10}$$

is well defined and is a homeomorphism in $L \setminus N_\alpha$. The topology in $H_\alpha$ is induced from $L$: if $\{V_i, \ i \in I\}$ is a base of neighborhoods of a point $Z$ in $L$, then $\{\gamma_\alpha(V_i), \ i \in I\}$ is a base of neighborhoods of the element $f_Z$ in $H_\alpha$. This procedure endows $H'_\alpha$ with the same topology as before. The operator inverse to $\gamma_\alpha$ describes the inverse of $\alpha$ with values in $L$:

$$\alpha^{-1} : H_\alpha \to L; \quad \alpha^{-1}(f_Z) = Z, \ Z \in L. \tag{A.2.11}$$

Since $\gamma_\alpha$ (see A.2.10)) is a homeomorphism, $H_\alpha$ is a simply connected two-dimensional manifold. We define the projection

$$\varphi_\alpha : H_\alpha \to \mathbb{C}; \quad \varphi_\alpha(f_Z) = \alpha(Z) \quad \forall \ f_Z \in H_\alpha \tag{A.2.12}$$

corresponding to the bundle $Ln^*$ and having the property

$$\varphi_\alpha \circ \gamma_\alpha = \alpha. \tag{A.2.13}$$

Since $\varphi_\alpha|_{H'_\alpha}$ is a local homeomorphism (see (A.2.6)), the mapping $\varphi_\alpha$ is a local coordinate for the set $H'_\alpha$. This fact will be used for the choice of an atlas on

$H_\alpha$. In the notation of formulas (A.2.8), (A.2.9), we put $\mathfrak{A}_\alpha = \{(U_W, \zeta_W),\ W \in L\}$, where

$$U_W = \{f_Z \in H_\alpha : Z \in f_W(V_W)\},\ \zeta_W := \varphi_\alpha : U_W \to V_W$$

if $W \in L \setminus N_\alpha$;

$$U_W = \{f_Z \in H_\alpha : Z \in f(S_\lambda)\},\ \zeta_W : U_W \to S_\lambda,\ \zeta_W(f_Z) = f^{-1}(Z) \quad \text{(A.2.14)}$$

if $W \in N_\alpha$. For instance, if $X \in L \setminus N_\alpha$, $W \in N_\alpha$, and $U_X \cap U_W \neq \varnothing$, then by (A.2.9), (A.2.13) we have

$$\zeta_X \circ \zeta_W^{-1}(t) = \varphi_a[\gamma_\alpha \circ f(t)] = \alpha(W) + t^\mu,\ t \in V, \quad \text{(A.2.15)}$$

where $\mu = \mu(W)$, and $V = \zeta_W(U_X \cap U_W)$ is a simply connected open subset of $S_\lambda \setminus \{0\}$. Therefore, the maps of the atlas $\mathfrak{A}_\alpha$ are biholomorphically compatible. Thus, we have proved the following fact.

**Property A.2.1.** *Let $\alpha$ be a polynomial of fractional order $\rho > 0$ of the form* (A.2.4). *The space $H_\alpha$ associated with $\alpha$ admits a complex structure making it a simply connected Riemann surface, and the inverse $\alpha^{-1}$ with values in $L$ is a biholomorphic mapping.*

**Property A.2.2.** *The mapping $\varphi_\alpha : H_\alpha \to \mathbb{C}$ is a holomorphic covering equivalent to $\alpha : L \to \mathbb{C}$, and $\varphi_\alpha \circ \gamma_\alpha = \alpha$* (see (A.2.10)-(A.2.13)), *where $\gamma_\alpha : L \to H_\alpha$ is the operator defined by* (A.2.1).

We mention some other properties of the covering $\varphi_\alpha : H_\alpha \to \mathbb{C}$ .

**Property A.2.3.** *There exists a cyclic, free, and totally disconnected group $G_H$ of conformal automorphisms that operates on $H_\alpha$ and is generated by*

$$B : H_\alpha \to H_\alpha,\ Bf_Z = f_{AZ}\ , \quad \text{(A.2.16)}$$

*where $A$ is the generator of the group $G_L$* (see item A.2.1), *and* (see (A.2.11))

$$\varphi_a(Bm) = e^{i2\pi\rho}\varphi_\alpha(m) \quad \forall\, m \in H_\alpha; \quad \text{(A.2.17)}$$

$$\alpha^{-1}(B^k f_Z) = A^k Z \quad \forall\, Z \in L, \quad k \in \mathbb{Z}\ . \quad \text{(A.2.18)}$$

Formula (A.2.17) shows that the automorphism $B$ is not fiberwise.

◀ From (A.2.9) and (A.2.14) it follows that $B$ is a conformal automorphism. Condition (A.2.18) is a consequence of (A.2.16) and (A.2.11). So, the operator $\alpha^{-1}$ agrees with the action of $G_H$ and $G_L$, and these groups are isomorphic. But $G_L$ is a totally disconnected group operating on $L$ freely. Consequently, so is $G_H$. Finally, formula (A.2.17) is a consequence of (A.2.16) and (A.2.12) ▶

We discuss yet another property of the covering $\varphi_\alpha : H_\alpha \to \mathbb{C}$. This property will be used below; it is related to uniformization of the set $\{(Z,p) \in L \times \mathbb{C} : \alpha(Z) = p\}$ in the limits $|Z| \to 0$, $|Z| \to \infty$.

**Propersty A.2.4.** *There exist nonintersecting open sets $U_i$ in $H_\alpha$, some numbers $\varepsilon_i > 0$, and homeomorphisms*

$$\Psi_i : U_i \to V_i\ , \quad V_i = (0, \varepsilon_i) \times \mathbb{R} \subset L, \quad i = 0, \infty,$$

*satisfying the equations*

$$\varphi_\alpha \circ \Psi_0^{-1} \equiv \pi(T^{\rho - n}),\ T \in V_0\ ,\ \varphi_\alpha \circ \Psi_\infty^{-1}(T) \equiv \pi(T^{-\rho}),\ T \in V_\infty; \quad \text{(A.2.19)}$$

*moreover, in the previous notation we have*

$$\Psi_0(B_m) = A\Psi_0(m),\ m \in U_0;\ \Psi_\infty(B_m) = A_\infty^{-1}(m),\ m \in U_\infty\ . \quad \text{(A.2.20)}$$

◀ We uniformize the set

$$\{(Z, p) \in l \times \mathbb{C} : \alpha(Z) - p = 0\}$$

as $|Z| \to \infty$. We have

$$\alpha(Z) = \pi(T^{-\rho}) \cdot P_n(1/z), \quad \text{(A.2.21)}$$

where $P_n$ is a polynomial on $\mathbb{C}$ of power $n$ and such that $P_n(0) = 1$.

Take $R > 0$ in such a way that $P_n(1/z) \neq 0$ whenever $Z \in V$, where $z = \pi(Z)$, $V = \{Z \in L : |Z| > R\}$. There exists $\varepsilon_\infty > 0$ with $\pi(T^{-\rho}) \subset \alpha(V)\ \forall\ T \in V_\infty$. We shall look for a solution $Z = Z(T)$, $T \in V_\infty$, of the equation

$$\alpha(Z) = \pi(T^{-\rho}), \quad T \in V_\infty\ , \quad \text{(A.2.22)}$$

such that, moreover, $Z(V_\infty) \subset V$. For this it suffices (see (A.2.21)) to find an inverse (defined on $V_\infty$) for the mapping

$$T = F(Z),\ Z \in V; F(Z) = g(z)Z^{-1},$$

where $g(z) = [P_n(1/z)]^{-1/\rho}$ is the branch of the multivalued function in question with the properties that $g$ is holomorphic near $\infty$ and satisfies $g(\infty) = 1$.

The mapping $F$ agrees with the action of $G_L$ (see Section A.1); consequently, the factor-mapping $t = f(z) := g(z)z^{-1}$ is holomorphic and univalent in the vicinity of $\infty$. The function $z = f_\infty(t)$ inverse to it is holomorphic in the punctured neighborhood $\sigma_\infty = \{t \in \mathbb{C} : 0 < |t| < \varepsilon_\infty\}$ of the point 0 (for sufficiently small $\varepsilon_\infty$); this inverse has a pole at zero, and $f_\infty(t) = k(t)/t$, $t \in \sigma_\infty$, where $k(0) = 1$. The required mapping is

$$Z = F_\infty(T) := T^{-1}k(t), \quad T \in V_\infty\ ,\ t = \pi(T), \quad \text{(A.2.23)}$$

(this is verified immediately). It has the property

$$F_\infty(A^{-1}T) = A \circ F_\infty(T), \quad T \in V_\infty\ , \quad \text{(A.2.24)}$$

where $A$ is the generator of the group $G_L$ (see (A.2.2)).

Consider the set $U_\infty = \{f_Z \in H_\alpha : Z \in F_\infty(V_\infty)\}$. Since $F_\infty$ is a univalent function, the mapping

$$\Psi_\infty : U_\infty \to V_\infty, \quad \Psi_\infty(f_Z) = F_\infty^{-1}(Z)$$

is a homeomorphism. Now (A.2.10), (A.2.22), (A.2.23), and the relation $\Psi_\infty^{-1} = \gamma_\alpha \cdot F_\infty$ imply that

$$\varphi_\alpha \circ \Psi_\infty^{-1}(T) = \alpha \circ F_\infty(T) \equiv \pi(T^{-\rho}), \quad T \in V_\infty\ ,$$

i.e., the second identity in (A.2.19) is valid. Since $\Psi_\infty = F_\infty^{-1} \circ \alpha{-1}$ (see (A.2.11)), from (A.2.18) and (A.2.24) we deduce that $\Psi_\infty(B_m) = A^{-1}\Psi_\infty(m)$, $m \in U_\infty$. The first identities in (A.2.19) and (A.2.20) are proved similarly ▶

**Remark.** Earlier we showed that there exists a solution $Z = F_\infty(T)$ of the equaltion $\alpha \circ Z(T) = \pi(T^{-\rho})$, $T \in V_\infty$ , such that in the notation of item A.2.1 we have

$$\lim_{|T|\to 0} T \cdot F_\infty(T) = \mathbb{I},$$

and the convergence is uniform with respect to $Arg\, T \in L$. Analysis of the above proof shows that every other solution of the same equation has the form $Z = F_\infty(A_\nu T)$, $T \in V_\infty$ , where $\nu = s/\rho$, $s \in \mathbb{Z}$, and $A_\nu T = (|T|, Arg\, T + 2\pi\nu)$ (see (A.2.3)).

Now we pass to the central result. Formula (A.2.18) means that $\alpha^{-1}$ agrees with the action of the groups $G_H$ and $G_L$ . From this, Properties A.2.1-A.2.4, and Proposition A.1.9 we deduce the following fact.

**Theorem A.2.5.** *Let $\mathfrak{M}'_\alpha$ be the space of orbits of the group $G_H$ acting on the Riemann surface $H_\alpha$ for the inverse $\alpha^{-1}$* (the values of these inverse are taken in $L$, see (A.2.11)). *We equip $\mathfrak{M}'_\alpha$ with the factor-topology. There exists a two-point extension $\mathfrak{M}_\alpha$ of $\mathfrak{M}'_\alpha$ such that $\mathfrak{M}_\alpha$ is a Riemann surface conformally equivalent to the Riemann sphere $\overline{\mathbb{C}}$. Homeomorphism of $\mathfrak{M}_\alpha$ and $\overline{\mathbb{C}}$ is established by the mapping*

$$\hat{\alpha}^{-1} : \mathfrak{M}_\alpha \to \overline{\mathbb{C}}\ , \tag{A.2.25}$$

*which is a continuous extension to $\mathfrak{M}_\alpha$ of the factor-mapping for $\alpha^{-1}$.*

On $H_\alpha$, passage to the factor-space identifies the germs that belong to $Ln^*$ and are taken by $\pi$ (see (A.2.1)) to the same point. If a regular germ admits analytic continuation, e.g., along the path $\{p = \alpha(r, \theta + t) \in \mathbb{C} : 0 \leqslant t \leqslant 2\pi\}$, where $(r, \theta) \in L$, then the resulting element is the same as the initial (up to the above identification). This path is not closed if $\rho$ is a noninteger. In this connection, it is useful to give the following definition (for integral $r$ it is equivalent to the classical definition of the Riemann surface of the inverse to a polynomial of order $\rho$).

**Definition A.2.6.** The holomorphic mapping $\widehat{\alpha}^{-1}$ (see (A.2.25)) is called the *factorized inverse* of the polynomial $\alpha$ (A.2.4), and the space $\mathfrak{M}_\alpha$ is called the *Riemann surface* of the mapping $\widehat{\alpha}^{-1}$.

**A.2.4. Auxiliary apparatus of $(\rho, n)$-covering.** To examine the question of equivalence of a given covering to the covering associated (in the manner indicated above) with a certain polynomial of fractional order, we need to study some properties of a class of coverings.

**Definition A.2.7.** Suppose $\rho > 0$ is a noninteger, $n > 0$ is an integer, $M$ is a Riemann surface, $G_M$ is a cyclic free and totally disconnected group of conformal automorphisms with generator $B$ operating on $M$, and $\varphi : M \to \mathbb{C}$ is a holomorphic covering. The triplet $(\varphi, M, \mathbb{C})$ is called a $(\rho, n)$-*covering* if the following conditions are fulfilled:

1) $\varphi(Bm) = \varphi(m)e^{i2\pi\rho} \quad \forall\, m \in M;$ (A.2.26)

2) there exist mutually disjoint sets $U_i \subset M$, numbers $\varepsilon_i > 0$, and homeomorphisms $\Psi_i : U_i \to V_i$, $i = 0, \infty$, where $V_i = (0, \varepsilon_i) \times \mathbb{R} \subset L$, such that

$$\varphi \circ \Psi_0^{-1}(T) \equiv T^{\rho-n}, \ T \in V_0; \ \varphi \circ \Psi_\infty^{-1}(T) \equiv T^{-\rho}, \ T \in V_\infty; \tag{A.2.27}$$

$$\Psi_0(Bm) = A\Psi_0(m), \ m \in U_0; \ \Psi_\infty(Bm) = A^{-1}\Psi_\infty(m), \ m \in U_\infty\,, \tag{A.2.28}$$

where $A$ is the automorphism of $L$ given by $A(r, \theta) = (r, \theta + 2\pi) \ \forall\, (r, \theta) \in L$.

Condition 2 implies that the covering $\varphi$ has two logarithmic branching points. We assume that the complex structure on $M$ is compatible (in a sense) with the mapping $\varphi$ and the group $G_M$. This complex structure is defined by a complex atlas

$$\mathfrak{A} = \{(U(x), \psi_x), \ x \in M\} \tag{A.2.29}$$

satisfying the following conditions:

1) $U(x)$ is a parametric disk with center at $x$, i.e., $\psi_x[U(x)]$ is an open disk in $\mathbb{C}$ with center at $\psi_x(x)$, and

$$U(gx) = gU(x), \ U(x) \cap U(gx) = \varnothing \quad \forall\, g \in G_M \setminus \{\mathbf{1}\}, \tag{A.2.30}$$

where $\mathbf{1}$ is the unit of $G_M$;

2) if $x$ is not a branching point of $\varphi$, then $\psi_x = \varphi|_{U}(x)$;

3) if $x$ is a branching point of $\varphi$, then there is an integer $\nu = \nu(x)$ (the multiplicity of $\varphi$ at $x$) and a disk $S_\lambda = \{t \in \mathbb{C} : |t| < \lambda\}$ such that

$$\varphi \circ \psi_x^{-1}(t) \equiv \varphi(x) + t^\nu, \ t \in S_\lambda, \tag{A.2.31}$$

and, in the notation of the Definition A.2.7,

$$U(x) \cap U_i = \varnothing, \ i = 0, \infty; \ U(x) \cap U(a) = \varnothing, \tag{A.2.32}$$

where $a$ is any branching point of $\varphi$ different from $x$.

The latter condition can be imposed because $\varphi : M \to \mathbb{C}$ is a holomorphic covering and, consequently, the branching points for $\varphi$ are isolated points in $M$.

**Proposition A.2.8.** *The Hausdorff space of orbits $\mathfrak{M}' := M/G_M$ equipped with the factor-topology admits a complex structure that makes $\mathfrak{M}'$ a Riemann surface.*

◀ First, we note that the canonical mapping $\gamma : M \to \mathfrak{M}'$ is open; if $U$ is an open set in $M$, then

$$\gamma^{-1} \circ \gamma(U) = \bigcup gU|\ g \in G_M \tag{A.2.33}$$

is an open set. Since $\gamma$ is a continuous mapping, $\mathfrak{M}'$ is connected.

The set $V(a) = \gamma[U(a)]$ is open in $\mathfrak{M}'$ for every $a \in M$ by condition (A.2.30), and $\gamma : U(a) \to V(a)$ is a homeomorphism. Let $\theta_a : V(a) \to U(a)$ be the inverse of $\gamma$. Consider the set (cf. (A.2.29))

$$\mathfrak{N}' := \{(V(a), \delta_a),\ \delta_a := \psi_a \circ \theta_a : V(a) \to \mathbb{C},\ a \in M\}, \tag{A.2.34}$$

which consists of complex charts on $\mathfrak{M}'$. We show that every two charts in $\mathfrak{N}'$ are biholomorphically compatible.

Let $V := V(a) \cap V(x) \neq \varnothing$, where $a, x \in M$. By (A.2.30), there is a unique element $g \in G_M$ such that $\theta_a(V) = U(a) \cap U(gx) \neq \varnothing$ (see A.2.34)). Then $\theta_a(V) = U(g^{-1}a) \cap U(x)$, where $g^{-a}$ is the element of $G_M$ inverse to $g$.

Consider the mapping $\delta_a \circ \delta_x^{-1} : \delta_x(V) \to \delta_a(V)$, i.e.,

$$\delta_x(V) \overset{\psi_x^{-1}}{\to} U(g^{-1}a) \cap U(x) \overset{\theta_x^{-1}}{\to} V \overset{\theta_a}{\to} U(a) \cap U(gx) \overset{\psi_a}{\to} \delta_a(V).$$

Since $B$ is a generator of the cyclic group $G_M$, by (A.2.30) and (A.2.32) we see that there is $k \in \mathbb{Z}$ such that $g = B^k$,

$$\theta_a \circ \theta_x^{-1}(m) = gm = B^k m, \quad m \in U(g^{-1}a) \cap U(x). \tag{A.2.35}$$

Therefore, by (A.2.26) we have

$$\delta_a \circ \delta_x^{-1}(z) = \psi_a \circ g \circ \psi_x^{-1}(z), \quad z \in \delta_x(V). \tag{A.2.36}$$

Two cases are possible.

1) The points $a$ and $x$ are not branching points for $\varphi$. Then $\psi_a = \varphi|_{U(a)}$ and $\varphi_x = \varphi|_{U(x)}$. Consequently, by (A.2.35) and (A.2.26) we have:

$$\delta_a \circ \delta_x^{-1} = \varphi \circ B^k \circ \varphi^{-1} = z^{i2\pi k\rho}, \quad z \in \delta_x(V).$$

2) One of the points (say, $x$) is a branching point for $\varphi$ ($a$ and $x$ cannot be branching points simultaneously because $V \neq \varnothing$, see (A.2.32)).

Then $\delta_X(V)$ is a simply connected set in $S_\lambda \setminus \{0\}$ (see (A.2.31)) and this time from (A.2.35), (A.2.36) snd (A.2.34) we deduce that

$$\delta_a \delta_x^{-1}(t) = \varphi \circ B^k \cdot \psi_x^{-1}(t) = [\varphi(x) + t^\nu] e^{i2\pi k\rho}, \quad t \in \delta_x(V).$$

The assertion is proved.

We compactify the manifold $\mathfrak{M}'$ locally. For this, consider its extension $\mathfrak{M} = \mathfrak{M}' \cup \{\Lambda_0, \Lambda_\infty\}$ and endow it with a topology and complex structure. From (A.2.28) it follows that

$$\Psi_0(B^k m) = A^k \Psi_0(m), \ m \in U_0; \ \Psi_\infty(B^k m) = A^{-k} \Psi_\infty(m), \ m \in U_\infty ,$$

where $k \in \mathbb{Z}$. Therefore,

$$\Psi_0 \circ \sigma = T_\sigma \circ \Psi_0, \ \Psi_\infty \circ \sigma = T_\sigma^{-1} \circ \Psi_\infty \quad \forall\, \sigma \in G_M . \tag{A.2.37}$$

It is easy to verify that the mappings $\sigma \mapsto T_\sigma$ and $\sigma \mapsto T_\sigma^{-1}$ are isomorphisms of the groups $G_M$ and $G_L$ (see Subsection A.2.1). Hence, the factor-mapping

$$\widehat{\Psi_i} : U_i/G_M \to V_i/G_L \approx S(\varepsilon_i) \setminus \{0\}, \ S(\varepsilon_i) = \{z \in \mathbb{C} : |z| < \varepsilon_i\} \tag{A.2.38}$$

is a homeomorphism for $i = 0, \infty$. We put

$$\widehat{\Psi}_0(\Lambda_0) = 0; \ \widehat{\Psi}_\infty(\Lambda_\infty) = 0; \ \omega_i := U_i/G_M \cup \{\Lambda_i\}, \ i = 0, \infty.$$

Then $\widehat{\Psi}_i : \omega_i \to S(\varepsilon_i)$, $i = 0, \infty$, are one-to-one mappings. For a base of neighborhoods of the point $\{\Lambda_i\}$ we take the preimage of a base of neighborhoods of $0 \in \mathbb{C}$ under $\widehat{\Psi}_i$, $i = 0, \infty$. In the topological space $\mathfrak{M}$, we consider the atlas

$$\mathfrak{N} = \mathfrak{N}' \cup \{(\omega_i, \widehat{\Psi}_i), \ i = 0, \infty\}.$$

We show that the maps in $\mathfrak{N}$ are biholomorphically compatible. Let $(V(x), \delta_x) \in \mathfrak{N}'$ (then $V := V(x) \cap \omega_\infty \neq \varnothing$; here $x$ is a fixed point in $M$ that is not a branching point for $\varphi$; see (A.2.32)). Consider the mapping $\delta_x \circ \widehat{\Psi}_\infty^{-1} : \widehat{\Psi}_\infty(V) \to \delta_x(V)$.

Schematically, we represent this mapping with the help of a commutative diagram (we employ the above notation, see the proof of Proposition A.2.8). Observing that $U := U(x) \cap U_\infty \neq \varnothing$, we use (A.2.30) and (A.2.37) to conclude that the domain $\Psi_\infty(U)$ in $L$ intersects each orbit of $G_L$ by at most one point. Therefore, the restriction of $\pi$ (see (A.2.1)) to $\Psi_\infty(U)$ is a homeomorphism. Denoting the inverse to this restriction by $\chi_x$, we represent the mapping $\delta_x \circ \widehat{\Psi}_\infty^{-1}$ (see Subsection A.2.1, (A.2.34), (A.2.38)) as follows:

$$\begin{array}{ccccccc}
\Psi_\infty(U) & \xrightarrow{\Psi_\infty^{-1}} & U & & & & \\
\uparrow \chi_x & & \downarrow \gamma & & & & \\
\widehat{\Psi}_\infty(V) & \xrightarrow{\widehat{\Psi}_\infty^{-1}} & V & \xrightarrow{\theta_x} & U & \xrightarrow{\psi_x} & \delta_x(V).
\end{array}$$

Now, $\chi_x = (\pi|_{\Psi_\infty(U)})^{-1}$ is a homeomorphism, $\theta_x \circ \gamma : U \to U$ is the identity mapping, and $\psi_x = \varphi|_{U(x)}$ (by condition 2 concerning the atlas $\mathfrak{A}$ in M defined by (A.2.9)). Therefore, by (A.2.27) we have:

$$\delta_x \circ \widehat{\Psi}_\infty^{-1}(t) \equiv [\pi(t)]^{-\rho}, \quad t \in \widehat{\Psi}_\infty(V).$$

Since $\Psi_\infty(V)$ is a simply connected domain in $S(\varepsilon_\infty) \setminus \{0\}$, the chart $(\omega_\infty, \widehat{\Psi}_\infty)$ is biholomorphically compatible with the charts of the atlas $\mathfrak{N}'$. We arrive at the following statement.

**Theorem A.2.9.** *The topological space* $\mathfrak{M}$ *is a Riemann surface.*

We give a constructive example of a $(\rho, 1)$-covering.

**Example A.2.10.** Let $\rho \in (1/2, 1)$, and let $P = \{p_j : j \in \mathbb{Z}\}$ be a broken line in $\mathbb{C}$ the vertices of which satisfy the condition

$$p_0 \neq 0; \; p_j = p_{j+1} e^{i2\pi\rho}; \quad j \in \mathbb{Z},$$

where $L_j = \{p_j \cdot t, t \geqslant 1\}$ is a radial ray with vertex at $p_j$. Consider two systems $\{M_j^-, M_j^+, j \in \mathbb{Z}\}$ of convex (and, respectively, concave) infinite triangles in $\mathbb{C}$ such that for every $j \in \mathbb{Z}$ the triangles $M_j^-$ and $M_j^+$ have common boundary

$$\Gamma_j = L_{j-1} \cup [p_{j-1}, p_j] \cup L_j \; .$$

We glue $M_j^-$ and $M_j^+$ along the segment $[p_{j-1}, p_j]$ so that the union $M_j = M_j^- \cup M_j^+$ is the plane with a cut along the rays $L_{j-1}, L_j$. Every point $m \in M_j$ will be represented by the pair $m = (p, j)$, where $p$ is the complex coordinate of $m$. On the union $\widetilde{M} = \bigcup M_j | j \in \mathbb{Z}$ , we define a mapping $\tilde{\varphi} : \widetilde{M} \to \mathbb{C}$ by setting $\tilde{\varphi}(m) = p$ if $m = (p, j) \in M_j$ , $j \in \mathbb{Z}$. We factorize $\widetilde{M}$ as follows. We glue each pair $M_{j-1}^-$ , $M_j^-$ of convex triangles along the ray $L_j$ , and do similarly for every pair $M_{j-1}^+$ , $M_j^+$ of concave triangles. Let $M$ be the factor-set obtained, and let $\varphi : M \to \mathbb{C}$ be the factor-mapping generated by $\tilde{\varphi}$. Then $\varphi : M \to \mathbb{C}$ is a covering all branching points of which have multiplicity two, and over $\infty$ there are two logarithmic branching points.

In a natural way, we introduce a complex structure on $M$, after which $M$ becomes a Riemann surface, and the covering $\varphi : M \to \mathbb{C}$ becomes holomorphic (see conditions 2, 3 related to the atlas $\mathfrak{A}$ (A.2.29)).

Now, we describe the group of automorphisms operating on $M$. Since the plane with the cuts $M_{j-1}$ is obtained from the plane with the cuts $M_j$ by rotation of the latter clockwise through the anlge of $2\pi\rho$, the folowing mapping is well defined on $M$:

$$B : M \to M; \; B(M_j) = M_{j-1}; \; B[(p, j)] = \left(pe^{i2\pi\rho}, j-1\right), \quad j \in \mathbb{Z}.$$

This automorphism satisfies (A.2.26) and generates a cyclic totally disconnected and free group $G_M$ acting on $M$.

In order to conclude that $\varphi : M \to \mathbb{C}$ is a $(\rho, 1)$-covering, it remains to verify conditions (A.2.27) and (A.2.28). Suppose

$$R > |p_0|;\ D_R = \{p \in \mathbb{C} :\ |p| > R\};\quad U_j^{\pm} = M_j^{\pm} \cap D_R\ ,\quad j \in \mathbb{Z}.$$

Then

$$U_0 = \bigcup U_j^- \mid j \in \mathbb{Z},\quad U_\infty = \bigcup U_j^+ \mid j \in \mathbb{Z}$$

are neighborhoods of logarithmic branching points of the covering $\varphi : M \to \mathbb{C}$. Polar coordinates in $U_0$ and $U_\infty$ yield homemorphisms

$$\varkappa_i : U_i \to (R, \infty) \times \mathbb{R},\ i = 0, \infty,$$

and (see (A.2.23))

$$\varkappa_\infty(Bm) = A_\rho \varkappa_\infty(m),\ m \in U_\infty;\quad \varkappa_0(Bm) = A_{1-\rho}^{-1} \varkappa_0(m),\quad m \in U_0;$$

this follows from the geometric properties of the mapping $B$ and the nature of the gluing. Suppose

$$\gamma > 0;\ f_\gamma : L \to L;\ f_\gamma(r, \theta) = (r^{-1/\gamma}, -\theta/\gamma),\ (r, \theta) \in L.$$

The mappings $\psi_0 = f_{1-\rho} \circ \varkappa_0$ and $\psi_\infty = f_\rho \circ \varkappa_\infty$ are required.

The interior of every set $M_i$ is a fundamental domain of the group $G_M$. A topological model of the completion $\mathfrak{M}$ of the space of orbits for $G_M$ can be obtained from $M_0$ as follows. Each of the triangles $M_0^-$ and $M_0^+$ is completed by adjoining the point at infinity in the topology of the Riemann sphere. Both in $M_0^-$ and in $M_0^+$, the rays $L_0$ and $L_1$ are glued at the points equidistant from the vertices $p_0, p_1$ (see Fig. 10).

**A.2.5. Characteristic property of the Riemann surface for the inverse of a polynomial of fractional order.** From Properties A.2.1 – A.2.4 and Theorem A.2.5 we deduce the "if" part of the following theorem, which describes the coverings associated with polynomials of fractional order (A.2.5).

**Theorem A.2.11.** *Let $\varphi : M \to \mathbb{C}$ be a $(\rho, n)$-covering* (Definition A.2.7), *and let $\mathfrak{M}$ be the completion* (by the method indicated in A.2.4) *of the space of orbits of the group $G_M$ operating on the simply connected manifold $M$. The manifold $\mathfrak{M}$ is a simply connected compact Riemann surface if and only if there is a polynomial $\alpha$ of the form* (A.2.5) *such that the covering $\varphi : M \to \mathbb{C}$ is equivalent to the covering $\alpha : L \to \mathbb{C}$ or to the covering $\varphi : H_\alpha \to \mathbb{C}$ associated with $\alpha$* (see Definition A.1.1, (A.2.12)) *and there exists a holomorphic fiberwise homeomorphism $\Phi : L \to M$ with the following properties:*

*a) the mapping $\Phi$ agrees with the action of $G_L$ on $L$ and of $G_M$ on $M$;*

*b) the mapping $\chi := \Phi \circ \alpha^{-1} : H_\alpha \to M$*(see (A.2.11)) *is a fiberwise homeomorphism compatible with the action of $G_H$ on $H_\alpha$ and of $G_L$ on $L$.*

◀ *The "only if" part.* 1. Let $\{\Lambda_0, \Lambda_\infty\} = \mathfrak{M} \setminus \mathfrak{M}'$ be the new points of the completion $\mathfrak{M}' = M/G_M$ (see A.2.4). By the Riemann theorem (see the proof of Theorem 4.3.3) the space $\mathfrak{M}$ is conformally equivalent to the Riemann sphere $\overline{\mathbb{C}}$, and the stock of the linear-fractional automorphisms allows us to choose the corresponding mapping $\widehat{\Phi} : \overline{\mathbb{C}} \to \mathfrak{M}$ in such a way that

$$\widehat{\Phi}(0) = \Lambda_0; \ \widehat{\Phi}(\infty) = \Lambda_\infty. \tag{A.2.39}$$

Since every orbit of the free and totally disconnected group $G_M$ acting on $M$ is discrete, the mapping $\widehat{\Phi} \circ \pi : L \to \mathfrak{M}$ and the canonical mapping $\gamma : M \to \mathfrak{M}'$ are unbranched unbounded coverings.

In the atlas $\mathfrak{N}$ on $M$, we take a complex chart $(\omega_0, \widehat{\Psi}_0)$, where $\omega_0$ is a coordinate neighborhood of the point $\Lambda_0$ on $M$ (see Theorem A.2.4). Then from (A.2.39) if follows that $D := \widehat{\Phi}^{-1}(\omega_0)$ is a neighborhood of 0 in $\mathbb{C}$. But $\widehat{\Phi}$ is a conformal mapping. Therefore, the function $\widehat{\beta} = \widehat{\Psi}_0 \circ \widehat{\Phi}$ (which is holomorphic in $D$) has the following properties for sufficiently small $\varepsilon_0$:

$$\begin{gathered}\widehat{\beta}(z) = zk(z), \ k(0) = b \in \mathbb{C}^*, \quad k(z) = b + O(z), \ z \to 0; \\ \{k(z) : \ z \in D\} \subset \{w \in \mathbb{C}^* : \ |w - b| < |b|\}.\end{gathered} \tag{A.2.40}$$

A.2. We denote by $\arg k(z)$, $z \in D$, a fixed branch of the multivalued function $\operatorname{Arg} k(z), z \in D$, and put $T = \pi^{-1}(D \setminus \{0\})$. If $z = \pi(Z)$, then

$$\beta(Z) = k(z)Z = Z|k(z)| \exp\{i \arg k(z)\}, \quad Z \in T.$$

We show that, in the notation of formula (A.2.28), the function $\Phi_1 = \Psi_0^{-1} \circ \beta$ satisfies

$$\Phi_1 \circ A = B \circ \Phi_1. \tag{A.2.41}$$

From the structure of $\beta$ we conclude that

$$\beta(AZ) = A\beta(Z), \quad Z \in T. \tag{A.2.42}$$

By (A.2.28), we have: $\Psi_0^{-1}(AZ) = B\Psi_0^{-1}(Z), \quad Z \in V_0$, and (A.2.42) implies the identities

$$\Phi_1(AZ) = \Psi_0^{-1} \circ \beta(AZ) = B\Psi_0^{-1} \circ \beta(Z) = B\Phi_1(Z), \quad Z \in V_0,$$

i.e., (A.2.41) is valid. So, the mapping $\Phi_1$ is compatible with the action of the groups $G_L$ on $T$ and $G_M$ on $U_0$, and the following diagram is commutative:

$$\begin{array}{ccccc} T & \xrightarrow{\Phi_1} & U_0 & \xrightarrow{\Psi_0} & V_0 \\ \downarrow \pi & & \downarrow \gamma & & \downarrow \pi \\ D & \xrightarrow{\widehat{\Phi}} & \omega_0 & \xrightarrow{\widehat{\Psi}_0} & S_{\varepsilon_0} \end{array}$$

Fix a point $Z_0 \in T$. Let $m_0 = \Phi_1(Z_0)$. Since $L$ and $M$ are simply connected Riemann surfaces, there exists (see Propositions A.1.10, A.1.11) only one homeomorphic and holomorphic lifting $\Phi : L \to M$ such that $\Phi(Z_0) = m_0$, $\widehat{\Phi} \circ \pi(Z_0) = \gamma(m_0)$. For the same reason, $\Phi_1 = \Phi|_T$ ($T, U_0$ are simply connected domains in $L$ and $M$, respectively). Together with the uniqueness theorem for holomorphic functions, this implies formula (A.2.41) for $\Phi$. Therefore, we also have

$$\Phi \circ A^k = B^k \circ \Phi \quad \forall\, k \in \mathbb{Z}. \tag{A.2.43}$$

3. Now we pass to finding the required polynomial. Consider the function

$$H(Z) = \varphi \circ \Phi(Z) \cdot Z^{-\rho}, \ Z \in L,$$

where $\varphi : M \to \mathbb{C}$ is a given $(\rho, n)$-covering. Since $\Phi : L \to M$ is a holomorphic lifting, $H$ is a holomorphic function on $L$. From (A.2.43) and (A.2.26) we deduce that

$$H(A^k Z) = H(Z), \ Z \in L, \ k \in \mathbb{Z}. \tag{A.2.44}$$

Therefore, the holomorphic function

$$f(z) = H \circ \pi^{-1}(z), \ z \in \mathbb{C}^*$$

is also well defined. We study its behavior near 0 and $\infty$.

A. $z = 0$. Using the notation of formula (A.2.42), we obtain the representation

$$\varphi \circ \Phi(Z) = (\varphi \circ \Psi_o^{-1}) \circ \beta(Z), \quad Z \in \Phi^{-1}(U_0). \tag{A.2.45}$$

But $\beta$ is a holomorphic "lifting" of $\hat{\beta}$ (see (A.2.41), i.e., $\pi\beta(Z) = \hat{\beta} \circ \pi(Z)$, $Z \in \Phi^{-1}(U_0)$. Consequently, taking (A.2.41) and (A.2.42) into account and restricting ourselves to the fundamental set $\{(r, \theta) \in L : r > 0, \ -\pi < \theta \leqslant \pi\}$ (see (A.2.44)), we find:

$$|f(z)| = |z|^{-\rho} \cdot |\hat{\beta}(z)|^{\rho - n} = |z|^{-n} |b(1 + O(z))|^{\rho - n}, \ z \to 0 \ (b \neq 0).$$

So, the function $f$ has a pole of order $n$ at 0.

B. $z = \infty$. Taking into account that $\widehat{\widehat{\Phi}}(\infty) = \Lambda_\infty$ and arguing as in the proof of formula (A.2.41), from (A.2.27) we deduce the relation

$$\widehat{\Psi}_\infty \circ \widehat{\Phi}(z) = \frac{a}{z}[1 + O(1/z)], \quad |z| \to \infty \ (a \neq 0).$$

Next, we use (A.2.27) and a representation of $\varphi \circ \Phi$ as $|z| \to \infty$ similar to (A.2.45) to conclude (in the same way as above) that $f$ is a function bounded near $\infty$. So, $f$ is a meromorphic function in $\overline{\mathbb{C}}$ with the only pole of order $n$ at 0. Hence (see Subsection A.2.1),

$$f(z) = P_n(1/z), \quad P_n(w) = b_0 + b_1 w + \cdots + b_n w^n, \quad w \in \mathbb{C},$$

$b_0, b_n \neq 0$ and, moreover,

$$\varphi \circ \Phi(Z) = Z^{\rho} \cdot P_n(Z^{-1}), \quad Z \in L.$$

Let $b_0^{-1/\rho}$ be a root of power $\rho$ from the number $1/b_0$. Denoting

$$\Phi^*(W) = \Phi(b_0^{-1/\rho} W), \quad W \in L,$$

we see that the function

$$\alpha(W) := \varphi \circ \Phi^*(W) \equiv W^{\rho} + a_1 W^{\rho-1} + \cdots + a_n W^{\rho-n}, \; W \in L,$$

is the required polynomial of fractional order. This is explained by the fact that (in the notation of Subsection A.2.3, see (A.2.11), (A.2.12)) the diagram

$$\begin{array}{ccccc} H_\alpha & \xrightarrow{\alpha^{-1}} & L & \xrightarrow{\Phi^*} & M \\ & \searrow \varphi_\alpha & & \swarrow \varphi & \\ & & \mathbb{C} & & \end{array}$$

is commutative ▶

The mapping $\Phi$ in Theorem A.2.11 may have some additional properties.

**Theorem A.2.12.** *Let $\varphi : M \to \mathbb{C}$ be a $(\rho, n)$-covering that satisfies the necessary condition of Theorem* A.2.11. *Suppose $A : L \to L$, $B : M \to M$, $\Psi_\infty : U_\infty \to V_\infty$ are homeomorphisms treated in Definition* A.2.7. *Then there exists a holomorphic homeomorphism $\Phi : L \to M$ such that*

1) *$\alpha := \varphi \circ \Phi$ is a polynomial on $L$ of the form* (A.2.5)*;*

2)

$$\Phi \circ A = B \circ \Phi; \tag{A.2.46}$$

3) *the function $f_\infty := \Phi^{-1} \circ \Psi_\infty^{-1} : V_\infty \to L$ satisfies the conditions $f_\infty \circ A^{-1} = A \circ f_\infty$ and*

$$\lim_{|T| \to 0} T \cdot f_\infty(T) = \mathbb{I}, \tag{A.2.47}$$

*where the convergence is uniform with respect to $Arg\, T$ in $\mathbb{R}$;*

4) *for the factor-mapping $\hat{f}_\infty$ we have*

$$\hat{f}_\infty(\tau) = \tau^{-1}(1 + O(\tau)), \; \tau \to 0. \tag{A.2.48}$$

◀ By Theorem A.2.11 there exists a mapping $\Phi_1 : L \to M$ having properties a) and b). Consider the function $f_1 = \Phi_1^{-1} \circ \Psi_\infty^{-1} : V_\infty \to L$. Since $B\Psi_\infty^{-1} = \Psi_\infty^{-1} \circ A^{-1}$ and $A \circ \Phi_1^{-1} = \Phi_1^{-1} \circ B$ (see (A.2.28), (A.2.40)), we see that

$$f_1 \circ A^{-1} = A \circ f_1. \tag{A.2.49}$$

Let $\alpha_1 = \varphi \circ \Phi_1$. Then $\alpha_1 \circ f_1(T) \equiv T^{-\rho}$, $T \in V_\infty$ (see (A.2.27)). From the remark to Property A.2.4 we deduce that there exists $s \in \mathbb{Z}$ such that

$$\lim_{|T|\to 0} A_\nu T \circ f_1(T) = \lim_{|T|\to 0} T \cdot A_\nu \circ f_1(T) = \mathbb{I}, \qquad \text{(A.2.50)}$$

where $\nu = s/\rho$, $A_\nu T = (|T|, Arg\, T = 2\pi\nu)$, and the convergence is uniform with respect to $Arg\, T$.

We show that the mapping $\Phi = \Phi_1 \circ A_\nu^{-1}$ is required. First, $\alpha := \varphi \circ \Phi = \alpha_1 \circ A_\nu^{-1}$ is a polynomial on $L$ of the form (A.2.5). Next, the mapping $f_\infty := \Phi^{-1} \circ \Psi_\infty^{-1} = A_\nu f_1$ satisfies condition (A.2.47) (see (A.2.50)). Since the automorphisms $A$ and $A_\nu$ commute, properties 1) – 3) are valid for $\Phi$ (see (A.2.49)). Finally, property 4) is a consequence of property 3) ▶

**A.2.6. Fundamental domain of the group $G_H$ on $H_\alpha$.** We examine the question of the number of sheets of the covering $\varphi_\alpha : D \to \mathbb{C}$, where $D_H$ is the fundamental domain of the group $G_H$ on $H_\alpha$. We confine ourselves to the case of $n > \rho$, which is important for further applications.

By (A.2.18) and the formula $\varphi_\alpha \circ \gamma_\alpha = \alpha$ (see (A.2.13)), the question under consideration is equivalent to the problem about the number of sheets of the covering $\alpha : D_L \to \mathbb{C}$, where $D_L$ is a fundamental domain of the group $G_L$ on $L$. In the case where $\rho$ is an integer, this number of sheets is equal to $n$, as follows from the main theorem of algebra. Below we shall see that no direct analog of the theorem is available if $\rho$ is a fractional number.

A simplest fundamental domain $D_L$ for $G_L$ is the half-strip of width $2\pi$ bounded by smooth Jordan curves

$$\Gamma_0 = \{Z = (t, \theta(t)) \in L,\ t > 0\}, \quad \Gamma_1 = A\Gamma_0,$$

where $A$ is the generator of the group $G_L$ (see (A.2.3)). Here we assume the existence of finite limits $\lim_{t\to\infty} \theta(t)$, $\lim_{t\to 0} \theta(t)$.

Let $T_i = \alpha(\Gamma_i)$, $i = 0, 1$. Since $n > \rho$, we see that $\{T_i\}_0^1$ are paths passing through $\infty$ and closed on the Riemann sphere; they may fail to be Jordan. We have

$$T_1 = \{we^{i2\pi\rho},\ w \in T_0\}. \qquad \text{(A.2.51)}$$

An analog of the main theorem of algebra for the polynomials of the form (A.2.5) looks like this.

**Theorem A.2.13.** *Let $p \notin T_0 \cup T_1$. The number $N_p(D_L)$ of p-points (with account of multiplicity ) of the mapping* (A.2.5) *in $D_L$ is equal to $n + \nu(D_L)$, where $n > \rho$ and*

$$\nu(D_L) = \frac{1}{2\pi_i} \int_{T_0} \frac{p(1 - e^{i2\pi\rho})\, dw}{(w - p)(w - pe^{-i2\pi\rho})} \qquad \text{(A.2.52)}$$

*is an integer belonging to the segment* $[-1,1]$.

◀ For $0<\varepsilon<R<\infty$, let

$$D_L(\varepsilon,R)=D_L\cap\{Z\in L:\ \varepsilon<|Z|<R\}.$$

We endow $D_L$ with positive orientation. The mapping $\pi$ (see (A.2.1)) is a homeomorphic inclusion of $D_L$ into $\mathbb{C}$, and $\pi(D_L)=\mathbb{C}^*\setminus\Gamma$ is the complex plane with a cut along the curve $\Gamma:=\pi(\Gamma_0)=(\Gamma_1)$. Put

$$\Phi(Z)=\alpha'(Z)[\alpha(Z)-p]^{-1}\,dZ=d[\pi(Z)];$$

$$\Gamma(\varepsilon,R)=\partial D_L(\varepsilon,R);\quad \Gamma_i(\varepsilon,R)=\Gamma(\varepsilon,R)\cap\Gamma_i,\quad i=0,1.$$

From the argument principle it follows that

$$N_p(D_L)=\lim_{\varepsilon\to0;R\to\infty}\frac{1}{2\pi i}\int\limits_{\partial D_L(\varepsilon,R)}\Phi(Z)\,dZ=\lim_{\varepsilon\to0;R\to\infty}\frac{1}{2\pi i}\left[\int\limits_{\Gamma_0(\varepsilon,R)}\Phi(Z)\,dZ-\right.$$

$$\left.-\int\limits_{\Gamma_1(\varepsilon,0)}\Phi(Z)\,dZ\right]+\frac{1}{2\pi}\left[\int\limits_{\theta(R)}^{\theta(R)+2\pi}\Phi(R,\theta)Re^{i\theta(R)}\,d\theta-\int\limits_{\theta(\varepsilon)}^{\theta(\varepsilon)+2\pi}\Phi(\varepsilon,\theta)\varepsilon e^{i\theta(\varepsilon)}\,d\theta\right].$$

As $\varepsilon\to0$ and $R\to\infty$, the difference of the first two integrals approaches the number $\nu_p(D_L)$ (see (A.2.53)). To see this, we make the holomorphic change of variables $w=\alpha(Z)$ and use formula (A.2.52): $pe^{-i2\pi\rho}\notin T_0$, because $p\notin T_1$. The asymptotic behavior of the function $\Phi(Z)$ as $|Z|\to0$ and as $|Z|\to\infty$, and condition (A.2.47) allow us to conclude that the last two integrals approach the numbers $\rho$ and $\rho-n$, respectively ▶

Theorem A.2.13 implies the estimate

$$|\nu_p(D_L)|=|N_p(D_L)-n|\leqslant1\quad\forall\,p\notin T_0\cup T_1\,.$$

The following example shows that this is sharp.

**Example A.2.14.** For $1/2<\rho<1$ and $\lambda>0$, we put

$$\alpha(Z)=Z^\rho-\frac{\lambda\rho}{\rho-1}Z^{\rho-1}.$$

A) $\Gamma_0=\{(t,0)\in L:\ t>0\}$. Then $T_0=\{p\in\mathbb{C}:\ \operatorname{Im}p=0\}$ is the real axis. From formula (A.2.53) we find

$$\nu_p(D_L)=\begin{cases}-1, & -2\pi(1-\rho)<\arg p<0,\\ 0, & 0<\arg p<\pi(2\rho-1),\ -\pi<\arg p<-2\pi(1-\rho),\\ 1, & \pi(2\rho-1)<\arg p<\pi.\end{cases}$$

B) $\Gamma_0 = \{(t, -\pi) \in L : t > 0\}$. Then

$$T_0 = \left\{ (t^\rho + \frac{\lambda\rho}{1-\rho} t^{\rho-1}) e^{-i\pi\rho} : t > 0 \right\} = \left\{ re^{-i\pi\rho} \in \mathbb{C} : r \geqslant \frac{\lambda^\rho}{1-\rho} \right\}$$

is a ray passed twice in different directions. Then

$$\nu_p(D_L) \equiv 0 \quad \forall\, p \notin T_0 \cup T_1 \ .$$

**Remark.** If the domain $D_L$ is chosen as in B), then

$$N_p(D_L) \equiv n \quad \forall\, p \notin T_0 \cup T_1 \ ,$$

i.e., $n$ is the number of sheets of the covering $\alpha : D_L \to \mathbb{C}$ .

# Appendix B

# Indicator and Conjugate Diagrams of Entire Function of a Given Proximate Order*)

In this appendix we deal with a generalization of the Polya theorem (see the Introduction, Theorem $A_1$; Theorem 3.2.3) to entire functions of a given proximate order with nonnegative generalized indicator, and to entire functions of order 1 and minimal type with generic indicator. The presentation is based on the asymptotic properties of analog of the Mittag-Leffler function (specifically, this analog is related to an analytic proximate order). The results are applied to partial fraction expansion for meromorphic functions of a special class.

## B.1. Necessary information about proximate orders. Indicator Diagram

**B.1.1. Basic properties of proximate orders.** A growth characteristic convenient for an entire function $f$ of order $\rho$ and of maximal or minimal type is given by the so-called *proximate order*, which can be chosen in such a way that the entire function $f$ is of *normal* type relative to this order.

**Definition B.1.1.** A nonnegative continuously differentiable function $\rho(r)$ defined on the set $(c, \infty)$, $c \geqslant 0$, where $c$ is a constant depending on $\rho(r)$, is called a *proximate order* if

$$\lim_{r\to\infty} \rho(r) = \rho < \infty; \quad \lim_{r\to\infty} r\rho'(r)\ln r = 0. \tag{B.1.1}$$

Two proximate orders $\rho_i(r)$, $i = 1, 2$, are said to be *equivalent* if

$$\lim_{r\to\infty} [\rho_1(r) - \rho_2(r)] \ln r = 0.$$

**Property B.1.2.** *For the function* $\rho(r)$ (see Definition B.1.1) *to be a proximate order it is necessary and sufficient that the function* $V(r) = r^{\rho(r)}$ *have the*

*) Initially, the research was supported by the Krasnoyarsk Territorial Science Foundation, grant 1F28 for 1993.

*properties*

$$\lim_{r\to\infty} \frac{\ln V(r)}{\ln r} = \rho; \quad \lim_{r\to\infty} \frac{rV'(r)}{V(r)} = \rho. \tag{B.1.2}$$

**Remark.** If $\ln V(r) = \rho(r)\ln r \to \infty$ as $r \to \infty$, then the first formula in (B.1.2) is a consequence of the second, by the L'Hostpital rule.

**Definition B.1.3.** Let $\rho(r)$ be a proximate order, and let $f$ be an entire function. If

$$0 < \sigma_f := \overline{\lim_{r\to\infty}}\, r^{-\rho(r)} \ln M_f(r) < \infty, \tag{B.1.3}$$

then $\rho(r)$ is called a *proximate order of* $f$, the quantity $\sigma_f$ is called the *type* of $f$ relative to $\rho(r)$, and the function

$$h_f(\theta) = \overline{\lim_{r\to\infty}}\, r^{-\rho(r)} \ln |f(re^{i\theta})|, \quad \theta \in \mathbb{R} \tag{B.1.4}$$

is called the *generalized indicator of* $f$.

The following theorem shows that a proximate order satisfying condition (B.1.3) exists.

**Theorem B.1.4.** *For every entire function* $f$ *of order* $\rho < \infty$ *there exists a proximate order* $\rho(r)$ *satisfying* (B.1.3) *and the inequality*

$$|\rho'(r)| \leqslant [(r+e)\ln(r+e)]^{-1} |\rho(r) - \rho|;$$

*moreover,* $\rho(r)$ *is a monotone function.*

**Theorem B.1.4′.** *Let* $f$ *be an entire function of order* $\rho \in (0,\infty)$. *If* $f$ *is of minimal (maximal) type, then* $f$ *has a proximate order* $\rho(r)$ *such that* $L(r) \downarrow 0$ $(L(r) \uparrow \infty)$ *as* $r \to \infty$, *where* $L(r) = r^{\psi(r)}$, $\psi(r) = \rho(r) - \rho$.

◀ Let $f$ be of minimal type, and let $\rho(r)$ be its proximate order with the properties described in Theorem B.1.4. Then there exists a sequence $r_n \to \infty$ as $n \to \infty$ such that $\lim_{n\to\infty} L(r_n) = 0$ and $\psi(r_n) < 0$, $\rho(r_n) < \rho$. Since $\rho(r)$ is a monotone function, we find: $\rho(r) \uparrow \rho$ as $r \to \infty$, $\psi(r) < 0$ for $r > R$. On the other hand, by the inequality in Theorem B.1.4 we have $\psi(r) \leqslant -\psi'(r)(r+e)\ln(r+e)$ if $r > R$, where $R > 0$ is a constant. Therefore,

$$\frac{d\ln L(r)}{d\ln r} = \psi(r) + \psi'(r) r\ln r \leqslant \psi(r)\frac{q(r) - r\ln r}{q(r)} < 0, \ r > R,$$

where $q(r) = (r+e)\ln(r+e)$. Thus, $L(r) \downarrow 0$. The second statement of the theorem is proved similarly ▶

**Theorem B.1.5.** *The generalized indicator of an arbitrary entire function* $f$ *of order* $\rho > 0$ *is a finite* $2\pi$*-periodic trigonometrically* $\rho$*-convex function. Every function* $h$ *of this sort has the following property: if* $(\alpha,\beta) \subset \mathbb{R}$, $\beta - \alpha > \pi/\rho$, *and* $h(\theta) \leqslant 0$, $\theta \in (\alpha,\beta)$, *then* $h(\theta) = 0$, $\theta \in (\alpha,\beta)$,

We mention some properties of a proximate order.

**Property B.1.6.** *The function* $L(r) := V(r)r^{-\rho}$ (see (B.1.2)) *satisfies the relation* $\lim\limits_{r\to\infty} L(tr)/L(r) = 1 \quad \forall\, t \in (0,\infty)$, *and the convergence is uniform in the variable* $t$ *ranging through an arbitrary fixed segment* $[a,b]$, *where* $0 < a < b < \infty$.

Any function $L$ with the above property is called *slowly varying.* The next property is a consequence of Property B.1.6.

**Property B.1.7.** *If* $\rho(r)$ *is a proximate order,* $V(r) = r^{\rho(r)}$, *and* $0 < a < b < \infty$, *then*

$$\lim_{r\to\infty} V(tr)/V(r) = t^{\rho} \quad \forall\, t \in (0,\infty),$$

*uniformly in the variable* $t$ *on an arbitrary segment* $[a,b]$.

**Property B.1.8.** *Let* $\rho(r)$, $r > c \geqslant 0$, *be a proximate order. Suppose that* $\rho > 0$ *in* (B.1.1). *Then* $V(r) := r^{\rho(r)}$ *is a monotone increasing function for* $r > R$, *where* $R$ *is a sufficiently large number. The inverse function* $r = U(s)$, $s > V(R)$, *has the property that* $\rho^*(s) = (\ln s)^{-1} \ln U(s)$ *is a proximate order such that* $\rho^*(s) \to 1/\rho$ *as* $s \to \infty$.

The function $\rho^*(s)$ considered in B.1.8 is called the proximate order *conjugate* to the proximate order $\rho(r)$.

**Theorem B.1.9.** *The type* $\sigma_f$ *of the function* $f$ *relative to its proximate order* $\rho(r)$ (see Definition B.1.3) *is found by the formula*

$$(\sigma_f e\rho)^{1/\rho} = \varlimsup_{r\to\infty} \varphi(k)|c_k|^{1/k}, \tag{B.1.5}$$

*where* $r = \varphi(t)$ *is the inverse to the function* $t = r^{\rho(r)}$, $r > R$; $R$ *is a sufficiently large number* (see Property B.1.8); $\{c_k\}$ *are the Taylor coefficients of the function* $f$.

**B.1.2. Canonical Weierstrass product.** Let $\{a_k\}$ be a sequence of complex numbers such that $0 < |a_1| \leqslant |a_2| \leqslant \dots$, and let $p \geqslant 0$ be the minimal integer for which the series $\sum\limits_{k=1}^{\infty} |a_k|^{-1-p}$ converges. We introduce the *Weierstrass primary factor* as follows:

$$E(z;p) = \begin{cases} 1-z, & p = 0,\\ (1-z)\exp\{\sum\limits_{k=1}^{p} z^k/k\}, & p \geqslant 1. \end{cases}$$

**Theorem B.1.10.** The *product* $f(z) = \prod\limits_{k=1}^{\infty} E(z/a_k;p)$ *converges absolutely and uniformaly on any compact subset of* $\mathbb{C}$ *and represents an entire function.*

The function $f$ is called the *Weierstrass canonical product of genus* $p$.

**Theorem B.1.11.** *Under the conditions of Theorem* B.1.10, *let* $\{a_k\}_1^{\infty}$ *be a sequence of positive numbers . We denote by* $n(r)$ *the sum of orders of the* *zeros of* $f$ *in the disk* $S(r) = \{z \in \mathbb{C} : |z| \leqslant r\}$. *Let* $\rho(r)$ *be a proximate order satisfying* $p < \rho < p+1$ (see (B.1.1)), *and let* $0 < \Delta < \infty$. *If*

$$n(r) \sim \Delta r^{\rho(r)} \quad as \quad r \to \infty, \quad i.e., \quad \lim_{r\to\infty} r^{-\rho(r)} n(r) = \Delta,$$

*then* $f$ *is an entire function of order* $\rho$ *and of proximate order* $\rho(r)$ *such that*

$$\lim_{r\to\infty} \frac{\ln f(-re^{i\theta})}{r^{\rho(r)}} = \frac{\pi\Delta}{\sin \pi\rho} e^{i\rho\theta}, \quad |\theta| < \pi; \quad \ln f(0) = 0. \tag{B.1.6}$$

*Here the convergence is uniform in the variable* $\theta$ *ranging over an arbitrary fixed segment* $[\alpha, \beta] \subset (-\pi, \pi)$.

**Theorem B.1.12.** *Let* $\rho(r)$ *be a proximate order. Suppose*

$$\rho(r) \to \rho \in \mathbb{N}; \;\; L(r) = r^{\rho(r)-\rho}; \;\; \Delta \in (0, \infty).$$

*If* $L(r) \downarrow 0$, *then there exists a proximate order* $\rho_1(r) \to \rho$ *and an entire function* $f$ *of order* $\rho(r)$ *with positive zeros such that* (see Theorems B.1.10, B.1.11)

$$f(z) = \prod_{k=1}^{\infty} E(z/a_k; \rho - 1); \;\; n(r) \sim \Delta r^{\rho_1(r)}.$$

*Moreover, the entire function* $g(z) = f(-z)$ *satisfies the condition*

$$\lim_{r\to\infty} \frac{\ln g(re^{i\theta})}{r^{\rho(r)}} = -\Delta e^{i\rho(\pi+\theta)}, \quad |\theta| < \pi, \quad \ln g(0) = 0, \tag{B.1.7}$$

*where the convergence is uniform in the variable* $\theta$ *ranging over an arbitrary fixed segment* $[a, b] \subset (-\pi, \pi)$.

*If* $L(r) \uparrow \infty$ *as* $r \to \infty$, *then the statement remains true with the only difference that the genus of the function* $f$ *is equal to* $p = \rho$ *and for* $g(z)$ *the minus sign on the right in* (B.1.7) *must be deleted.*

**Theorem B.1.13.** *Let* $f$ *be an entire function of proximate order* $\rho(r) \to \rho \in (0, \infty)$; *let* $h_f = h(\cdot, f)$ *be the generalized indicator of* $f$ (see (B.1.4)). *Then* $\rho(r)$ *is a proximate order of the derivative* $f'$, *and for the generalized indicator* $h(\cdot, f')$ *we have:*

1) $h(\cdot, f') \leqslant h(\cdot, f)$; $h^+(\cdot, f') = h^+(\cdot, f)$, $h^+ = \max\{0, h\}$;

2) *if* $h(\cdot, f)$ *is a trigonometric indicator, i.e., the* $2\pi$ *periodic extension to* $\mathbb{R}$ *of the trigonometric function*

$$H(\theta) = a\cos \rho\theta + b \sin \rho\theta, \;\; \theta \in [\phi, \phi + 2\pi], \;\; H(\phi) = H(\phi + 2\pi)$$

*(here* $\phi \in \mathbb{R}$ *is a constant), then* $h(\cdot, f') = h(\cdot, f)$.

◀ Statement 1) is well known. We prove statement 2).

Let $\rho > 1/2$. The first relation in 1) shows that $h(\theta, f') \leqslant H(\theta)$, $\theta \in [\phi, \phi + 2\pi]$. Since $h(\theta, f') - H(\theta)$ is a nonpositive trigonometrically $\rho$-convex function, we apply Theorem B.1.5 to deduce the identity $h(\cdot, f') = h(\cdot, f)$.

Let $\rho \leqslant 1/2$. Since in this case an arbitrary $2\pi$-periodic trigonometrically $\rho$-convex function is nonnegative, the second relation in 1) implies 2) ►

**B.1.3. Plane indicator diagram of an entire function of a given proximate order and with nonnegative indicator.** This notion can be introduced by analogy with Definition 3.1.17, on the basis of Theorem B.1.5.

**Definition B.1.14.** Let $h = h_f$ be a generalized indicator of the entire function $f$ of finite order $\rho > 0$ (see Definition B.1.3), and $h \geqslant 0$; let $\Phi_\rho$ be the function defined by formula (3.1.6):

$$\Phi_\rho(z) = \begin{cases} |z|^\rho \cos(\rho \arg z), & |\arg z| \leqslant \gamma_\rho; \\ 0, & \rho \geqslant 1/2; \quad \gamma_\rho \leqslant |\arg z| \leqslant \pi \end{cases} \tag{B.1.8}$$

We define the *indicator diagram of the function* $f$ to be the set

$$X_h = \{z \in \mathbb{C} : \ \Phi_\rho(ze^{i\theta}) \leqslant h(\theta) \quad \forall\, \theta \in \mathbb{R}\}. \tag{B.1.9}$$

## B.2. Analytic proximate orders and associated functions

For $\gamma \in (0, \pi]$, let $L(\gamma, R) = \{z = re^{i\theta} \in \mathbb{C} : \ |\theta| < \gamma; \ r > R\}$ be an angle on the plane $\mathbb{C}$ without a sector, and let $L(\gamma) = \{z \in \mathbb{C} : \ |\arg\ z| < \gamma\}$. Suppose $\rho(r)$, $r > c$, is a positive function such that $\rho(r) \to \rho$ as $r \to \infty$. The function $\rho(r)$ is called an *analytic proximate order* if the function $V(r) = r^{\rho(r)}$, $r > c$, has an analytic extension $V(z)$ to a domain containing the set $L(\gamma, R)$ for some $\gamma, R > 0$ and, moreover,

$$\lim_{t\to\infty} [V(t)]^{-1} V(tz) = z^\rho, \quad \forall\ z \in L(\gamma), \tag{$B$.2.0}$$

where the convergencee is uniform on every compact subset of $L(\gamma)$. Here and in what follows we mean the principal value of the symbol $z^\rho$: $z^\rho = r^\rho e^{i\rho\theta}$, where $z = re^{i\theta} \in \mathbb{C}$. Condition (B.2.0) is stronger than the limit condition in Proposition B.1.7 for the *usual* proximate order. The term is justified by the following statement.

**Property B.2.1.** *Any analytic proximate order* $\rho(r)$, $r > c$, *is a proximate order in the sense of Definition* B.1.1.

◄ As $r \to \infty$, the sequence of analytic functions

$$\{f_r(z) := V(rz)/V(r), \ z \in L(\gamma, R) : \ r > c\}$$

converges uniformly on every compact subset of $L(\gamma)$, in particular, on every segment $[a, b] \subset (0, \infty)$. Consequently, the sequence of their derivatives $\{f'_z(\cdot, r) : \ r > c\}$ has a similar property. Therefore,

$$\lim_{r\to\infty} \frac{rV'(rt)}{V(r)} = \rho t^{\rho-1} \quad \forall\, t \in (0, \infty).$$

For $t = 1$ this means that condition (B.1.2) (see Property B.1.2) is fulfilled. Thus, $\rho(r)$ is a proximate order ▶

In this section we prove the existence of an analytic proximate order $\rho(r)$ for any entire function of finite order. The function $V(z)$ occuring in the definition of $\rho(r)$ admits the representation

$$V(z) = z^{\rho(z)}, \text{ where } \rho(z) \to \rho \text{ as } z \to \infty, \ |\arg z| \leqslant \delta,$$

and $\delta$ is an arbitrary number in the interval $(0, \gamma)$. It turns out that the same properties are also true for $\gamma > \pi$, i.e., when the sets $L(\gamma, R)$ and $L(\gamma)$ are situated on the Riemann surface of the logarithm (this is important for the sequel). Besides, for the purposes of presentation in Section B.3, we study various properties of certain functions associated with the analytic proximate orders mentioned above.

**B.2.1. Existence of an analytic proximate order for entire functions of finite order.** It turns out that proximate orders of entire functions can be chosen in the classes of restrictions to the positive semiaxis of the functions holomorphic in an angle of sufficiently large opening.

**Lemma B.2.2.** *Let $f$ be the entire function introduced in Theorem* B.1.11 *($f$ is of order $\rho \in (0,1)$, and of proximate order $\rho(r)$); let $\Delta > 0$, and let $\delta = (\pi\Delta)^{-1} \sin \pi\rho$. We put $L(\pi) = \{z \in \mathbb{C} : |\arg z| < \pi\}$ and consider the branch $G(z)$ of $\ln f(-z)$ such that $\ln f(0) = 0$. Then*

$$\lim_{r\to\infty} \frac{G(rz)}{G(r)} = z^{\rho}, \tag{B.2.1}$$

*and the convergence is uniform in the variable $z$ on every compact set $K \subset L(\pi)$. Moreover, the function*

$$\gamma(r) = (\ln r)^{-1} \ln \delta G(r) \tag{B.2.2}$$

*is a proximate order equivalent to $\rho(r)$.*

◀ First, by (B.1.6) we have:

$$\lim_{r\to\infty} \frac{\delta G(re^{i\theta})}{r^{\rho(r)}} = e^{i\rho\theta}, \quad |\theta| < \pi, \tag{B.2.3}$$

and the convergence is uniform in $\theta$ on every segment $[\alpha, \beta] \subset (-\pi, \pi)$. Next, Property B.1.7 shows that

$$(rt)^{\rho(tr)} r^{-\rho(r)} \to t^{\rho} \quad \text{as} \quad r \to \infty \tag{B.2.4}$$

uniformly in $t$ on every segment $[a, b] \subset (-\pi, \pi)$ if $0 < a < b < \infty$. Therefore, from (B.2.3), (B.2.4) we deduce that the sequence of holomorphic functions $g_r(z) = [G(r)]^{-1} G(rz)$, $z \in L(\pi)$, satisfies condition (B.2.1). Consequently, the sequence $g'_r(z)$ converges to the function $\rho z^{\rho-1}$ in the same sense. For $z = 1$ we obtain the relation

$$\lim_{r\to\infty} \frac{rG'(r)}{G(r)} = \rho. \tag{B.2.5}$$

Condition (B.2.5) implies the last statement of the lemma by (B.1.2), (B.2.3) ▶

In a sense, the following result is a converse to Proposition 1.4.22.

**Lemma B.2.3.** *If the derivative $f'(x)$ of a function $f(x)$ is continuous and monotone nonincreasing on the interval $(a, b)$, then $f(x)$ is concave. If, moreover, the derivative $f'(x)$ is strictly monotone decreasing, then $f(x)$ is strictly concave.* *)

◀ Suppose $x_0 \in (a, b)$, $a < x < y < b$, $z = \lambda x + \mu y$, where $\lambda \in (0, 1)$ and $\mu = 1 - \lambda$. By the Newton-Leibniz formula,

$$f(x) = f(x_0) + \int_{x_0}^{x} f'(t)dt.$$

Invoking the mean value theorem and the conditions of the lemma, we obtain

$$f(z) - f(x) = \int_{x}^{z} f'(t)dt = (z - x)f'(c) \geqslant (z - x)f'(z)$$

for some $c \in (x, z)$. The inequality $f(y) - f(z) \leqslant (y - z)f'(z)$ is proved similarly. Therefore, observing that $\lambda + \mu = 1$, we have

$$\lambda[f(z) - f(x)] + \mu[f(z) - f(y)] \geqslant [\lambda(z - x) - \mu(y - z)]f'(z) = 0.$$

Consequently, $f(z) \geqslant \lambda f(x) + \mu f(y)$, i.e., $f$ is a concave function. If the derivative $f'(x)$ is strictly monotone decreasing on the interval $(a, b)$, then the inequalities in question are strict. Then $f$ is strictly concave ▶

**Theorem B.2.4.** *Let $\rho > 0$, and let $L(\gamma) = \{(r, \theta) \in L : r > 0;\ |\theta| < \gamma\}$ be an angle on the Riemann surface of the logarithm. For every function $f(z)$ of finite order $\rho > 0$ there exists a proximate order $\rho_0(r)$ such that the function $V(r) = r^{\rho_0(r)}$ has the following properties:*

*1) the function $V(r)$ has an analytic extension to $L(\gamma)$ such that*

$$\lim_{r\to\infty} \frac{V(rW)}{V(r)} = W^\rho, \quad W \in L(\gamma), \tag{B.2.6}$$

*where $W^\rho = r^\rho e^{i\rho\theta}$ for $W = (\rho, \theta) \in L$, and the convergence is uniform on every compact set $K \subset L(\gamma)$;*

*) I.e., $-f(x)$ is a strictly convex function.

2) *the complex conjugate function* $\overline{V}$ *satisfies* $\overline{V(W)} = V(\overline{W})$ *for all* $W \in L(\beta; R)$, *where* $\overline{W} = (|W|, -\operatorname{Arg} W)$;

3) *the function* $V(r)$ *is positive on the positive semiaxis, and increasing and strictly convex relative to* $\ln r$; *moreover,* $V(0) = 0$;

4) $\ln V(r)$, $r > 0$, *is a strictly convex function of the variable* $r$.

◀ We fix some numbers $\nu$, $\alpha$, and $\gamma$ such that $0 < \nu < \alpha < \pi/\gamma$, and construct a proximate order $\rho(r)$ for $f(z)$ in accordance with Theorem B.1.4. The function

$$\mu(t) = \alpha^{-1}\nu(t^{1/\alpha}), \tag{B.2.7}$$

where $\nu(r) = \rho(r) - \rho + \nu$, $\nu(r) \to \nu$, is also a proximate order such that $\mu(t) \to \mu < 1$ as $t \to \infty$ ($\mu := \nu\alpha^{-1}$, see (B.1.1)). We consider the Weierstrass canonical product $f_0(z)$ of genus 0 (see Theorem B.1.11) with positive zeros $1 \leqslant a_1 \leqslant a_2 \leqslant \dots$ such that $n(t) \sim \Delta t^{\mu(t)}$ as $t \to \infty$, where $\Delta = \pi^{-1}\sin\pi\mu$. By Lemma 2.2, $(\ln t)^{-1}\ln G(t)$ is a proximate order equivalent to $\mu(t)$. Consequently, for the function $G(w) = \ln f_0(-w)$, where $\ln f_0(0) = 0$, we have (see (B.1.6), (B.2.1))

$$z^\mu = \lim_{t\to\infty}\frac{G(tz)}{t^{\mu(t)}} = \lim_{t\to\infty}\frac{G(tz)}{G(t)}. \tag{B.2.8}$$

Here the convergence to both limits is uniform on every compact set $K \subset L(\pi)$. The function $\Phi(W) = G(W^\alpha)$ is analytic in the angle $L(\pi/\alpha)$. By (B.2.8), (B.2.7), for it we find:

$$\lim_{r\to\infty}\frac{G(r^\alpha W^\alpha)}{r^{\alpha\mu(r^\alpha)}} = \lim_{r\to\infty}\frac{\Phi(rW)}{r^{\nu(r)}} = \lim_{r\to\infty}\frac{\Phi(rW)}{\Phi(r)} = W^\nu, \tag{B.2.9}$$

where the convergence is uniform on every compact subset $K$ of $\overline{L(\gamma)}$ with $\gamma < \pi/\alpha$. This implies that the function

$$V(W) = W^{\rho-\nu}\Phi(W), \quad W \in L(\pi/\alpha)$$

satisfies condition (B.2.6). Since $(\ln t)^{-1}\ln G(t)$ is a proximate order equivalent to $\mu(t)$, Property B.1.2 and formulas (B.2.7), (B.2.9) allow us to conclude that $\rho_0(r) = (\ln r)^{-1}\cdot\ln V(r)$, $r \geqslant a > 0$, where $V(a) \geqslant 1$ is the required proximate order equivalent to $\rho(r)$.

Property 2) follows from the structure of $V$.

Property 3) for $V$ is a consequence of a similar property for the function $G(t) - \ln f_0(-t) = \ln M_{f_0}(t)$ ($G$ does possess this property by the Hardamard three disks theorem, see (1.10.2); Subsection 7.1.1), and of the following statement (which is true by the definition of a strictly convex function and the inequality $\rho > \nu$).

*Let* $\mathfrak{M} = \{V : \mathbb{R} \to \mathbb{R}\}$ *be the class of positive monotone increasing strictly convex functions, and let* $V_1, V_2 \in \mathfrak{M}$. *Then* $V := V_1V_2 \in \mathfrak{M}$.

By Lemma B.2.3, for $c > 0$, $\alpha \in (0,1)$, the functions $\varphi_1(r) = \ln r$ and $\varphi_2(r) = \ln(1 + cr^\alpha)$ are strictly concave for $r > 0$. Therefore (see Exercise 1.4.9), the restriction to the positive semiaxis of the function $V$ in question has the property that

$$\ln V(r) = (\rho - \nu)\ln r + \ln\left(\lim_{n\to\infty}\sum_{\nu=1}^{n}\ln\left(1 + \frac{r^\alpha}{a_\nu}\right)\right)$$

is a stricitly concave function, i.e., condition 4) is fullfilled for $V$ ▶

**Remarks.** 1. If $\gamma \leqslant \pi$, the angle $L(\gamma)$ may be regarded as the angle $\{z \in \mathbb{C} : |\arg z| < \gamma\}$ in $\mathbb{C}$.

2. If $1/2 < \rho < 1$, $\gamma \leqslant \pi$, then, putting $\nu = \rho$, $\alpha = 1$ in the above proof, we see that the function $G(w) = \ln f_0(-w)$ satisfies the assumptions of the theorem.

Now, we study the behavior near $t = 0$ of the function occuring in the proof of Theorem B.2.4.

**Property B.2.5.** *Suppose* $\gamma > 0$ *and* $\alpha \in (0, \pi/\gamma)$. *Under the conditions of Theorem* B.2.4, *for every* $\beta \in (\rho, \rho + \alpha)$ *there is a function* $V$ *admitting the representation*

$$V(t) = \sum_{k=0}^{\infty}(-1)^k B_{k+1} t^{\beta + k\alpha}, \quad t \in [0,1],$$

*where* $B_l > 0 \ \forall\, l \in \mathbb{N}$, *and the series converges absolutely and uniformly for all* $t \in [0,1]$.

◀ Using Theorem B.1.10 and the notation of the proof of Theorem B.2.4, we find that the series

$$G(w) = \sum_{k=1}^{\infty}\ln\left(1 + \frac{w}{a_k}\right)$$

converges absolutely and uniformly on every compact subset of the disk $S = \{w \in \mathbb{C} : |w| \leqslant \delta\}$ with $0 < \delta < a_1$. Therefore, the function $G$ admits a power series expansion of the form

$$G(w) = \sum_{l=1}^{\infty}(-1)^{l-1}\frac{A_l}{l}w^l, \ w \in S; \ A_l = \sum_{k=1}^{\infty}\frac{1}{a_k^l}.$$

Putting $\beta = \rho - \nu + \alpha$ and $V_1(t) = t^{\rho-\nu}G(t^\alpha)$, we conclude that the function $V(t) = V_1(\delta t)$ has the required property ▶

**B.2.2. On a class of functions associated with an analytic proximate order.** Let $\mathfrak{B}(\gamma; \rho(r)) = \{r^{\rho_0(r)}\}$ be the class of functions satisfying conditions 1) - 4) of Theorem B.2.4, where, moreover, $\rho_0(r)$ is an analytic proximate order equivalent to $\rho(r)$ ($\rho(r) \to \rho > 0$ as $r \to \infty$).

Put $\gamma_\rho = \min\{\pi, \pi/2\rho\}$, $\Pi_\theta = \{(r,\varphi) \in L : r > 0; |\varphi - \theta| < \gamma_\rho\}$, and $L(\beta; R) = \{(r,\varphi) \in L : r \geqslant R > 0, |\varphi| \leqslant \beta\}$. On the elements of the Riemann surface $L$ of the logarithm, the operations described in Subsection A.2.1, are well defined. The proof of the main result of this section will require the knowledge of certain properties of the functions belonging to $\mathfrak{B}(\gamma; \rho(r))$. We begin with the following simplest properties. In the formulas below (except for (B.2.12)) the convergence is of the same nature as in (B.2.6), and $K$ is an arbitrary compact subset of $L(\gamma)$.

For every function $V \in \mathfrak{B}(\gamma; \rho(r))$ the following statements are true.

a) $\lim_{t\to\infty} [V(t)]^{-1} tV'(tW) = \rho W^{\rho-1}, \quad W \in L(\gamma),$ (B.2.10)

where $V'(Z) = e^{-i\theta}\frac{\partial V}{\partial r}$, $Z = (r,\theta)$ (see Subsection A.2.1). In particular, $(\ln t)^{-1} \ln V(t)$ is a proximate order (see (B.1.1), (B.1.2)):

$$\rho = \lim_{t\to\infty} [V(t)]^{-1} tV'(t) = \lim_{t\to\infty} (\ln t)^{-1} \ln V(t). \tag{B.2.11}$$

b) $\lim_{|W|\to\infty} [V(W)]^{-1} \pi(W) V'(W) = \rho, \quad |\operatorname{Arg} W| \leqslant \beta < \gamma,$ (B.2.12)

where $\pi : L \to \mathbb{C}$, $\pi(r,\theta) = re^{i\theta}$.

c) The function $v(W) = W^{-\rho} V(W)$, $W \in L(\gamma)$, is an analytic extension of the slowly varying function $v(t) = V(t)t^{-\rho}$, $t > 0$ (see Property B.1.6), and, moreover,

$$\lim_{t\to\infty} [v(t)]^{-1} v(tW) = 1, \quad W \in L(\gamma).$$

d) $\lim_{|W|\to\infty} [V(W)]^{-1} V(ZW) = Z^\rho; \ \lim_{|W|\to\infty} [v(W)]^{-1} v(ZW) = 1$ (B.2.13)

if $|\operatorname{Arg} W| \leqslant \beta < \gamma$ and $Z \in K$, where the number $\beta > 0$ and the compact set $K \subset L(\gamma)$ are such that $ZW \in L(\gamma)$ for $Z \in K$, $|\operatorname{Arg} W| \leqslant \beta$.

Putting $K = \{(1,\varphi) \in L : |\varphi| \leqslant \beta < \gamma\}$, from (B.2.6) we deduce that for every $\beta \in (0,\gamma)$ there is a number $Q_\beta > 0$ and a value of the argument of the function $V$ such that $V(W) \neq 0$ and $|\operatorname{Arg} V(W) - \rho \operatorname{Arg} W| < \pi/2$ for $W \in L(\beta; Q_\beta)$. Accordingly, we define the mapping

$$V : D \to L;\ \tilde{V}(w) = (|V(w)|, \operatorname{Arg} V(w)); D = \bigcup_{0<\beta<\gamma} L(\beta; Q_\beta) \tag{B.2.14}$$

which is a lifting of $V : D \to L$ relative to $\pi : L \to \mathbb{C}$, $\pi(r,\theta) = re^{i\theta}$ (see Subsection A.1.3).

**Theorem B.2.6.** *Let $\rho > 0$, and let $V \in \mathfrak{B}(\gamma; \rho(r))$. Let $t = U(s)$, $s > 0$, be the inverse to the function $s = V(t)$, $t > 0$; let $\rho^*(s)$ be the proximate order conjugate to $\rho(r)$* (see Property B.1.8). *Then $\ln U(s)/\ln s$ is an analytic proximate order equivalent to $\rho^*(s)$, and the function $U(s)$ admits analytic continuation to a function $U(W)$ in a domain $T \subset L(\rho\gamma)$ such that $T$ is symmetric relative to the real axis and, for every $\beta < \gamma$, contains the set $L(\rho\beta, T_\beta) = \{z = (r,\theta) \in L : |\theta| < \rho\beta;\ r > T_\beta\}$. Furthermore, the function $U$*

*possesses properties* 1)-4) *of the traces of functions of class* $\mathfrak{B}(\rho\gamma;\rho^*(s))$ *on the set* $D$.

*If* $\rho > 1/2$ *and* $\gamma \leqslant \min\{\pi, \pi/\rho\}$, *then* $L(\gamma) \subset L$ *and* $T \subset L(\rho\gamma) \subset \mathbb{C}$ (see Remark 1 to Theorem B.2.4). *Next, the function* $z = U(w)$, $w \in T$, *has an inverse coinciding with the function* $w = V(z)$ *for* $z \in U(T) \subset L(\gamma)$. *If* $\rho \leqslant 1/2$, *then the function* $U$ *has a lifting* $\chi: T \to L$ *relative to* $\pi: L \to \mathbb{C}$, *and this lifting is equal to* $\widetilde{V}^{-1}$, *where* $\widetilde{V}$ *is the one-to-one holomorphic mapping defined on the set* $U(T) \subset D$ *by formula* (B.2.14).

◀ 1. Fixing arbitrarily $\beta \in (0,\gamma)$ and putting $K = \{(1,\varphi) \in L : |\varphi| \leqslant \beta\}$ in (B.2.10), we see that $V'(W) \neq 0 \;\forall\, W \in L(\beta; P_1)$ if the number $P_1$ is siffuciently large. We show that for every $c \in (0,1)$ and for $W \in L(\beta; P)$ there is a number $\alpha \in (0,1)$ such that

$$|V(Z) - V(W)| > c|V'(W)| \cdot |z - w| \quad \forall\, Z \in \Omega_\alpha(W) \tag{B.2.15}$$

for some $P \geqslant P_1$, where $z = \pi(Z)$, $w = \pi(W)$, $\Omega_\alpha(W) = \{Z \in \Pi_{Arg\, W} : |z - w| \leqslant \alpha|w|\}$ (we observe that $\Omega_\alpha(W) \subset L(\gamma)$). Consider the function

$$G(W,Z) = [(z-w)V'(W)]^{-1} \times [V(Z) - V(W)],$$

where $W \in L(\beta; P_1)$ and $Z \in L(\gamma) \cap \Pi_{Arg\, W}$. Let

$$A(W) = [wV'(W)]^{-1}V(W), \quad \tau = (z-w)w^{-1}.$$

Then $G(W,Z) = G_1(W,Z) + G_2(W,Z)$ with

$$G_1(W,Z) = A(W)[(1+\tau)^\rho - 1]\tau^{-1};$$

$$G_2(W,Z) = A(W)(1+\tau)^\rho\tau^{-1}\{[v(W)]^{-1}v(Z) - 1\}, v(W) = \frac{V(W)}{W^\rho}.$$

By (B.2.12), for every $\delta_1 \in (c,1)$ there exist numbers $\alpha \in (0,1)$ and $P_2 \geqslant P_1$ such that

$$|G_1(W,Z)| > \delta_1 \quad \forall\, W \in L(\beta; P_2); \quad Z \in \Omega_\alpha(W) \subset L(\gamma). \tag{B.2.16}$$

For every $W \in L(\beta; P_2)$, the function $X(W,\tau) = v((1+\tau)W)$ is holomorphic in $\tau$ in the vicinity of the disk $\{\tau \in \mathbb{C} : |\tau| \leqslant \alpha < 1\}$. By the Schwarz lemma, since $Z = (1+\tau)W$, we obtain

$$|G_2(W,Z)| \leqslant |A(W)|(1+\alpha)^\rho\alpha^{-1}M_\alpha(W) \;\forall\, W \in L(\beta; P_2),\; Z \in \Omega_\alpha(W),$$

where $M_\alpha(W) = \sup\limits_{|\tau|=\alpha} |[v(W)]^{-1}X(W,\tau) - 1|$. Recalling (B.2.12) and (B.2.13), we see that

$$|G_2(W,Z)| < \delta_1 - c \quad \forall\, W \in L(\beta; P), \quad Z \in \Omega_\alpha(W) \tag{B.2.17}$$

for some $P \geqslant P_2$. Now, we deduce inequality (B.2.15) from (B.2.16) and (B.2.17). Thus,

$$\widetilde{V}(Z) \neq \widetilde{V}(W) \quad \forall\, Z \in \Omega_\alpha(W) \setminus \{W\},\ W \in L(\beta; P). \tag{B.2.18}$$

2. Now we prove that

$$\widetilde{V}(Z) \neq \widetilde{V}(W) \quad \forall\, Z \in L(\beta; P) \setminus \Omega_\alpha(W),\ W \in L(\beta; P). \tag{B.2.19}$$

In the notation of formula (B.2.15), let $W \in L(\beta; P)$, and let

$$\omega_\delta(W) = \{Z = \tau W \in L:\ 1 - \delta < |\tau| < 1 + \delta;\ |\,Arg\, \tau| < \delta\},$$

where the number $\delta \in (0, 1)$ is chosen in such a way that $\omega_\delta(W) \subset \Omega_\alpha(W)$ (see (B.2.15)). If $Z \notin \omega_\delta(W)$, three cases are possible.

A. $|\tau| \geqslant 1+\delta$. If $P > Q_\beta$ (see (B.2.14)), then $V(W) \neq 0$. Invoking (B.2.13), we see that the following inequality is true as $|W| \to \infty$, $W \in L(\beta; P)$ :

$$|V(Z) - V(W)| \geqslant |V(W)| \left| \frac{|V(\tau W)|}{|V(W)|} - 1 \right| \geqslant |V(W)|[(1+\delta)^{\rho+o(1)} - 1] > 0.$$

B. $|\tau| \leqslant 1 - \delta$. In a similar way, we arrive at the following inequality as $|W| \to \infty$, $W \in L(\beta; P)$ :

$$|V(Z) - V(W)| \geqslant |V(W)|[1 - (1-\delta)^{\rho+o(1)}] > 0.$$

C. $|\,Arg\, \tau| \geqslant \delta$. We introduce the additional condition $\rho\delta < \pi/2$. If $\rho|\,Arg\, \tau| = \rho|\,Arg\, Z - Arg\, W| \geqslant \pi/2$, then $Arg\, \widetilde{V}(Z) \neq Arg\, \widetilde{V}(W)$, $\widetilde{V}(Z) \neq \widetilde{V}(W)$ (see (B.2.14)). Let $\rho|\,Arg\, \tau| < \pi/2$. Then, using the inequality $|re^{i\theta} - 1| \geqslant (r+1)\sin\theta/2$ and arguing as in case A, we see that

$$|V(Z) - V(W)| \geqslant |V(W)|(|\tau|^{\rho+o(1)} + 1)\sin[\delta(\rho + o(1))/2] > 0,$$

where $o(1)$ is an infinitesimal as $|W| \to \infty$, $W \in L(\beta; P)$. So, formula (B.2.19) is true. After this, (B.2.18) and (B.2.19) imply the existence of the required number $R_\beta$ : the inverse function $\chi = \widetilde{V}^{-1}$ exists on the set $O_\beta := \widetilde{V}(L(\beta; R_\beta))$.

Let $U(s)$, $s > 0$, be the inverse of the function $s = V(t)$, $t > 0$. Then $y(s) = (\ln s)^{-1} \ln U(s) \to 1/\rho$ as $s \to \infty$, and $y(s)$ is a proximate order (see (B.2.11)). For every $\beta \in (0, \gamma)$ the set $O_\beta$ contains the infinite segment $\{(s, 0) \in L\ :\ s \geqslant V(R_\beta)\}$. Therefore, the function $\chi = \widetilde{V}^{-1}$ is an analytic extension of the function $\chi(s) = (U(s), 0)$, $s > 0$. By (B.2.6), for sufficiently small $\varepsilon > 0$ we have

$$1 - \varepsilon \frac{|V(W)|}{V(|W|)} < 1 + \varepsilon;\ |\,Arg\, \widetilde{V}(W) - \rho\, Arg\, W| < \rho\varepsilon\ \forall\, W \in L(\beta; S), \tag{B.2.20}$$

where $S = S(\varepsilon) > R_\beta$. This implies that

$$L((\beta-\varepsilon)\rho; T^{(1)}) \subset O_\beta \subset L(\rho(\beta+\varepsilon); T^{(2)})$$

for some $T^{(i)} = T^{(i)}(\delta, \beta)$, $i = 1, 2$. In particular, the domain of the function $\chi$ includes the set described in the theorem.

3. We retain the notation of step 2 of the proof. We claim that

$$\lim_{R\to\infty} \frac{U(RZ)}{U(R)} = Z^{1/\rho}, \quad U = \pi \circ \chi, \tag{B.2.21}$$

uniformly in $Z$ ranging over an arbitrary compact subset $K$ of $L(\rho\gamma)$.

There exist numbers $R_1 > 0$, $0 < \beta < \gamma$, and a sufficiently small number $\varepsilon \in (0, \beta)$ such that

$$RZ \in L[\rho(\beta-\varepsilon), T^{(1)}] \subset O_\beta \quad \forall\, R > R_1; \quad Z \in K$$

(see above). By construction, the holomoprhic fuction $\chi = \widetilde{V}^{-1}$ is defined on $O_\beta$. Let $(t,\theta) = \chi(Rs, \varphi)$, $R > R_1$, and let $Z = (s, \varphi) \in K$. By (B.2.20), we conclude that

$$\widetilde{V}(t,\theta) = (Rs, \varphi) = (V(t), \rho\theta)a, \quad a = 1 + \alpha(W),\ W = (t,\theta), \tag{B.2.22}$$

where $\alpha(W) \to 0$ as $|W| \to \infty$ and the convergence is uniform in $\theta \in [-\beta, \beta]$. Therefore,

$$Rs = |a|V(t), \quad t = U(sR|a|^{-1}),$$

where $U = V^{-1}$ is a monotone increasing function on $(0,\infty)$. Since $(\ln r)^{-1} \ln U(r)$ is a proximate order, we have $t \to \infty$ as $R \to \infty$ and, by Property B.1.7,

$$t = s^{1/\rho}(1 + \alpha_1(R))U(R),$$

where $\alpha_1(R) \to 0$ as $R \to \infty$ and the convergence is uniform in $Z \in K$. On the other hand, (B.2.22) implies that $\theta = \varphi/\rho + \alpha_2(R)$, where $\alpha_2(R) \to 0$ (as $R \to \infty$) uniformly in $Z \in K$. Thus, condition (B.2.21) is true.

4. The structure of the domain $D_{\widetilde{V}}$ and of the range $E_{\widetilde{V}}$ of the function $\widetilde{V}$ imply that these two sets are symmetric in $L$ relative to the ray $\{(r, 0) : r > 0\}$. Since $E_{\widetilde{V}} = D_\chi$, where $D_\chi$ is the domain of the function $\chi$, property 2) of the function $V$ shows that the function $U$ satisfies a similar condition:

$$U(\overline{Z}) = \overline{U(Z)} \quad \forall\, Z \in D_U = D_\chi.$$

5. Suppose $\lambda > 0$, $\mu > 0$, $\lambda + \mu = 1$, $0 < s < p$, $t = U(s)$, and $r = U(p)$, where $U$ is the inverse of $V$ (and $V(0) = U(0) = 0$). From property 3 of $V$ we deduce the inequality $V(t^\lambda r^\mu) < \lambda V(t) + \mu V(r)$. Since the inverse function $t = U(s)$, $s > 0$, is also monotone increasing, we see that

$$\lambda \ln U(s) + \mu \ln U(p) < \ln U(\lambda s + \mu p),$$

i.e., $\ln U$ is a strictly concave function on the interval $(0,\infty)$. Thus, the function $U$ possesses property 4).

Moreover, property 4) for the function $V$ yields $[V(t)]^\lambda \cdot [V(r)]^\mu < V(\lambda t + \mu r)$. Therefore, for the inverse function $U$ of $V$ we have $U(s^\lambda p^\mu) < \lambda U(s) + \mu U(p)$, i.e., property 3) is also fulfilled for $U$ ▶

**Remark.** Let $\rho > 0$, let $m > 1$ be a natural number, and let $\rho(r) \to \rho$ be a proximate order. Then $\rho_m(r) := m\rho(r^m) \to m\rho$ is also a proximate order, and the operator

$$A: \mathfrak{B}(\gamma;\rho(r)) \to \mathfrak{B}(\gamma/m;\rho_m(r)),\ A[V(r)] = V_m(r) := V(r^m),\ r \geqslant 0$$

is a bijection. Next, if $U, F$ are the inverse functions to $V, V_m$, respectively, then $F = U^m$.

In the sequel this property will be used for $\rho < 1/2$ and $m\rho > 1/2$.

### B.2.3. Integral properties of the functions belonging to $\mathfrak{B}(\gamma;\rho(r))$

**Theorem B.2.7.** *Suppose $\rho > 1/2, \gamma = \min\{\pi,\pi/\rho\}$, $V \in \mathfrak{B}(\gamma;\rho(r))$, and $\beta = \min\{\pi,\pi\rho\}$. Let $U = V^{-1}$ be the inverse function of $V$ holomorphic in $L(\beta;T_\beta)$. Fixing $\varphi_0 \in (\pi/2,\beta)$, we choose a number $a > 0$ so as to ensure that*

$$\Gamma_\varphi = \Gamma(\varphi;a) = \{a + te^{i\varphi} :\ t \geqslant 0\} \subset L(\pi;T_\beta) \tag{B.2.23}$$

*for all $\varphi \in [-\varphi_0,\varphi_0]$. Finally, we put*

$$N_\varphi(\zeta) = \int_{\Gamma_\varphi} \exp\{z - V(\zeta U(z))\}dz,\ G_\varphi = \{\zeta \in L(\alpha) : \operatorname{Re}(\zeta^\rho - 1)e^{i\varphi} > 0\},$$

*where $0 < \rho\alpha < \beta - \varphi_0$. Then*

*1) the function $N_\varphi$ is holomorphic in the domain $G_\varphi$ for every $\varphi \in [-\varphi_0,\varphi_0]$;*

*2) there exists a function $N$ holomorphic in the domain $D = L(\alpha)\setminus\{1\}$ and such that $N(\zeta) = N_\varphi(\zeta)$, $\zeta \in D \cap G_\varphi$;*

*3) the point $\zeta = 1$ is a pole of $N(\zeta)$, and*

$$\lim_{\zeta\to 1} \rho(\zeta - 1)N(\zeta) = 1. \tag{B.2.24}$$

◀ 1. Since $w \equiv V(U(w))$ and

$$G_\varphi = \{\zeta \in L(\alpha) : \operatorname{Re}(\zeta^\rho - 1)e^{i\varphi} > 0\} \neq \varnothing, \tag{B.2.25}$$

relations (B.2.6), (B.2.13) imply that

$$B(z;\zeta) := \exp\{z - V(\zeta U(z))\} = \exp\{z[(1 - \zeta^\rho) + \varkappa(z,\zeta)]\}, \tag{B.2.26}$$

where $\varkappa(z,\zeta) \to 0$ (as $|z| \to \infty$) uniformly in $\zeta$ on every compact set $K \subset \{\zeta \in L(\alpha) : \text{Re}\,(\zeta^\rho - 1)e^{i\varphi} \geqslant b > 0\}$. Then there exist $t_0$, $A > 0$ such that

$$N_\varphi(\zeta) \leqslant \int_0^{t_0} |B(a + te^{i\varphi}, \zeta)|\, dt + A \exp\{a\text{Re}\,(1 - \zeta^\rho)\} \quad \forall\, \zeta \in K.$$

Consequently, the integral $N_\varphi(\zeta)$ converges uniformly in $\zeta$ on $K$. Therefore, the function $N_\varphi(\zeta)$ is holomorphic in $G_\varphi$ for all $\varphi \in [-\varphi_0, \varphi_0]$ (see (B.2.25)). If $0 \in G_\varphi$ (for $\varphi \in (\pi/2, \varphi_0)$), then the function $N_\varphi(\zeta)$ is continuous at 0 on $G_\varphi$.

2. Let $0 < \varphi_2 - \varphi_1 < \pi$, $\zeta \in G_{\varphi_1} \cap G_{\varphi_2}$, $C_t = \{z \in \mathbb{C} : z = V(r) \cdot (a + te^{i\varphi}), \varphi_1 \leqslant \varphi \leqslant \varphi_2\}$. Then $\zeta \in G_\varphi$ for every $\varphi \in [\varphi_1, \varphi_2]$, and $\max\{\text{Re}\,(1 - \zeta^\rho)e^{i\varphi}, \varphi \in [\varphi_1, \varphi_2]\} < 0$. Arguing as above (see (B.2.26)), we conclude that $\int_{C_t} B(z,\zeta)\, dz \to 0$ as $t \to \infty$. Consequently, $N_{\varphi_1}(\zeta) \equiv N_{\varphi_2}(\zeta)$, $\zeta \in G_{\varphi_1} \cap G_{\varphi_2}$. Therefore, for $\varphi \in (\pi/2, \varphi_0)$ the functions $N_\varphi$ and $N_{-\varphi}$ are analytic extensions of the function $N_0$ to the domain $G := G_\varphi \cap G_{-\varphi}$ along the paths $\Gamma = \{\zeta = 1 + e^{i\theta}/2 : 0 \leqslant \theta \leqslant \pi\}$ and $\{z = \overline{\zeta} : \zeta \in \Gamma\}$, respectively.

We show that $N_\varphi(\zeta) = N_{-\varphi}(\zeta)$ $\forall\, \zeta \in G$. By the uniqueness theorem it suffices to verify this identity on the interval $(0,1) \subset G$. Using the integral representation of the function $N_\varphi$, property 2) of the function $V$ of the class $\mathfrak{B}(\gamma; \rho(r))$, and a similar property for the function $U$ (see (B.2.23) and Theorem B.2.6), after elementary transformations we obtain

$$N_\varphi(\zeta) = \overline{N_{-\varphi}(\zeta)} \quad \forall\, \zeta \in (0,1).$$

Such a relationship between two analytic functions is possible only if they are real on (0, 1) and coincide there. Thus, statements 1) and 2) of Theorem B.2.7 are true.

3. We represent the function $B(z,\zeta)$ (see (B.2.26)) in the form

$$\exp\{z(1 - \zeta^\rho)\} \cdot \exp\{\zeta^\rho z - V(\zeta U(z))\},$$

integrate by parts, and use the formula $U'(z) = [V'(U(z))]^{-1}$ to obtain

$$N_\varphi(\zeta) = (\zeta^\rho - 1)^{-1}[B(a,\zeta) + J_\varphi(\zeta)], \quad \zeta \in G_\varphi, \tag{B.2.27}$$

where

$$J_\varphi(\zeta) = \int_{\Gamma_\varphi} B(z,\zeta) H(z,\zeta)\, dz; \;\; H(z,\zeta) = \zeta^\rho - \zeta V'(\zeta U(z))[V'(U(z))]^{-1}.$$

By (B.2.27), $J_\varphi$ is a holomorphic function in $G_\varphi$ for every $\varphi \in [\varphi_0, \varphi_0]$, and $J_{\varphi_1}(\zeta) \equiv J_{\varphi_2}(\zeta)$, $\zeta \in G_{\varphi_1} \cap G_{\varphi_2}$, $0 < \varphi_2 - \varphi_1 < \pi$ (see (B.2.10)). The relation

$$\lim_{\zeta \to 1} \rho(\zeta - 1) N_\varphi(\zeta) = 1 \quad \forall\, \varphi \in [-\varphi_0, \varphi_0],$$

equivalent to (B.2.24), is fullfilled if and only if

$$\lim_{\zeta\to 1} J_\varphi(\zeta) = 0, \quad |\varphi| \leqslant \varphi_0. \tag{B.2.28}$$

4. For the proof of (B.2.28), we split the ray $\Gamma_\varphi$ into two parts $\Gamma_\varphi^1$, $\Gamma_\varphi^2$ by a point $z_0$ ($z_0$ corresponds to the value of the parameter $t = t_0$) in such a way that $\Gamma_\varphi^1 = [a, z_0]$, $\Gamma_\varphi^2 = \Gamma_\varphi \setminus \Gamma_\varphi^1$. Then (see (B.2.28))

$$J_\varphi(\zeta) = J_1(\zeta) + J_2(\zeta), \quad J_i(\zeta) = \int_{\Gamma_\varphi^i} B(z,\zeta)H(z,\zeta)\,dz, \quad i = 1,2. \tag{B.2.29}$$

From (B.2.6), (B.2.10), we deduce that

$$\lim_{z\in\Gamma_\varphi;|z|\to\infty} H(z,\zeta) = 0, \quad \zeta \in K, \tag{B.2.30}$$

and the convergence is uniform relative to the variable $\zeta$ ranging over a compact neighborhood $K$ of the point 1. Therefore, $H$ is a uniformly bounded function on the set $\Gamma_\varphi \times K$. Since $H(z,1) \equiv 0$, $z \in \Gamma_\varphi$, the Schwarz lemma yields the following estimate of the holomorphic function $H$:

$$|H(z,\zeta)| \leqslant M|\zeta - 1|, \quad (z,\zeta) \in \Gamma_\varphi \times K; \quad M = \text{const} > 0.$$

Consequently (see (B.2.29)),

$$|J_1(\zeta)| \leqslant M|\zeta - 1| \int_0^{t_0} |B(a + te^{i\varphi}, \zeta)|\,dt, \quad \zeta \in K.$$

Putting $S(1,\delta) = \{\zeta \in \mathbb{C} : \ |\zeta - 1| < \delta\}$, we conclude that for every $\varepsilon > 0$ there exists $\delta = \delta(\varepsilon; t_0)$ such that

$$|J_1(\zeta)| < \varepsilon/2 \quad \forall\, \zeta \in S(1,\delta). \tag{B.2.31}$$

Next, as before, taking (B.2.30) into account, we find a constant $N(t_0)$ such that

$$|H(z,\zeta)| < N(t_0)|\zeta - 1|, \quad (z,\zeta) \in \Gamma_\varphi^2 \times K; \quad \lim_{t_0\to\infty} N(t_0) = 0 \tag{B.2.32}$$

Arguing as above, we estimate the function $\varkappa$ occurring in (B.2.26):

$$|\varkappa(z,\zeta)| < P(t_0)|\zeta - 1|, \quad (z,\zeta) \in \Gamma_\varphi^2 \times K; \quad \lim_{t_0\to\infty} P(t_0) = 0. \tag{B.2.33}$$

Let $\lambda \in (0, \pi/2)$ be a fixed number. It suffices to prove (B.2.28) under the condition

$$\zeta \to 1; \quad \zeta \in G_\varphi(\lambda) := G_\varphi \cap \{\zeta \in L(\alpha) : \ |\varphi + \arg(\zeta^\rho - 1)| < \lambda\}.$$

Now we estimate the integral $J_2(\zeta)$ (see (B.2.29), (B.2.32), (B.2.33)):

$$|J_2(\zeta)| < AN(t_0)|\zeta - 1|(|\zeta^\rho - 1|\cos\lambda - P(t_0)|\zeta - 1|)^{-1}\exp\{a\mathrm{Re}\,(1 - \zeta^\rho)\},$$

where $\zeta \in G_\varphi(\lambda)$, $A = \mathrm{const} > 0$. Since the quantity $|\zeta^\rho - 1|^{-1} \cdot |\zeta - 1|$ is bounded near 1, taking the number $t_0$ sufficiently large, we deduce that

$$|J_2(\zeta)| < \varepsilon/2, \quad \zeta \in G_\varphi(\lambda) \cap S(1, \delta).$$

Combined with (B.2.31) this proves (B.2.28) ▶

In the following statement we estimate the asymptotic behavior of the function $N$ occurring in Theorem B.2.7.

**Theorem B.2.8.** *Suppose $\rho > 1/2$, $\alpha \in (0, \pi/2\rho)$. There exist positive constants $b, A, R$ such that under the conditions and in the notation of Theorem* B.2.7 *we have*

$$|N(\zeta)| \leqslant A\exp\{-bV(|\zeta|)\} \quad \forall\, \zeta \in L(\alpha; R). \tag{B.2.34}$$

◀ If $R > (\cos\rho\alpha)^{-1/\rho}$, then $L(\alpha; R) \subset G_0$ (see Theorem B.2.7); therefore,

$$N(\zeta) = N_0(\zeta) = \int_a^\infty \exp\{t - V[\zeta U(t)]\}\, dt, \quad \zeta \in L(\alpha; R).$$

Since $2 \in G_0$, for $\zeta \in L(\alpha; R)$ we obtain

$$|N(\zeta)| < B\exp\{\sup_{s \geqslant U(a)} \Phi(s, \zeta)\},\ B = \int_a^\infty \exp\{t - V[2U(t)]\}\, dt, \tag{B.2.35}$$

where $\Phi(s, \zeta) = \mathrm{Re}\,[V(2s) - V(\zeta s)]$.

Let $\zeta = \tau e^{i\lambda}$. Since $V(t)$ is a positive function for $t > 0$, formula (B.2.6) implies that for every $\delta \in (0, 1)$ there is a constant $T_\delta > 0$ such that

$$\Phi(s, \zeta) < H(s, \zeta) := V(2s) - (1 - \delta)V(\tau s)\cos\rho\lambda,\ s \geqslant 0, \tau > T_\delta/a, \tag{B.2.36}$$

uniformly in $\lambda \in [-\alpha, \alpha]$. Since $V(t)$ is a strictly convex function on $(0, \infty)$, for $\tau > 2[(1 - \delta)\cos\rho\alpha]^{-1}$ we have

$$H_s'(s, \zeta) = 2V'(2s) - (1 - \delta)\tau V'(\tau s)\cos\rho\lambda < 0.$$

Hence, putting $R_1 := \max\{2[(1 - \delta)\cos\rho\alpha]^{-1}, T_\delta/a\}$, we see that

$$\sup_{s \geqslant U(a)} H(s, \zeta) = H[U(a), \zeta];\ \zeta = \tau e^{i\lambda};\ \tau > R_1;\ |\lambda| \leqslant \alpha. \tag{B.2.37}$$

Together with (B.2.6), (B.2.35)–(B.2.37), this implies (B.2.34) with $R > \max\{R_2, (\cos\rho\alpha)^{-1/\rho}\}$, $A = B\exp\{V[2U(a)]\}$ ▶

The following statement is important for the sequel.

**Property B.2.9.** *Suppose* $\gamma \geqslant \pi/\rho$, $V \in \mathfrak{B}(\gamma; \rho(r))$, $\alpha \in (0, \pi/2\rho)$. *Then there exist constants* $b > 0$, $R_0 > 0$ *such that*

$$\operatorname{Re} V(Z) \geqslant bV(|Z|) \quad \forall\, Z \in L(\alpha; R_0). \tag{B.2.38}$$

◀ Let $k_\rho = \min\{1, \pi/2\rho - \alpha\}$, and let $\delta \in (0, k_\rho)$. By property 1) of the function $V$ (see (B.2.6)), we have

$$V(Z) = V(|Z|)e^{i\rho\theta[1+\varepsilon(Z)]}, \tag{B.2.39}$$

where $\theta = Arg\, Z$, and $\varepsilon(Z) \to 0$ (as $|Z| \to \infty$) uniformly in $\theta \in [-\alpha, \alpha]$. Therefore, there are real functions $\varepsilon_i(r), i = 1, 2,$ such that $\varepsilon(r) \to 0$ as $r \to \infty$. Consequently, for some $R_0 > 0$ we have

$$1 + \varepsilon_1(r) > 1 - \delta; \quad |\varepsilon_2(r) + \theta| \leqslant \alpha + \delta$$

for $r > R_0$. Combining this with (B.2.39), we obtain

$$\operatorname{Re} V(Z) \geqslant V(|Z|)(1 - \delta)\cos\rho(\alpha + \delta), \quad Z \in L(\alpha; R_0).$$

So, relation (B.2.38) is true ▶

## B.3. Entire functions associated with an analytic proximate orders

In this section we study a certain function associated with an analytic proximate order and similar to the Mittag-Leffler function. Specifically, we consider the entire function

$$E_\rho(z; V) = \sum_{k=0}^{\infty} \frac{z^k}{\Delta(k+1)}, \quad z \in \mathbb{C}, \tag{B.3.1}$$

where $V \in \mathfrak{B}(\gamma; \rho(r))$ (see Subsection B.2.2),

$$\Delta(\lambda) = \Delta_\rho(\lambda; V) = \rho \int_0^\infty t^{\lambda-1} \exp\{-V(t)\}dt, \quad \operatorname{Re}\lambda > 0. \tag{B.3.2}$$

The function $\Delta(\lambda) = \Delta_\rho(\lambda; V)$ is an analog of the Euler gamma-function for the analytic proximate order $\ln V(t)/\ln t$. The function $E_\rho(\cdot; V)$ will play a fundamental role in Section B.4 in the study of the integral transforms of entire functions of a given proximate order.

fundamental role in Section B.4 in the study of the integral transforms of entire functions of a given proximate order.

**B.3.1. An analog of the Hankel formula.** The Hankel formula for the gamma-function looks like this:

$$\frac{1}{\Gamma(\lambda)} = \frac{1}{2\pi i} \int\limits_{\gamma(R;\varphi)} e^{w} w^{-\lambda} dw, \qquad \lambda \in \mathbb{C}. \tag{B.3.3}$$

Here $R > 0$, $\varphi \in (\pi/2, \pi]$, and $\gamma(R;\varphi)$ is the contour formed by the rays $\{w \in \mathbb{C} : |w| \geqslant R;\ \arg w = \pm\varphi\}$ and the circular arc $\{w \in \mathbb{C} : |w| = R;\ |\arg w| \leqslant \varphi\}$, and oriented in the growth direction for $\arg w$. The symbol $w^{-\lambda}$ is understood as the principal value:

$$w^{-\lambda} = e^{-\lambda \ln w} = \exp\{-\lambda(\ln|w| + i\arg w)\}.$$

A similar formula is true for the function $\Delta(\lambda)$ involved in the definition of the function $E_\rho(\cdot; V)$ (see (1)).

**Theorem B.3.1.** *Suppose* $\rho > 0$, $m$ *is the smallest natural number satisfying* $m\rho > 1/2$, $\gamma \geqslant \min\{\pi/\rho, \pi m\}$, $V \in \mathfrak{B}(\gamma; \rho(r))$, $\gamma_\varphi = \{w = a + se^{i\varphi} : s \geqslant 0\}$, *where* $\pi/2 < \varphi < \min\{\pi, \pi\rho m\}$ *and the number* $a > 0$ *is chosen as in Theorem* B.2.7. *Let also* $L_\varphi = \gamma_\varphi \cup \gamma_{-\varphi}$ *be a contour oriented in the growth direction for the argument and placed in such a way that an analytic extension of the function* $t = U(s)$ $(s > 0)$ *inverse to the function* $s = V(t)$ $(t > 0)$ *is defined on* $L_\varphi$. *Let* $\Delta$ *be the function defined by formula* (B.3.2). *Then the function* $1/\Delta(\lambda)$ *extends analytically to* $\mathbb{C}$, *and this extension is given by the formula*

$$\frac{1}{\Delta(\lambda)} = \frac{1}{2\pi i} \int\limits_{L_\varphi} \frac{e^{w}}{[U(w)]^{\lambda}} dw, \qquad \lambda \in \mathbb{C}. \tag{B.3.4}$$

*Formula* (B.3.4) *remains true if the contour* $L_\varphi$ *is replaced by a contour* $\gamma(R;\varphi)$ (see (B.3.3)) *included in the domain of the function* $U$.

The theorem is proved by the integral transformation method. The proofs of (B.3.3) known to the author are based on the formula

$$\Gamma(z)\Gamma(1-z) = \frac{\pi}{\sin \pi z}, \qquad z \in \mathbb{C} \setminus \mathbb{Z}, \tag{B.3.5}$$

the verification of which is not elementary. We prove formula (B.3.4) on the basis of the definition of the function $\Delta(\lambda)$ (see (B.3.2)), by using standard means of complex analysis. As a trivial consequence of (B.3.3) in the case where $V(t) = t$, $\Delta(\lambda) = \Gamma(\lambda)$, we obtain a new proof of formula (B.3.5).

◀ 1. First, we consider the case where $\rho > 1/2$.

By $I(\lambda)$ we denote the integral on the right in formula (B.3.4). This is an entire function. So by the uniqueness theorem it suffices to prove formula (B.3.4) for $\lambda \geqslant 1$. Now we represent the integral $I(\lambda)$ in the following form:

$$I(\lambda) = \frac{1}{2\pi i}[I_{\varphi}(\lambda) - I_{-\varphi}(\lambda)], \qquad I_{\theta}(\lambda) = \int\limits_{\gamma_\theta} e^{w}[U(w)]^{-\lambda} dw, \tag{B.3.6}$$

where $\theta = \pm\varphi$ and $\gamma_\theta$ is the ray occurring in the statement of Theorem B.3.1. We choose numbers $\alpha > 0$, $\delta > 0$ in such a way that

$$\alpha\rho \in (\varphi - \pi/2, \pi/2), \ \rho\delta < \frac{\pi}{2} + \min\{\rho\alpha - \varphi, -\rho\alpha\}. \tag{B.3.7}$$

The asymptotic properties of the function $U$ (see Theorem B.2.6) imply that we can choose the number $a$ involved in the definition of the rays $\gamma_{\pm\varphi}$ so as to ensure the inequalities

$$\begin{gathered} -\delta < \arg U(w) < \frac{\varphi}{\rho} + \delta, \ w \in \gamma_\varphi; \\ -\delta - \frac{\varphi}{\rho} < \arg U(w) < \delta, \ w \in \gamma_{-\varphi}. \end{gathered} \tag{B.3.8}$$

Let

$$\theta(w) = \arg U(w) - \alpha, \ w \in \gamma_\varphi. \tag{B.3.9}$$

By inequalities (B.3.7), (B.3.8), we have:

$$0 < \beta_- \leqslant \theta(w) \leqslant \beta_+ < \frac{\pi}{2\rho}, \qquad \beta_\pm = \frac{\varphi}{\rho} \pm \delta - \alpha. \tag{B.3.10}$$

Therefore, $\theta(w) \in (0, \pi/2\rho)$ for every $w \in \gamma_\varphi$, and from Property B.2.9 we deduce that the integral

$$\int\limits_0^{\infty(\arg t = \theta(w))} t^{\lambda - 1} \exp\{-V(t)\} dt$$

converges absolutely for $w \in \gamma_\varphi$. Applying Property B.2.9 (once again) and the Cauchy integral formula (see (B.3.2)) we obtain,

$$\frac{\Delta(\lambda)}{[U(w)]^\lambda} = \frac{\rho}{[U(w)]^\lambda} \int\limits_0^{\infty(\arg t = \theta(w))} t^{\lambda-1} \exp\{-V(t)\} dt = \rho \int\limits_{T_\alpha} z^{\lambda - 1} \exp\{-V[zU(w)]\} dz,$$

where $T_\alpha = \{z \in \mathbb{C} : \ \arg z = -\alpha)\}$. Therefore (see (B.3.6)),

$$\Delta(\lambda) I_\varphi(\lambda) = \int\limits_{\gamma_\varphi} dw \int\limits_{T_\alpha} \Phi(z, w) dz, \tag{B.3.11}$$

where $\Phi(z,w) = \rho z^{\lambda-1}\exp\{w - V[zU(w)]\}$. The integral $J(w) := \int_{T_\alpha} \Phi(z,w)dz$ converges uniformly in the variable $w$ on every segment $K$ included in $\gamma_\varphi$. Indeed, by Property B.2.9 and inequality (B.3.10), there exist constants $b > 0$, $R > 0$ such that (see (B.3.9))

$$\text{Re } V[zU(w)] = \text{Re } V(r|U(w)|)e^{-i\theta(w)}] > bV[r|U(w)|,$$

where $r > R$, $w \in \gamma_\varphi$. Consequently,

$$\int_R^\infty |\Phi(z,w)|d|z| < \frac{\Delta(\lambda,b)}{|U(w)|^\lambda}, \ \Delta(\lambda,b) = \rho \int_0^\infty t^{\lambda-1}\exp\{-bV(t)\}dt, \qquad \text{(B.3.12)}$$

implying the claim about the integral $J(w)$.

Similarly, the integral

$$\int_{\gamma_\varphi} \Phi(z,w)dw = z^{\lambda-1}N_\varphi(z)$$

(see Theorem B.2.7 and part 1 of its proof for the notation) also converges uniformly in the variable $z$ on every segment $F \subset T_\alpha$.

We estimate the double integral

$$Q_\varphi(\lambda) = \int_{\gamma_\varphi} dw \int_{T_\alpha} |\Phi(z,w)|d|z|. \qquad \text{(B.3.13)}$$

Since, by inequalities (B.3.10) and (B.3.12), the integral

$$\Pi(\theta,\lambda) = \rho\int_0^\infty t^{\lambda-1}\exp\{-\text{Re } V(te^{i\theta})\}dt$$

converges uniformly in the variable $\theta$ on the segment $[\beta_-,\beta_+]$ , it is a continuous function there. Hence, after elementary transformations, we deduce the following inequality (see (B.3.13)):

$$|Q_\varphi(\lambda)| \leqslant \max\{\Pi(\theta,\lambda) : \theta \in [\beta_-\beta_+]\} \cdot \int_{\gamma_\varphi} \left|\frac{e^w}{U(w)^\lambda}\right| d|w|.$$

Thus, for $\lambda \geqslant 1$ the integral $Q_\varphi(\lambda)$ converges and it is legal to change the order of integration. Consequently, by (B.3.6), (B.3.11), in the notation of Theorem B.2.7, for all $\lambda \geqslant 1$ we have

$$\Delta(\lambda)I_\varphi(\lambda) = \int_0^{\infty(\arg z = -\alpha)} F(z)dz, \ F(z) = \rho N(z)z^{\lambda-1}, \qquad \text{(B.3.14)}$$

where $N(z)$ is a function holomorphic in the domain $L(\alpha)\setminus\{1\}$, continuous in $\overline{L(\alpha)}\setminus\{1\}$ and having a pole of order 1 at the point $z=1$. Similar arguments lead to the formula

$$\Delta(\lambda)I_{-\varphi}(\lambda) = \int\limits_{\gamma_{-\varphi}} e^w dw \int\limits_0^{\infty(\arg z=\alpha)} z^{\lambda-1}\exp\{-V[zU(w)]\}dz = \int\limits_0^{\infty(\arg z=\alpha)} F(z)dz \quad \text{(B.3.15)}$$

(see (B.3.11) and (B.3.14)). For $r>1$ we put $C_r=\{z=re^{i\theta}\in\mathbb{C}:\ |\theta|\leqslant\alpha\}$ and denote by $G_r$ the circular sector that has vertex at zero and is bounded by the arc $C_r$ and the rays $\{z\in\mathbb{C}:\ \arg z=\pm\alpha\}$, with the boundary $\partial G_r$ oriented positively. For $\alpha\in(0,\pi/2\rho)$ and $\lambda\geqslant 1$, Theorem B.2.8 implies the identity

$$\lim_{r\to\infty}\int\limits_{C_r} F(z)dz = 0.$$

Combining this with (B.3.6), (B.3.14), (B.3.15), and the Cauchy integral formula, we obtain on the basis of (B.2.24):

$$\Delta(\lambda)I(\lambda) = \lim_{r\to\infty}\frac{1}{2\pi i}\int\limits_{\partial G_r} F(z)dz = \operatorname*{res}_{z=1} N(z) = 1 \qquad \forall\,\lambda\geqslant 1.$$

So, (B.3.4) is true.

2. Suppose $\rho\in(0,1/2]$, $m>1$ is the number satisfying the assumptions of the theorem, and $V_m(\tau)=V(\tau^m)$, $\tau>0$. Then $V_m\in\mathfrak{B}(\gamma/m,\rho_m(r))$ (see the remark to Theorem B.2.6). In the notation of formula (B.3.2), after elementary transformations we obtain

$$\Delta(\lambda)=\Delta_\rho(\lambda;V)=\Delta_{m\rho}(m\lambda;V_m),\quad \operatorname{Re}\lambda>0. \qquad \text{(B.3.16)}$$

Let $\tau=F(s)$, $s>0$, be the inverse of the function $s=V_m(\tau)$. Then $F=U^{1/m}$, where $U$ is the inverse of the function $V$. Recalling that $m\rho>1/2$, we apply formula (B.3.4) to the function $V_m$; then (B.3.16) implies the relation

$$\frac{1}{\Delta(\lambda)}=\frac{1}{\Delta_{m\rho}(m\lambda;V_m)}=\frac{1}{2\pi i}\int\limits_{L_\varphi}\frac{e^w}{[F(w)]^{m\lambda}}dw=\frac{1}{2\pi i}\int\limits_{L_\varphi}\frac{e^w}{[U(w)]^{\lambda}}dw.$$

Here $\lambda\in\mathbb{C}$, $\pi/2<\varphi<\min\{\pi,\pi m\rho\}$ ▶

By formula (B.3.3), $\Delta(\lambda)$ is an analytic function in $\mathbb{C}$ except for at most countably many poles. The following statement (based on Theorem B.2.5) shows that poles may occur only on the negative real semiaxis.

**Proposition B.3.2.** *In the notation of Property* B.2.5, *the following formula is true:*

$$\Delta(\lambda)=\Delta_\rho(\lambda;V)=\frac{\rho}{\lambda}-\rho\sum_{k=0}^{\infty}\frac{(-1)^kB_{k+1}}{\lambda+\beta+k\alpha}+A(\lambda),$$

*where* $B_l = const > 0$ *for all* $l \in \mathbb{N}$, *and* $A(\lambda)$ *is a function analytic in the half-plane* $\operatorname{Re}\lambda > 2\beta$.

◀ Let $P(t) = \rho e^{-V(t)} t^{\lambda-1}$. Then

$$\Delta(\lambda) = \Delta_1(\lambda) + \Delta_2(\lambda), \quad \Delta_1(\lambda) = \int_0^1 P(t)dt, \; \Delta_2(\lambda) = \int_1^\infty P((t)dt,$$

where $\Delta_2(\lambda)$ is an entire function. The function $\Delta_1(\lambda)$ admits the representation

$$\Delta_1(\lambda) = \rho \int_0^1 t^{\lambda-1} dt - \rho I_1(\lambda) + \rho I_2(\lambda),$$

where

$$I_1(\lambda) = \int_0^1 V(t) t^{\lambda-1} dt, \quad I_2(\lambda) = \int_0^1 [e^{-V(t)} - 1 + V(t)] t^{\lambda-1} dt.$$

By Property B.2.5, the function $\Phi(t) = V(t) t^{-\beta}$ has the following uniformly convergent series expansion:

$$\Phi(t) = \sum_{k=0}^{\infty} (-1)^k B_{k+1} t^{k\alpha}, \quad t \in [0,1].$$

Therefore,

$$I_1(\lambda) = \sum_{k=0}^{\infty} \frac{(-1)^k B_{k+1}}{\lambda + \beta + k\alpha}.$$

Using the Taylor formula for the function $e^{-u}$ with remainder term in the Lagrange form and recalling the above series expansion for $\Phi$, we see that

$$I_2(\lambda) = \int_0^1 e^{-cV(t)} V^2(t) t^{\lambda-1} dt, \quad 0 < c = c(t) < 1$$

is a function analytic in the half-plane $\operatorname{Re}\lambda > 2\beta$. Now, we put $A(\lambda) = \Delta_2(\lambda) + I_2(\lambda)$ and observe (see Theorem B.2.4) that $V(0) = 0$. This proves Proposition B.3.2 ▶

**B.3.2. The order and the type of the function $E_\rho(\cdot; V)$.** We study the simplest growth characteristics of the function $E_\rho(\cdot; V)$. Keeping in mind further applications, we consider a more general function of the form (cf. (B.3.1))

$$E_\rho(z; V; \mu) = \sum_{k=0}^{\infty} \frac{z^k}{\Delta(k + \mu + 1)}, \quad z \in \mathbb{C}; \quad \mu \geqslant 0. \tag{B.3.1'}$$

**Theorem B.3.3.** *For every* $\mu \geqslant 0$ *the function* $E_\rho(z;V;\mu)$, *where* $V \in \mathfrak{B}(\gamma;\rho(r))$, *is an entire function of order* $\rho$ *and type* 1, *and of proximate order* $\rho(r) = (\ln r)^{-1} \ln V(r)$.

For the proof we formulate the following auxiliary theorem.

**Theorem B.3.4.** *Let* $\Phi(x)$ *be a positive two times continuously differentiable function satisfying the following conditions:*

1) $x^2\Phi''(x) \to +\infty$ *and* $\Phi'(x) \to +\infty$ *as* $x \to +\infty$;

2) $\Phi''(\zeta) \sim \Phi''(x)$ *as* $x \to +\infty$, $|\zeta - x| \leqslant A[\Phi''(x)]^{-1/2}$ ( $A > 0$ *may be fixed arbitrarily*).

*Then for* $\lambda$ *sufficiently large the function* $\Psi(x,\lambda) = x\lambda - \Phi(x)$ *has a unique maximum at the point* $x = c(\lambda)$; *as* $\lambda \to +\infty$, *the following asymptotic formula is true:*

$$\int_a^\infty \exp\{x\lambda - \Phi(x)\}\,dx \sim \sqrt{\frac{2\pi}{\Phi''[c(\lambda)]}} \exp\{\lambda c(\lambda) - \Phi[c(\lambda)]\}.$$

**Proof of Theorem B.3.3.** By (B.3.2), we have

$$\Delta(\lambda+\mu+1) = \rho\int_0^\infty t^{\lambda+\mu}\exp\{-V(t)\}dt = \rho\int_{-\infty}^\infty \exp\{\Psi(x,\lambda)\}\,dx,$$

where

$$\Psi(x,\lambda) = x(\lambda+\mu) - V(e^x), \quad |x| < \infty; \quad \lambda > 0.$$

We verify that the function $\Phi(x) = V(e^x) - \mu x$ satisfies the conditions of Theorem B.3.4. By (B.2.10), (B.2.11), we have

$$\Phi''(\ln t) = tV'(t) + t^2V''(t) \sim \rho^2 V(t), \quad t \to \infty,$$

because (see the proof of Lemma B.2.1)

$$\lim_{t\to\infty} [V(t)]^{-1} t^2 V''(t) = \rho(\rho-1).$$

This implies that conditions 1), 2) of Theorem B.3.4 are indeed satisfied. By (B.2.10), we find

$$\Psi'_x(x,\lambda) = [\lambda+\mu-\rho V(e^x)][1+\varepsilon(x)], \quad \varepsilon(x) \to 0, \quad x \to \infty.$$

Consequently, if $x = c(\lambda)$ is a point of maximum for the function $\Psi(x,\lambda)$ (this function is concave in the variable $x$), then

$$t = \exp\{c(\lambda)\} \sim U[(\lambda+\mu)/\rho], \quad \lambda \to \infty; \quad U = V^{-1}.$$

Now, by Theorem B.3.4 and (B.2.21), we conclude that

$$\sqrt[\lambda]{\Delta(\lambda+\mu+1)} \sim U(\lambda)(e\rho)^{-1/\rho}, \quad \lambda \to \infty.$$

It remains to refer to relation (B.1.5) ▶

**B.3.3. An integral representation of the function $E_\rho(\cdot; V)$.** An integral representation of the Mittag-Leffler function for $\rho > 1/2$ is a direct consequence of (B.3.3). Similarly, the analog (B.3.4) of the Hankel formula for the function $\Delta$ (see (B.3.2)) allows us to obtain an integral representation for the function $E_\rho(z; V)$ and for the more general function $E_\rho(z; V; \mu)$ (see $(B.3.1')$).

**Theorem B.3.5.** *Under the assumptions and in the notation of Theorem* B.3.1, *let* $s_a = \inf\{|U(w)| : w \in L_\varphi\}$. *Then for every* $z$ *with* $|z| < s_a$ *we have*

$$E_\rho(z; V; \mu) = \frac{1}{2\pi i} \int\limits_{L_\varphi} \frac{e^w}{(U(w) - z)[U(w)]^\mu} dw; \tag{B.3.17}$$

*moreover,* $s_a \to \infty$ *as* $a \to \infty$. *Formula* (B.3.17) *remains true if the contour* $L_\varphi$ *is replaced by the contour* $\gamma(R; \varphi)$ (see (B.3.3)) *included in the domain of the function* $U$. *In this case* $|z| < d_R := \inf\{|U(w)| : w \in \gamma(R; \varphi)\}$.

◀ By formulas (B.3.1) and (B.3.4), we have

$$E_\rho(z; V) = \frac{1}{2\pi i} \sum_{k=0}^{\infty} \int\limits_{L_\varphi} \frac{e^w z^k}{[U(w)]^{k+\mu+1}} dw. \tag{B.3.18}$$

Theorem B.2.6 shows that $U(s) = s^{\rho(s)}$, $s > 0$, where $\rho(s)$ is an analytic proximate order and $\rho(s) \to 1/\rho$ as $s \to \infty$. Therefore, $|U(w)| \sim U(|w|)$ as $|w| \to \infty$, and this equivalence is uniform in $\arg w \in [-\varphi_0, \varphi_0]$, where $\varphi < \varphi_0 < \min\{\pi, \pi\rho\}$. This implies the required asymptotic property of the quantity $s_a$. The series

$$\sum_{k=0}^{\infty} \frac{e^w z^k}{[U(w)]^{k+\mu+1}}$$

converges uniformly in the variable $w \in L_\varphi$ for fixed $z$ with $|z| < s_a$. Therefore, it admits termwise integration over the contour $L_\varphi$. Consequently, by formula (B.3.18), for $\varphi > \pi/2$ we obtain

$$E_\rho(z; V) = \frac{1}{2\pi i} \int\limits_{L_\varphi} \sum_{k=0}^{\infty} \frac{e^w z^k}{[U(w)]^{k+\mu+1}} dw,$$

which proves formula (B.3.17).

Since $d_R \to \infty$ as $R \to \infty$, the Cauchy integral formula implies the last statement of the theorem ▶

**B.3.4. Asymptotic properties of the function $E_\rho(\cdot;V)$.** We study the asymptotics of the function $E_\rho(\cdot;V)$ using Theorem B.3.5 and the notation introduced in it.

1. Let $\rho > 1/2$, let $\delta = \delta(\varphi) > 0$ be a sufficiently small number satisfying

$$\pi/2 < \varphi \pm \rho\delta < \min\{\pi, \pi\rho\} \tag{B.3.19}$$

(see Theorem B.3.1), and let $R_0$ be a number such that for $R \geqslant R_0$ the contour $\gamma(R;\varphi)$ is included in the domain of the function $U$, and the inequalities

$$|\arg U(se^{i\varphi}) - \varphi/\rho| < \delta, \ |\arg U(se^{-i\varphi}) + \varphi/\rho| < \delta \tag{B.3.20}$$

are true for all $s \geqslant R_0$. We choose $\varepsilon > 0$ so that $\rho\varepsilon < \varphi - \rho\delta - \pi/2$. Let $|z| > r_0$, where

$$r_0 = \max\{U(R_0e^{i\theta}) : |\theta| \leqslant \varphi\}, \ |\arg z| \leqslant \frac{\pi}{2\rho} + \varepsilon. \tag{B.3.21}$$

Then $|\arg z| < \min\{\pi, \pi/\rho\}$. There exists a number $R > R_0$ such that $|z| < d_R$ (see Theorem B.3.2). The oriented contour $\gamma(R;\varphi)$ determines an orientation of the Jordan contour $\Gamma(R;\varphi) := \{U(w) \in \mathbb{C} : w \in \gamma(R;\varphi)\}$; the latter bounds a positively oriented domain $D_R^+ \ni z$. Then by Theorem B.3.2 we have:

$$E_\rho(z;V) = \frac{1}{2\pi i}\int\limits_{\gamma(R;\varphi)} \frac{e^w}{U(w)-z}\,dw. \tag{B.3.22}$$

Now, using standard complex analysis tricks, from formula (B.3.21) and the asymptotic properties of the function $U$ (see Theorem B.2.6) we deduce the formula

$$E_\rho(z;V) = \frac{1}{2\pi i}\int\limits_{\gamma(R_0;\varphi)} \frac{e^w}{U(w)-z}\,dw + \operatorname*{res}_{w=V(z)} \frac{e^w}{U(w)-z}.$$

Assuming (as in Theorem B.3.1) that $V \in \mathfrak{B}(\gamma;\rho(r))$ with $\gamma \geqslant \min\{\pi/\rho, \pi\}$, by Theorem B.2.6 we obtain

$$E_\rho(z;V) = \frac{1}{2\pi i}\int\limits_{\gamma(R_0;\varphi)} \frac{e^w}{U(w)-z}\,dw + V'(z)e^{V(z)}.$$

By (B.3.19)-(B.3.21), there exists a constant $m > 0$ such that $|U(w)/z - 1| \geqslant m$ for all $w \in \Gamma(R_0;\varphi)$ and all $z$ satisfying inequalities (B.3.21) and the additional condition $|z| \geqslant r$, where $r > r_0$. Now, by (B.3.19), we arrive at the following asymptotic formula as $z \to \infty$:

$$E_\rho(z;V) = V'(z)e^{V(z)} + O\left(\frac{1}{|z|}\right), \quad |\arg z| \leqslant \frac{\pi}{2\rho} + \varepsilon. \tag{B.3.23}$$

Let $0 < \rho\varepsilon < \rho\pi - \pi/2$. We choose the parameters $\varepsilon, \delta$ so as to ensure condition (B.3.19) and the inequalities $0 < \varphi + \rho\delta - \pi/2 < \rho\varepsilon$. Let

$$|z| > r_0, \quad \frac{\pi}{2\rho} + \varepsilon \leqslant |\arg z| \leqslant \pi$$

(see (B.3.21)). In the same way as in the proof of (B.3.23), we use formula (B.3.22) to obtain the asymptotic relation

$$E_\rho(z;V) = \frac{1}{2\pi i} \int\limits_{\gamma(R_0;\varphi)} \frac{e^w}{U(w)-z} dw = O\left(\frac{1}{|z|}\right). \tag{B.3.24}$$

Thus, the following uniform asymptotic estimate in $\mathbb{C}$ for the function $E_\rho(z;V)$ as $|z| \to \infty$ is true:

$$E_\rho(z;V) = \begin{cases} V'(z)e^{V(z)} + O(1/|z|) & |\arg z| \leqslant \dfrac{\pi}{2\rho} + \varepsilon, \\ O(1/|z|), & \dfrac{\pi}{2\rho} + \varepsilon \leqslant |\arg z| \leqslant \pi, \end{cases} \tag{B.3.25}$$

where $0 < \rho\varepsilon < \pi \min\{1/2, \rho - 1/2\}$.

Estimate (B.3.24) can be refined. For this, we transform the integral in formula (B.3.24) with the aid of Theorem B.3.1 and the elementary formula

$$\frac{1}{q-1} = \frac{q^n}{q-1} - \sum_{k=0}^{n-1} q^k, \quad q \in \mathbb{C} \setminus \{1\},$$

obtaining

$$E_\rho(z;V) = \frac{1}{2\pi i} \int\limits_{\gamma(R_0;\varphi)} \frac{e^w U^n(w)}{z^n(U(w)-z)} dw - \sum_{k=1}^{n-1} \frac{1}{\Delta(-k)z^{k+1}},$$

where $n \in \mathbb{N} \setminus \{1\}$. This implies the following uniform asymptotic formula as $|z| \to \infty$:

$$E_\rho(z;V) = O\left(\frac{1}{|z|^{n+1}}\right) - \sum_{k=1}^{n-1} \frac{1}{\Delta(-k)z^{k+1}}, \quad \frac{\pi}{2\rho} + \varepsilon \leqslant |\arg z| \leqslant \pi. \tag{B.3.26}$$

where $\varepsilon + \pi/2\rho \leqslant |\arg z| \leqslant \pi$. The asymptotics of the function $E_\rho(\cdot;V;\mu)$ (see (B.3.1$'$), Subsection B.3.2) is analysed much in the same way. The following generalization of formula (B.3.25) is true:

$$E_\rho(z;V;\mu) = \begin{cases} z^{-\mu}V'(z)e^{V(z)} + O\left(\dfrac{1}{|z|}\right), & |\arg z| \leqslant \dfrac{\pi}{2\rho} + \varepsilon, \\ O\left(\dfrac{1}{|z|}\right), & \dfrac{\pi}{2\rho} + \varepsilon \leqslant |\arg z| \leqslant \pi, \end{cases} \tag{B.3.27}$$

where the symbol $z^{-\mu}$ is understood as the principal value, and

$$0 < \rho\varepsilon < \pi \min\{1/2, \rho - 1/2\}.$$

2. Now, let $\rho \leqslant 1/2$, and let $m$ be the smallest *odd* natural number such that $m\rho > 1/2$. We suppose that $\gamma \geqslant \min\{\pi/\rho, \pi m\}$ and $V \in \mathfrak{B}(\gamma; \rho(r))$ (compare with the conditions of Theorem B.3.1). In this case the function $V$ extends analytically to the angle of opening greater than $2\pi$ on the Riemann surface of the logarithm. Later on we shall need the following analog of Dzhrbashyan's formula.

**Proposition B.3.6.** *In the notation of formulas* (B.3.1), (B.3.2) and (B.3.1′) *the following identity is true:*

$$mE_\rho(z; V) = \sum_{n=-\nu}^{\nu} E_{m\rho}\left(z^{1/m}\exp\{i\pi n/m\}; V_m; \mu\right), \tag{B.3.28}$$

*where* $z \in \mathbb{C}$, $\nu = (m-1)/2$, $\mu = m - 1$, $z = re^{i\theta}$, $z^{1/m} = r^{1/m}\exp\{i\theta/m\}$, $|\theta| \leqslant \pi m$, *and* $V_m(\tau) := V(\tau^m)$ *for* $\tau \geqslant 0$.

◀ We observe that $V_m \in \mathfrak{B}(\gamma/m; \rho_m(r))$ (see the remark to Theorem B.2.6). Changing the variable $t = \tau^m$ in the integral (B.3.2), we find

$$\Delta(\lambda) = m\rho \int_0^\infty \tau^{m\lambda - 1}\exp\{-V_m(\tau)\}d\tau = \Delta(m\lambda; V_m), \quad \operatorname{Re}\lambda > 0.$$

Therefore, we have the representations

$$E_\rho(w^m; V) = \sum_{s=0}^{\infty} \frac{w^{sm}}{\Delta(m(s+1); V_m)}, \quad w \in \mathbb{C}. \tag{B.3.29}$$

We use the elementary identity

$$S_k := \sum_{n=-\nu}^{\nu} \exp\{i2\pi kn/m\} = \begin{cases} m, & k \equiv 0 \pmod{m} \\ 0, & k \not\equiv 0 \pmod{m}. \end{cases}$$

By (B.3.29), this implies that

$$E_\rho(w^m; V) = \frac{1}{m}\sum_{k=0}^{\infty} \frac{S_k w^k}{\Delta(k+1+\mu)} = \frac{1}{m}\sum_{n=-\nu}^{\nu}\sum_{k=0}^{\infty} \frac{w^k e^{i2\pi nk/m}}{\Delta(k+1+\mu)}, \quad w \in \mathbb{C}.$$

This means that formula (B.3.28) is true if the summands in it are understood as functions on the Riemann surface of the function $z^{1/m}$ ▶

**Remark.** It suffices to consider the summands in formula (B.3.28) only on one sheat of the Riemann surface of the function $z^{1/m}$, restricting ourselves, e.g., to the values of the parameter $\theta \in [-\pi, \pi]$.

Denoting $N_m = \{0, \pm 1, ..., \pm \nu\}$, for $n \in N_m$ we introduce the set $I_n(R) = \{\theta \in [-\pi, \pi] : \ |\theta + 2\pi n| \leqslant R\}$. If $R = \pi/2\rho + \varepsilon$, where

$$0 < \varepsilon\rho < \pi \min\{1/2, m\rho - 1/2\}, \tag{B.3.30}$$

then the system of segments $\{I_n(R) : n \in N_m\}$ covers the segment $[-\pi, \pi]$, and $I_n(R) \neq \varnothing$ for all $n \in N_m$, $I_0(R) = [-\pi, \pi]$. For every $r > 0$, $\theta \in I_n(R)$, and $n \in N_m$, the pair $(r, \theta + 2\pi n)$ lies in $L(\gamma)$, i.e., in the domain of $V$ (see (B.3.30)). In the notation of Proposition B.3.6, from formulas (B.3.27) and (B.3.28) we deduce the following uniform asymptotic expansion in $\mathbb{C}$ as $|z| \to \infty$:

$$E_\rho(re^{i\theta}; V) = \sum_{n=-\nu}^{\nu} F_n(z_n), \tag{B.3.31}$$

where $z_n = r^{1/m} \exp\{i(\theta + 2\pi n/m)\}$ and

$$F_n = F_n(r, \theta) = \begin{cases} \dfrac{e^{V_m(z_n)} V_m'(z_n)}{m z_n^{m-1}} + O\left(\dfrac{1}{r^{1/m}}\right), & \theta \in I_n(R), \\ O\left(\dfrac{1}{r^{1/m}}\right), & \theta \in [-\pi, \pi] \setminus I_n(R). \end{cases}$$

Here we consider the analytic extension of the function $V_m$ to the angle

$$L(\alpha) = \{z \in \mathbb{C} : \ |\arg z| < \alpha\}, \quad \alpha m = \min\{\pi m, \pi/\rho\} \leqslant \gamma.$$

By the uniqueness of an analytic extension, we have

$$V_m(z) = V(|z|^m, m \arg z), \ V_m'(z) = m z^{m-1} V'(|z|^m, m \arg z),$$

where $z \in L(\alpha)$; consequently, we arrive at the following asymptotic formula in $\mathbb{C}$ as $|z| \to \infty$ :

$$F_n = \begin{cases} \exp\{V(r, \theta + 2\pi n)\} V'(r, \theta + 2\pi n) + S(r), & \theta \in I_n(R), \\ S(r), \quad \theta \in [-\pi, \pi] \setminus I_n(R), \end{cases} \tag{B.3.32}$$

where $S(r) = O(1/r^{1/m})$. We find the principal term of the asymptotic expansion of the function $E_\rho(\cdot; V)$ (see (B.3.31), (B.3.32)) on the segment $[-\pi + \delta, \pi - \delta]$, where $\delta > 0$ is fixed but may be arbitrarily small. Let $n \in N_m \setminus \{0\}$. From formulas (B.2.6), (B.2.10) we deduce the relations

$$\Phi_n(\theta) := \lim_{r \to \infty} \frac{\text{Re}\ [V(r, \theta + 2\pi n) - V(r, \theta)]}{V(r)} = \cos \rho(\theta + 2\pi n) - \cos \rho\theta,$$

$$\lim_{r \to \infty} |[V'(r, \theta)]^{-1} V'(r, \theta + 2\pi n)| = 1,$$

where the convergence to the limits is uniform on the segment $[-\pi+\delta, \pi-\delta]$. Since for sufficiently small $\varepsilon = \varepsilon(\delta)$ we have

$$\sup\{\Phi_n(\theta) : \theta \in [-\pi+\delta, \pi-\delta]\} < 0, \quad \forall\, n \in N_m \setminus \{0\},$$

relations (B.3.31), (B.3.32) imply the uniform asymptotic formula (as $z \to \infty$)

$$E_\rho(re^{i\theta}) = e^{V(r,\theta)}V'(r,\theta) + o(1), \quad \theta \in [-\pi+\delta, \pi-\delta]. \tag{B.3.33}$$

**B.3.5. The indicator of the function $E_\rho(\cdot, V)$.** Asymptotic properties of the function $E_\rho(\cdot, V)$ described in Subsection B.3.4 allow us to find the nonnegative cutoff by zero $h_E^+ = \max\{0, h_E\}$ of the indicator $h_E$ of this function, and in some cases the indicator $h_E$ itself, if $h_E$ is related to the proximate order $\rho(r) = \ln V(r)/\ln r$.

**Theorem B.3.7.** *In the notation introduced above and in formula* (B.1.8), *the following statements are true.*

1. *For every $\rho > 0$ we have*

$$h_E^+(\theta) = \Phi_\rho(e^{i\theta}), \quad \theta \in [-\pi, \pi]. \tag{B.3.34}$$

2. *If $\rho > 1$, then $h_E(\theta) = h_E^+(\theta)$, $\theta \in \mathbb{R}$.*
3. *If $\rho \leqslant 1/2$, then $h_E(\theta) = \cos\rho\theta$, $|\theta| \leqslant \pi$.*
4. *If $1/2 < \rho \leqslant 1$ and there exists $k \in \mathbb{N}$ such that* (see (B.3.2))

$$\Delta(-k) \neq 0, \tag{B.3.35}$$

*then $h_E(\theta) = h_E^+(\theta)$, $\theta \in \mathbb{R}$.*

◀ Statements 1 and 4 are direct consequences of the asymptotic formulas (B.3.25), (B.3.26), (B.3.33) and of Property (B.2.0) for the analytic proximate order $\rho(r)$ under consideration (see Theorem B.2.4). Proposition B.2.5 shows that condition (B.3.35) is fullfilled for an appropriate choice of the function $V \in \mathfrak{B}(\gamma; \rho(r))$. Statement 3 is a consequence of statement 1 and Property 1.9.9.

Next, statement 1 implies that $h_E(\theta) \leqslant 0$, $\gamma_\rho \leqslant |\theta| \leqslant \pi$, where $\gamma_\rho = \min\{\pi, \pi/2\rho\}$. Since $h_E$ is a $2\pi$-periodic function, we conclude that the length $l$ of every component of the set $\{\theta \in \mathbb{R} : h_E(\theta) \leqslant 0\}$ is equal to $2\pi(1-1/\rho)$. Therefore, $l > \pi/\rho$ if $\rho > 1$. Now, statement 2 follows from Property 1.9.8 ▶

## B.4. Entire functions of a given proximate order and with nonnegative indicator, and their generalized Laplace-Borel transforms

On the basis of the results exposed in the preceding sections, we turn to construction of an analog of Theorem 3.2.3 for entire functions of a given proximate order and with nonnegative indicator.

**B.4.1. Generalized Borel transformation and its integral representation.** For $\rho > 0$, let $V \in \mathfrak{B}(\gamma; \rho(r))$, and let $\mathfrak{M}_\rho(V) = \{f\}$ be the class of entire functions in $\mathbb{C}$ such that for some $M = M(f)$, $m = m(f) > 0$, we have

$$|f(z)| \leqslant M \exp\{mV(|z|)\} \quad \forall\, z \in \mathbb{C}. \tag{B.4.1}$$

We consider the following generalized Borel transformation (see (1.10.7), (1.10.8)).

**Definition B.4.1.** Let

$$\Delta(k+1) = \rho \int\limits_0^\infty t^k \exp\{-V(t)\}\, dt, \quad f(z) = \sum_{k=0}^\infty \frac{a_k}{\Delta(k+1)} z^k \tag{B.4.2}$$

(see (B.3.2)), and let $f$ be an entire function of class $\mathfrak{M}_\rho(V)$. A funciton defined by the formula

$$F(p) = \sum_{k=0}^\infty \frac{a_k}{p^{k+1}} \tag{B.4.3}$$

is called the *Borel $V$-transform* of $f$.

The following statement is an analog of Proposition 1.10.4.

**Proposition B.4.2.** *Let $H_0(\infty)$ be the class of functions holomorphic in a neighborhood of $\infty \in \overline{\mathbb{C}}$ and vanishing at the point $\infty$. Then the operator*

$$\nu : \mathfrak{M}_\rho(V) \to H_0(\infty); \quad \nu(f) = F$$

(see (B.4.2), (B.4.3)) *is an isomorphism.*

◀ We show that the entire function $f$ (see (B.4.2)) satisfies condition (B.4.1) if and only if the series $F$ (see (B.4.3)) converges for all $|p| > m^{1/\rho}$. In the course of the proof of Theorem B.3.3, it was shown for $\mu = 0$ that

$$\sqrt[\lambda]{\Delta(\lambda+1)} \sim U(\lambda)(e\rho)^{-1/\rho}, \quad \lambda \to \infty,$$

where $t = U(s)$ $(s \geqslant 0)$ is the inverse to the function $s = V(t)$, $t \geqslant 0$. Therefore, the following two inequalities are equivalent:

$$\varlimsup_{k\to\infty} U(k) \sqrt[k]{|c_k|/\Delta(k+1)} \leqslant (me\rho)^{1/\rho}; \quad \varlimsup_{k\to\infty} \sqrt[k]{|c_k|} \leqslant m^{1/\rho}.$$

It only remains to use formula (B.1.5) and the Cauchy formula for the radius of divergence of the series (B.4.3), and also to observe that $\nu$ is a linear operator ▶

Let $P_\rho^+$ be the class of finite nonnegative $2\pi$-periodic trigonometrically $\rho$-convex functions, and let $h \in P_\rho^+$. In accordance with Theorem 3.1.15, the set

$$X_h = \{z \in \mathbb{C} : \Phi_\rho(ze^{i\theta}) \leqslant h(\theta)\ \forall\, \theta \in \mathbb{R}\} \tag{B.4.4}$$

(where

$$\Phi_\rho(z) = \begin{cases} |z|^\rho \cos(\rho \arg z), & |\arg z| \leqslant \gamma_\rho; \\ 0, \quad \rho \geqslant 1/2; & \gamma_\rho \leqslant |\arg z| \leqslant \pi, \end{cases} \tag{B.4.5}$$

$\gamma_\rho = \min\{\pi, \pi/2\rho\}$) is a unique $\rho$-convex compact set having $h$ as its $\rho$-support function (see (3.1.6), (3.1.9)). Using the class $P_\rho^+$, we may represent the set $\mathfrak{M}_\rho(V)$ of entire functions described above as follows:

$$\mathfrak{M}_\rho(V) = \bigcup P_\rho(V; h)|h \in P_\rho^+, \tag{B.4.6}$$

where $P_\rho(V, h) = \{f\}$ is the class of entire functions satisfying the inequality

$$|f(re^{i\theta})| \leqslant A \exp\{(h(\theta) + \varepsilon)V(r)\}, \quad re^{i\theta} \in \mathbb{C}, \tag{B.4.7}$$

for every $\varepsilon > 0$ and for some $A = A(f, \varepsilon) > 0$. The Borel $V$-transforms of entire functions of class $P_\rho(V, h)$ admit a specific integral representation.

**Theorem B.4.3.** *Suppose $\rho > 0$, $h \in P_\rho^+$, $V \in \mathfrak{B}(\gamma; \rho(r))$, where $\gamma \geqslant \gamma_\rho$* (see (B.4.5) ). *We denote $D_\rho(\theta; b) = \{p \in \mathbb{C} : \Phi_\rho(pe^{i\theta}) > b\}$ for $\theta \in \mathbb{R}$, $b \geqslant 0$. Let $f \in P_\rho(V, h)$. Then for every $\theta \in \mathbb{R}$ the function*

$$F_\theta(p) = \rho \int_0^{\infty(\arg z = \theta)} \exp\{-V(pz)\} f(z)\, dz \tag{B.4.8}$$

*is holomorphic in the domain $D_\rho(\theta, h(\theta))$. In formula* (B.4.8) *we mean an analytic extension of $V$ to the angle $L(\gamma_\rho) \subset \mathbb{C}$, and the system of functions*

$$\{F_\theta(p) : p \in D_\rho(\theta, h(\theta))\}, \quad \theta \in \mathbb{R}$$

*determines the analytic extension of the Borel $V$-transform of $f$ to the exterior of the $\rho$-convex compact set $X_h$* (B.4.4).

◄ Let $p = re^{i\varphi}$, $z = te^{i\theta}$. For every $\varepsilon > 0$ there exists a number $t_0 = t_0(\varepsilon) > 0$ such that (see (B.2.0), (B.2.13))

$$\operatorname{Re} V(pz) = \operatorname{Re} V(rte^{i(\varphi+\theta)}) > r^\rho V(t)[\cos \rho(\varphi + \theta) - \varepsilon] \quad \forall\, t > t_0,\ p \in K,$$

where $K$ is an arbitrarily fixed compact subset of the domain $D_\rho(\theta, h(\theta))$. By inequality (B.4.7), the integral $F_\theta$ converges uniformly on $K$. Consequently, the function $F_\theta$ is holomorphic on $D_\rho(\theta, h(\theta))$.

As usual (see the proof of Proposition 1.10.5), expanding the function $f$ in the series (B.4.2) and integrating termwise we prove that

$$F_\theta(p) \equiv F(p), \quad |p| > R \tag{B.4.9}$$

for some $R$. So, $F_\theta(p)$ is the analytic extension of the function $F(p)$ to $D_\rho(\theta; h(\theta))$. Since $X_h$ is a star-like compact set, the complement

$$\mathbb{C} \setminus X_h = \bigcup D_\rho(\theta, h(\theta)) | \theta \in \mathbb{R}$$

is a simply connected domain on the Riemann sphere ($F(\infty) = 0$). Now, Theorem B.4.3 follows from (B.4.9) and the monodromy theorem ▶

**B.4.2. Integral representation of an entire function of a given proximate order, and an analog of the Polya theorem.**

There is also a reciprocal integral formula relating a function $f \in \mathfrak{M}_\rho(V)$ and its Borel $V$-transform $F$. Specifically,

$$f(z) = \frac{1}{2\pi i} \int\limits_{\Gamma} E_\rho(zp; V) F(p)\, dp; \quad E_\rho(w; V) = \sum_{k=0}^{\infty} \frac{w^k}{\Delta_k}, \tag{B.4.10}$$

where $\Gamma$ is a positively oriented Jordan contour surrounding the singularities of $F$ (see Definition B.4.1, $\{\Delta(k) : k = 0, 1, 2, ...\}$ is the sequence defined by formula (B.4.2), $V \in \mathfrak{B}(\gamma; \rho(r))$, and the number $\gamma$ satisfies the conditions of Theorem B.3.1. The proof of this formula is entirely similar to the classical case, where $V(t) = t$, $\rho = 1$, $E_\rho(w; V) = e^w$ (see the proof of Proposition 1.10.5). In order to establish an analog of the Polya theorem for the entire functions of a given proximate order and with nonnegative indicator, we need the following fact.

**Theorem B.4.4.** *Let $K$ be a $\rho$-convex compact set and $h := h_K$ its $\rho$-support function* (see Definition 3.1.11). *Suppose $V \in \mathfrak{B}(\gamma; \rho(r))$ and $f \in \mathfrak{M}_\rho(V)$. If the Borel $V$-transform of $f$ extends analytically outside $K$, then $f \in P_\rho(V; h)$.*

◀ For $\varepsilon > 0$ let $K_\varepsilon = X_{h+\varepsilon}$ be the $\rho$-convex compact set associated with the function $h+\varepsilon$ by formula (B.4.4) (see Theorem 3.1.15). The system of sets $\{K_\varepsilon : \varepsilon > 0\}$ has the following properties: a) $\bigcap K_\varepsilon | \varepsilon > 0 = K$; b) $K_{\varepsilon_1} \subset K_{\varepsilon_2}$ if $\varepsilon_1 < \varepsilon_2$; c) $K \subset K_\varepsilon$, $\varepsilon > 0$; d) the $\rho$-support function of the compact set $K_\varepsilon$ is $h + \varepsilon$. For any $\delta > 0$ there exist $\varepsilon > 0$ and $\gamma > 0$ such that

$$\varepsilon + \gamma \sup\{|p|^\rho,\ p \in K_\varepsilon\} < \delta. \tag{B.4.11}$$

By Theorem B.3.7, the nonnegative part $h_E^+(\theta)$ of the generalized indicator of the function $E_\rho(\cdot; V)$ is equal to $\Phi_\rho(e^{i\theta})$ (see (B.4.5)). Therefore, for every $\gamma > 0$ there exists a constant $A_\gamma$ such that

$$|E_\rho(w; V)| < A_\gamma \exp\{V(|w|)[\Phi_\rho(e^{i \arg w}) + \gamma]\}, \quad w \in \mathbb{C}.$$

Since $0 \notin \partial K_\varepsilon$, Property B.1.7 implies the following asymptotic estimate as $r \to \infty : V(r|p|) = V(r)|p|^\rho(1 + o(1))$ uniformly in $p \in \partial K_\varepsilon$. Consequently, for some $R > 0$ and all $p \in \partial K_\varepsilon$ we have the following inequality:

$$|E_\rho(re^{i\theta}p;V)| < A_\gamma \exp\{V(|r|)[\Phi_\rho(pe^{i\theta}) + \gamma|p|^\rho](1+o(1))\},\ r > R,\ \theta \in \mathbb{R}. \tag{B.4.12}$$

Together with (B.4.11), this implies the theorem, because, by assumption, in formula (B.4.10) we may choose the boundary $\partial K_\varepsilon$ for the role of $\Gamma$ ▶

Now we formulate a simple analog of Theorem 3.2.3 for entire functions of a given proximate order and with nonnegative generalized indicator.

**Theorem B.4.5.** *Suppose* $\rho > 0, V \in \mathfrak{B}(\gamma;\rho(r))$, $h \in P_\rho^+$, *and* $f$ *is an entire function of the class* $\mathfrak{M}_\rho(V)$ (see (B.4.1)) *of proximate order* $(\ln r)^{-1}\ln V(r)$ *and with indicator* $h_f$. *Let* $h(\cdot) = \max\{0, h_f(\cdot)\}$, *and let* $K$ *be the conjugate diagram of* $f$, *i.e., the smallest* $\rho$-*convex compact set such that the Borel* $V$-*transform of* $f$ (see Definition B.4.1) *extends analytically to the exterior of* $K$ . *Then*

1) *the plane indicator diagram* $X_h$ *of* $f$ (see Definition B.1.14) *coincides with* $K$;

2) *there exists an analytic functional* $\mu$ *such that for its Laplace* $V$-*transform we have*

$$\mu_p[E_\rho(zp;V)] = f(z),\ z \in \mathbb{C}, \tag{B.4.13}$$

*and* $X_h$ *is a* $\rho$-*convex support of* $\mu$.

◀ Applying Theorem B.4.3 to the function $f$, we see that $K \subset X_h$. Let $h_K$ be the $\rho$-support function of the compact set $K$. Then $h_K \in P_\rho^+$ and, by Theorem B.4.4, $f \in P_\rho(V;h_K)$. Since $h_K \geqslant 0$, we obtain $h \leqslant h_K$. Consequently, $X_h \subset K$. Thus, $X_h = K$.

Let $H$ be the space of all entire functions on $\mathbb{C}$. We endow $H$ with the topology of uniform convergence on compact sets. Formula (B.4.10) implies that the analytic functional

$$\mu(\varphi) = \frac{1}{2\pi i}\int_\Gamma \varphi(p)F(p)dp, \quad \varphi \in H, \tag{B.4.14}$$

satisfies condition (B.4.13). We consider the collection $\{K_\varepsilon : \varepsilon > 0\}$ of $\rho$-convex compact sets occurring in the proof of Theorem B.4.4; for $K$ we take the conjugate diagram of $f$ (then, by statement 1) above and Theorem 3.1.15, the $\rho$-support function of $K$ is $h$). Then $K \subset \operatorname{int}K_\varepsilon\ \forall\, \varepsilon > 0$. As the contour $\Gamma$ in (B.4.14) we take the boundary $\partial K_\varepsilon$ of the compact set $\omega := K_\varepsilon$. By (B.4.13), there is a constant $A_\omega$ such that $|\mu(\varphi)| \leqslant A_\omega \sup\{|\varphi(p)| : p \in \omega\}$. This means that $K$ is a $\rho$-convex determining set for the functional $\mu$ (see Definition 3.2.2).

Let $L \subset K$ be a $\rho$-convex compact set determining for $\mu$, and let $h_L$ be the $\rho$-support function of $L$. Recalling formula (B.4.13) and estimating the function $E_\rho(re^{i\theta}p;V)$ asymptotically for $p \in L$ (much as before; see (B.4.12)), we see that $f \in P_\rho(V;h_L)$. Consequently, $h \leqslant h_L$ and $X_h \subset L \subset K = X_h$, i.e., $L = K = X_h$ is a $\rho$-convex support of the functional $\mu$ ▶

## B.5. Analytic proximate orders generated by entire or meromorphic functions

In this section we consider a special class of analytic proximate orders generated by the logarithms of entire or meromorphic functions. An entire function $g$ of proximate order $\rho(r) \to \rho > 1/2$ is called a *generator* of an analytic proximate order $\rho_0(r)$ equivalent to $\rho(r)$ and a *generator* of the function $V(r) = r^{\rho_0(r)}$ if $V \in \mathfrak{B}(\pi, \rho(r))$ and either $V(r) = \ln g(r)$ or $V(r) = \ln 1/g(r)$ for $r \geqslant 0$. The object under study in this section is the subclass $\mathfrak{A}(\rho(r)) = \{V\}$ of $\mathfrak{B}(\pi, \rho(r))$ distinguished by the following properties: the generator of every function $V \in \mathfrak{A}(\rho(r))$ is an entire function $g = g(V)$ of proximate order $\rho(r)$ such that $g$ is a canonical Weierstrass product, as described in Theorems B.1.11, B.1.12. In particular, the zeros of this function are located on the negative semiaxis only. For simplicity, in the definition of the class $\mathfrak{A}(\rho(r))$ we retain only properties 1)–3) of the functions belonging to $\mathfrak{B}(\pi, \rho(r))$ (see Theorem B.2.6), because property 4) of these functions will not be used below. If $\rho(r) \equiv \rho \in \mathbb{N}$, we agree that the class $\mathfrak{A}(\rho(r))$ consists of a single function, namely, $\exp\{z^\rho\}$. In this case, the results presented below are trivial, and, mostly, we omit them without mention.

Let $\rho(r) \to \rho \in (0, 1/2]$ be a proximate order. In Sections B.3–B.4 we considered an equivalent analytic proximate order $\rho_0(r)$ such that $V(r) = r^{\rho_0(r)} \in \mathfrak{B}(\gamma, \rho(r))$, where $\gamma \geqslant \pi m$, and $m$ is a natural number satisfying $m\rho > 1/2$. Therefore, in this case we modify the definition of the class $\mathfrak{A}(\rho(r))$. An entire function $g$ of order $m\rho$ and of proximate order $\rho_m(r) := m\rho(r^m)$ is called a *generator* of an analytic proximate order $\rho_0(r)$ (and at the same time a *generator* of the function $V(r) = r^{\rho_0(r)}$) if $V \in \mathfrak{B}(\pi m, \rho(r))$ and either $V(r) = \ln g(r^{1/m})$ or $V(r) = \ln 1/g(r^{1/m})$ for $r \geqslant 0$. In particular, if $\rho(r) \equiv 1/m$, $m \in \mathbb{N} \setminus \{1\}$, then the function $e^z$ is a generator for the function $V(r) = r^{1/m}$. This new notion of a generator is the only distinction of the definition of $\mathfrak{A}(\rho(r))$ from the above definition in the case where $\rho > 1/2$.

We study supplementary properties of the entire function $E_\rho(\cdot; V)$ (see (B.3.1)), assuming that $V \in \mathfrak{A}(\rho(r))$. This will enable us to obtain new information (compared to Theorem B.3.7) concerning the indicator of the function $E_\rho(\cdot; V)$ relative to the proximate order $\rho(r)$ if $\rho \in (1/2, 1]$.

**B.5.1. The structure of the functions belonging to $\mathfrak{A}(\rho(r))$.** We show that the class $\mathfrak{A}(\rho(r))$ is nonempty for every proximate order $\rho(r) \to \rho \in (0, \infty)$. Taking Theorems B.1.4′, B.1.12 into account, we restrict ourselves to the case where $\rho > 1/2$, and $L(r) = r^{\rho(r) - \rho}$ is a monotone function if $\rho \in \mathbb{N}$.

**Theorem B.5.1.** 1. *For the existence of an analytic proximate order $\rho_0(r)$ equivalent to $\rho(r)$ and such that $V(r) := r^{\rho_0(r)} \in \mathfrak{A}(\rho(r))$, and the generator $g$ of the function $V$ satisfies the relation $V(r) = \ln g(r)$, it suffices that one of the following conditions be fulfilled:*

a) $\rho = 2n+1,\ n = 0, 1, ...;\ L(r) \downarrow 0;$
b) $\rho = 2n,\ n = 1, 2, ..;\ L(r) \uparrow \infty;$
c) *$\rho$ is not an integer and* $\rho \in (2n, 2n+1),\ n = 0, 1, ....$

2. *In the same notation, a similar statement obtained by replacement of the property* $V(r) = \ln g(r)$ *with* $V(r) = \ln 1/g(r)$ *is true if one of the following conditions is fulfilled :*
a) $\rho = 2n,\ n = 1, 2, ...;\ L(r) \downarrow 0;$
b) $\rho = 2n+1,\ n = 0, 1, ...;\ L(r) \uparrow \infty;$
c) *$\rho$ is not an integer and* $\rho \in (2n-1, 2n),\ n = 1, 2, ....$

◀ Suppose condition 1 a) is fullfilled. Let $g(z) = \prod_{k=1}^{\infty} E(-z/a_k; 2n)$ be a function satisfying (B.1.7) under the condition $\Delta = 1$. Arguing as in the proof of Lemma B.2.2, we see that the function

$$\rho_0(r) = (\ln r)^{-1} \ln V(r), \qquad V(r) = \ln g(r),\ \ln g(0) = 0, \tag{B.5.1}$$

is an analytic proximate order equivalent to $\rho(r)$. However,

$$V(r) = \sum_{k=1}^{\infty} F\left(\frac{r}{a_k}\right), \quad F(t) = \ln(1+t) + \sum_{l=1}^{2n} \frac{(-t)^l}{l},\ t > 0. \tag{B.5.2}$$

Differentiating $F$ relative to the variable $\ln t$, we obtain

$$\frac{dF(t)}{d\ln t} = \frac{t^{2n+1}}{1+t}; \quad \frac{d^2F(t)}{d(\ln t)^2} = \frac{2nt^{2n+2} + (2n+1)t^{2n+1}}{(1+t)^2}.$$

This means that $F$ is a monotone increasing convex function of $\ln t$. Consequently (see (B.5.2)), the function $V$ posesses a similar property. Since $\ln g(z)$ is an analytic extension to $L(\pi)$ of the real-analytic function $V$, we conclude that $V \in \mathfrak{A}(\rho(r))$ (see Section B.2).

Let condition 2 a) be fullfilled. By Theorem B.1.12, there exists an entire function $g$ of proximate order $\rho(r)$ and such that

$$\lim_{r\to\infty} r^{-\rho(r)} \ln g(r) = -1.$$

In this case the function given by formula (B.5.1) with $V(r) = -\ln g(r)$ is an analytic proximate order equivalent to $\rho(r)$ (see (B.1.7)). Therefore, $V$ admits an analytic extension $V(z)$ to $L(\pi)$, and $\exp\{V(z)\} = 1/g(z)$ is a meromorphic function. Next, $g$ is a function of genus $2n-1$, and

$$G(t) = \sum_{l=1}^{2n-1} \frac{(-1)^{l-1}t^l}{l} - \ln(1+t),\ t > 0,$$

is a monotone increasing convex function of $\ln t$; so, arguing as above we prove that $V \in \mathfrak{A}(\rho(r))$.

In cases 1 b) and 2 b) the theorem is proved by the same procedure, and the proof is based on Theorem B.1.12.

Let $\rho$ be a noninteger, and let $\Delta = |\sin \pi\rho|/\pi$ in Theorem B.1.11. Then, the entire function $g$ of proximate order $\rho(r)$ discussed in that theorem satisfies the relation $\lim\limits_{r\to\infty} r^{-\rho(r)} \ln g(r) = \operatorname{sgn} \sin \pi\rho$. The remaining arguments for cases 1 c) and 2 c) are similar to the above. It suffices to observe that the genus of the function $g$ is $[\rho]$, where $[\rho]$ is the integral part of the number $\rho$ ▶

**Remarks.** 1. In cases 1 c), 2 c) the theorem is true , in particular, for $\rho(r) \equiv \rho$. For instance, the entire function

$$g(z) = \prod_{k=1}^{\infty} E(-z/k^{1/\rho}; [\rho]), \quad z \in \mathbb{C} \, ,$$

of order $\rho$ and of normal type satisfies the conditions of the theorem.

2. Let $g$ be an *arbitrary fixed* real-analytic entire function of proximate order $\rho(r)$, which is representable as a Weierstrass canonical product and belongs to the class of functions treated in Theorems B.1.11, B.1.12. In the course of the proof it was shown that the function $g$ decreases on the positive semiaxis if one of conditions 2 a)-c) is fulfilled. Therefore, in these cases no function $V \in \mathfrak{B}(\pi, \rho(r))$ can extend analytically to the function $\ln g(z)$, $\ln g(0) = 0$.

In the sequel we shall need the following well-known concept (see Definition 1.10.11).

An entire function $f$ of proximate order $\rho(r)$ and with generalized indicator $h_f$ is called a *function of completely regular growth* (c.r.g.) if the following limit exists along the complement of $E$:

$$h_f(\theta) = \lim_{r\to\infty} r^{-1} \ln |f(re^{i\theta})|, \quad \theta \in \mathbb{R}; \tag{B.5.3}$$

here $E$ is a $E^0$-set independent of $\theta$ (see Definition 1.10.11), and the convergence is uniform in the variable $\theta \in \mathbb{R}$.

The property of an entire function to be a generator of an analytic proximate order is invariant under differentiation.

**Theorem B.5.2.** *If an entire function $g$ of proximate order $\rho(r) \to \rho > 1/2$ is a generator of an analytic proximate order $\rho_0(r)$, then the derivative $g'$ has the same property relative to an analytic proximate order $\rho_1(r)$ equivalent to $\rho(r)$ and $\rho_0(r)$. Moreover, $g, g'$ are entire functions of completely regular growth.*

◀ Let $V(r) = r^{\rho_0(r)}$ and, for instance, $V(r) = \ln g(r)$. Then

$$\lim_{r\to\infty} \frac{\ln g(rz)}{V(r)} = z^{\rho}, \quad \lim_{r\to\infty} \frac{rg'(rz)}{g(rz)V(r)} = \rho z^{\rho-1},$$

and the convergence is uniform in $z \in K$, where $K$ is an arbitrary compact subset of $L(\pi)$. Taking the logarithm of the second limit, we obtain

$$\lim_{r\to\infty}\frac{\ln g(rz)}{V(r)}=\lim_{r\to\infty}\frac{\ln g'(rz)}{V(r)}=z^{\rho}, \tag{B.5.4}$$

where the convergence is uniform in the same sense. By Theorem B.1.13, $\rho(r)$ is a proximate order of $g'$. As in the proof of Theorem B.5.1, it is not difficult to show that $V_1(r) = \ln g'(r)$ is a convex monotone increasing function for $r \geqslant 0$. So, $V_1 \in \mathfrak{B}(\pi, \rho(r))$ and $\rho_1(r) = (\ln r)^{-1}\ln V_1(r)$ is the required analytic proximate order.

A similar result is true if $V(r) = -\ln g(r)$. From (B.5.3) and (B.5.4) we deduce that, for every $\theta \in (-\pi,\pi)$, $\{re^{i\theta} : r \geqslant 0\}$ is a ray of completely regular growth of the functions $g$ and $g'$. Since such rays constitute a closed set, the last statement of the theorem is true ▶

**B.5.2. "Gamma-functions" associated with the functions of class $\mathfrak{A}(\rho(r))$.** We study the properties of the analog of the gamma-function (see (B.3.1) and (B.3.2)) which determines the structure of the function $E_\rho(\cdot;V)$. Theorem B.3.1 is supplemented with the following results.

**Theorem B.5.3.** *Let $V \in \mathfrak{A}(\rho(r))$, where $\rho(r) \to \rho > 1/2$ is a proximate order. Suppose $g$ is an entire function serving as a generator of the function $V$, $\Delta(\lambda)$ is the function defined by formula* (B.3.2), *$\Lambda_\Delta$ is the collection of all poles of the meromorphic extension of $\Delta$ to $\mathbb{C}$. Let $\mathbb{Z}_- = \{0,-1,-2,...\}$ be the set of all nonpositive integers. Then*

$$\Delta(\lambda)=\Delta_\rho(\lambda;g)=\rho\int_0^\infty t^{\lambda-1}/g(t)dt,\quad \operatorname{Re}\lambda>0, \tag{B.5.5}$$

*if $V(r) = \ln g(r)$, $r \geqslant 0$, and*

$$\Delta(\lambda)=\Delta_\rho(\lambda;g)=\rho\int_0^\infty g(t)t^{\lambda-1}dt,\quad \operatorname{Re}\lambda>0, \tag{B.5.6}$$

*if $V(r) = -\ln g(r)$, $r \geqslant 0$. Moreover, the following statements are true:*

1) *every element of the set $\Lambda_\Delta$ is a simple pole of $\Delta$, and $\Lambda_\Delta \subset \mathbb{Z}_-$;*

2) *$\Lambda_\Delta \cap \{-1,-2,...,-p\} = \varnothing$ if $g$ is an entire function of genus $p > 0$;*

3) *if $V(r) = \ln g(r)$, $r \geqslant 0$, where $g$ is an entire function of order $\rho \in (1/2, 1]$ and of minimal type for $\rho = 1$*[*)] *, then $\Lambda_\Delta = \mathbb{Z}_-$, and*

$$\Delta(k+1)=\rho\int_0^\infty t^k/g(t)dt=\frac{k!}{g^{(k+1)}(0)},\quad k=0,1,2,.... \tag{B.5.7}$$

---

[*)] That is, for $n = 0$ either condition 1 a) or 1 c) of Theorem B.5.1 is fulfilled (see Theorem B.1.4′).

◄ Let, for instance, $V(r) = \ln g(r)$, $r \geqslant 0$. Then formula (B.3.2) is equivalent to formula (B.5.5). We expand the function $1/g$ in a Taylor series on the segment $[0, \delta]$ with $\delta < a_1$, where $-a_1$ is the zero of $g$ nearest to the origin: $1/g(t) = \sum_{k=0}^{\infty} b_k t^k$. Now, on the same segment, integrating the series $t^{\lambda-1} g(t)$ termwise, we arrive at the following identity:

$$\Delta(\lambda) = \rho\delta^{\lambda} \sum_{k=0}^{\infty} \frac{b_k \delta^k}{\lambda + k} + \rho \int_{\delta}^{\infty} t^{\lambda-1}/g(t)dt, \quad \mathrm{Re}\ \lambda > 0. \tag{B.5.8}$$

Here the series converges uniformly in the variable $\lambda$ ranging over an arbitrary compact subset of $\mathbb{C} \setminus \mathbb{Z}_-$, and the second summand is an entire function. Therefore, formula (B.5.8) determines a meromorphic extension of the function $\Delta(\Lambda)$ to $\mathbb{C} \setminus \mathbb{Z}_-$, and statement 1) is true. If $g$ is an entire function of genus $p > 0$, then, since $g$ is a Weierstrass canonical product, we conclude that $b_1 = b_2 = \cdots = b_p = 0$. Consequently, statement 2) is also true. In the case where $V(r) = -\ln g(r)$, $r \geqslant 0$, the proofs of statements 1) and 2) are similar and are based on formula (B.5.6).

Suppose the assumptions of statement 3) are fulfilled. In order to determine the values of the entire function $1/\Delta(\lambda)$ at the points of the set $\mathbb{Z}$, we use the analog of the Hankel formula, in which we change the variables $w = \ln g(\zeta), \ln g(0) = 0$ (see Theorem B.3.1):

$$\frac{1}{\Delta(n)} = \frac{1}{2\pi i} \int_{\gamma_\varphi} \frac{e^w}{[U(w)]^n} dw = \frac{1}{2\pi i} \int_{T_\varphi} \frac{g'(\zeta)}{\zeta^n} d\zeta, \quad n \in \mathbb{Z}, \tag{B.5.9}$$

where $\zeta = U(w)$ is the inverse of the function $w = \ln g(\zeta)$ and $T_\varphi$ is the image of the contour $\gamma_\varphi$ under the map $\zeta = U(w)$, $\pi/2 < \varphi < \pi\rho$. The contour $T_\varphi$ is symmetric relative to the abscissa axis because $\overline{U(w)} = U(\overline{w})$ (see Theorem B.2.6). We consider another contour $T_\varphi(S)$, which possesses the same axial symmetry and consists of the curves

$$T_{\pm\varphi}(S) = \{\zeta = U(a + s\exp\{\pm i\varphi\}) : \ s \geqslant S > 0\}$$

and the circular arc

$$\{\zeta = Re^{i\theta} : \ \psi \leqslant \theta \leqslant 2\pi - \psi\}, \quad Re^{i\psi} = U(a + Se^{i\varphi});$$

$T_\varphi(S)$ is oriented in the direction of increase of $\arg\zeta$. By Theorem B.2.6, here $Re^{i\psi} = U(S)\exp\{i\varphi/\rho\}(1 + o(1))$ as $S \to \infty$. Consequently, for $\delta > 0$ sufficiently small there exists $S_0 = S_0(\delta)$ such that

$$\frac{\pi}{2\rho} < \frac{\varphi}{\rho} - \delta < |\arg\zeta| \leqslant \pi, \quad \zeta \in T_\varphi(S), \ S > S_0. \tag{B.5.10}$$

By the structure of the contour $T_\varphi$, the domain bounded by $T_\varphi(S)$ and positively oriented relative to it contains 0. Therefore, by (B.5.8), for any sufficiently large $S > S_0$ we find

$$\frac{1}{\Delta(n)} = \frac{1}{2\pi i} \int\limits_{T_\varphi(S)} \frac{g'(\zeta)}{\zeta^n} d\zeta + \operatorname*{res}_{\zeta=0} \frac{g'(\zeta)}{\zeta^n} \quad \forall\, n \in \mathbb{Z}. \tag{B.5.11}$$

By Theorem B.1.13, the function $g'$ has the indicator $\cos\rho\theta$ relative to the proximate order $\rho(r)$. For every $\varepsilon > 0$ we have

$$\ln|g'(re^{i\theta})| < V(r)(\cos\rho\theta + \varepsilon), \; r > R_0(\epsilon).$$

Next, choosing $\varepsilon < -\cos(\varphi - \rho\delta)$ (see (B.5.10)), we conclude that in formula (B.5.11) the integral tends to 0 as $S \to \infty$. Therefore, (B.5.7) is true and

$$1/\Delta(\;k) = 0, \quad k = 0, 1, 2, \ldots.$$

Thus, statement 3) is proved ▶

**Theorem B.5.3′.** *Let $\rho(r) \to \rho \in (0, 1/2]$ be a proximate order such that $L(r) = r^{\rho(r)-\rho} \downarrow 0$ if $\rho = 1/2$. Suppose $V(r) \in \mathfrak{A}(\rho(r))$, $m$ is the smallest natural number satisfying $m\rho > 1/2$, and $g$ is an entire function of order $m\rho$ and of proximate order $\rho_m(r) := m\rho(r^m)$ that is a generator of $V(r)$. Then, in the notation of Theorem* B.5.3, *we have*

$$\Delta(\lambda) = \Delta_\rho(\lambda; V) = m\rho \int\limits_0^\infty t^{m\lambda-1}/g(t)dt, \quad \operatorname{Re}\lambda > 0, \tag{B.5.12}$$

*and the following statements are true.*

1) *Every element of the set $\Lambda_\Delta$ is a simple pole for the function $\Delta$, and $\Lambda_\Delta = \{z = -k/m : k = 0, 1, 2, \ldots\}$.*

2) *For every $k \in \mathbb{N} \cup \{0\}$ we have*

$$\Delta(k+1) - \Delta_\rho(k+1; V) = \frac{(km+\mu)!}{g^{m(k+1)}(0)}, \quad \mu = m - 1. \tag{B.5.13}$$

◀ By assumption, the proximate order $\rho_m(r) \to m\rho \in (1/2, 1]$ satisfies either condition 1 a) or 1 c) of Theorem B.5.1. Therefore, $V(r) = \ln g(r^{1/m})$ and, by formula (B.3.2), relation (B.5.12) is fullfilled. Since $V_1(r) := V(r^m) = \ln g(r) \in \mathfrak{A}(\rho_m(r))$, this implies that

$$\Delta(\lambda) = \Delta_\rho(\lambda; V) = \Delta_{m\rho}(m\lambda; V_1) \quad \operatorname{Re}\lambda > 0.$$

To complete the proof, it only remains to apply statement 3) of Theorem B.5.3 ▶

**B.5.3. "Exponentials" associated with functions of class $\mathfrak{A}(\rho(r))$.** For the analog $E_\rho(\cdot;V)$ (see (B.3.1)) of the Mittag-Leffler function, we study the specific features of its Taylor coefficients and its asymptotics. In this investigation, identities (B.5.7) and (B.5.13) will allow us to avoid application of the integral formula (B.3.17) in the appropriate cases.

**Theorem B.5.4.** 1. *Suppose $\rho(r)\to\rho\in(1/2,1]$ is a proximate order such that $L(r)=r^{\rho(r)-1}\downarrow 0$ if $\rho=1$. Let $V\in\mathfrak{A}(\rho(r))$, $V(r)=\ln g(r)$, where $g$ is the generator of $V$. Then*

$$E_\rho(z;g):=E_\rho(z;V)\equiv g'(z),\ \ z\in\mathbb{C}, \tag{B.5.14}$$

*and we have the following asymptotic estimate as $r\to\infty$ :*

$$\ln E_\rho(re^{i\theta};g)=V(r)e^{i\rho\theta}[1+o(1)],\quad |\theta|<\pi. \tag{B.5.15}$$

*This estimate is uniform in the variable $\theta$ on every segment $[a,b]\subset(-\pi,\pi)$.*

2. *In the notation of statement 1, suppose $\rho\in(0,1/2]$, $L(r)=r^{\rho(r)-\rho}\downarrow 0$ if $\rho=1/2$, and $m$ is the smallest natural number satisfying $m\rho>1/2$. Let $g$ be an entire function of order $m\rho$ and of proximate order $\rho_m(r):=m\rho(r^m)$. If $g$ is the generator for the function $V(r)$, then*

$$E_\rho(w^m;g):=E_\rho(w^m;V)\equiv\frac{1}{mw^{m-1}}\left[\sum_{\nu=0}^{m-1}g(we^{i2\pi\nu/m})\right]',\ \ w\in\mathbb{C}. \tag{B.5.16}$$

◄ Identity (B.5.14) is a consequence of formula (B.5.7). Formula (B.5.15) is true by Theorem B.5.2 and formula (B.2.6).

We prove statement 2 of the theorem. Let $\mu=m-1$, and let

$$S_n:=\sum_{\nu=1}^{\mu}\exp\{i2\pi\nu n/m\}=\begin{cases} m, & n\equiv 0\ (\mathrm{mod}\ m),\\ 0, & n\not\equiv 0\ (\mathrm{mod}\ m).\end{cases}$$

Since the assumptions of statement 2) of Theorem B.5.3′ are fulfilled, from formulas (B.3.1), (B.3.2), and (B.5.12) we deduce that

$$mw^\mu E_\rho(w^m;g)=\sum_{n=0}^{\infty}\frac{[g^{(n+1)}(0)]w^nS_n}{n!}=\sum_{\nu=0}^{\mu}\sum_{n=0}^{\infty}\frac{w^n}{n!}(e^{i2\pi\nu/m})^n[g^{(n+1)}(0)]$$

$$=\sum_{\nu=0}^{\mu}\sum_{n=0}^{\infty}\frac{w^n}{n!}\left[g'(we^{i2\pi\nu/m})\right]_w^{(n)}|_{w=0}=\sum_{\nu=0}^{\mu}[g(we^{i2\pi\nu/m})]'_w,\quad w\in\mathbb{C}.$$

So, formula (B.5.16) is proved ►

Thus, in the cases under consideration, the asymptotics of the function $E_\rho(\cdot;V)$ is entirely determined by the asymptotics of the generator $g$ of $V$. In the other cases the asymptotics of $E_\rho(\cdot;V)$ is described by another theorem, which follows from formulas (B.3.25), (B.3.26) and statement 2 of Theorem B.5.3.

**Theorem B.5.5.** *Suppose $\rho(r) \to \rho \in [1,\infty)$ is a proximate order such that $L(r) = r^{\rho(r)-1} \uparrow \infty$ if $\rho = 1$. Let $V \in \mathfrak{A}(\rho(r))$, and let an entire function $g$ of genus $p$ be the generator of $V$. Then we have the following asymptotic estimate of the function $E_\rho(z;V)$ as $|z| \to \infty$, which is uniform in* $\arg z$ :

$$E_\rho(z;V) = \begin{cases} F'(z) + O\left(\dfrac{1}{|z|^2}\right), & |\arg z| \leqslant \dfrac{\pi}{2\rho} + \varepsilon\ , \\ O\left(\dfrac{1}{|z|^2}\right), & \dfrac{\pi}{2\rho} + \varepsilon \leqslant |\arg z| \leqslant \pi, \end{cases} \tag{B.5.17}$$

*where $0 < \rho\varepsilon < \pi \min\{1/2, \rho - 1/2\}$, $F = g^{\pm 1}$, and the sign is the same as in the formula $V(r) = \pm \ln g(r)$. Next,*

$$O\left(\frac{1}{|z|^2}\right) = O\left(\frac{1}{|z|^{p+2}}\right) - \sum_{k=1}^{p} \frac{1}{\Delta(-k) z^{k+1}}. \tag{B.5.18}$$

*Here*

$$\Delta(-k) = \int_0^\delta \left[\frac{1}{F(t)} - 1\right] t^{-k-1} dt - \frac{\delta^{-k}}{k} + \int_\delta^\infty t^{-k-1} 1/F(t) dt \neq 0 \tag{B.5.19}$$

*for some $\delta > 0$ and for $k = 1, 2, ..., p$* (see (B.5.4), (B.5.5)).

**B.5.4. The indicator of the function $E_\rho(\cdot, V)$, $V \in \mathfrak{A}(\rho(r))$.** Now everything is ready for calculating the generalized indicator $h_E$ of the function $E_\rho(\cdot, V)$ (and not merely its nonnegative part $h_E^+$) relative to the proximate order $\rho(r) \to \rho$ for the case where $V \in \mathfrak{A}(\rho(r))$. The next statement supplements Theorem B.3.7 and describes some cases in which the indicator $h_E$ may take negative values.

**Theorem B.5.6.** *In the above notation, the following statements are true.*

1. *Let $\rho \in (0, 1]$, and, moreover, let $L(r) = r^{\rho(r)-1} \downarrow 0$ if $\rho = 1$. Then the indicator of the function $E_\rho(\cdot, V)$ relative to $\rho(r)$ is the $2\pi$-periodic extension to $\mathbb{R}$ of the function $h_E(\theta) = \cos \rho\theta$, $|\theta| \leqslant \pi$.*

2. *Let $\rho \in [1, \infty)$, and, moreover, let $L(r) = r^{\rho(r)-1} \uparrow \infty$ if $\rho = 1$. Then* (see (B.1.8)) *$h_E(\theta) = \Phi_\rho(e^{i\theta})$, $\theta \in [-\pi, \pi]$.*

◀ Statement 1 for $\rho > 1/2$ follows from (B.5.15). Relation (B.5.19) implies condition (B.3.35). Therefore, the remaining part of the theorem follows from Theorem B.3.7 ▶

## B.6. Applications of the preceding results

This section is devoted to two applications of the results of Section B.5. On the one hand, we prove a complete analog of Theorem $A_1$ for an entire function of order 1 and of minimal type (see the Introduction). On the other hand, we study the partial fraction expansion for the meromorphic function $1/g$ and its derivative, where $g$ is an entire function of order 1 and of maximal type that serves as the generator for a given function $V \in \mathfrak{A}(\rho(r))$ (see Section B.5).

**B.6.1. The indicator diagram and the conjugate diagram for an entire function of order 1 and of minimal type.** We prove an analog of Theorem B.4.5 for an entire function $f$ of order 1 and of minimal type with indicator that may take negative values. We use the notation and terminology of Sections B.4–B.5.

Let $P_1$ be the class of finite $2\pi-$periodic trigonometrically 1-convex functions on $\mathbb{R}$. For $h \in P_1$ we put

$$X_h = \{p \in \mathbb{C} : \ \mathrm{Re}\, pe^{i\theta} \leqslant h(\theta) \quad \forall\, \theta \in \mathbb{R}\}. \tag{B.6.1}$$

If $h = h_f$ is the indicator of the function $f$, then the set $I_h = \{w = \overline{p} : \ p \in X_h\}$ is called the *indicator diagram* of $f$. We formulate an analog of Theorem B.4.3.

**Theorem B.6.1.** *Suppose* $\rho(r) \to 1$ *is a proximate order,* $L(r) = r^{\rho(r)-1} \downarrow 0$, $V \in \mathfrak{A}(\rho(r))$, $V(r) = \ln g(r)$, *where* $g$ *is the generator of the function* $V$, $h \in P_1$, *and* $f \in P_\rho(V,h)$ (see (B.4.7)). *Then the system of functions* $\{F_\theta : \ |\theta| \leqslant \pi\}$, *where*

$$F_\theta(p) = \int\limits_0^{\infty(\arg z=\theta)} f(z)/g(pz)dz, \tag{B.6.2}$$

*determines an analytic extension of the Borel* $V$*-transform* $F$ *of the function* $f$ *to the complement* $\mathbb{C} \setminus X_h$ *of the compact convex set* $X_h$ (B.6.1).

◀ For $\theta \in [-\pi,\pi]$ we put $\Pi_\theta = \{p \in \mathbb{C} : \ \mathrm{Re}\, pe^{i\theta} > h(\theta)\}$ and $L_\theta = \{p = -re^{-i\theta} : \ r \geqslant 0\}$. Let $K$ be an arbitrary compact subset of $T_\theta = \Pi_\theta \setminus L_\theta$. There exists a number $\delta = \delta(K) > 0$ such that

$$\mathrm{Re}\, pe^{i\theta} > h(\theta) + \delta \quad \forall\, p \in K.$$

Choosing $\varepsilon \in (0, \delta/2)$ and recalling that $V(r) = \ln g(r) \in \mathfrak{B}(\pi, \rho(r))$ (see (B.2.6)), we conclude that there is $t_0 = t_0(\varepsilon)$ satisfying

$$\ln|g(pte^{i\theta})| > V(t)(\mathrm{Re}\, pe^{i\theta} - \varepsilon), \quad t > t_0,\ p \in K.$$

Together with (B.4.7), this implies the following estimate of the integral $F_\theta$ (see (B.6.2)):

$$|F_\theta(p)| < \int\limits_0^{t_0} |g(pte^{i\theta})|dt + M\int\limits_0^{\infty} \exp\{V(t)[2\varepsilon - \delta]\}dt < \infty,\ p \in K.$$

Thus, the integral $F_\theta$ converges uniformly on $K$ and, consequently, represents a function holomorphic in $T_\theta$.

Assume that $\Pi_\theta \cap L_\theta \neq \varnothing$, $p_0 \in \Pi_\theta \cap L_\theta$ , and, for instance, that $-\pi < \theta \leqslant \pi$. Since the function $h$ is continuous, there exists a number $\psi < \theta$ such that $p_0 \in T_\psi$. Therefore,

$$\bigcup T_\theta | \theta \in [-\pi, \pi] = \bigcup \Pi_\theta | \theta \in [-\pi, \pi] = \mathbb{C} \setminus X_h.$$

The remaining part of the proof follows the outline of the proof of Theorem B.4.3 and is based on formula (B.5.5) ▶

Suppose the function $V$ satisfies the conditions of Theorem B.6.1. Then, by (B.5.14), the inverse integral formula expressing a function $f \in \mathfrak{M}_\rho(V)$ in terms of its Borel $V$-transform $F$ looks like this:

$$f(z) = \frac{1}{2\pi i} \int_\Gamma g'(zp)F(p)dp, \quad z \in \mathbb{C}, \tag{B.6.3}$$

where $g$ is the generator of $V$, and $\Gamma$ is a positively oriented Jordan contour enclosing the singularities of the function $F$ (see Definition B.4.1).

In this case the following statement is an analog of Theorem B.4.4.

**Theorem B.6.2.** *Suppose a function* $V \in \mathfrak{A}(\rho(r))$ *satisfies the conditions of Theorem* B.6.1, $K$ *is a compact convex set,* $h_K(\theta) = \sup\{\operatorname{Re} pe^{i\theta} : p \in K\}$, $\theta \in \mathbb{R}$ *is its "support" function, and* $f \in \mathfrak{M}_\rho(V)$. *If the Borel* $V$*-transform of* $f$ *admits analytic continuation beyond* $K$, *then* $f \in P_\rho(V; h_K)$ (see (B.4.7)).

◀ Let $K_\varepsilon = K + S_\varepsilon$, where $S_\varepsilon = \{p \in \mathbb{C} : |p| \leqslant \varepsilon\}$, be the $\varepsilon$-neighorhood of $K$. If $0 \notin K$, then we choose $\varepsilon > 0$ such that $0 \notin K_\varepsilon$. The remaining arguments are the same as in the proof of Theorem B.4.4 and are based on Theorem B.5.2 and formula (B.6.3). The functions $\Phi_\rho(pe^{i\theta})$ and $E_\rho(\cdot; V)$ are replaced with $\operatorname{Re} pe^{i\theta}$ and $g,'$ respectively ▶

We pass to the relationship between the indicator and conjugate diagrams for an entire function of order 1 and of minimal type.

**Theorem B.6.3.** *Suppose* $f$ *is an entire function of order* 1 *and of minimal type,* $\rho(r)$ *is its proximate order,* $h = h_f$ *is the generalized indicator of* $f$, $V \in \mathfrak{A}(\rho(r))$, *and* $V(r) = \ln g(r)$, *where* $g$ *is the generator of* $V$. *Let* $K$ *be the conjugate diagram of* $f$, *i.e., the smallest compact convex set to the exterior of which the Borel* $V$*-transform of* $f$ *extends analytically. Then*

1) *the set* $K$ *coincides with the "indicator diagram"* $X_h$ *of the function* $f$ (see Definition B.4.1, (B.6.1));

2) *there exists an analytic functional* $\mu$ *such that its Laplace* $V$*-transform* $\mu_p[g'(zp)]$ *coincides with* $f(z)$, $z \in \mathbb{C}$; *moreover,* $X_h$ *is a convex support of* $\mu$.

◀ By Theorem B.1.4′, there is no loss of generality in assuming that $L(r) = r^{\rho(r)-1} \downarrow 0$. The proof of statement 1) is based on Theorems B.6.1, B.6.2 and follows a classical outline (see the proof of Theorem B.4.5) ▶

**B.6.2. On the expansion of a meromorphic function in partial fractions.** We shall need the following convergence criterion for the series of principal parts of the Laurent expansion of a meromorphic function, and a criterion for the existence of a partial fraction expansion.

**Theorem B.6.4.** *Let $\Phi$ be a meromorphic function in $\mathbb{C}$, and let $\Lambda = \{b_k\}_1^\infty$ be the sequence of all its poles arranged in the order of increase of the absolute values and tending to $\infty$. Next, let*

$$\left\{\Phi_k(z) = \sum_{n=1}^{N_k} \frac{A_{k,n}}{(z-b_k)^n} : k = 1, 2, ...\right\} \tag{B.6.4}$$

*be the sequence of principal parts of the Laurent expansion of the function $\Phi$ near the poles. Then the series*

$$\sum_{k=1}^{\infty} \Phi_k(z) = \lim_{R\to\infty} \sum_{|b_k|<R} \Phi_k(z)$$

*converges for all $z \in \mathbb{C}\setminus\Lambda$ if and only if the series*

$$R(z) = \sum_{k=1}^{\infty} \operatorname*{res}_{w=b_k} \Psi(w,z), \;\; z \in \mathbb{C}\setminus\Lambda; \;\; \Psi(w,z) = \frac{\Phi(w)}{w-z},$$

*converges. Moreover,*

$$\Phi_k(z) = -\operatorname*{res}_{w=b_k} \Psi(w,z), \;\; z \in \mathbb{C}\setminus\Lambda, \;\; k \in \mathbb{N}; \;\; R(z) = -\sum_{k=1}^{\infty} \Phi_k(z). \tag{B.6.5}$$

*If the series $R(z)$ converges uniformly on every compact set in $\mathbb{C}\setminus\Lambda$, then $\alpha(z) = \Phi(z) - \sum\limits_{k=1}^{\infty} \Phi_k(z)$ is an entire function equal to the total sum of residues of the function $\Psi(w,z)$ in the variable $w$.*

◀ For the function $\Phi$ we fix its pole $b = b_k$ of multiplicity $N_k$ for some $k \in \mathbb{N}$. We find the residue of the function $\Psi(w,z)$ for fixed $z \neq b$ with the aid of its Laurent series expansion in the variable $w$ near the point $w = b$. The principal part of this series can be calculated in terms of the function $\Psi_k(w,z) = \Phi_k(w)(w-z)^{-1}$ (see (B.6.4)):

$$\Psi_k(w,z) = -\sum_{k=1}^{N_k} A_{k,n} \sum_{r=0}^{\infty} \frac{(w-b)^{r-n}}{(z-b)^{r+1}} = -\sum_{s=-N_k}^{\infty} (w-b)^s \sum_{n=1}^{N_k} \frac{A_{k,n}}{(z-b)^{n+s+1}}.$$

Now, assuming that $s = -1$ and using (B.6.4), we verify (B.6.5), which implies also the second statement of the theorem ▶

**Definition B.6.5.** We shall say that a meromorpic function $\Phi$ *expands in partial fractions* if, in the notation of Theorem B.6.4, we have

$$\Phi(z) = \lim_{R\to\infty} \sum_{|b_k|<R} \Phi_k(z) + \alpha(z), \quad z \in \mathbb{C} \setminus \Lambda, \tag{B.6.6}$$

where $\alpha$ is an entire function and the series converges uniformly on every compact subset of $\mathbb{C} \setminus \Lambda$.

The simplest example of such a function is the Gamma-function. It is representable in the form

$$\Gamma(z) = \sum_{k=0}^{\infty} \frac{(-1)^k}{k!(z+k)} + \alpha(z), \quad \alpha(z) = \int_1^{\infty} e^{-t} t^{z-1} dt.$$

Here $\alpha(z)$ is an entire function of order 1.

We apply the criterion (see Theorem B.6.4) to the partial fraction expansion of a meromorphic function of the form $1/g$, where $g$ is the generator of an analytic proximate order $\rho(r)$ satisfying $\rho(r) \to 1$ and $L(r) = r^{\rho(r)-1} \uparrow \infty$ as $r \to \infty$.

**Definition B.6.6.** Suppose that $g$ is an entire of order $\rho > 0$ and of proximate order $\rho(r)$. Let $0 \leqslant R_1 < R_2 < \cdots < R_k < \ldots$ be sequence of radii of all circumferences with center at 0 containing zeros of $g$. The function $g$ is called a function of *locally regular growth* if there exists an increasing sequence of positive numbers $\{r_n\}_1^\infty$ such that for some $k \in \mathbb{N}$ we have

$$R_{k+n} < r_n < R_{k+n+1} \quad \forall\, n \in \mathbb{N} \tag{B.6.7}$$

and

$$\lim_{n\to\infty} r_n^{-\rho(r_n)} \log |g(r_n e^{i\theta})| = h_g(\theta) \quad \theta \in \mathbb{R}, \tag{B.6.8}$$

where $h_g$ is a generalized indicator of $g$ and convergence in (B.6.8) is uniform in $\theta \in \mathbb{R}$.

**Theorem B.6.7.** *Let* $V(r) = r^{\rho(r)} = -\ln g(r)$, *where the generator* $g$ *of* $\rho(r)$ *is a function of locally regular growth* (see Definition B.6.6) *) . *Denote by* $Z_g = \{b_1, b_2, ...,\}$ *the sequence of all zeros of* $g$ *such that* $0 > b_1 > b_2 > \ldots$. *Then the function* $\Phi = 1/g$ *and its derivative* $\Phi'$ *expand in partial fractions, i.e.,* (B.6.6) *is true with* $\Lambda = Z_g$, *and it remains true after termwise differentiation. Here* $\alpha, \alpha'$ *are entire functions of proximate order* $\rho(r)$, *and the generalized indicator of each of them is equal to* $\max\{0, \cos\theta\}$, $\theta \in \mathbb{R}$. *Moreover* (see (B.3.1)), $\alpha'(z) \equiv E_\rho(z; V)$, $z \in \mathbb{C}$.

◀ We verify the assumptions of Theorem B.6.5. Let $S_R = \{z \in \mathbb{C} : |z| < R\}$, and let $\gamma := \gamma(R, \pi - \delta)$ (where $R > 0$ and $\delta \in (0, \pi/2)$) be the

*) This requirement means that the zeros of $g$ with sufficiently large moduli cannot be arbitrarily close to one another.

contour considered in (B.3.3). Suppose $K$ is an arbitrary compact subset of $S_R \cap (\mathbb{C} \setminus Z_g)$. Consider the integral

$$\alpha(z) = \frac{1}{2\pi i} \int_\gamma \Psi(w,z)dw, \quad z \in K, \tag{B.6.9}$$

where

$$\Psi(w,z) = \frac{\Phi(w)}{w-z}, \quad \Phi = g^{-1}.$$

By (B.2.6), putting $\theta = \arg z$, we find:

$$\ln g(z) = -V(z) = -V(|z|)e^{i\theta}[1 + o(1)], \quad |\theta| < \pi, \tag{B.6.10}$$

as $|z| \to \infty$; furthermore, this asymptotic estimate is uniform in the variable $\theta$ if $\theta \in [a,b] \subset (-\pi, \pi)$. Therefore, for every $\varepsilon \in (0, \cos\delta)$ there exist $R_0$ such that

$$\ln |g(re^{\pm i(\pi-\delta)})| > -V(r)[\cos(\pi - \delta) + \varepsilon], \quad r > R_0.$$

Now by (B.6.10) we conclude that the integral $\alpha(z)$ converges uniformly in the variable $z \in K$. Since $R > 0$ is an arbitrary fixed number, $\alpha(z)$ is an entire function.

We observe that $g$ is an entire function of locally regular growth. Then there exists a sequence $\{r_n > 0 : n \in \mathbb{N}\}$ satisfying conditions (B.6.7), (B.6.8) of Definition B.6.6. Consequently, for every $\varepsilon \in (0, \cos\delta)$ we have

$$\ln |g(r_n e^{i\theta})| > -V(r_n)(\cos\theta + \varepsilon), \ n > n_0(\varepsilon) \ \ \forall\, \theta \in [-\pi, \pi].$$

This estimate implies that, in the notation of formula (B.6.9), we have

$$\lim_{n\to\infty} \int_{T_n} \Psi(w,z)dw = 0, \ z \in K; \ T_n = \{w = r_n e^{i\theta} : \ \theta \in [\pi - \delta, \pi + \delta]\},$$

where the convergence is uniform in $z \in K$. Combining this with the same formula (B.6.9), by standard methods of complex analysis we obtain:

$$\alpha(z) = \Phi(z) + \sum_{k=1}^{\infty} \operatorname*{res}_{w=b_k} \Psi(w,z), \ z \in \mathbb{C} \setminus Z_g; \ \Phi = 1/g.$$

Now we see that formula (B.6.5) implies (B.6.6) for the function $\Phi = 1/g$. By the well-known properties of the analytic functions, this implies the following partial fraction expansion for the derivative $(1/g)'$ :

$$\left(\frac{1}{g(z)}\right)' = \sum_{k=1}^{\infty} \Phi_k'(z) + \alpha'(z), \quad z \in \mathbb{C} \setminus Z_g \, . \tag{B.6.11}$$

On the other hand, the following integral representation is true (see Theorem B.3.5):

$$E_\rho(z;V) = E_\rho(z;\ln 1/g) = \frac{1}{2\pi i}\int\limits_{\gamma(R,\varphi)} \frac{e^w}{U(w)-z}dw, \quad |z| < d_R\ , \tag{B.6.12}$$

where $\varphi \in (\pi/2, \pi)$, and $\zeta = U(w)$ ($w \in \gamma(R,\varphi)$) is the inverse of the function $w = V(\zeta) = -\ln g(\zeta)$. Next, $d_R \to \infty$ as $R \to \infty$. Let $M(R,\varphi)$ be the image of the contour $\gamma(R,\varphi)$ under the mapping $\zeta = U(w)$. Substituting the variable $w = -\ln g(\zeta)$ in the integral formula (B.6.12) and integrating by parts, we obtain:

$$E_\rho(z;\ln 1/g) = \frac{1}{2\pi i}\int\limits_{M(R,\varphi)} \frac{[1/g(\zeta)]'}{\zeta - z}d\zeta = \frac{1}{2\pi i}\int\limits_{M(R,\varphi)} \frac{1/g(\zeta)}{(\zeta-z)^2}dw = \beta'(z)$$

for $|z| < d_R$, where

$$\beta(z) = \frac{1}{2\pi i}\int\limits_{M(R,\varphi)} [1/g(\zeta)](\zeta - z)^{-1}d\zeta, \quad |z| < d_R.$$

Now, by the asymptotic estimate (B.6.10) and the Cauchy integral formula, (B.6.9) implies that $\alpha(z) = \beta(z)$, $|z| < \min\{R, d_R\}$. By the uniqueness theorem, we conclude that $\alpha'(\cdot) \equiv E_\rho(\cdot;V)$. Finally, Theorems B.3.3 and B.1.9 show that $\alpha$ is an entire function of proximate order $\rho(r) \to 1$. Now, the required property of the generalized indicator of the function $\alpha$ follows from Theorems B.3.7, B.5.6, and B.1.13 ▶

## Comments

### On the Introduction

Borel transforms of this kind were studied by D. Polya [152]. The original approach of E. Borel is exposed in Section 3.3 (see (3.3.1)). Theorem $A_2$ and its version for compact convex sets (see Theorem 1.10.9) are due to A. Martineau [139-140] and L. Ehrenpreis [35]. Concerning applications of Theorems $A_1$, $A_2$ we refer the reader, e.g., to the monographs [18, 99-103, 148]. In [148, Ch.3, §12] the proofs of Theorems $A_2$ and 1.10.9 can be found. The relationship between the regularity of growth of an entire function of exponential type and the arrangement of the singularities of its Borel transform was studied by N. V. Govorov and N. M. Chernykh [32-33]. As was noted by Ch. O. Kiselman in [79], L. Ehrenpreis deserves the credit for a far-reaching generalization of Theorem $A_2$ to the class of solutions of a partial linear differential equation with constant coefficients in a domain $D \subset \mathbb{R}^n$ (holomorphic functions in a domain $D \subset \mathbb{C}$ are solutions of the equation $\dfrac{\partial f}{\partial \bar z} := \dfrac{\partial f}{\partial x} + i\dfrac{\partial f}{\partial y} = 0$, $z = x + iy \in D$). In an equivalent way, the transformation (1) was introduced by A. I. Markushevich [138]: $f(\lambda) = \mu_z[K(\lambda, z)]$, $\lambda \in \mathbb{C}$ ; here $\mu \in H^*(\mathbb{C})$ is defined by

$$\mu(\varphi) = \sum_{k=0}^{\infty} a_k \varphi^{(k)}(0)/k!, \qquad \overline{\lim_{k\to\infty}} |a_k|^{1/k} < \infty.$$

However, the situation considered in [138] is more general: $K$ is a function biholomorphic in a bidisk (see also [31, p. 4]). The generalized Borel transformation corresponding to the function $E_\rho(z;\mu)$ for $\rho > 0$, $\mu > 0$ was introduced by M. F. Subbotin [182] (see also [105]). M. M. Dzhrbashyan studied it for $\rho > 0$, Re $\mu > 0$ (see [31, 34]). In the case of $K(\lambda, z) = A(\lambda z)$, where $A$ is an entire function, the transformation (1) was studied by M. A. Evgrafov [40, p. 36], Yu. A. Kaz'min [73-75], A. Yu. Popov [153] (see also [104, p. 5: 81]). A rather general form of an analog of the Fourier and Borel transformations was proposed recently by Yu. F. Korobeinik [85]. The order in the totality of variables was defined by E. Borel in [23, p. 80].

### On Chapter 1

When writing Chapter 1, the author used the following literature.
**Monographs**:
N. Bourbaki [25, 26] (Sections 1.3 and 1.4) *); V. S. Vladimirov [190] (1.4); A. A. Goldberg, I. V. Ostrovsky [59] (1.1); J. Cassels [27] (1.7), B. Ya. Levin [103] (1.9-1.10); A. A. Raikov [154] (1.3); P. Rockafellar [157] (1.2-1.6); L. I.

*) In parentheses, we indicate the sections of Chapter 1 where a particular reference was used.

Ronkin [164] (1.1, 1.2, 1.8); V. P. Khavin [76] (1.10.3); B. V. Shabat [179] (1.8); L. S. Maergoiz [118] (1.1-1.8).

**Articles:**

A. Bronsted [28] (1.5); B. A. Vertgeim, G. Sh. Rubinstein [189] (1.4); A. D. Ioffe, V. M. Tikhomirov [68] (1.5); C. O. Kiselman [77] (1.8, 1.9); L. S. Maergoiz [117, 122] (1.9).

Now we give references for the proofs of some statements.

Theorem 1.4.27: see [157]; Proposition 1.6.1: see [157, Ch. 2]; Theorems 1.8.3, 1.8.6-1.8.8, 1.8.11, 1.9.18: see [164, pp. 123-128; 53; 289; 56]; Theorems 1.8.5, 1.8.17: see [172, part 2, p. 259; part 1, p. 320]; Properties 1.9.20, 1.9.27, Theorems 1.10.2, 1.10.3, 1.10.12, 1.10.15, 1.10.16: see [103, pp. 77-78, 108, 13, 74, 124, 186, 258, 219]; Proposition 1.10.6: see [156]; Theorem 1.10.7 (the well-known Köthe-Grothendieck-Sebastião e Silva theorem): see [88]; Proposition 1.10.10: see [199]; Theorem 1.10.13: see [45]; Theorem 1.10.14: see [122]; Theorem 1.10.17: see [15, p. 297], and also [8], [122].

An assertion more general than Proposition 1.10.1 is proved in Ch. 8 (Property 8.1.1). The proof of Theorem 1.10.8 presented here was proposed by O. V. Epifanov.

## On Chapter 2

The term "relaxation" (regaining the equilibrium) is used in physics and chemistry to describe the behavior of various equilibrium systems subject to external impact (see, e.g. [5] *) ). The study of homeostasis systems in various extreme states subject to external impact can be found, e.g., in t he monographs [1, 36, 37]. The problem of presenting initial data of an experiment as a sum of a finite number of exponentials, which are groupped subsequently in the processing, occurs in many fields of natural science and technology, and there are quite a few mathematical lines of attack - see, e.g., the survey [15] and the papers [8, 12, 18]. This problem was studied systematically in the theory of chemical relaxation in [5] (see also [16]). The role of quasipolynomials and other entire functions in applied mathematics, physics, and technology was discussed in the monographs [10, 11, 42].

About two hundred years ago, Gasparo Riche, Baron de Prony [39] proposed a method to interpolate a quasipolynomial starting with the data of a physical experiment that are measured in equal time lapses. Under this, a process is modeled by a sum of $n$ exponentials of the form $A\exp\{\alpha t\}$, where $A, \alpha \in \mathbb{R}$; the parameters of this model are identified on the basis of $2n$ uniform counts, and in the general case this procedure is unstable relative to small errors of the measurements [17, 20]. The concept of a well-posed algorithm, as it is adopted in mathematical physics [41, p. 16], demands finding conditions of existence, uniqueness, and stability of the algorithm. Quite a general problem similar to

*) In this section we mean the list of references to Ch. 2.

that of finding a quasipolynomial by its moments (see (1.2)) was considered by A. O. Gelfond [7, p. 216] in the class of entire functions of exponential type when the sequence of moments is infinite. Its solution is not always stable under small changes of the moments (V. K. Ivanov [14]). An estimate of the error of extrapolating entire functions that belong to the Wiener class in the case of a prescribed finite sequence of moments was obtained by the author in [24, 25]. In the case of inaccurate data, a similar problem was treated in [34] in a more general situation. The proofs of Theorems 2.2.1, 2.2.2 can be found in [7, Ch. 5]. The properties of skew-symmetric entire functions resemble the properties of skew-symmetric polylinear mappings [4, pp. 192-195]. The principal results of Ch. 2 are based on the author's paper [22]. In the terminology due to M. M. Lavrent'ev [19], formula (2.2.19) yields a solution of the inner problem for equation (2.2.1) with boundary conditions on a uniform grid. Unlike the class of signals considered in the well-known theorem of V. A. Kotelnikov [11, p. 152], the signals in $\Pi_n(d)$ are synthesized on the basis of a finite number of uniform counts. On the other hand, formula (2.2.19) can be regarded as a solution of the inverse problem for equation (2.2.1) with boundary conditions (2.2.3). The inverse problem for evolution equations in a multidimensional space or with two-dimensional time was considered in [2, 3]. The use of the functional $\varphi_F$ is typical for the generalized Prony method, when it is required only to approximate a sequence of $N$ uniform counts by $n$ exponentials, where $N \gg n$ (see S. M. Kay, S. L. Marple Jr. [15],[43, Ch. 11] and A. A. Kondrat'ev [17]). Cases 5)-10) in Property 2.4.3 were studied by A. I. Karapetyan. Property 2.4.2$'$ and a property similar to Property 2.4.5 for $n = 3$ were established by B. N. Varava in [31], where an outline of the proof can also be found. A complete proof of Property 2.4.2$'$ was given in [32]. The results of Section 2.5 are taken from [29]. The mathematical analysis of relaxation characteristics of wave type given here is based on applying a modification of the Prony harmonic expansion method (see [15]). The case of approximating uniform counts by one sinusoid was considered in detail by M. I. Levin and Yu. I. Semko [21]. For other modifications of the generalized Prony method, see the papers of V. V. Golubkov [8], M. N. Zav'yalov [44] (the case of systems of linear differential equations with constant coefficients) and G. Price [38]. In [33], the author and I. O. Egorushkin studied an example of applying the Prony method to the case of a nonuniform grid. The author's paper [23] present a method of mathematical modeling for an indicator of a homeostasis system in the case where the latter depends not only on time, but also on other variables. The input information about such a system is a set of curves or surfaces specifying, e.g., the space distribution of the parameter in question at equidistant time moments. Its behavior dynamics is assumed to be described by a homogeneous linear differential evolution equation of finite order with analytic coefficients depending only on the space variables. Methods of the theory of entire functions were used in [23] to study direct and inverse problems for this equation. Among other things, to solve the problem of identifying an indicator to be modeled, a multidimensional version of the Prony algorithm was developed; conditions of its existence and uniqueness were studied.

### On Chapter 3

Definition 3.1.1 and the definitions of the $\rho$-support function and the $\rho$-convex hull of a set are due to A. E. Avetisyan and M. M. Dzhrbashyan [34, pp. 333-334]. In a related form, the equivalence of assertions 1), 2) of Theorem 3.2.3 was proved by M. F. Subbotin [182], I. F. Lokhin [104], A. E. Avetisyan [7], and M. M. Dzhrbashyan [34, pp. 335]. M. M. Dzhrbashyan even considered a more general case: the generalized Borel transformation corresponding to the Mittag-Leffler function $E_\rho(z, \mu)$, $\rho > 0$, $\operatorname{Re}\mu > 0$ (see the comments to the Introduction). For $\rho = 1$, Theorem 3.2.3 in an equivalent form was proved by A. Martineau and L. Ehrenpreis (see [139, Ch. 3, §12]; [35, p. 137]). Lemma 3.2.5 and formula (3.2.6) are due to M. M. Dzhrbashyan [34, p. 330].

About the Borel polygon and its properties for $\rho = 1$, see [22], [66, pp. 231-237]. Among other things, E. Borel proved [66, p. 232] the possibility of analytic continuation of the power series (3.3.1) to $\bigcup\{z \in \mathbb{C} : \operatorname{Re}\zeta/z > 1\}|\zeta \in B$; this set coincides with $M \setminus \{0\}$. For $\theta = 0$ the integral formula (3.3.3) was used is the questions of summing a power series in its Mittag-Leffler star (see M. M. Dzhrbashyan [34, pp. 157-159]).

### On Chapter 4

Theorem 1 is due to the author [117]. A different proof of this theorem was outlined by O. V. Epifanov [37]. The construction of the Riemann surface associated with the indicator of an entire function of order $\rho > 0$ and of normal type was proposed by V. Bernstein [17]. A formalization of this construction is represented by the surfaces $(\varphi_l, E_l)$ and $(\varphi_l, E_h)$ $(h \geqslant l)$ considered in Ch. 4. A similar assertion is valid for the polygons $\mathcal{A}_l$ (see Definition 4.2.11) if $\rho$ is an integer. Many-sheeted polygons and their conformal mappings were studied by G. Unkelbach [187]. The existence of Riemann surfaces with prescribed boundary curve were studied by S. R. Nasyrov and F. G. Akhvadiev [149], [5].

We treated the geometry of $(\rho, \alpha)$-convex sets with the use of the Minkowsky duality pattern developed by S. S. Kutateladze and A. M. Rubinov [94, p. 21].

### On Chapter 5

V. Bernstein [17] considered the generalized Borel transformation that takes each function of the form (5.0.1) to the holomorphic function

$$F(P) = \sum_{k=0}^{\infty} b_k P^{-(k+1)/\rho}, \ P = (r, \varphi) \in U,$$

where $P^\nu = r^\nu e^{i\nu\varphi}$, $U = \{(r,\varphi) \in L : r > T; \varphi \in \mathbb{R}\}$, $L = (0,\infty) \times \mathbb{R}$ is the Riemann surface of the logarithm, and $T = T(F) > 0$ is a certain number. By using with this transformation and the construction of the many-sheeted "indicator diagram" (see the comments to Ch. 4), V. Bernstein [17] constructed an entire function of order $\rho > 0$, of normal type, and with prescribed indicator of general form (see also [117]). An even more general transformation of the same nature has also found applications in complex analysis [122, 37]. For $\alpha(\zeta) = \zeta^\rho$ the operator $B_\alpha$ turns into the generalized Borel transformation which, among other authors, was studied by M. F. Subbotin and M. M. Dzhrbashyan (see the comments to the Introudction). The integral formula (5.1.4) is a generalization of the result obtained by A. Macintyre [105] and M. M. Dzhrbashyan [34, p. 327]: if $p = \alpha(\zeta) = \zeta^\rho$ and $g(\zeta) = \widehat{f}(\zeta)$ (see (5.0.2)), then $\Phi(\lambda, \psi_1, \psi_2) = f(\lambda)$ (see (5.0.1) and the arguments in Example 5.1.3). The integral formula (5.2.3) is a particular case of formula (1) of A. I. Markushevich (see the Introduction), its advantage is that in the case where $K = K_\alpha$ there is an inverse integral representation (namely, (5.2.2)) relating $\hat{f}$ and $f$. For $\rho = 1$, Proposition 5.2.2 and Theorem 5.2.3 were proved by A. Martineau [140] if $h$ is a finite function, and by O. V. Epifanov [36] under the condition that $h \in P_1$. The case of $\rho > 0$, $h \geqslant 0$ was treated by O. V. Epifanov, A. A. Lenev [39] and Yu. F. Korobeinik [82]. Theorems 5.2.3 and 5.2.6 are analogs of Theorems $A_1$ and $A_2$ for $n = 1$ (see the Introduction).

Another realization of the space $[\rho, h(\theta)]$ in the case where $\rho > 0$ and $h$ is a positive finite function in $P_\rho$ was proposed by V. A. Tkachenko [184, Lemma 4.1].

Let $\rho^* = \rho/(\rho - 1)$, and let $M(z) := r^\rho h(\theta)$, $z = re^{i\theta} \in \mathbb{C}$, be a convex function. We denote by $h^*(\theta) r^{\rho^*} = M^*(z) = \sup\{\text{Re } zw - M(w) : w \in \mathbb{C}\}$ the Young transform of $M$. Then, the Laplace transformation

$$\mu_w(e^{zw}) = f(z), \ \mu \in [\rho^*, H^*(\theta)]^*; \ f \in [\rho, h(\theta)]$$

is an isomorphism of the spaces $[\rho, h(\theta)]$ and $[\rho^*, h^*(\theta)]^*$. Without the requirement of convexity, this assertion fails.

If the set $G_h := \{\theta \in \mathbb{R} : h(\theta) < 0\}$ is not empty, then the function $M$ is not convex. Indeed, let $\theta_0 \in G_h$. Then

$$M^*(z) \geqslant \sup_{t \geqslant 0} [\text{Re } ze^{i\theta_0} - t^{\rho-1} h(\theta_0)] = \infty \ \forall \ z \in \mathbb{C} \ .$$

Therefore, $M^{**}(z) \equiv -\infty$ . On the other hand, supposing that $M$ is convex and using the continuity of $M$ and the fact that the Young transformation is an involution, we would obtain the relation $M = M^{**}$. This is a contradiction. Moreover, if $h \in P_\rho$ and $G_h \neq \varnothing$, then $e^{zw} \notin [\rho, h(\theta)]$ for every $z \in \mathbb{C}$. So it is reasonable to apply the generalized Laplace transformation in the situation where $h$ may take negative values (see also, e.g., Theorem 3 in [84]).

V. A. Tkachenko studied the operator $\Lambda^*$ and showed that for $h \geqslant 0$ it can be realizes as the Gelfond-Leont'ev operator [184, p. 452]. The definition of the generalized derivative can be found in [101, p. 60]. Generalized convolution operators for $\rho \neq 1$ were studied, e.g., by V. A. Tkachenko [185] and O. V. Epifanov [38]. Theorems 5.3.1 and 5.3.4 extend the results of V. A. Tkachenko in [185, §4], where the case of $h \geqslant 0$ was considered. The work of V. V. Napalkov (see [146-149]) and I. F. Krasichkov (see [89]) was devoted to various aspects of the theory of convolution operators and its generalizations (also multidimensional), and to spaces of entire functions. Representing systems for the classes $H(\overline{\mathcal{D}}), H(\mathcal{D})$, where $\mathcal{D}$ is a convex or a $\rho$-convex domain, can be found, e.g., in the monographs by A. F. Leont'ev [99, 101], the papers by Yu. F. Korobeinik [82], A. V. Abanin [2], and S. N. Melikhov [142]. Theorem 5.3.5 is an analog of Theorems 2.3.1 and 2.3.3 in [101] for the case of the space $H(\overline{\mathcal{D}})$, where $\mathcal{D}$ is a bounded $\rho$-convex domain. Theorem 5.3.6 is a generalisation to $(\rho, \alpha)$-convex domains of the results by A. F. Leont'ev and A. V. Abanin [1] about "continuability inwards" of a representing system for the space $H(\mathcal{D})$, where $\mathcal{D}$ is a convex [86] or $\rho$-convex [1] domain.

Theorem 5.3.5 admits a generalization, which was presented in the author's paper [133]. To formulate the result, we need several additional concepts.

Let $\Lambda = \{\lambda_k\}_1^\infty$ be an infinite set of points in $\mathbb{C}$ such that

$$0 < |\lambda_k| \leqslant |\lambda_{k+1}|, \; k \in \mathbb{N}; \quad \lim_{k\to\infty} |\lambda_k| = \infty \,. \tag{1}$$

We denote by $n(r)$ be the number of points of $\Lambda$ in the disk $\{z \in \mathbb{C} : |z| < r\}$. Assume that the set $\Lambda$ has a finite upper density with respect to some proximate order $\rho(r)$, i.e.,

$$\overline{\lim_{r\to\infty}}\; n(r) \cdot r^{-\rho(r)} < \infty.$$

On the algebra $\sigma(S^1)$ of Borel sets in $S^1 := \{z \in \mathbb{C} : \; |z| = 1\}$, we define a system of measures $\{s_r : r > 0\}$ as follows.

Let $n(E)$ be the number of points of $\Lambda$ beloning to $E \subset \mathbb{C}$. For $\Omega \in \sigma(S^1)$, we denote by $K_r(\Omega) = \{te^{i\theta} \in \mathbb{C} : t \in (0, r); \; e^{i\theta} \in \Omega\}$ the sector of the disk of radius $R$ with the "base" $\Omega \in \sigma(S^1)$. Also, we put

$$s_r(\Omega) = r^{-\rho(r)} n[K_r(\Omega)], \quad \Omega \in \sigma(S^1). \tag{2}$$

In the spirit of the research of V. S. Azarin [10, 11] and L. I. Ronkin [165], we define the angular density of the set $\Lambda$ (this definition is equivalent to the classical one [103, p. 119]).

**Definition 1.** The set $\Lambda$ *has angular density with index* $\rho(r)$ if on $\sigma(S^1)$ there exists a measure $s$ such that (see (2)) $\lim\limits_{r\to\infty} s_r = s$ in the topology of the space $\mathcal{D}^*(\mathcal{S}^\infty)$, i.e.,

$$\lim_{r\to\infty} \int_{S^1} \varphi \, ds_r = \int_{S^1} \varphi \, ds, \quad \varphi \in \mathcal{D}(\mathcal{S}^\infty), \tag{3}$$

where $\mathcal{D}(S^{\infty})$ is the space of infinitely differentiable functions on $S^1$ equipped with the standard topology.

On $\sigma(S^1)$, consider the following measure associated with a finite $2\pi$-periodic trigonometrically $\rho$-convex function $h$ on $\mathbb{R}$. Let

$$s(\theta) = h'_{-}(\theta) + \int_0^{\theta} \rho^2 h(t)\,dt, \quad \theta \in \mathbb{R}, \tag{4}$$

where $h'_{-}$ is the left derivative of $h$ and $\mathcal{P}$ is the class of arcs of the form $\Omega(\psi, \theta) = \{e^{it} : t \in [\psi, \theta)\}$, where $0 < \theta - \psi \leqslant 2\pi$.

On $\mathcal{P}$, we introduce a set function $\nu$ (see (4)) by

$$\nu[\Omega(\psi, \theta)] = s(\theta) - s(\psi) = h'_{-}(\theta) - h'_{-}(\psi) + \int_{\psi}^{\theta} \rho^2 h(t)\,dt. \tag{5}$$

The $\sigma$-ring of subsets of $S^1$ generated by $\mathcal{P}$ coincides with the algebra $\sigma(S^1)$ of Borel subsets of $S^1$. On $\sigma(S^1)$ there exists a unique measure $\nu_h$ such that (see (5)) $\nu_h(E) = \nu(E)\ \forall\ E \in \mathcal{P}$ [65, Ch. II, §8]. In particular,

$$\nu_h([\psi, \theta]) = h'_{+}(\theta) - h'_{-}(\psi) + \int_{\psi}^{\theta} \rho^2 h(t)\,dt,$$

$$\nu_h((\psi, \theta)) = h'_{-}(\theta) - h'_{+}(\psi) + \int_{\psi}^{\theta} \rho^2 h(t)\,dt,$$

where $\psi < \theta < \psi + 2\pi$ and $\psi \in \mathbb{R}$ .

Now we formulate the following supplement to Definition 1.

**Definition 2.** In the previous notation, let $\mu_h := \nu_h / 2\pi\rho$. The set $\Lambda$ *has angular density (with index $\rho(r)$) associated with the function $h$* if $s = \mu_h$ in (3).

Definition 2 makes it possible to formulate a "minimal" criterion for completely regular growth (see Definition 1.10.11, [103, Ch. 3, §3, Theorems 3, 4]; [3, Lemma 1]).

**Theorem 3.** *Let $F$ be an entire function of order $\rho$ and proximate order $\rho(r)$. Suppose $F$ has a prescribed generalized indicator $h$. Let $N$ be the (countable) set of all zeros of $F$. Then $F$ is a function of completely regular growth if and only if $N \setminus \{0\}$ has an infinite subset $\Lambda$ possessing the angular density $\mu_h$ with index $\rho(r)$* (see Definition 2).

Now we pass to the result about an absolutely representing system for the space $H(D)$ of holomorphic functions in an arbitrary bounded $(\rho, \alpha)$-convex domain $D$.

Let $\Pi$ be a sequence of pairwise different complex numbers with the only limit point at $\infty$; let $h$ be the $(\rho, \alpha)$-support function of the domain $D$; finally, let $M(\Pi; h)$ be the class of entire functions of order $\rho$ and normal type with prescribed subset $\Pi$ of simple zeros and with indicator $h$. In the case of $\rho$-convex domains, the following result was proved by A. V. Abanin [3, Theorem 1] in a more general form.

**Theorem 4.** *If $L \in M(\Pi; h)$ and there exists a subsequence $\Lambda = \{\lambda_k\}_1^\infty \subset \Pi \setminus \{0\}$ satisfying condition* (5.3.8) *and having angular density with index $\rho$ assocated with the function $h$* (see Definition 2), *then, in the notation of Theorem* 1.3.5, *the set*

$$\mathfrak{M}(\Lambda) = \{K_\alpha(\lambda_k, z), \; z \in \mathbb{C}\}_1^\infty$$

*is an absolutely representing system for $H(D)$.*

Quite a substential role in the proof of this theorem was played this theorem was played by the paper [85] by Yu. F. Korobeinik.

## On Chapter 6

The idea to study the asymptotic properties of the function $M_f(r)$ on the basis of its convexity as a function of $\ln r_1, \dots, \ln r_n$, and the first results date back to J. Valiron [188]. This line of research was developed by L. I. Ronkin [162] and the author [106]. The class $\mathfrak{A}$ was introduced by L. I. Ronkin. V. S. Vladimirov [191-192] used methods of convex analysis to investigate the asymptotics of a plurisubharmonic function $u$ defined in a tubular radial domain $\Omega = \mathbb{R}^n + iC$; the growth of the function $u$ was compared to the growth of its majorant $v_u(y) = \sup\{u(x + iy) : x \in \mathbb{R}^n\}$, $y \in C$. In [192] V. S. Vladimirov completely described the indicator $h_u(y) = \overline{\lim\limits_{t\to\infty}} \; t^{-\rho} v_u(ty)$, $y \in C$, of the function $u = \ln|f|$, where $f$ is a function holomorphic in the domain $\Omega$, and $p > 0$ is the order of $v_u$ in the totality of variables in the cone $C \subset \mathbb{R}^n$ with vertex at 0.

The definition of the $x$-order (Definition 2.3) and Proposition 6.2.5 for the class $\mathfrak{A}_0$ (see (7.2.1)) are due to A. A. Goldberg, in [56-57] it was shown that the $(G, x)$-type, i.e., the type of the function $M_\Phi(t; G; x)$, depends on the domain $G$. For the class $\mathfrak{A}_0$ and for $x = \mathbb{I}$, these notions were introduced by L. I. Ronkin [162]. In some particular cases of exhausting $\mathbb{R}^n_+$, for the class $\mathfrak{A}_0$ these notions have been considered by many authors. For instance, E. Borel [23, Ch. 6] and J. Sire [176] treated the case where $G = \{r \in \mathbb{R}^n_+ : r \leqslant \mathbb{I}\}$ is a cube; P. Lelong [96] took $G = \{r \in \mathbb{R}^n_+ : |r| < 1\}$, A. A. Temlyakov [183] took $G = \{r \in \mathbb{R}^n_+ : \|r\| \leq 1\}$. Some generalizations of these notions can be found in the papers of F. I. Geche [49-50] and M. N. Sheremeta [173]. Proposition 6.2.6 generalizes the result of J. Valiron [188] about the convexity of the function $\varphi(k) = \rho_\phi(1, k)$, where $\Phi \in \mathfrak{A}_0$, $n = 2$. Some properties of the function $M_\phi(t;G;x)$ (see Definition 6.2.3), e.g., its convexity as a function of $\ln t$, were studied by Sh. I. Strelits [181, §26] in the case when $\Phi \in \mathfrak{A}_0$.

E. Borel was the first who proved Theorem 6.2.7 for the class $\mathfrak{A}_0$ [23, Ch. 6]. Later, J. Valiron proved it in a somewhat more general form [188]; for the class $\mathfrak{A}$ the result was established by L. I. Ronkin [164, p 180]. The method of the proof of Lemma 6.2.8 is close to Ronkin's method of the proof of Theorem 6.2.7 [164, p. 180]. For the class $\mathfrak{A}_0$ the content of Proposition 6.2.10 was obtained by E. Borel [164, Ch. 6] and later by J. Valiron in a somewhat more general form [188].

The notion of the hypersurface of conjugate orders for an entire function of two variables was introduced by L. Baumgartner [13] in 1914. The definition of conjugate types is also due to him; here the presentation follows the book [164, p.178] by L. I. Ronkin.

Let $(\gamma_1, \ldots, \gamma_n)$ be the system of positive conjugate orders of a function $\Phi \in \mathfrak{M}_n$ (see (6.2.12)). Denote by $B_\sigma$ the set of all points $\tau \in \mathbb{R}^n_+$ for which, as $|r| \to \infty$, we have the asymptotic inequality $\Phi(r) < \tau_1 r_1^{\gamma_1} + \cdots + \tau_n r_n^{\gamma_n}$. The boundary $S_\sigma$ of $B_\sigma$ is called the hypersurface of conjugate types for the conjugate orders $(\gamma_1, \ldots, \gamma_n)$ of $\Phi$.

Until recently the work of L. Baumgartner was neglected by mathematicians. For various subclasses these and similar notions associated with various methods of exhausting $\mathbb{R}^n_+$ were considered by M. M. Dzhrbashyan [32], L. I. Ronkin [159-161], and A. A. Goldberg [56-57] (see also F. Gross [64]). A detailed survey of these results can be found in [118, Ch. 1, §2, 3].

Theorem 6.2.12, except for formula (6.2.9), is due to L. I. Ronkin [164, p. 177]. J. Krishna and I. Rao [161] considered comparison functions of the form $Ar_1^{y_1} \ldots r_n^{y_n}$ close to the order function, see (6.2.21): if $\varphi \in Y$, then $\exp\{\varphi(\ln r_1, \ldots, \ln r_n)\} = \sup\{r_1^{y_1} \ldots r_n^{y_n} : y \in K_\varphi\}$, where $K_\varphi$ is the compact convex set in $\mathbb{R}^n_+$ associated with $\varphi$ (see Theorem 1.3.19). Assertion 2) of Theorem 6.2.19 is due to L. I. Ronkin [161]. The type $\sigma_n(r_1, \ldots, r_{n-1})$ (Definition 6.3.17) was studied by L. I. Ronkin [162]. Theorem 6.3.18, except for formula (6.3.28), is also due to L. I. Ronkin [162].

If the hyperasymptote $\Pi$ for the set $\operatorname{epi} V$ (Definition 6.4.4) does not intersect $\operatorname{epi} V$, then it is the same as the $n$-dimensional asymptote (see Klee, [80]) for the closed convex set $\operatorname{epi} V$. Theorems 6.3.11, 6.4.3 are due to the author: see [113, Theorem 2, Theorem 4]; [114, Theorem 1, Proposition 8, Theorem 5]; [109, Proposition 4]; [116, Theorem 3].

"Geometric" versions of Theorems 6.3.11, 6.4.3 are also valid for the functions of class $\mathfrak{A}$ with prescribed order hypersurface or hypersurface of conjugate orders, or with prescribed hypersurface of conjugate types. These facts are due to the author, G. Shopf, and F. I. Geche (see Theorems 1.4.13, 1.5.10, 1.5.10′, 1.5.11 in [118] and comments on them).

**On Chapter 7**

Theorem 7.1.1 is due to J. Valiron [188]. Lemma 7.1.2 can be found in E. Borel's monograph [23, Ch. 6], and in a more general form and with a complete proof in M. M. Dzhrbasyan's paper [32]. A close version of the inequality in Lemma 7.1.3 can be found, e.g., in the paper [56] of A. A. Goldberg. The proof of Lemma 7.1.3, particularly, of its first assertion does not differ essentially from the proof of the corresponding inequalities for $n = 1$ (see, e.g., B. Ya. Levin [103, p. 13]). The formulas in Theorems 7.1.4, 7.1.5 for $x \in \mathbb{R}^n_+$ are due to A. A. Goldberg [57].

The proof of Theorem 7.1.6 was presented in the author's paper [125]. Theorems 7.2.2, 7.2.3 are due to the author [108, 113]. Their proofs are based, in particular, on the studies close to the results of G. A. Friedman [47] about the relationship between the asymptotics of entire functions and their Young transforms. Construction of entire functions in various classes distinguished by indication of other growth characteristics (conjugate orders, conjugate types, etc.) were considered by L. I. Ronkin [164, Theorems 3.13, 3.15], G. Shopf [166-169], and B. N. Ginzburg [53-55]. In the papers of E. I. Yakovlev [194-196, 204] and in the author's papers [136-137] the main results of Ch. 6, Sections 7.1 and 7.2 were carried over to entire functions of infinite order in the totality of variables with nonempty order hypersurface, and to related classes of functions holomorphic in unbounded Reinhardt domains in $\mathbb{C}^n$.

A weaker version of Theorem 7.3.2 is implicit in the paper [188] by J. Valiron:

*For $n = 2$ and in the notation and under the conditions of Theorem 7.3.2, let $G = \{u \in \mathbb{R}^2 : \gamma_j(u) = \infty\}$ (see (7.3.3)). Then condition (7.3.4) is fulfilled if $\overline{K} \setminus \{0\} \subset \overset{\circ}{G}$.*

Corollary 7.3.6 is due to J. Krishna [91-92]. A result similar to Theorem 7.3.2 was proved by I. F. Bitlyan and A. A. Goldberg [19].

*For every $\varepsilon > 0$ there exists a constant $A = A(\varepsilon)$ such that*

$$M_f(r) \leqslant A\mu_f[\ln \mu_f(r)]^{3/2+\varepsilon} \quad \forall\, r \in \mathbb{R}^2_+ \setminus Q$$

*for $n = 2$, where the set $Q$ is such that its intersection with every ray emanating from the point $(0,0)$ has finite logarithmic measure.*

The proof of Theorem 7.3.2 for $n = 1$ turns into the well-known elementary proof of Proposition 7.3.1 (see, e.g., M. A. Evgrafov [44]). Several other results extending some facts of the Wieman-Valiron theory to the multidimensional case were obtained by Sh. I. Strelits [181, §26].

The statement of Theorem 7.4.2 can be found in P. Lelong [97, p. 20] and in L. I. Ronkin [164, p. 138]. L. I. Ronkin obtained a result similar to the assertion of Theorem 7.4.3 concerning the hypersurface of conjugate orders (see Theorem 1.2.19, assertion 2)):

*Let $B$ be a convex domain complete in $\mathbb{R}^n_+$, and let $B^* = \{(b_1^{-1}, \dots b_n^{-1}) : b \in B \cap \mathbb{R}^n_0\}$. Then there exists an entire function $f$ such that the hypersurface* *of conjugate orders of the function*

$$N_f(r) = \frac{1}{(2\pi)^n} \times \int_0^{2\pi} \cdots \int_0^{2\pi} \ln|f(r_1 e^{i\varphi_1}, \ldots, r_n e^{i\varphi_n})|\, d\varphi_1 \ldots d\varphi_n$$

*coincides with* $\partial B^*$.

G. Shopf [170] constructed an entire function $f$ for which the function $N_f(r)$ has a prescribed order function and prescribed type function in the fixed direction of growth. Similarly, G. Shopf [171] studied another real function associated with an entire function $f$ and characterizing the zero set of $f$. Inequality (7.4.1) is due to A. A. Goldberg. For $n = 1$, it turns into an inequality proved by R. Nevanlinna (see [150]). A cruder inequality (as compared to (7.4.1)) had been found before that by L. I. Ronkin [164, p. 407].

We use the term "growth category" following B. Ya. Levin [103]. Theorem 7.4.6 is a multidimensional analog of a known result (see [59, p. 64]). Lemma 7.4.7 is a modification of a result of L. I. Ronkin [164, p. 189], who considered the case of $x = e_n$. For $n = 1$ N. V. Govorov proved an inequality finer than (7.4.5) ($\rho(x) = \max\{0, \gamma x\}$, see [59, the Appendix]).

Some results about the orders and types of entire functions of several variables related somewhat to particular questions of Ch. 6 and 7 are due to S. Dalal [30], J. P. Singh [175], V. Sreenivasulu [178], R. K. Srivactava and V. Kumar [179].

### On Chapter 8

In the case of $n$ complex variables ($n > 1$), in order to prove the multidimensional analog of Theorem $A_1$, V. K. Ivanov [70] used a certain geometric image (in $\mathbb{R}^n$) of an entire function of exponential type. This geometric image reflects the growth of this function along a submanifold of $\mathbb{C}^n$ obtained by fixing the arguments of the coordinates. S. G. Gindikin [52] considered a more general geometric image of such a function, corresponding to a convex cone in $\mathbb{R}^n_+$. L. I. Ronkin [164, p. 270] found an analog of the above result of V. K. Ivanov [164, p. 254] for an entire function with a given system of positive conjugate orders. In the multidimensional version of Theorem 1.10.8′, the role of the indicator diagram for the convex radial indicator is played by a compact convex set $K$ in $\mathbb{C}^n$. As was noted by the author in [119], Ch. O. Kiselman [77] proved a multidimensional analog of the Polya theorem. In that analog, the geometric image $\omega_h$ of the radial indicator $h$ lies in the projective space $\mathbb{P}^n$ :

$$\omega_h = \bigcup \Pi_h(\tau) | \tau \in \mathbb{C},\ \Pi_h(\tau) = \{(z_1, \ldots, z_n, t) \in \mathbb{P}^n : h(\tau z < \mathrm{Re}\ \ t\tau\},$$

where $(z_1, \ldots, z_n, t)$ is the direction vector of a complex line passing thorugh 0 in $\mathbb{C}^{n+1}$ (such a line uniquely determines an element of $\mathbb{P}^n$). "Analytic" multidimensional analogs of Theorem $A_1$ are due to N. S. Krasikova, V. M. Trutnev [90] and E. I. Yakovlev [196]; E. I. Yakovlev considred a multidimensional version of a "spiral" indicator (dating back to A. Macintyre [105]; see also [12]) for an entire function with prescribed system of positive conjugate orders. This indicator is an immediate generalization of the $\rho$-indicator introduced by the author (see Definition 1.4, [120]). I. Kh. Musin [144] studied a certain subclass of entire functions of order $\rho > 0$ and of finite type in the totality of variables. For this class, the role of indicator diagrams is played by the so-called "elementary $r$-convex sets in $\mathbb{C}^n$". For $x = \mathbb{I}$, the set $\Omega_h$ (see (8.1.4)) was studied by Ch. O. Kiselman [77]. Circular sets in $\mathbb{C}^n$ and their generalizations naturally emerge in the summation theory for $n$-fold power series. For $n = 1$, in this connection we refer the reader to Section 3.3 and comments to Ch. 3, and to the author's paper [85]; for $n > 1$, see the papers by V. K. Ivanov [71] and by L. A. Aizenberg and V. M. Trutnev [4], [186]. See also the author's paper [119] and the paper [196] by E. I. Yakovlev. As in Ch. 3, 4, the Minkowsky duality (see [94]) is used to prove Theorem 8.2.3. Theorem 8.2.12 for $x = \mathbb{I}$ is due to Ch. O. Kiselman [77, Theorem 8.3.1]. It is proved in accordance with a pattern exposed in [77]. The proof of Lemma 8.2.10 can be found in the monograph by V. S. Vladimirov [190, p. 109].

Under the conditions of Theorem 8.3.1, if $h$ is a convex function and $K$ is a compact convex set in $\mathbb{C}^n$ such that $h = h_K$, then $\Omega_h = \tilde{K}$ (see Lemma 8.2.5 and Theorem 3.3.1). In this form, the theorem is implicit in the studies by A. Martineau and L. Ehrenpreis (see [147, Ch. 3, §12]), V. M. Trutnev [186]. About the conjugate orders belonging to the class of the so-called linear-convex sets in $\mathbb{C}^n$, see the paper [202] by S. V. Znamensky, containing a bibliographical survey. The transformation (8.3.2) was proposed by E. I. Yakovlev [196]. Theorems 8.3.1 and 8.4.5 are due to the author [129]; Corollary 8.4.2 is a weak version of the Martineau theorem (see [164, p. 294]). The part of Theorem 8.4.5 concerning the relationship between the "Borel polygon" $B = B_0$ and the indicator of the entire function associated with the series (8.4.1) was found earlier for $x = \mathbb{I}$ by V. M. Trutnev [186], and in the general case by E. I. Yakovlev [196]. The formula $S = \{r \subset \mathbb{R}_0^n : \sigma_f(r; x) = 1\}$ was obtained by E. I. Yakovlev [196] by a different method.

### On Appendix A

Mostly, the results of Section A.1 can be found in the monograph of P. Forster [46]. In addition, to write A.1 the author used the monographs of B. P. Apanasov [6] (Propositions A.1.7 and A.1.9, Definition A.1.8), W. Massi and J. Stollings [141] (Propositions A.1.11 and A.1.13, Definition A.1.12), V. A . Rokhlin and D. B. Fux [158, p. 427] (Theorem A.1.18), T. Bröcker and L. Lander [27] (Theorem A.1.17). The construction of the Riemann surface for the inverse of a polynomial of a fractional order is based on the well-known procedure used to construct the Riemann surface of an algebraic function (see, e.g., [150], [134]).

### On Appendix B

The first results related to generalization of Theormes $A_1$, $A_2$ to the class of entire functions of one variable with finite order $\rho > 0$ and with minimal or maximal type are due to V. Bernstein [16]. He studied asymptotic properties of the function

$$E(z;W) = \sum_{k=0}^{\infty} \frac{z^n}{\theta(n)}, \ z \in \mathbb{C}^n; \quad \theta(\zeta) = \int_0^{\infty} e^{-W(t)} t^{\zeta} dt, \ \mathrm{Re}\, \zeta > 1, \qquad (1)$$

where $W(z)$ is a function analytic in the angle $|\,Arg\, z| < \pi/\rho$ and satisfying the asymptotic formula $W(re^{i\theta}) = W(r)e^{i\rho\theta}[1 + o(1)]$ as $r \to \infty$. Here the asymptotics is uniform with respect to $\theta \in [-\delta, \delta]$, where $\delta < \pi/\rho$, and $\rho_0(r) = \ln W(r)/\ln r$ is a proximate order equivalent to the given one. If $W(r) = r^{\rho}$, then $\rho^{-1}E(\cdot; W) = E_{\rho}$ is the Mittag-Leffler function (see (3.2.3)).

This Appendix is a modified version of the author's paper [135] (the latter was written on the basis of the note [134] and the paper [124]).

In Section B.1, the principal facts about proximate orders are taken from the monographs [103], [59]. Theorem B.1.4 is due to A. F. Grishin and T. I. Malyutina [63], and Theorem B.1.4′ is a modification of another result of theirs (see Remark 2 to Theorem 3 in [16]). Theorem B.1.12 can be found in the paper [58] by A. A. Goldberg.

In Section B.2, we further develop the results of Vl. Bernstein [16] about analytic proximate orders equivalent to a given one. In particular, Theorem B.2.6 is an immediate generalization of one of Bernstein's results in [16].

The proof of Theorem B.3.1 is based on the methods of integral transformations in the paper [105] by A. Macintyre. For the functions of the form (1), Theorem B.3.3 was proved by Vl. Bernstein [16]. Theorem B.3.4 can be found in the monograph [44, p.21]. Asymptotic expansions of the functions of the form $E(\cdot; W)$ (see (1)) were obtained by the arguments used in [34, p. 133], [59, p.113] to investigate the asymptotics of the Mittag-Leffler function. Proposition B.3.6 is entirely parallel to one of Dzhrbashyan's results in [34, p. 120]. Vl. Bernstein obtained a lower bound for the generalized indicator of the function $E(\cdot; W)$. Statement 2 of Theorem B.3.7 and its proof are due to M. A. Evgrafov [43]. Other statements of that theorem were proved independently by M. A. Evgrafov [43] and by the author [121], [124].

Theorem B.4.5, which is an analog of Theorems $A_1$, $A_2$ for entire functions of arbitrary proximate order with nonnegative generalized indicator, is due to the author [115], [121], [124]. In those papers, a function somewhat similar to $E(\cdot; W)$ (see (1)) played the role of the kernel of the generalized Laplace transformation. Assertion 1 of that theorem was proved also by M. A. Evgrafov [43], independently and by a different method.

In Section B.5 we study analytic proximate orders generated by entire or meromorphic functions. The asymptotic analysis of "gamma-functions and exponentials" associated with them (in terms of the paper [48]), i.e., of analogs for the Mittag-Leffler functions of the form (1), leads to deeper results as compared to Section B.3, especially for the functions of order $\rho \in (0,1]$. In particular, the generalized indicator of an "exponential" of order $\rho \in (1/2,1)$ and of order 1 and minimal type turns out to be trigonometric on the interval $[-\pi,\pi]$ (Theorem B.5.4). The first example of this kind was announced by M. A. Evgrafov in [43]. Assertion 1) of Theorem B.5.3 for a more general situation was also proved there.

In Section B.6 we deal with applications of results of Section B.5. There, on one hand, we deduce a *complete* analog of Theorem A for the entire functions of order 1 and minimal type with an arbitrary generalized indicator (Theorem B.6.3). On the other hand, we give a convergence criterion for a series formed by the main parts of the Laurent expansion of a meromorphic function, and discuss partial fraction expansion for such functions (Theorem B.6.4). This criterion is verified for the meromorphic function $1/g$ and its derivative, where $g$ is an entire function of order 1 and maximal type with trigonometric generalized indicator (Theorem B.6.5). For the entire functions of class $A$, such an expansion was obtained by M. G. Krein [103, p. 333].

* *

*

The results not mentioned in the comments on Chapters 1-8 and Appendices A, B are due to the author.

# Bibliography *)

## References to Chapters 1, 3-8


1. Abanin A. V. *Representing systems in ρ-convex domains.* Izvestiya Sev.-Kavkaz. Nauch. Tsentra Vysshei Shkoly (1980), **no. 4**, 3-5. (In Russian).
2. Abanin A. V. *Distribution of the indices of representing systems of generalized exponentials.* Investigations in complex analysis: Collection of papers, Ufa, BFAN SSSR, 1987, pp. 5-14. (In Russian).
3. Abanin A. V. *Characterization of minimal systems of indices for representing systems of generalized exponentials.* Mat. Zametki (1991), **no. 2 (345)**, pp. 3-12. (In Russian).
4. Aizenberg L. A., Trutnev V. M. *A Borel-type summation method for n-fold power series.* Sibirsk. Mat. Zh. (1971), **Vol. 12., no. 6**, pp. 1895-1901. (In Russian).
5. Akhvadiev F. T., Hasyrov S. R. *Construction of a Riemann surface by its boundary.* Izv. VUZ, Matem. (1986), **no. 5**, pp. 3-11. (In Russian).
6. Apanasov B. N. *Discrete transformation groups and manifold structures.* Nauka, Siberian Branch, 1983. 242 pp. (In Russian).
7. Avetisyan A. E. *Generalization of a theorem of G.Polya.* Dokl. AN SSSR (1955), **Vol. 105, no. 4,** pp. 885-888. (In Russian).
8. Avetisyan A. E. *Two theorems on functions analytic in angular domains.* Dokl. AN Arm. SSR (1959), **Vol. 29, no. 5**, pp. 193-202. (In Russian).
9. Azarin V. S. *Theory of growth of subharmonic functions. Part* 1. Kharkov State University, 1978. 73 pp. (In Russian).
10. Azarin V. S. *Theory of growth of subharmonic functions. Part* 2. Kharkov State University, 1982. 74 pp. (In Russian).
11. Azarin V. S. *On the asymptotic behavior of subharmonic functions of finite order,* Mat. Sb. (1979), **Vol. 108 (150), no. 2,** pp. 147-167. (In Russian).
12. Balashov S. K. *On entire funcions of completely regular growth along curves with regular rotation.* Abstract of the Ph. D. Thesis. Rostov-on-Don, 1972. 16 pp. (In Russian).
13. Baumgartner L. *Beitrage zur Theorie der ganzen Functionen von zwei komplexsen Veranderlichen.* Monatshefte fur Math. und Phis. (1914), **Vol. 25**, pp. 3-70.
14. Bernstein V. *Sur une généralisation de la métode de sommation exponentielle de M. Borel.* C.R. Acad. Sci. Paris (1932), **Vol. 194**, pp. 1887-1889.
15. Bernstein V. *Leçons sur les progrès recents de théorie des series de Dirichlet.* Paris: Gaunter-Villars, 1933. 320 pp.
16. Bernstein V. *Sulla crescenza delle transcendenti intere di ordine finito.* Reale Accad. d'Italia. Memoire della classe di scien. fis. mat. e natur. (1933), **Vol. 4**, pp. 339-401.
17. Bernstein V. *Sulle proprieta caratteristiche delle indicatrici di crescenza delle transcendenti intere d'ordine finito.* Ibid. (1936), **Vol. 6**, pp. 131-189.
18. Bieberbach L. *Analytishe Fortsetzung.* Springer-Verlag, Berlin-Gottingen-Heidelberg, 1955.


*) Many Russian journals and books are translated into English cover-to-cover. Below, the author refers only on the original sources.

19. Bitlyan I. F., Goldberg A. A. *The Viman-Valiron theorem for entire functions of several complex variables.* Vestnik Leningradskogo Univ., ser. matem., mekhan., astron. (1959), **no. 13**, issue 2, pp. 27-41 (In Russian).

20. Boas R. P. *Entire functions.* New York: Acad. Press, 1954. 276 pp.

21. Boas R. P., Buck R. G. *Polynomial expansions of analytic functions.* Berlin a.o.: Springer-Verlag, 1958. 77 pp.

22. Borel E. *Leçons sur les séries divergentes.* Paris: Gaunter-Villars, 1901. 184 pp.

23. Borel E. *Leçons sur les séries a termes positifs.* Paris, 1902. 91 pp.

24. Borisovich Yu. G., Bliznyakov M. M., Izrailevich Ya. A., Fomenko T. N. *Introduction to topology.* Vyssh. Shk., Moscow, 1980. 295 pp. (In Russian).

25. Bourbaki N. *Espaces vectoriels topologiques.* Hermann & C[ie], Éditeurs, Paris.

26. Bourbaki N. *Fonctions d'une variable réelle.* Hermann & C[ie], Éditeurs, Paris.

27. Bröcker Th., Lander L. *Differentiable germs and catastrophes.* Cambridge University Press, 1975.

28. Bronsted A. *Conjugate convex functions in topological vector spaces.* Mat. Fys. Medd. Dansk. Vid. Selsk. (1964), - **Vol. 34, no. 2**, pp. 1-26.

29. Cassels J. W. S. *An introduction to the geometry of numbers.* Springer-Verlag, Berlin-Gottingen-Heidelberg, 1959.

30. Dalal S. S. *On the order and type of integral functions of several complex variables.* Trans. Indian. Math. Soc. (1970), **Vol. 33, no. 2-4**, pp. 215-220.

31. Dzhrbashyan M. M. *Representation and uniqueness theorems for analytic functions.* Izv. AN SSSR, ser. matem. (1952), **Vol. 16**, pp. 225-252. (In Russian).

32. Dzhrbashyan M. M. *Concerning the theory of certain classes of entire functions of several complex variables.* Izv. AN Arm. SSR, ser. fiz-mat., estestv. i technich. nauk (1955), **Vol. 8**, pp. 1-29. (In Russian).

33. Dzhrbashyan M. M. *Integral expansion and generalized Taylor series expansion for entire functions of several complex variables.* Mat. Sb., (1957), **Vol. 41(83), no. 2**, pp. 257-276. (In Russian).

34. Dzhrbashyan M. M. *Integral transformations and expansions in the complex domain.* Nauka, Moscow, 1966. 672 pp. (In Russian).

35. Ehrenpreise L. *Fourier analysis in several complex variables.* New York: Wiley-Intersciense publishers, 1970. 506 pp.

36. Epifanov O. V. *On the solvability type of a differential equation of infinite order.* In: Mat. Analyz i ego Prilozh., Rostov-on-Don University Publishers, Rostov-on-Don, 1969, pp. 72-82. (In Russian).

37. Epifanov O. V. *A differential operator with polynomial coefficients on classes of entire functions with a prescribed estimate for the indicator.* Mat. Sb. (1981), **Vol. 114, no. 1**, 85-109. (In Russian).

38. Epifanov O. V. *Multiplication operators on spaces of entire functions of finite order, and convolution-type operators.* Mat. Sb., 120 (1983), **Vol. 120, no. 1**, pp. 505 528. (In Russian).

39. Epifanov O. V., Lenev A. A. *On the solvability of a certain integral equation in the space of analytic functions.* In: Mat. Analyz i ego Prilozh., Rostov-on-Don University Publishers, Rostov-on-Don, 1974, pp. 258-261. (In Russian).

40. Evgrafov M. A. *The Abel-Goncharov interpolation problem.* GITTL, Moscow, 1954. 127 pp. (In Russian).

41. Evgrafov M. A. *The method of close systems in spaces of analytic functions, and its applications to interpolation.* Trudy Mosk. Mat. Obshch. (1956), **Vol. 5**, pp. 89-201. (In Russian).

42. Evgrafov M. A. *On some aspects of the stationary phase method.* Institute of Applied Math. of the Soviet Academy of Science / Preprint (1975), **no. 35**. 68 pp. (In Russian).

43. Evgrafov M. A. *Generalized Borel transformation.* Institute of Applied Math. of the Soviet Academy of Science / Preprint (1976), **no. 35**. 57 pp. (In Russian).

44. Evgrafov M. A. Asymptotic estimates and entire functions. Nauka, Moscow, 1979. 320 pp. (In Russian).

45. Firsakova O. S. *Some problems of interpolation by entire functions.* Dokl. AN SSSR (1958), **Vol. 120, no. 3**, pp. 477-480. (In Russian).

46. Forster O. *Riemannsche Flächen.* Springer-Verlag, Berlin-Heidel berg-New York, 1977.

47. Fridman G. A. *On the asymptotic behavior of Young dual functions.* Mat. Sb. (1966), **Vol. 70, no. 4**, pp. 518-525. (In Russian).

48. Gelfand I. M., Graev M. I., Retakh V. S. *General gamma-functions, exponentials, and hypergeometric functions.* Uspekhi Matem. Nauk (1998), **Vol. 53, no. 1**, pp. 3-60. (In Russian).

49. Geche F. I. *On growth characteristics for entire functions of several complex variables.* Dokl. AN SSSR, (1965), **Vol. 164, no. 33**, pp. 487-490 (In Russian).

50. Geche F. I. *On proximate growth characteristics for entire functions of several complex variables.* Litovsk. Mat. Sb. (1968), **Vol. 8, no. 3**, pp. 461-487. (In Russian).

51. Geche F. I. *Growth control for entire functions of several complex variables with the help of proximate characteristics.* Dopovidi AN Ukr. SSR, ser. A (1975), **no. 2**. pp. 38-49. (In Ukrainian).

52. Gindikin S. G. *The Borel transform and the indicator of an entire function of several variables.* Uspekhi Mat. Nauk (1964), **Vol. 19, no. 2**, pp. 202-204. (In Russian).

53. Ginzburg B. N. *On the growth of entire characteristic functions of probability laws.* In: Theory of Functions, Functional Analysis, and its Applications, **no. 20**. Izd. Khar'kov Univ., Kharkov, 1974, pp. 38-49. (In Russian).

54. Ginzburg B. N. *On the growth of entire ridge functions of several variables.* In: Theory of Functions, Functional Analysis, and its Applications, **no. 26.**, Izd. Khar'kov Univ., Kharkov, 1976, pp. 9-11. (In Russian).

55. Ginzburg B. N. *On the characteristic properties of some hypersurfaces for entire functions of several variables.* In: Theory of Functions, Functional Analysis, and its Applications, **no. 28**. Izd. Kharkov Univ., Kharkov, 1978, pp. 9-15. (In Russian).

56. Goldberg A. A. *Elementary remarks about the formulas for the order and type of entire functions of several variables.* Dokl. AN Arm. SSR (1959), **Vol. 29**, pp. 145-152. (In Russian).

57. Goldberg A. A. *On the formulas for the order and type of an entire functions of several variables.* Dokl. i Soobshch. Uzhgorod. Gos. Univ., ser. fiz.-mat. nauk (1961), **no. 4**, pp. 101-103. (In Russian).

58. Goldberg A. A. *Integral representation of monotone slowly varying functions.* Izv. VUZ., Matem. (1988), **no. 4**, pp. 21-27. (In Russian).

59. Goldberg A. A., Ostrovskii I. V. *Distribution of values of meromorphic functions.* Nauka, Moscow, 1970. 592 pp. (In Russian).

60. Goldberg A. A., Ostrovskii I. V. *On the derivatives and primitives for entire functions of completely regular growth.* In: Theory of Functions, Functional Analysis, and its Applications, **no. 18**. Izd. Khar'kov Univ., Kharkov, 1973, pp. 70-81. (In Russian).

61. Govorov N. V., Chernykh N. M. *On the completely regular growth of the Borel transform with finitely many singularities for the analytic extension of the associated function.* Izv. AN SSSR, ser. matem. (1978), **Vol. 42, no. 5**, pp. 965-971. (In Russian).

62. Govorov N. V., Chernykh N. M. *Completely regular growth for some classes of analytic functions of exponential type that are representable by Borel integrals or power series.* Izv. AN SSSR, ser. matem. (1985), **Vol. 42, no. 6**, pp. 1155-1176. (In Russian).

63. Grishin A. F., Malyutina T. I. *On the proximate order.* In: Complex analysis and mathematical physics. Krasnoyarsk Gos. Univ., Krasnoyarsk, 1998, pp. 10-24. (In Russian).

64. Gross F. *Generalized Tailor series, and orders and types of entire functions of several complex variables.* Trans. Amer. Math. Soc. (1965), **Vol. 120, no. 1**, pp. 124-144.

65. Halmos P. R. *Measure theory*. New York, 1950.

66. Hardy G. H. *Divergent series*. Oxford, 1949.

67. Hörmander L. *An introduction to complex analysis in several variables.* D. Van Nostrand Company, Inc. Princeton, New Jersey, Toronto-New York-London, 1966. 208 pp.

68. Ioffe A. D., Tikhomirov V. M. *Duality of convex functions and extremal problems,* Uspekhi Mat. Nauk, (1968), **Vol. 23, no. 6(144)**, pp. 51-116. (In Russian).

69. Ivanov V. K. *Distribution of singularities of a potential, and a spatial analog of the Polya theorem.* Mat. Sb. (1956), **Vol. 40, no.3**, pp. 319-338. (In Russian).

70. Ivanov V. K. *The relationship between the growth of an entire function of several variables and the distribution of singularities of the associated function.* Mat. Sb. (1957), **Vol, 43, no. 3**, pp. 367-375. (In Russian).

71. Ivanov V. K. *A growth characteristic for entire functions of two variables, and its applications to summation of double power series.* Mat. Sb. (1959), **Vol. 47, no. 1**, pp. 131-140. (In Russian).

72. Kamthan P. K., Gupta M. *Space of entire functions of several complex variables having finite order point.* Math. Japonicae. (1975), **Vol. 20, no. 1**, pp. 7-19.

73. Kaz'min Yu. A. *Methods of interpolation of analytic functions and their applications.* Abstract of the Doctor Thesis, Moscow, 1972. 22 pp. (In Russian).

74. Kaz'min Yu. A.*On a problem of A. O. Gel'fond.* Mat. Sb., (1973), **Vol. 90, no. 4**, pp. 521-544. (In Russian).

75. Kaz'min Yu. A.*Comparison of functions.* In: Mathematical Ecyclopaedy, Sov. entsiklopediya, 1985, p. 159. (In Russian).

76. Khavin V. P. *Spaces of analytic functions.* In: Itogi Nauki, Mat. Analiz. VINITI, 1964, 76-164. (In Russian).

77. Kiselman C. O. *On entire functions of exponential tape and indicators of analytic functionals.* Acta math. (1967), **Vol. 117**, pp. 1-35.

78. Kiselman C. O. *Existence of entire functions of one variable with prescribed indicator.* Arkiv math. (1968), **Vol. 7, no. 35**, pp. 505-508.

79. Kiselman C. O. *Functionals on the space of solutions to a differential equation with constant coefficients. The Fourier and Borel transformations.* Math. Scand. (1968), **Vol. 23**, pp. 27-53.

80. Klee V. *Asymptotes and projections of convex sets.* Ibid. (1960), **Vol. 8**, pp. 356-362.

81. Korobeinik Yu. F. On a converse to the Liouville theorem, Izv. VUZ, Matem. (1965), **no. 1**, pp. 81-90. (In Russian).

82. Korobeinik Yu. F. *Representing systems.* Uspekhi Mat. Nauk (1981), **Vol. 36, no. 1(217)**, pp. 73-126. (In Russian).

83. Korobeinik Yu. F. *Inductive and projective topologies. Sufficient sets and representing systems.* Izv. AN SSSR, ser. matem. (1986), **Vol. 50, no. 3**, pp. 539-565. (In Russian).

84. Korobeinik Yu. F. *Absolutely representing families.* Mat. Zametki (1987), **Vol. 42, no. 5**, pp. 670-680. (In Russian).

85. Korobeinik Yu. F. *Absolutely representing systems and realization of the dual space.* Izv. VUZ., Matem. (1990), **no. 2**, pp. 68-76. (In Russian).

86. Korobeinik Yu. F., Leont'ev A. F. *On the property of continuability inwards for representing systems of exponentials.* Mat. Zametki (1980), **Vol. 28, no. 2**, pp. 243-254. (In Russian).

87. Korobeinik Yu. F., Melikhov S. N. *Realization of the conjugate space with the help of the generalized Fourier-Borel transformation. Applications.* In: Complex analysis and mathematical physics. IF SO AN SSSR, Krasnoyarsk, 1988, pp. 62-73. (In Russian).

88. Köthe G. *Dualität in der Funktionen Theorie.* G. Reine und Angew. Math. (1953), **Vol. 191, no. 1-2**, pp. 30-49.

89. Krasichkov I. F. *Spectral synthesis of analytic functions.* Abstract of the Doctor Thesis, Ufa, 1974. 10 pp. (In Russian).

90. Krasikova N. S., Trutnev V. M. *Multidimensional analog of the Polya theorem.* In: On holomorphic functions of several complex variables. IF SO AN SSSR, Krasnoyarsk, 1988, pp. 62-73. (In Russian).

91. Krishna G. *Maximum term of power series in one and several complex variables.* Pacif. J. Math. (1969), **Vol. 29, no. 3**, pp. 609-622.

92. Krishna G. *Probabilistic techniques leading to a Valiron type theorem in several complex variables.* Ann. Math. Statist. (1970), **Vol. 41, no. 6**, pp. 2126-2129.

93. Krishna J. G., Rao J. H. N. *On orders and types of an entire function over* $\mathbb{C}^n$. J. Australian Math. Soc. (1973), **Vol. 15, no. 4**, pp. 393-408.

94. Kutateladze S. S., Rubinov A. M. *Minkowski duality and its applications.* Nauka, Siberian Branch, Novosibirsk, 1976. 251 pp. (In Russian).

95. Lavrent'ev M. A. *Sur une classe de representations continues.* Mat. Sb. (1935), **Vol. 42, no. 4**, pp. 407-424.

96. Lelong P. *Sur quelques problèmes de la théorie des fonctions de deux variables complexes.* Ann. École Notm. Sup. (1948), **Vol. 58**, pp. 83-177.

97. Lelong P. *Fonctions plurisousharmoniques et formes differéntielles positives.* Paris a. o.: Gordon and Breach, 1968. 79 pp.

98. Lelong P. Gruman L. *Entire functions of several complex variables.* Berlin a. o.: Springer-Verlag, 1986. 270 pp.

99. Leont'ev A. F. *Series of exponentials.* Nauka, Moscow, 1976. 536 pp. (In Russian).

100. Leont'ev A. F. *Sequences of polynomials of exponentials.* Nauka, Moscow, 1980. 384 pp. (In Russian).

101. Leont'ev A. F. *Generalizations of series of exponentials.* Nauka, Moscow, 1981. 320 pp. (In Russian).

102. Leont'ev A. F. *Entire functions. Series of exponentials.* Nauka, Moscow, 1983. 176 pp. (In Russian).

103. Levin B. Ya. *Distribution of zeros of entire functions.* Gostekhizdat, Moscow, 1956. 632 pp. (In Russian).

104. Lokhin I. F. *On functions associated with entire functions of finite order.* Uchen. Zapiski Gor'k. Univ. (1953), **no. 28,** pp. 20-23. (In Russian).

105. Macintyre A. I. *Laplace's transformation and integral functions.* Proc. London Math. Soc. (1938), **Vol. 45, no. 2**, pp. 1-20

106. Maergoiz L. S. *On the relationship between various definitions of the orders of entire functions of several variables.* Sibirsk. Mat. Zh. (1966), **Vol. 7, no. 6**, pp. 1268-1292. (In Russian).

107. Maergoiz L. S. *Some properties of convex sets and their applications to the growth theory of convex and entire functions.* Sibirsk. Mat. Zh. (1968), **Vol. 9, no. 3**, pp. 577-591. (In Russian).

108. Maergoiz L. S. *On growth scales for entire functions of several variables.* Dokl. AN SSSR (1970), **Vol. 192, no. 3**, pp. 495-498 . (In Russian).

109. Maergoiz L. S. *A boundary value problem for convex functions, and its applications to the study of the asymptotics of functions.* Dokl. AN SSSR, (1971), **Vol. 198, no. 4**, pp. 762-765. (In Russian).

110. Maergoiz L. S. *On the asymptotics of convex functions of several variables.* Optimizatsiya (1971), **no. 1(18)**, pp. 67-81. (In Russian).

111. Maergoiz L. S. *The order function and growths scales for entire functions of several variables.* Sibirsk. Mat. Zh., (1972), **Vol. 13, no. 1**, pp. 118-132. (In Russian).

112. Maergoiz L. S. *Plane ρ-convex sets and some their applications.* In: Holomorphic functions of several complex variables, IF SO AN SSSR, Krasnoyarsk, 1972, pp. 75-91. (In Russian).

113. Maergoiz L. S. *On types and the related growth scales for entire functions.* Dokl. AN SSSR (1973), **Vol. 213, no. 5**, pp. 1025-1028. (In Russian).

114. Maergoiz L. S. *The type functions of entire function of several complex variables on directions of its growth.* Sibirsk. Mat. Zh., (1973), **Vol. 14, no. 5**, pp. 1037-1056. (In Russian).

115. Maergoiz L.S. *An analog of the Polya theorem for entire functions of a given proximate order.* In: Holomorphic functions of several complex variables, IF SO AN SSSR, Krasnoyarsk, 1973, pp. 109-121. (In Russian).

116. Maergoiz L. S. *On a multidimensional analog for the type of an entire function.* Uspekhi Matem. Nauk (1975), **Vol. 30, no. 5(185)**, pp. 215-216. (In Russian).

117. Maergoiz L. S. *On the structure of the indicator of an entire function of finite order and normal type.* Sibirsk. Mat. Zh. (1975), **Vol. 16, no. 2**, pp. 301-313. (In Russian).

118. Maergoiz L. S. *Elementas of the growth theory for convex and entire functions of one or several variables.* Krasnoyarsk University Publishers, Krasnoyarsk, 1976. 142 pp. (In Russian).

119. Maergoiz L. S. *On analogs of the Polya theorem for entire functions of several variables.* In: Holomorphic functions of several complex variables, IF SO AN SSSR, Krasnoyarsk, 1976, pp. 193-199. (In Russian).

120. Maergoiz L. S. *On geometric multidimentional analogs of the growth order and type of an entire function. In: Problems of mathemetics.* Collection of papers. Tashkent, 1976, **no. 510**, pp. 37-38. (In Russian).

121. Maergoiz L. S. *An analog of the Polya theorem for entire functions of a given proximate order, and applications.* Funktsion. analiz i ego prilozh., (1977), **Vol. 11, no. 3**, pp. 84-85. (In Russian).

122. Maergoiz L. S. *An analog of the Polya theorem.* Izv. VUZ, Matem. (1977) **no. 12,** pp. 54-64. (In Russian).

123. Maergoiz L. S. *On a rezult of Valiron.* In: Theory of Functions, Functional Analysis, and its Applications, **no. 29.**, Izd. Khar'kov Univ., Kharkov, 1978, pp. 89-98. (In Russian).

124. Maergoiz L. S. *An analog of the Polya theorem and its applications.* Izv. AN Arm. SSR, (1980), **Vol. 15, no. 1**, pp. 45-62. (In Russian).

125. Maergoiz L. S. *On regularized types of convex and entire functions with strictly convex order hypersurface.* Izv. VUZ, Matem. (1980) **no. 11**, pp. 60-66. (In Russian).

126. Maergoiz L. S. *Analytic realization of operators commuting with the generalized differentiation operator.* Uspekhi Matem. Nauk, (1982), **Vol. 37, no. 3**, pp. 191-192. (In Russian).

127. Maergoiz L. S. *On representing systems of entire functions.* Dokl. AN SSSR, (1985), **Vol. 285, no. 5**, pp. 1058-1061. (In Russian).

128. Maergoiz L. S. *The plane indicator diagram for an entire function of integral order* $\rho \neq 1$. Sibirsk. Matem. Zh., (1987), **Vol. 28, no. 2**, pp. 107-124. (In Russian).

129. Maergoiz L. S. *Indicator diagram for an entire function of several variables with nonnegative indicator.* Dokl. AN SSSR, (1988), **Vol. 299, no. 3**, pp. 551-554. (In Russian).

130. Maergoiz L. S. *The plane indicator diagram for an entire function of nonintegral order.* In: Complex analysis and mathematical physics, School-seminar (abstracts). IF SO AN SSSR, Krasnoyarsk, 1987, p. 69 . (In Russian).

131. Maergoiz L. S. *Construction of the Riemann surface for the inverse of a polynomial of fractional order.* All-union conference on geometric theory of functions (abstracts), IM SO AN SSSR, Novosibirsk, 1988, pp. 63. (In Russian).

132. Maergoiz L. S. *On applications of the Lagrange interpolation series.* In: Republic meeting-seminar on complex analysis and applied control problems (abstracts). Math. Inst. Ukrainian Acad. Sci., Kiev, 1989, **Vol. 30**, p. 30. (In Russian).

133. Maergoiz L. S. *Representing systems for the space of holomorphic functions in a* $(\rho, \alpha)$*-convex domain.* Zapiski Nauchn. Semin. POMI, **Vol. 206** (1993), pp. 91-106. (In Russian).

134. Maergoiz L. S. *An analog of the Mittag-Leffler function associated with an analytic proximate order.* In: Complex analysis and mathematical physics, Krasnoyarsk State University, Krasnoyarsk, 1998, pp. 101-111. (In Russian).

135. Maergoiz L. S. *Indicator diagram and generalized Borel-Laplace transforms for entire functions of a given proximate order.* Algebra i Analiz (2000), **Vol. 12, no. 2**, pp. 1-63. (In Russian).

136. Maergoiz L. S., Yakovlev E. I. *On the growth of convex or entire functions of infinite order in the totality of variables.* In: Some questions of multidimensional complex analysis, IF SO AN SSSR, Krasnoyarsk, 1980, pp. 79-88. (In Russian).

137. Maergoiz L. S., Yakovlev E. I. *The Dirichlet problem for convex functions in an unbounded domain, and its applications to the asymptotic study of functions of several variables.* Optimizatsiya (1985), **no. 35**, pp. 79-88. (In Russian).

138. Markushevich A. I. *On a basis in a space of analytic functions.* Mat. Sb., 17 (1945), **Vol. 17, no. 2**, pp. 211-252. (In Russian).

140. Martineau A. *Equations differéntielles d'ordre infini*. Bull. Math. France, (1967), **Vol. 95**, pp. 109-154.

141. Massey W. S., Stallings J. *Algebraic topology: An Itroduction*. Yale University Press, New Haven and London, 1971.

142. Melikhov S. N. *Expansion of analytic functions in a series of exponentials*. Izv. AN SSSR, ser. matem. (1988), **Vol. 52, no. 5**, pp. 991-1004. (In Russian).

143. Muraev E. B. *Borel summation of n-fold series, and related entire functions*. Sibirsk. Matem. Zh. (1978), **Vol. 19, no. 6**, pp. 1332-1340. (In Russian).

144. Musin I. Kh. *Analytic functionals in ρ-convex domains of a multidimensional space, and some functonal equations*. Ph. D. Thesis abstract. Ufa, 1987. 14 pp. (In Russian).

145. Napalkov V. V. *On discrete sufficient sets in certain spaces of entire function*. Dokl. AN SSSR (1980), **Vol. 250, no. 4**, pp. 809-812. (In Russian).

146. Napalkov V. V. *On discrete weakly sufficient sets in certain spaces of entire function*. Izv. AN SSSR, ser. matem. (1981), **Vol. 45, no. 5**, pp. 1082-1099. (In Russian).

147. Napalkov V. V. *Convolution equations in multidimensional spaces*. Nauka, Moscow, 1982. 240 pp. (In Russian).

148. Napalkov V. V. *Spaces of analytic functions of prescribed growth near the boundary*. Izv. AN SSSR, ser. matem. (1987), **Vol. 51, no. 2**, pp. 287-305. (In Russian).

149. Nasyrov S. R. *Construction of finite Riemann surfaces by the boundary curve*. Dokl. AN SSSR (1987), **Vol. 297, no. 6**, pp. 1311-1314. (In Russian).

150. Nevanlinna R. *Uniformisierung*. Berlin, 1953.

151. Pasechnik G. M. *Geometric characterization of the set of indices for a representing system of exponentials in H(G)*. Dokl. AN SSSR (1980), **Vol. 252, no. 3**, pp. 553-555. (In Russian).

152. Polya G. *Untersuchungen über Lücken and Singularitäten von Potenzsrihen*. (1929), **Bd. 29**, S. 630-549.

153. Popov A. Yu. *Inversion of the generalazed Borel transformations*. Fundamental'naya i Prikladnaya Matematika (1999), **Vol. 5, no. 3**, pp. 817-841. (In Russian).

154. Raikov D. A. *Vector spaces*. Fizmatgiz, Moscow, 1962. 212 pp. (In Russian).

155. Riekstynsh A. Ya. *Asymptotic expansion of integrals, Vol. 2*. Zinatne, Riga, 1977. 460 pp. (In Russian).

156. Robertson A. P., Robertson W. *Topological vector spaces*. Cambridge University Press, 1964.

157. Rockafellar P. T. *Convex analysis*. Princeton University Press, Princeton, New Jersey, 1070.

158. Rokhlin V. A., Fuks D. B. *Initial course of topology. Geometric chapters*. Nauka, Moscow, 1977. 488 pp. (In Russian).

159. Ronkin L. I. *On types of entire functions of two complex variables*. Mat. Sb. (1956), **Vol. 39(81), no. 2**, pp. 253-266. (In Russian).

160. Ronkin L. I. *On a certain growth characteristic for entire function of several variables*. Zap. Mat. Otdel. Fiz.-Mat. Fak. Khar'kovsk. Gos. Univ. Khar'kovsk. Mat. Obshch. (1961), ser. 4, **Vol. 27**, pp. 59-65. (In Russian).

161. Ronkin L. I. *On conjugate orders and types for an entire function of several variables*. Ukr. Mat. Zh. (1964), **Vol. 14, no. 3**, pp. 408-413. (In Russian).

162. Ronkin L. I. *On the growth of the function $\Phi(r_1, \ldots, r_n)$ that is convex relative to* $\ln r_1, \ldots, \ln r_n$. Dokl. AN SSSR (1967), **Vol. 172, no. 5**, pp. 1028-1031. (In Russian).

163. Ronkin L. I. *An analog of the Weierstrass canonical product for entire functions of several variables.* Dokl. AN SSSR (1967), **Vol. 175, no. 4**, pp. 767-770. (In Russian).

164. Ronkin L. I. *Introduction to the theory of entire functions of several variables.* Nauka, Moscow, 1971. 432 pp. (In Russian).

165. Ronkin L. I. *Functions of Completly Regular Growth. Mathematics and its Applications.* Kluwer Academic Publishers. 1992. 392 pp.

166. Schopf G. *The dependence of the hypersurface of conjugate types for a system of conjugate orders on the choice of the conjugate orders.* Izv. VUZ, ser. matem. (1974) **no. 12**, pp. 35-46. (In Russian).

167. Schopf G. *The dependence of the hypersurfaces of conjugate types on the conjugate orders for a certain class of entire functions of several variables.* Izv. VUZ, ser. matem. (1976) **no. 4**, pp. 105-121. (In Russian).

168. Schopf G. *On conjugate types for entire characteristic functions of several variables.* Ukr. Mat. Zh. (1977), **Vol. 29, no. 4**, pp. 489-498. (In Russian).

169. Schopf G. *On conjugate orders and conjugate types for the decay of a multivariate probabilistic law.* Ukr. Mat. Zh. (1978), **Vol. 30, no. 5**, pp. 701-707. (In Russian).

170. Schopf G. *Construction of an entire function of several variables with a given asymptotics for the distribution of zeros.* Ukr. Mat. Zh. (1981), **Vol. 33, no. 4**, pp. 476-481. (In Russian).

171. Schopf G. *On analytic functions for a divisor on $\mathbb{C}^n$ with minimal order function.* Berlin, 1982. 38 pp. (Preprint / Institut für Mathematic Academie der Wissenshaften der DDR. P-MATH 11/82).

172. Shabat B. V. *Introduction to complex analysis.* Nauka, Moscow, 1985. (In Russian). Part 1: Functions of one variable, 336 pp. Part 2: Functions of several variables, 464 pp.

173. Sheremeta M. N. *The relationship between the growth of the maximum of the modulus of an entire function and absolute values of the coefficients of its power series expansion.* Izv. VUZ, ser. matem. (1967) **no. 2**, pp. 100-108. (In Russian).

174. Sigurgson R. *Growth properties of analytic and plurisubharmonic functions of finite order.* Lund University, 1984. 109 pp.

175. Singh J. P. *On the order and type of entire functions of several complex variables.* Riv. mat. Univ. Parma. (1969), **Vol. 10**, pp. 111-121.

176. Sire J. *Sur les fonctions entiéres de deux variables d'ordre apparent total fini.* Rend. Circ. Math. Palermo (1911), **Vol. 31**, pp. 1-91.

177. Springer G. *Introduction to Riemann surfaces.* Addicon-Wesley Publishing Company, Inc. Reading, Massachusetts, U.S.A., 1957.

178. Sreenivasulu V. *A theorem on the order and type of entire function of several complex variables.* Indian J. Pure and Appl. Math. (1970), **Vol. 2, no. 2**, pp. 312-317.

179. Sreevactava R. K., Kumar V. *On the order and type of integral functions of several complex variables.* Comp. Math. (1966) **Vol. 17, no. 2**, pp.161-166.

180. Stoilow S. *Teoria Functiilor de o variabiliă complexă, Vol. 1.* Editura Academiei Republlich Populare Române, 1954.

181. Strelits Sh. I. *Asymptotic properties of analytic solutions of differential equations.* Mintis, Vilnius, 1972. 467 pp. (In Russian).

182. Subbotin M. Th. *Sur les propriétés-limites du module des fonctions entières d'ordre fini.* Math. Ann. (1930), **Vol. 104**, pp. 377-386.

183. Temlyakov A. A. *Entire functions of two complex variables.* Uchenye Zapiski Mosk. Obl. Ped. Inst. (1954), **Vol. 20**, pp. 7-16. (In Russian).

184. Tkachenko V. A. *On operators commuting with the generalized differentiation on spaces of analytic functionals with a given growth indicator.* Mat. Sb. (1977), **Vol. 182, no. 3**, pp. 435-456. (In Russian).

185. Tkachenko V. A. *Equations of convolution type in spaces of analytic functionals.* Izv. AN SSSR, ser. matem. (1977), **Vol. 41, no. 2**, pp. 378-392. (In Russian).

186. Trutnev V. M. *The radial indicator in the Borel summation theory, and some applications.* Sib. Mat. Zh. (1972), **Vol. 13, no. 3**, pp. 659-664. (In Russian).

187. Unkelbach H. *Die Konforme Abbildung echter Polygone.* Math. Ann. (1952), **Vol. 125**, pp. 82-118.

188. Valiron G. *Sur un théorème de M. Hadamard.* Bull. Sci. (1922), **Vol. 47, no. 1**, pp. 177-191.

189. Vertgeim B. A., Rubinshtein G. Sh. *Concerning the definition of quasiconvex functions.* In: Mathematical Programming. Nauka, Moscow, 1966, pp. 121-134. (In Russian).

190. Vladimirov V. S. *Methods of the theory of functions of several complex variables.* Nauka, Moscow, 1964. 411 pp. (In Russian).

191. Vladimirov V. S. *On plurisubharmonic functions in tubular domains.* Izv. AN SSSR, ser. matem. (1965), **Vol. 29, no. 6**, pp. 1123-1146. (In Russian).

192. Vladimirov V. S. *Homogeneous convex functions as growth indicators for holomorphic functions.* Mat. Zametki (1967), **2, no. 2**, pp. 167-174. (In Russian).

193. Volkovysskii L. I. *Quasiconformal mappings.* Izdatel'stvo L'vovskogo Universiteta, Lvov, 1954. 155 pp. (In Russian).

194. Yakovlev E. I. *Asymptotic behavior and methods of analytic continuation of multiple power series.* Ph. D. Thesis abstract, Sverdlovsk, 1986. 23 pp. (In Russian).

195. Yakovlev E. I. *On the existence of a function holomorphic in an unbounded Reinhardt domain in $\mathbb{C}^n$ and having prescribed growth characteristics.* Izv. VUZ, ser. matem. (1985), **no. 6**, pp. 71-76. (In Russian).

196. Yakovlev E. I. *Construction of a function holomorphic in an unbounded Reinhardt domain in $\mathbb{C}^n$ under a matching conditions for type functions.* In: Multidimensional complex analysis, IF SO AN SSSR, Krasnoyarsk, 1985, pp. 268-271. (In Russian).

197. Yanushauskas A. I. *Direct methods of the study of analytic continuability for the sum of a double power series.* Litov. Mat. Sb. (1976), **Vol. 16, no. 3**, pp. 216-227. (In Russian).

198. Yanushauskas A. I. *Double series.* Nauka, Siberian branch, Novosibirsk, 1980. 224 pp. (In Russian).

199. Yocida K. *Functional analysis.* Springer-Verlag, Berlin- Gottingen-Heidelberg, 1965.

200. Yulmukhametov R. S. *Entire functions of several variables with prescribed behavior at infinity.* Izv. RAN, ser. matem. (1996), **Vol. 60, no. 4**, pp. 206-224. In Russian).

201. Yulmukhametov R. S. *On entire functions with given asymptotic behavior.* Funkt. Anal. i ego Prilozh. (1998), **Vol. 32, no. 3**, pp. 50-61. (In Russian).

202. Znamenskii S. V. *Strong linear convexity. I. Duality of spaces of holomorphic functions.* Sibirsk. Mat. Zh. (1985), **Vol. 26, no. 3**, pp. 32-43. (In Russian).

203. Maergoiz L. S. *A constructive procedure yielding Laurent polynomials associated with the indicator of an entire function of order $\rho \neq 1$: new examples.* In: Multidimensional complex analysis, Krasnoyarsk, KGU, 2002; 120-138. (In Russian).

204. Yakovlev E. I. *Liouville theorem and descriptions of $n$-circular domains of bounded holomorphy.* Izv. VUZov, Matematica, 1985, **no. 9**, 52-55. (In Russian).

## References to Chapter 2

1. Akhtunin V. M., Nefedov V. P., Sakharov M. P. et al. *Engineer physiology and modeling of systems of an organism.* Nauka, Siberian Branch, Novosibirsk, 1987. 240 pp. (In Russian).

2. Anikonov Yu. E. *New mathematical models and inverse problems.* In: Mathematical Simulation in Geophysics. VTS SO AN SSSR, Novosibirsk, 1987. pp. 3-42. (In Russian).

3. Anikonov Yu. E., Bubnov B. A. *On a method of the study of multidimensional inverse problems.* In: Methods of Study of Inverse Problems and Ill-posed Problems. VTS SO AN SSSR, Novosibirsk, 1987, pp. 22-40. (In Russian).

4. Cartan H. *Calcul différentiel. Formes différentielles.* Hermann, Paris,1967.

5. Eigen M., Maeyer L. D. *Theoretical base of the relaxation spectroscopy.* In: Investigation of Rates and Mechanisms of Reactions. Third edition. John Wiley & Sons, New York-London-Sydney-Toronto, 1974.

6. Eykhoff P. *System identification.* John Wiley and Sons Ltd., 1974.

7. Gelfond A. O. *Calculus of finite differences.* Nauka, Moscow, 1967. 376 pp. (In Russian).

8. Golubkov V. V., Shcherbov S. Ya. *Exponential approximation.* All-Union Institute of System Research, preprint. Moscow, 1980. 36 pp. (In Russian).

9. Hamming R. W. *Numerical methods for scientists and engineers.* Mc Grow-hill Book Company, Inc. New York, San Francisko, Toronto, London, 1962.

10. Hurgin Ya. I., Yakovlev V. P. *Methods of the theory of entire functions in radiophysics.* Fizmatgiz, Moscow, 1962. 220 pp. (In Russian).

11. Hurgin Ya. I., Yakovlev V. P. *Finitary functions in physics and engineering.* Nauka, Moscow, 1971. 408 pp. (In Russian).

12. Ivakhnenko A. G., Krotov G. I., Cherebkus V. I. *Harmonic and exponential-harmonic algorithms of the cluster analysis method for long-time prediction of oscillatory processes. Part* 1. *One-channal algorithms.* Authomatics (1981), **no. 3**, pp. 17-30. (In Russian).

13. Ivakhnenko A. G., Krotov G. I., Yurachkovskii Yu. P. *Exponential-harmonic algorithms of the cluster analysis method.* Authoma tics (1981), **no. 2**, pp. 23-30. (In Russian).

14. Ivanov V. K. *Stability of the complex moment problem.* Mat. Sb. (1965), **Vol. 68(110), no. 4**, pp. 570-585. (In Russian)

15. Kay S. M., Marple S. L., Jr. *Spectrum Analysis - A Modern Perspective.* Proc. of the IEEE (1981), **Vol. 49, no. 11**, pp. 1380-1419.

16. Komarov A. I. *Computer based methods of analysis of kinetical information in biophisical research.* Abstract of the Ph. D. Thesis, Moscow, 1983. 25 pp. (In Russian).

17. Kondrat'ev V. P. *Approximation by exponential sums. Optimization programs (approximation of functions).* In: Sb. Nauch. Tr. Inst. Mat. Mech., UNTS AN SSSR (1975), **no. 6**, pp. 3-17. (In Russian).

18. Kritskii V. G. *Separation of two components in the exponential representation of functions.* In: Analysis and Interpretation of Physical Experiments. Moscow University Publishers, Moscow, 1980, pp. 104-115. (In Russian).

19. Lavrent'ev M. M. *Interior problems for differential equations. In: Topical Problems of Computational Mathematics and Mathematical Simulation.* Nauka, Siberian Branch, Novosibirsk, 1985, pp. 119-124. (In Russian).

20. Lanczos C. *Applied Analysis.* Prentice Hall, Inc, 1956.

21. Levin M. I., Semko Yu. I. *Finding the parameters of periodic signals by their instant values.* Autometriya (1966), **no. 1**, pp. 33-40. (In Russian).

22. Maergoiz L. S. *A method of identification of relaxation characteristics for homeostasis processes, and the domain of its stability.* Physics Institute, USSR Academy of Sciences, Siberian Branch, Preprint no. 69B, Krasnoyarsk, 1987, 23 pp. (In Russian).

23. Maergoiz L. S. *A multidimensional version of the Prony algorithm.* Sibirsk. Mat. Zh. (1994), **35, no. 2**, pp. 389-406. (In Russian).

24. Maergoiz L. S. *Extreme properties of entire functions in the Wiener class, with applications.* Doklady RAN (1997), **Vol. 356, no. 2**, pp. 161-165. (In Russian).

25. Maergoiz L. S. *Optimal estimate for extrapolation in the Wiener class.* Sibirsk. Mat. Zh. (2000), **Vol. 41, no. 6**, pp. 1363-1375. (In Russian).

26. Maergoiz L. S., Solovei L. I., Avramenok A. K. *On mathematical prediction of the glycemia dynamics.* Cybernetics (1984), **no. 1**, pp. 106-109. (In Russian).

27. Maergoiz L. S., Varava B. N. *On a system of principles in mathematical prediction of homeostasis processes.* In: Homeostasis Mechanisms in Isolated Systems in an Organism, IF SO AN SSSR, Krasnoyarsk, 1984, pp. 134-138. (In Russian).

28. Maergoiz L. S., Varava B. N. *A method of mathematical prediction of the dynamics of homeostasis-type processes.* In: Cybernetics and Computational Technics. Collected Research papers. V. M. Glushkov Cybernetics Institute, Ukrainian Academy of Sciences, **no. 70**, Kiev, 1986, pp. 46-50. (In Russian).

29. Maergoiz L. S., Varava B. N. *On Prony's method of harmonic expansion.* In: Complex Analysis and Differential Equations, Krasnoyarsk State University, Krasnoyarsk, 1996, pp. 86-94. (In Russian).

30. Maergoiz L. S., Varava B. N. *Relationship between linear third order differential and difference equations with constant coefficients.* In: Complex Analysis and Differential Operators, Izd. Krasnoyarsk State University, Krasnoyarsk, 2000, pp. 80-84. (In Russian).

31. Maergoiz L. S., Varava B. N., Manchuk V. T. *Mathematical simulation and prediction for the homeostasis of glucose and sugar in blood.* Avtomatica (1985), **no. 2**, pp. 60-65. (In Russian).

32. Maergoiz L. S., Varava B. N., Karapetyan A.I. *Constant level stabilized relaxation indices and their identification algorithm.* AMSE Transactions "Scientific Siberian". Series A Dynamics (1992), **Vol. 4**, pp. 95-103.

33. Maergoiz L. S., Zakharova L. B., Egorushkin I. O., Kondrat'eva V. P. *On mathematical prediction of the growth dynamics for the ferment activity in ontogenesis.* Byull. Sib. Otd. AMN SSSR (1984), **no. 1**, pp. 92-95. (In Russian).

34. Maergoiz L. S., Fedotov A. M. *Optimal discrepancy for analytic extension from a finite set in the case of inaccurate data, for Hilbert spaces of entire functions,* Sibirsk. Math. Zh. (2001), **Vol. 42, no. 5**, pp. 1106-1116. (In Russian).

35. Mirolyubov A. A., Soldatov M. A. *Linear homogeneous difference equations.* Nauka, Moscow, 1981, 208 pp. (In Russian).

36. Nefedov V. P., Dorer G. A., Yarygina N. V. *Perfusion parameters for isolated organs.* Nauka, Siberian Branch, Novosibirsk, 1983. 229 pp. (In Russian).

37. Novosel'tsev V. N. *Control theory and biological systems.* Nauka, Moscow, 1978. 319 pp. (In Russian).

38. Price H. I. *An improved Prony algorithm for exponential analysis.* In: International Simposium on Electromagnetic Compatibility. San Diego: IEEE, 1979, pp. 310-313.

39. Prony G. R. B. *Essai expérimental et analytique, etc.* J. de L'École Polytechnique (1795), **Vol. 1, no. 2**, pp. 24-76.

40. Romanovskii R. M., Stepanova A. V., Chernavskii D. S. *Mathematical biophysics.* Nauka, Moscow, 1984. 303 pp. (In Russian).

41. Tikhonov A. N., Arsenin V. Ya. *Methods of solution of ill-posed problems.* Nauka, Moscow, 1979. 288 pp. (In Russian).

42. Vasilenko V. A., Zyuzin M. V., Kovalkov A. V. *Splines and digital filters.* VTS SO AN SSSR, Novosibirsk, 1984. 156 pp. (In Russian).

43. Marple S. L., Jr. *Digital spectral analysis with applications.* Prentice-Hall, Inc., Englewood Cliffs, New Jersey, 1987.

44. Zav'yalov M. N. *Modification of the Prony algorithm for a system of ordinary linear differential equations with constant unknown coefficients.* In: Multidimensional complex analysis, Krasnoyarsk, KGU, 2002; 37-47. (In Russian).

## Notation

$\mathbb{R}^n$ is the $n$-dimensional Euclidean space; $\mathbb{R}^1 = \mathbb{R}$;
$\mathbb{C}^n$ is the $n$-dimensional complex space; $\mathbb{C}^1 = \mathbb{C}$;
$u = (u_1, \ldots, u_n)$; $r = (r_1, \ldots, r_n)$; $z = (z_1, \ldots, z_n)$;
$\mathbb{R}^n_+ = \{r \in \mathbb{R}^n : r_1, \ldots, r_n \geqslant 0\}$; $\mathbb{R}^n_0 = \{r \in \mathbb{R}^n : r_1, \ldots, r_n > 0\}$
$\langle u, y\rangle = \sum\limits_{i=1}^{n} u_i y_i$;
$\mathbb{I} = (1, \ldots, 1)$; $e_1 = (1, 0, \ldots, 0), \ldots, e_n = (0, \ldots, 0, 1) \in \mathbb{R}^n$;
$\varphi(e^u) = \varphi(e^{u_1}, \ldots, e^{u_n})$; $\|r\| = r_1 + \cdots + r_n$; $|r| = \sqrt{r_1^2 + \cdots + r_n^2}$;
$xt + u = (x_1 t + u_1, \ldots, x_n t + u_n)$; $rt^x = (r_1 t^{x_1}, \ldots, r_n t^{x_n})$;

$$\ln^+ \Phi(r) = \begin{cases} \ln \Phi(r), & \Phi(r) \geqslant 1, \\ 0, & \Phi(r) < 1; \end{cases}$$

$\mathbb{Z}$ is the set of integers;
$\mathbb{N}$ is the set of natural numbers;
$\mathbb{Z}_+ = \mathbb{N} \cup \{0\}$;
$\overline{\mathbb{C}}$ is the Riemann sphere;
$\overline{z}$ is the conjugate number for a complex number $z$;
$K + a = \{x + a : x \in K\}$;
$V^*(y) = \sup\limits_{u \in S}[\langle u, y\rangle - V(u)]$, $y \in \mathbb{R}^n$, is the Young transform of a function $V : S \to [-\infty, \infty]$, $S \subset \mathbb{R}^n$;
$\operatorname{dom} V = \{u \in S : V(u) < \infty\}$, $S \subset \mathbb{R}^n$;
$\operatorname{epi} V = \{(u, u_{n+1}) \in \mathbb{R}^n \times \mathbb{R} : u \in \operatorname{dom} V;\ u_{n+1} \geqslant V(u)\}$ is the epigraph of a function $V$;
$VO^+(x) = \lim\limits_{t \to \infty} t^{-1}[V(xt + u) - V(u)]$, $u \in \operatorname{dom} V$, is the asymptotic function of a convex function $V$ such that $-\infty \notin V(S)$;
$\partial V(x) = \{y \in \mathbb{R}^n : V(u) \geqslant V(x) + \langle u - x, y\rangle \quad \forall\, u \in \mathbb{R}^n$ is the subdifferential of a convex function $V : \mathbb{R}^n \to (-\infty, \infty]$ at the point $x \in \operatorname{dom} V$;
$\operatorname{conv} V$ is the convex envelope of a function $V$;
$\overset{\circ}{K} = \operatorname{int} K$ is the collection of inner points of a set $K \subset \mathbb{R}^n$;
$\operatorname{ri} K$ is the interior of a convex set $K$ with respect to its affine hull;
$\partial K$ is the boundary of a set $K \subset \mathbb{R}^n$;
$\overline{K}$ is the closure of a set $K \subset \mathbb{R}^n$;
$p_K(x) = \inf\{\alpha > 0 : \alpha^{-1} x \in K\}$ is the Minkowski functional of a set $K \subset \mathbb{R}^n$ containing 0;
$SH(D)$ is the class of subharmonic functions in a domain $D \subset \mathbb{C}$; $PSH_n(\Omega)$ is the class of plurisubharmonic functions in a domain $\Omega \subset \mathbb{C}^n$; $\mathfrak{P} = PSH(\mathbb{C}^n) - PSH_n(\mathbb{C}^n)$,
$L = (0, \infty) \times \mathbb{R}$; $Z = (|Z|, Arg\, Z)$ is a generic point of $L$, where $|Z|$ is the modulus of $Z$; $Arg\, Z$ is the argument of $Z$;
$\pi : L \to \mathbb{C}^*$; $\pi(Z) = |Z| \exp\{i\, Arg\, Z\}$ is the distinguished local homeomorphism; $z = \pi(Z)$; $\mathbb{C}^* = \mathbb{C} \setminus \{0\}$;

$\mathfrak{A} = \{\Phi\}$ is the class of functions defined on $\mathbb{R}^n_+$ and such that $\Phi(z) := \Phi(|z_1|, \ldots, |z_n|) \in \mathfrak{P}$;

$\mathfrak{G}$ is the subclass of the class $\mathfrak{A}$ consisting of all functions nonnegative on $\mathbb{R}^n_+$ and logarithmically convex as functions of the variables $\ln r_1, \ldots, \ln r_n$ on $\mathbb{R}^n_0$;

$P_\rho$ is the class of lower semicontinuous trigonometrically $\rho$-convex $2\pi$-periodic functions defined on $\mathbb{R}$ and taking values in $(-\infty, \infty]$; $M_\rho$ is the subclass of minimal elements of the class $P_\rho$ partially ordered in a natural way;

$\rho_\Phi(u) = \overline{\lim\limits_{t\to\infty}} (\ln t)^{-1} \cdot \ln^+ \Phi(t^u)$, $u \in \mathbb{R}^n$ is the order function of a function $\Phi \in \mathfrak{A}$;

$\rho(\Phi) = \rho_\Phi(\mathbb{I})$ $(\rho_i(\Phi) = \rho_\Phi(e_i))$ is the order of a function $\Phi \in \mathfrak{A}$ in the totality of variables (respectively, in the variable $r_i$, $i = 1, \ldots, n$);

$S(\Phi)$ is the hypersurface of conjugate orders of a function $\Phi \in \mathfrak{A}$; $S(\Phi) \cap \mathbb{R}^n_0 = \{\gamma \in \mathbb{R}^n_0 : \rho_\Phi(\gamma_1^{-1}, \ldots, \gamma_n^{-1}) = 1\}$;

$T_\Phi = \{u \in \mathbb{R}^n : \rho_\Phi(u) = 1\}$ is the order hypersurface of a function $\Phi \in \mathfrak{A}$;

$\mathfrak{M}_n = \{\Phi \in \mathfrak{A} : 0 < \rho(\Phi) < \infty\}$;

$Y$ is the class of sublinear nonnegative functions monotone nondecreasing in each variable and defined on $\mathbb{R}^n$ (we agree that $\phi_0(u) \equiv 0 \notin Y$);

$T^\rho = \{u \in \mathbb{R}^n : \rho(u) = 1\}$, $\rho \in Y$;

$D_\rho = \{u \in \mathbb{R}^n : \rho(u) > 0\}$;

$\mathfrak{M}_n(\rho) = \{\Phi \in \mathfrak{A} : \rho_\Phi(u) \equiv \rho(u),\ u \in \mathbb{R}^n\}$, $\rho \in Y$;

$\sigma_\Phi(x) = \overline{\lim\limits_{t\to\infty}}\, t^{-\rho_\Phi(x)} \cdot \Phi(t^x)$ is the $x$-type of a function $\Phi \in \mathfrak{A}$, where $x \in D_{\rho_\Phi}$;

$\mathfrak{N}^x_n(\rho) = \{\Phi \in \mathfrak{M}_n(\rho) :\ 0 < \sigma_\Phi(x) < \infty\}$;

$\sigma_\Phi(r; x) = \overline{\lim\limits_{a\to r}}\, \overline{\lim\limits_{t\to\infty}}\ t^{-\rho_\Phi(x)} \cdot \Phi(at^x)$, $\quad r \in \mathbb{R}^n_+$ , is the $x$-type function of a function $\Phi \in \mathfrak{A}$;

$A_\Phi = \{\sigma_\Phi(r; x),\ x \in D_{\rho_\Phi}\}$ is the type of a function $\Phi \in \mathfrak{M}_n$;

$\varphi|_U$ is the restriction of a function $\varphi$ to a subset $U$ of the domain of $\varphi$;

$[a, b]$, $(a, b)$, $(a, b]$, and $[a, b)$ denote, respectively, the segment, the interval, and the half-intervals in $\mathbb{C}$ with ends $a$ and $b$; if $a, b \in \mathbb{R}$, it is assumed that $a < b$;

the symbols $h_k \downarrow h$; $h_k \uparrow h$ indicate strictly monotone pointwise convergence of the sequence $h_k$ to $h$ as $k \to \infty$ at each point of the common domain of the $h_k$.

MIX
Papier aus verantwortungsvollen Quellen
Paper from responsible sources
FSC® C105338

If you have any concerns about our products,
you can contact us on
**ProductSafety@springernature.com**

In case Publisher is established outside the EU,
the EU authorized representative is:
**Springer Nature Customer Service Center GmbH**
**Europaplatz 3, 69115 Heidelberg, Germany**

Printed by Libri Plureos GmbH
in Hamburg, Germany